Genome-Scale Algorithm Design

Biological Sequence Analysis in the Era of High-Throughput Sequencing

High-throughput sequencing has revolutionized the field of biological sequence analysis. Its application has enabled researchers to address important biological questions, often for the first time.

This book provides an integrated presentation of the fundamental algorithms and data structures that power modern sequence analysis workflows. The topics covered range from the foundations of biological sequence analysis (alignments and hidden Markov models), to classical index structures (k-mer indexes, suffix arrays, and suffix trees), Burrows–Wheeler indexes, graph algorithms, and a number of advanced omics applications. The chapters feature numerous examples, algorithm visualizations, exercises, and problems, each chosen to reflect the steps of large-scale sequencing projects, including read alignment, variant calling, haplotyping, fragment assembly, alignment-free genome comparison, transcript prediction, and analysis of metagenomic samples. Each biological problem is accompanied by precise formulations and complexity analyses, providing graduate students and researchers in bioinformatics and computer science with a powerful toolkit for the emerging applications of high-throughput sequencing.

The book is accompanied by a website (www.genome-scale.info) offering LaTeX source files for the exercises, along with relevant links.

Veli Mäkinen is a Professor of Computer Science at the University of Helsinki, Finland, where he heads a research group working on genome-scale algorithms as part of the Finnish Center of Excellence in Cancer Genetics Research. He has taught advanced courses on string processing, data compression, biological sequence analysis, along with introductory courses on bioinformatics.

Djamal Belazzougui is a postdoctoral researcher at the University of Helsinki. His research topics include hashing, succinct and compressed data structures, and string algorithms.

Fabio Cunial is a postdoctoral researcher at the University of Helsinki. His research focuses on string algorithms and genome analysis.

Alexandru I. Tomescu is a postdoctoral researcher at the University of Helsinki. His current research interests lie at the intersection of computational biology and computer science.

Genome-Scale Algorithm Design

Biological Sequence Analysis in the Era of High-Throughput Sequencing

VELI MÄKINEN
DJAMAL BELAZZOUGUI
FABIO CUNIAL
ALEXANDRU I. TOMESCU
University of Helsinki, Finland

CAMBRIDGE
UNIVERSITY PRESS

CAMBRIDGE
UNIVERSITY PRESS

University Printing House, Cambridge CB2 8BS, United Kingdom

Cambridge University Press is part of the University of Cambridge.

It furthers the University's mission by disseminating knowledge in the pursuit of
education, learning and research at the highest international levels of excellence.

www.cambridge.org
Information on this title: www.cambridge.org/9781107078536

First published 2015

Printed in the United Kingdom by TJ International Ltd. Padstow Cornwall

A catalogue record for this publication is available from the British Library

Library of Congress Cataloging in Publication data
Mäkinen, Veli, author.
Genome-scale algorithm design : biological sequence analysis in the era of high-throughput sequencing /
Veli Mäkinen, Djamal Belazzougui, Fabio Cunial, and Alexandru I. Tomescu.
    p. ; cm.
Includes bibliographical references and index.
ISBN 978-1-107-07853-6 (hardback : alk. paper)
I. Belazzougui, Djamal, author. II. Cunial, Fabio, author. III. Tomescu, Alexandru I., author.
IV. Title.
[DNLM: 1. Genomics. 2. Algorithms. 3. Sequence Analysis–methods. QU 460]
QH447
572.8'629–dc23   2014045039

ISBN 978-1-107-07853-6 Hardback

Contents

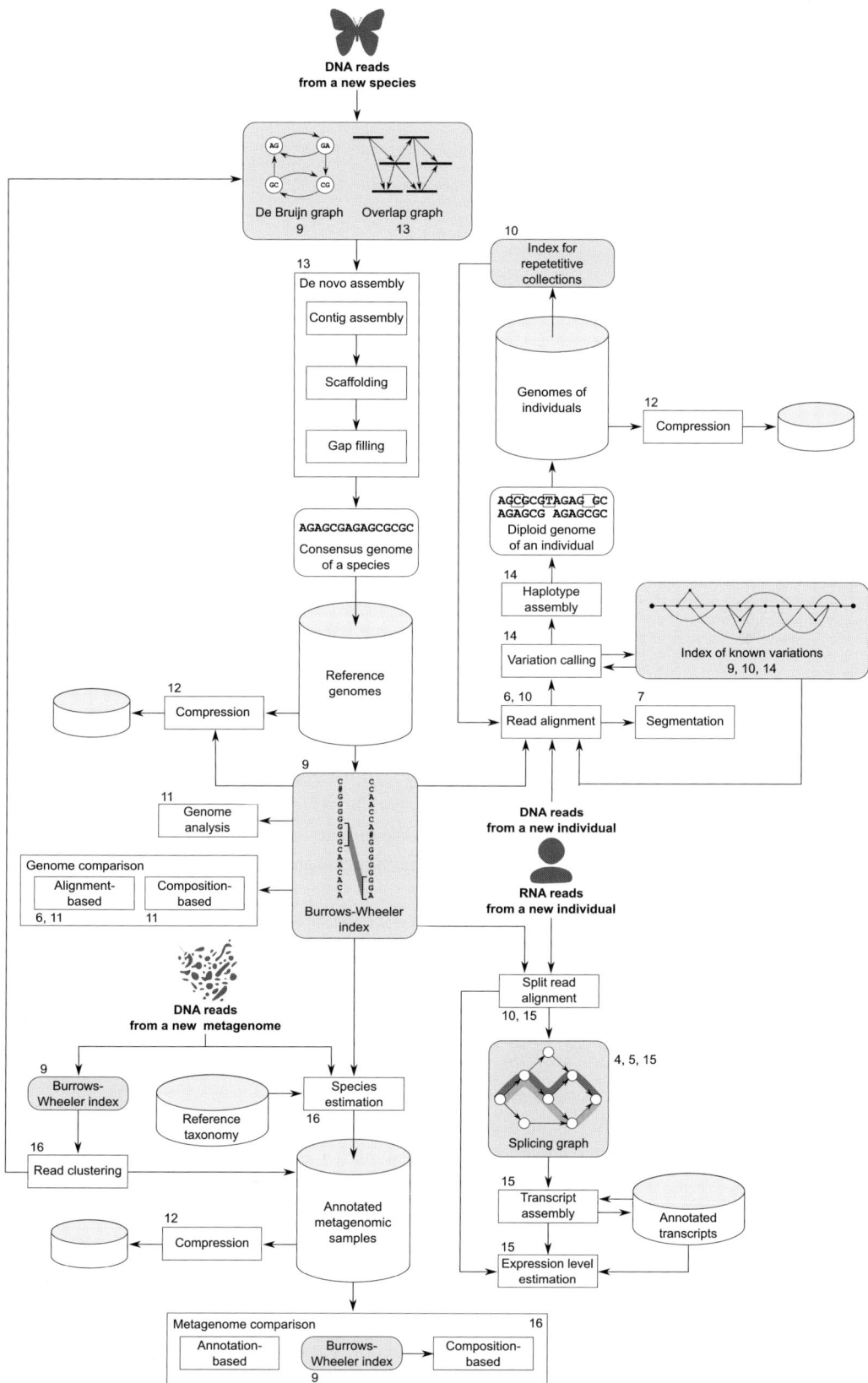

Figure 1 High-level summary of the main computational steps in high-throughput sequencing. Key data structures are highlighted in gray. Cylinders represent databases. Numbers indicate the chapters that cover each step.

Notation

Strings, arrays, sets

i, j, k, n	Integer numbers.
$[i..j], \|[i..j]\|$	The set of integers $\{i, i + 1, \ldots, j - 1, j\}$ and its cardinality $j - i + 1$.
$\mathbb{I}(x, y), \mathbb{I}(x)$	A function that returns the set of integers $[i..j]$ associated with the pair of objects (x, y). We use $\mathbb{I}(x)$ when y is clear from the context.
$\overleftrightarrow{x}, \overleftarrow{x}, \overrightarrow{x}, \|\overleftrightarrow{x}\|$	Assume that $\mathbb{I}(x, y) = [i..j]$, where both function $\mathbb{I}$ and object y are clear from the context. Then, $\overleftrightarrow{x} = [i..j], \overleftarrow{x} = i, \overrightarrow{x} = j$, and $\|\overleftrightarrow{x}\| = \|[i..j]\|$.
$\Sigma = [1..\sigma]$	Alphabet of size σ. All integers in Σ are assumed to be used.
$\Sigma \subseteq [1..u]$	An ordered alphabet, in which not all integers in $[1..u]$ are necessarily used. We denote its size $\|\Sigma\|$ with σ.
a, b, c, d	*Characters*, that is, integers in some alphabet Σ. We also call them *symbols*.
$T = t_1 t_2 \cdots t_n$	A *string*, that is, a concatenation of characters in some alphabet Σ, with character t_i at position i. We use the term *sequence* in a biological context.
$T \cdot S, t \cdot s$	Concatenation of strings T and S or multiplication of integers t and s, with the operation type being clear from the context. We sometimes omit $\cdot$ if the operands of the concatenation/multiplication are simple.
$T = \text{ACGATAGCTA}$	A *string*, with characters given explicitly and represented as letters of the English alphabet.
$\overleftarrow{T}$	The *reverse* of a string T, i.e. string T read from right to left.
$\overleftarrow{\mathcal{T}}$	The *reverse complement* of a DNA sequence T, that is, string T read from right to left, replacing A with T and C with G, and vice versa.
$T_{i..j}$	The *substring* $t_i t_{i+1} \cdots t_{j-1} t_j$ of string T induced by the indexes in $[i..j]$.
$T[i..j]$	Equivalent to $T_{i..j}$, used for clarity when i or j are formulas rather than variables.

subsequence	A string $t_{i_1} t_{i_2} \cdots t_{i_k}$ obtained by selecting a set of positions $1 \leq i_1 < i_2 < \cdots < i_k \leq n$ and by reading the characters of a string $T = t_1 t_2 \cdots t_n$ at those positions. In a biological context, subsequence is used as a synonym of substring.
$S = \{T^1, T^2, \ldots, T^n\}$	Set of strings, with T^i denoting the ith string.
$\Sigma^*, \Sigma^+, \Sigma^n$	The set of all strings over alphabet Σ, the set of all non-empty strings over alphabet Σ, and the set of strings of length n over alphabet Σ, respectively. We use shorthand $A = a^n$ for $A \in \{a\}^n$, that is, for a string consisting of n occurrences of a.
$\delta(i..j, c)$	A function that maps an interval $[i..j]$ and a character $c \in \Sigma$ onto exactly one interval $[i'..j']$.
$\ldots, \#_2, \#_1, \#_0$	Shorthands for non-positive integers, with $\#_0 = 0$ and $\#_i = -i$.
$\#$	Shorthand for $\#_0$.
$\$_1, \$_2, \ldots$	Shorthands for positive integers greater than σ, with $\$_i = \sigma + i$.
$\$$	Shorthand for $\$_1$.
$A[1..n]$	*Array A* of integers, indexed from 1 to n.
$A[i..j], A[\vec{x}]$	The *subarray* of array A induced by the indexes in $[i..j]$ and in $\vec{x}$, respectively.
$X = (p, s, v)$	*Triplet* of integers, with primary key $X.p$, secondary key $X.s$, and value $X.v$.
$D[1..m, 1..n]$	An *array/matrix* with m rows and n columns.
$D_{i_1..j_1, i_2..j_2}$	*Subarray* of D.
$D[i_1..j_1, i_2..j_2]$	Same as above.
$d_{i,j} = D[i,j]$	An element of the array D.

Undirected graphs

an *edge* between vertex a and vertex b, denoted as (a, b) or as (b, a)

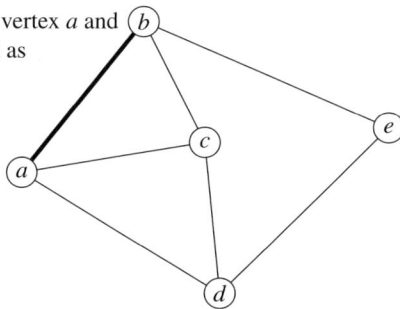

one *connected component* of G

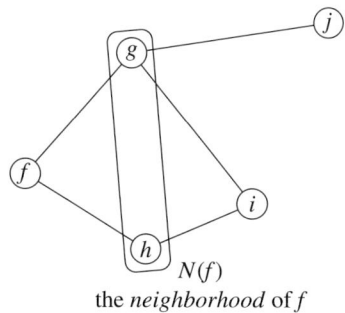

$N(f)$
the *neighborhood* of f

another connected component of G

Figure 2 An *undirected graph* $G = (V, E)$, with *vertex* set V and *edge* set E.

$V(G)$ Set V of *vertices* of a graph $G = (V, E)$.
$E(G)$ Set E of *edges* of an undirected graph $G = (V, E)$.
$(x, y) \in E(G)$ An edge of an undirected graph G; the same as (y, x).
$N(x)$ The *neighborhood* of x in G, namely the set $\{y \mid (x, y) \in E(G)\}$.

Directed graphs

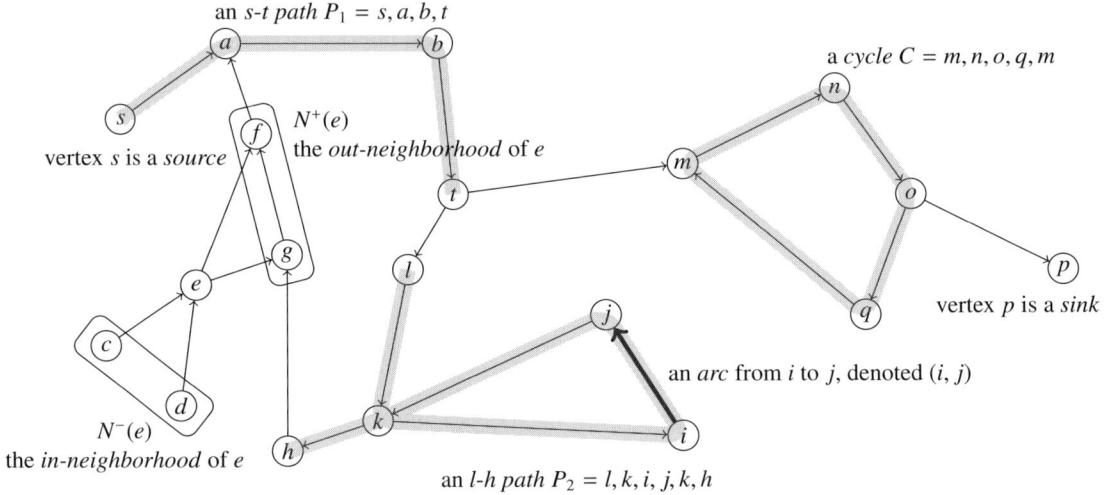

Figure 3 A *directed graph* $G = (V, E)$, with *vertex* set V and *arc* set E.

$(x, y) \in E(G)$ An arc of a directed graph G; the arc (x, y) is different from (y, x).
$N^-(x)$ The in-neighborhood of x in G, namely the set $\{y \mid (y, x) \in E(G)\}$.
source   A vertex v is a source if $N^-(v) = \emptyset$.
$N^+(x)$ The out-neighborhood of x in G, namely the set $\{y \mid (x, y) \in E(G)\}$.
sink   A vertex v is a sink if $N^+(v) = \emptyset$.
$P = v_1, \ldots, v_k$ A path in G, namely a sequence of vertices of G connected by arcs with the same orientation, from v_1 to v_k; depending on the context, we allow or not P to have repeated vertices.
s-t path   Path from vertex s to vertex t.
$C = v_1, \ldots, v_k, v_1$ A cycle in G, namely a path in G in which the first and last vertex coincide; depending on the context, we allow or do not allow C to have other repeated vertices than its first and last elements.
$(x, y) \in S$ Arc $(x, y) \in E(G)$ appears on S, where S is a path or cycle of G.

Trees

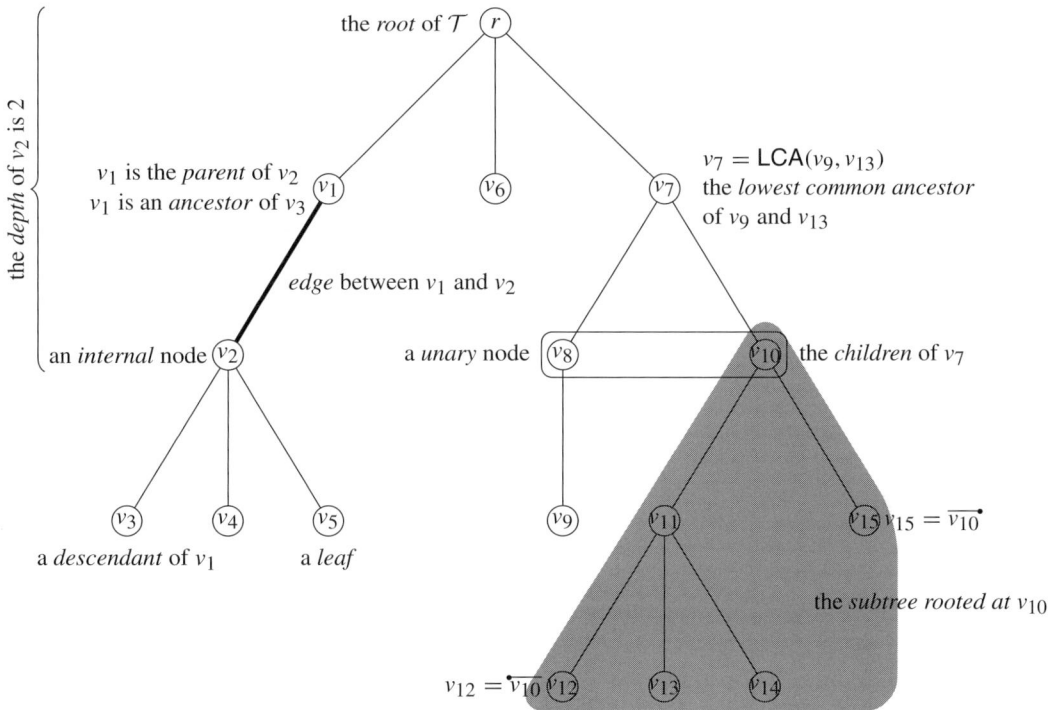

Figure 4 A tree $\mathcal{T} = (V, E)$, with *node* set V and *edge* set E. Unless stated otherwise, we assume all trees to be *ordered*, that is, we assume that there is a total order on the children of every node.

v_2 is a child of v_1	If there is an edge between v_1 and v_2 and v_1 appears on the path from the root to v_2.
v_1 is the parent of v_2	If v_2 is the child of v_1.
degree of a node v	The number of children of v.
leaf	A node with degree 0.
internal node	A node with degree at least 1.
unary node	A node with degree 1.
depth of v	The number of edges of the path from the root to v.
subtree rooted at v	The subtree of $\mathcal{T}$ having root v and consisting of all nodes reachable through a path starting at v made up only of nodes of depth at least the depth of v.
v_2 is descendant v_1	If $v_2 \neq v_1$ belongs to the subtree rooted at v_1.
v_1 is ancestor v_2	If $v_2 \neq v_1$ belongs to the subtree rooted at v_1.
$\mathsf{LCA}(v_1, v_2)$	The *lowest common ancestor* of v_1 and v_2, that is, the deepest node which is an ancestor of both v_1 and v_2.
$\overleftarrow{v}$	The left-most leaf of the subtree rooted at a node v.
$\overrightarrow{v}$	The right-most leaf of the subtree rooted at a node v.

Preface

Background

High-throughput sequencing has recently revolutionized the field of biological sequence analysis, both by stimulating the development of fundamentally new data structures and algorithms, and by changing the routine workflow of biomedical labs. Most key analytical steps now exploit index structures based on the Burrows–Wheeler transform, which have been under active development in theoretical computer science for over ten years. The ability of these structures to scale to very large datasets quickly led to their widespread adoption by the bioinformatics community, and their flexibility continues to spur new applications in genomics, transcriptomics, and metagenomics. Despite their fast and still ongoing development, the key techniques behind these indexes are by now well understood, and they are ready to be taught in graduate-level computer science courses.

This book focuses on the rigorous description of the *fundamental algorithms and data structures* that power modern sequence analysis workflows, ranging from the foundations of biological sequence analysis (like alignments and hidden Markov models) and classical index structures (like *k*-mer indexes, suffix arrays, and suffix trees), to Burrows–Wheeler indexes and to a number of advanced *omics* applications built on such a basis. The topics and the computational problems are chosen to cover the actual steps of large-scale sequencing projects, including read alignment, variant calling, haplotyping, fragment assembly, alignment-free genome comparison, compression of genome collections and of read sets, transcript prediction, and analysis of metagenomic samples: see Figure 1 for a schematic summary of all the main steps and data structures covered in this book. Although strongly motivated by high-throughput sequencing, many of the algorithms and data structures described in this book are general, and can be applied to a number of other fields that require the processing of massive sets of sequences. Most of the book builds on a coherent, self-contained set of algorithmic techniques and tools, which are gradually introduced, developed, and refined from the basics to more advanced variations.

The book is accompanied by a website

www.genome-scale.info

that provides references to implementations of many index structures and algorithms described here. The website also maintains a list of typos and mistakes found, and we encourage the reader to send corrections as requested therein.

This book introduces a number of significant novelties in presenting and organizing its content. First, it raises to a central role the *bidirectional Burrows–Wheeler index*: this powerful data structure is so flexible as to be essentially the only index needed by most sequence analysis primitives, like maximal repeats, maximal unique and exact matches, and alignment-free sequence comparison. In this book we use k-mer indexes, suffix arrays, and suffix trees mostly as *conceptual tools* to help the reader learn the bidirectional Burrows–Wheeler index and formulate problems with its language.

Another key concept that recurs in a large fraction of the book is *minimum-cost network flow*, a flexible model in combinatorial optimization that can be solved in polynomial time. We use this model as a "Swiss Army knife", both by providing a unified presentation of many well-known optimization problems in terms of minimum-cost flow (like maximum-weight bipartite matching and minimum-weight minimum path cover), and by showing that a number of key problems in fragment assembly, transcriptomics, and metagenomics can be elegantly solved by reductions to minimum-cost flow.

Finally, the book spices up the presentation of classical bioinformatics algorithms by including a number of advanced topics (like Myers' bitparallel algorithm), and by presenting inside a unifying framework the dynamic programming concepts that underlie most such algorithms. Specifically, many seemingly unrelated problems in classical bioinformatics can be cast as *shortest-path problems on a directed acyclic graph* (DAG). For example, the book describes the Viterbi algorithm for hidden Markov models as a special case of the Bellman–Ford algorithm, which can itself be interpreted as a solution to the shortest-path problem on a DAG created by layered copies of the input graph. Even the gap-filling problem in fragment assembly is solved through a similar reduction to a DAG problem. The book contains a number of other problems on DAGs, like aligning paths in two labeled DAGs, indexing labeled DAGs using an extension of the Burrows–Wheeler transform, and path covering problems on weighted DAGs arising from alternative splicing.

The book is designed so that key concepts keep reappearing throughout the chapters, stimulating the reader to establish connections between seemingly unrelated problems, algorithms, and data structures, and at the same time giving the reader a feeling of organic unity.

Structure and content

The book adopts the style of theoretical computer science publications: after describing a biological problem, we give precise problem formulations (often visually highlighted in a frame), algorithms, and pseudocode when applicable. Finally, we summarize our results in one or more theorems, stating the time and space complexity of the described algorithms. When we cannot obtain a positive result in the form of a polynomial-time algorithm, we give an NP-hardness proof of the problem, thereby classifying each problem as either tractable or intractable.

Nonetheless, ad-hoc optimizations and heuristics are important in practice, and we do explain a number of such strategies for many problems. We choose to present these methods inside a special frame, called *insight*. An insight can also contain additional background information, methods explained only by example, mathematical and statistical derivations, and practical bioinformatics issues. Visually separating insights from the main text puts the key algorithmic concepts in the foreground, helping the reader to focus on them and to potentially skip marginal details.

Every chapter ends with a collection of exercises, of varying difficulty. Some exercises ask the reader just to practice the topics introduced in the chapter. Other exercises contain simple, self-contained parts of long proofs, stimulating the reader to take an active part in the derivations described in the main text. Yet other exercises introduce new problem formulations, or alternative strategies to solve the same problems, or they ask the reader to play with variants of the same data structure in order to appreciate its flexibility. This choice makes some exercises quite challenging, but it allows the main text to stay focused on the key concepts. By solving the majority of the exercises, the reader should also gain a broad overview of the field.

A small number of sections describe advanced, often technical concepts that are not central to the main flow of the book, and that can be skipped safely: such sections are marked with an asterisk. The book is designed to be self-contained: apart from basic data structures such as lists and stacks, and apart from basic notions in algorithm complexity, such as the big-oh notation, every chapter of the book builds only on data structures and algorithms that have been described in previous chapters. Therefore, a reader could potentially implement every algorithm we describe, by just reading this book: using the computer science jargon, we could say that the book is entirely "executable". For pedagogical reasons we choose sometimes not to present the most time- or space-efficient algorithm for a problem. In all such cases, we briefly sketch the more efficient variants in the main text, leave them as exercises for the reader, or cite them in the literature section of the chapter.

The book focuses on *algorithm design*. This means that we mainly focus on combinatorial strategies that can be used to solve a variety of different problems, and that we try to find analogies between the solution of every problem and the solution of other problems described in previous chapters. Our focus on design also implies that we do not include in the book any algorithm that requires advanced mathematical tools for analyzing its performance or for proving its correctness: a basic understanding of amortized analysis and of combinatorics is enough to follow the derivations of all worst-case bounds. No average- or expected-case analysis is included in the book, except for a small number of insights that describe algorithms whose worst-case complexity is not interesting.

A significant part of the book focuses on applications of space-efficient data structures. Research on such structures has been very active over the last 20 years, and this field would deserve a textbook on its own. Our goal was not to delve into the technical data structure fundamentals, but to select the minimal setup sufficient to keep the book self-contained. With this aim, we avoided fully dynamic data structures, entropy-compressed data structures, and data structures to support constant time range minimum queries. These choices resulted in some new insights on algorithm design. Namely, we observed that one can still obtain optimal solutions to a number of suffix tree problems by resorting to amortized analysis on batched queries solvable without the need for any advanced basic data structure. A noteworthy new result in the book exemplifying this sufficiency is that Lempel–Ziv factorization can be obtained space-efficiently in near-linear time with the chosen minimal data structure setup.

Owing to the wide adaptation of the algorithms appearing in the literature to our self-contained minimal data structure setup, we decided to gather all of the references into a literature section appearing at the end of each chapter. This enables an undisturbed text flow and lets us explain the difference between our description and the original work. Some exercises are also directly related to some published work, and in such cases we mention the reference in the literature section. With regard to new results developed for the book or work under preparation, we added some future references in the literature sections.

Finally, almost all algorithms presented in the book are sequential, and are designed to work in random access main memory. By *genome-scale* we mean both a first logical layer of space-efficient and near-linear-time algorithms and data structures that process and filter the *raw data* coming from high-throughput sequencing machines, and a second layer of polynomial-time algorithms that work on the inherently smaller output of the first layer and that solve *semantical problems* closer to biology (for example in transcriptomics and haplotype assembly). The concepts and methodologies detailed in this book are, however, a vantage point for designing secondary-memory algorithms, parallel shared-memory algorithms, GPU algorithms, and distributed algorithms, whose importance is bound to increase in high-throughput sequencing, and in particular in metagenomics. Some exercises explicitly ask the reader to explore such directions, and the last chapter of the book suggestively ends by referring to an existing distributed algorithm, whose sequential version is described in the book.

Target audience

Since the book has a clear focus on algorithmic sequence analysis for high-throughput sequencing, the main audience consists in graduate students in bioinformatics, graduate students in computer science with a strong interest in molecular biology, and bioinformatics practitioners willing to master the algorithmic foundations of the field. For the latter, the insights scattered throughout the book provide a number of techniques that can be of immediate use in practice. The structure of the book is strongly focused on applications, thus the book could be used as an introduction to biological sequence analysis and to high-throughput sequencing for the novice. Selected parts of the book could even be used in an introductory course in bioinformatics; however, such basic topics were not chosen to cover the whole of bioinformatics, but just to give the minimal foundations required to understand the more advanced concepts that appear in later chapters. Our fresh presentation of recent theoretical topics in succinct data structures, and in biological applications of minimum-cost flow and dynamic programming, might also appeal to the specialist in algorithm design.

Acknowledgements

First and foremost, we wish to warmly thank our families and friends for their constant support throughout the sometimes laborious process of writing a book. We apologize once again for the long nights spent at work, and for frequent periods of pressure and discouragement that they helped us endure. We are also profoundly grateful to the editors, Katrina Halliday and Megan Waddington, for their constant encouragements – particularly important to keep us motivated at the beginning of the writing process. Our deep gratitude goes also to Gonzalo Navarro, Nadia Pisanti, and Romeo Rizzi, for volunteering to read major parts of the book, and for providing invaluable comments on it. Selected chapters were also reviewed by Carlo Comin, Emanuele Giaquinta, Anna Kuosmanen, Paul Medvedev, Martin Milanič, Alberto Policriti, Nicola Prezza, Kristoffer Sahlin, and Leena Salmela. Some chapters were in a quite early phase during such reviews, and we probably introduced new errors during the last weeks of intensive writing, so the reviewers should not be held responsible for any error that slipped through. Veli Mäkinen wishes also to thank the Finnish Cultural Foundation for supporting the sabbatical year that enabled his full concentration on the book.

The selection of content and structure on edit distances, alignments, hidden Markov models, and text indexing was highly influenced by the *String Processing Algorithms*, *Data Compression Techniques*, and *Biological Sequence Analysis* courses taught repeatedly at the Department of Computer Science, University of Helsinki, starting in the nineties with Esko Ukkonen, Jorma Tarhio (now at Aalto University), and Juha Kärkkäinen, and occasionally inherited by other colleagues, including Veli Mäkinen. The seed of this book consisted in a lecture script of some 150 pages by Veli, which gradually emerged in multiple rounds of lectures for the *Biological Sequence Analysis* course at the University of Helsinki, for the *Genome-Scale Algorithmics* course at the University of Bielefeld, and for a tutorial on *Genome-Scale Sequence Analysis Using Compressed Index Structures* at ISMB 2009. The students of the 2015 edition of the Biological Sequence Analysis course gave valuable feedback on a few chapters of the preprint of this book.

However, this initial draft was then heavily rewritten, restructured, and extended, capitalizing on the heterogeneous educational backgrounds of the authors to broaden its original scope. In the end, 11 out of the 16 chapters of this book were written from scratch, while the other 5 were heavily reworked. Some of the new chapters are based on some distinctive expertise that each one of us brought to the book. For example, Djamal brought in competence on succinct data structures, Fabio on genome analysis methods, and Alexandru on graph theory and computational complexity. As unexpected as it might sound, the very experience of writing a book together contributed to establishing a genuine friendship. We also thank our academic advisors, as well as all our teachers, research colleagues, and students, for their invaluable contribution in enriching our knowledge – as well as our lives. Finally, our understanding of high-throughput sequencing, and of the computational problems that underpin it, had been greatly increased by collaborating with colleagues inside the Finnish Center of Excellence in Cancer Genetics Research, led by Lauri Aaltonen, inside the *Melitaea cinxia* sequencing project, led by Ilkka Hanski, and inside the HIIT bio-focus area on metagenomics, led by Antti Honkela.

Sometimes chance events have long-term consequences. While visiting our research group at the University of Helsinki, Micheaël Vyverman, from Ghent University, gave a talk on maximal exact matches. Veli's lecture script already contained Algorithm 11.3, which uses the bidirectional BWT index to solve this problem, but without any analysis of its time and space complexity. Micheaël's talk inspired us to study this algorithm with more care: soon we realized that its running time is $O(n \log \sigma)$, but the use of a stack caused a space problem. Luckily, Juha Kärkkäinen happened to hear our discussions, and directed us to a classical stack-trick from the quicksort algorithm, which solved our space problem. After even more thinking, we realized that the algorithm to solve maximal exact matches could be adapted to solve a large number of seemingly unrelated problems in sequence analysis, essentially becoming Algorithm 9.3, one of the center-pieces of a book that was still in its infancy. Later we learnt that the related strategy of implicit enumeration of the internal nodes of a suffix tree with bidirectional Burrows–Wheeler index had already been developed by Enno Ohlebusch and his colleagues.

Another random event happened when we were working on a problem related to RNA transcript assembly. We had sent Romeo Rizzi a submitted manuscript containing an NP-hardness proof of a certain problem formulation. After some time, Romeo replied with a short message claiming he had a polynomial-time algorithm for our problem, based on minimum-cost flows. After a stunned period until we managed to get into contact, it of course turned out that we were referring to different problems. The NP-hard formulation is now presented in Exercise 15.11, and the problem Romeo was initially referring to is Problem 15.5. Nonetheless, this initial connection between assembly problems and minimum-cost flow solutions led to the many applications of minimum-cost flow throughout this book.

While we were working hard on this project, Travis Gagie and Simon Puglisi were working even harder to keep the research productivity of the group at a high level. Simon kept the group spirit high and Travis kindly kept Daniel Valenzuela busy with spaced suffix arrays while there was yet another book meeting.

Many students got the opportunity to implement our ideas. These implementations by Jarno Alanko, Jani Rahkola, Melissa Riesner, Ahmed Sobih, and many others can be found at the book webpage.

Part I

Preliminaries

1 Molecular biology and high-throughput sequencing

In this chapter we give a minimalistic, combinatorial introduction to molecular biology, omitting the description of most biochemical processes and focusing on inputs and outputs, abstracted as mathematical objects. Interested readers might find it useful to complement our abstract exposition with a molecular biology textbook, to understand the chemical foundations of what we describe.

1.1 DNA, RNA, proteins

Life consists of fundamental units, called *cells*, that interact to form the complex, emergent behavior of a colony or of a multicellular organism. Different parts of a multicellular organism, like organs and tissues, consist of specialized cells that behave and interact in different ways. For example, muscle cells have a fundamentally different function and structure from brain cells. To understand such differences, one should look inside a cell.

A cell is essentially a *metabolic network* consisting of a mixture of molecules that interact by means of chemical reactions, with new "output" molecules constantly produced from "input" molecules. A set of reactions that are connected by their input and output products is called *pathway*. Each reaction is facilitated or hindered by a specific set of molecules, called *enzymes*, whose regulation affects the status of the entire network inside a cell. Enzymes are a specific class of *proteins*, linear chains of smaller building blocks (called *amino acids*) which can fold into complex three-dimensional structures. Most proteins inside a human cell consist of amino acids taken from a set of 20 different types, each with peculiar structural and chemical properties.

The environment in which a cell is located provides input molecules to its metabolic network, which can be interpreted as signals to activate specific pathways. However, the cell can also regulate its own behavior using a set of instructions *stored inside itself*, which essentially specify how to assemble new proteins, and at what rate. In extreme simplification, this "program" is immutable, but it is executed dynamically, since the current concentration of proteins inside a cell affects which instructions are executed at the next time point. The conceptual network that specifies which proteins affect which instruction is called the *regulation network*. To understand its logic, we need to look at the instruction set more closely.

In surprising analogy to computer science, the material that stores the instruction set consists of a collection of *sequential*, chain-like molecules, called *deoxyribonucleic acid (DNA)*. Every such molecule consists of the linear concatenation of smaller building blocks, called *nucleotides*. There are just four different types of nucleotides in a DNA molecule, called *bases*: adenine, cytosine, guanine, and thymine, denoted by A, C, G, and T, respectively. Every such sequence of nucleotides is called a *chromosome*, and the whole set of DNA sequences in a cell is called its *genome*. A DNA molecule is typically *double-stranded*, meaning that it consist of two chains (called *strands*), bound around each other to form a double helix. The sequence of nucleotides in the two strands is identical, up to a systematic replacement of every A with a T (and vice versa), and of every C with a G (and vice versa): such pairs of bases are called *complementary*, and the attraction of complementary base pairs, among other forces, holds the two strands together. The redundancy that derives from having two identical sequences for each chromosome up to replacement is an ingenious mechanism that allows one to *chemically copy* a chromosome: in extreme simplification, copying happens by separating the two strands, and by letting the bases of each isolated strand spontaneously attract complementary, unbound bases that float inside the cell. The ability to copy the instruction set is one of the key properties of life. Owing to the chemical mechanism by which consecutive nucleotides in the same strand are connected to each other, each strand is assigned a direction. Such directions are symmetrical, in the sense that if a strand is read from left to right, its complementary strand is read from right to left.

Some subsequences of a chromosome *encode* proteins, in the sense that there is a molecular mechanism that *reads* such subsequences and that interprets them as sequential instructions to concatenate amino acids into proteins. Such program-like subsequences are called *genes*. Specifically, a gene consists of a set of *exons*, contiguous, typically short subsequences of DNA that are separated by large subsequences called *introns* (see Figure 1.1). Exons and introns are ordered according to the direction of the strand to which they belong. A *transcript* is an ordered subset of the exons of a gene. The same gene can have multiple transcripts.

Transcription is the first step in the production of a protein from a gene. In extreme simplification, the two strands of a chromosome are first separated from each other around the area occupied by the gene, and then a *chain of ribonucleic acid (RNA)* is produced from one of the strands by a molecular machine that reads the gene *sequentially* and "prints" a complementary copy. RNA is like DNA, but thymine (T) is replaced by *uracil* (U), which becomes the complement of adenine (A). This first RNA copy of a gene is called the *primary transcript*, or *pre-mRNA*, where the letter "m" underlines the fact that it encodes a message to be converted into proteins, and the prefix "pre" implies that it is a precursor that still needs to be processed.

Further processing consists in extracting subsets of exons from the pre-mRNA, and in combining them into one or more predefined transcripts, discarding introns altogether. This biochemical process is called *splicing*, and the creation of multiple transcripts from the same gene is called *alternative splicing* (see Figure 1.1). Alternative splicing is heavily exploited by complex organisms to encode a large number of alternative proteins from a small set of genes. The beginning and the end of introns are typically marked by special subsequences that signal the positions at which introns must be cut out of the primary transcript (for example dinucleotide GT at the beginning, and dinucleotide AG

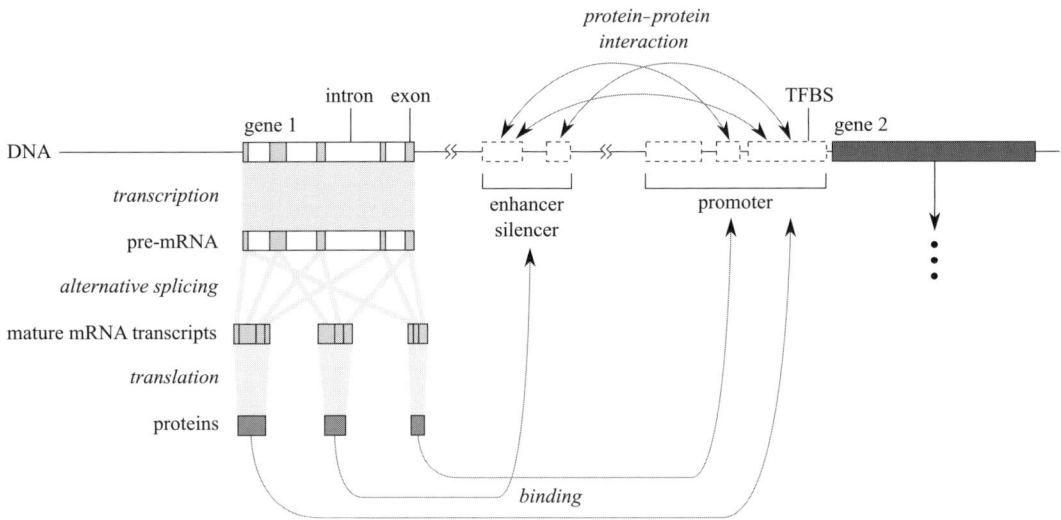

Figure 1.1 A schematic illustration of the central dogma. Gene 1 has three alternatively spliced transcripts. The relative expression of such transcripts affects the regulatory modules of gene 2, and eventually its expression. Definitions are given in Section 1.1.

at the end of an intron). The output of splicing is called *mature messenger RNA (mRNA)*. Introns are mostly absent from simple forms of life, like bacteria, and the entire process of transcription is much simpler in such cases.

The final step to convert mRNA into protein is called *translation*: in extreme simplification, a complex molecular machine reads a mature mRNA transcript *sequentially*, from left to right, three bases at a time, converting every such triplet (called a *codon*) into a specific amino acid determined by a fixed translation table, and chaining such amino acids together to form a protein. This universal table, shared by most organisms, is called *the genetic code*. There are $4^3 = 64$ distinct codons, so different codons are converted to the same amino acid: this redundancy is typically exploited by organisms to tolerate deleterious mutations in genes, or to tune the efficiency of translation. Specific *start* and *stop codons* signal the beginning and the end of the translation process. In addition to signalling, the start codon codes for a specific amino acid. Table 1.1 shows the genetic code used by the human genome.

This unidirectional flow of information from DNA to mRNA to protein, in which sequential information is neither transferred back from protein to RNA or DNA, nor from RNA to DNA, is called the *central dogma* of molecular biology, and it is summarized in Figure 1.1. The actual biological process is more complex than this dogma, and includes exceptions, but this abstraction is accurate enough to help understand the topics we will cover in the following chapters.

As mentioned earlier, proteins can work as enzymes to facilitate or inhibit specific chemical reactions in the metabolic network of a cell. Proteins also make up most of the structural parts of the cell, contributing to its shape, surface, and mobility. Finally, specific proteins, called *transcription factors*, control the rate of the transcription process

Table 1.1 The genetic code used by the human genome. Amino acids (AA) are abbreviated by letters. Symbols ▷ and ◁ denote the start and stop signals, respectively.

AA	Codons	AA	Codons
A	GCT, GCC, GCA, GCG	N	AAT, AAC
C	TGT, TGC	P	CCT, CCC, CCA, CCG
D	GAT, GAC	Q	CAA, CAG
E	GAA, GAG	R	CGT, CGC, CGA, CGG, AGA, AGG
F	TTT, TTC	S	TCT, TCC, TCA, TCG, AGT, AGC
G	GGT, GGC, GGA, GGG	T	ACT, ACC, ACA, ACG
H	CAT, CAC	V	GTT, GTC, GTA, GTG
I	ATT, ATC, ATA	W	TGG
K	AAA, AAG	Y	TAT, TAC
L	TTA, TTG, CTT, CTC, CTA, CTG	◁	TAA, TGA, TAG
M, ▷	ATG		

itself: indeed, the rate at which a gene is transcribed, called the *expression level* of the gene, is controlled by the binding of transcription factors to specific *regulatory regions* associated with the gene. Such proteins form molecular complexes called *enhancer* or *silencer modules*, which respectively facilitate or inhibit the formation of the molecular machine required to start transcription. In extreme simplification, this is the way in which the concentration of proteins in the cell at a given time point affects which instructions will be executed at the next time point.

The subsequences of DNA to which transcription factors bind are called *transcription factor binding sites* (TFBS), and they can be located both in a *promoter area* close to the beginning of a gene and at a large sequential distance before (*upstream* of) the beginning of a gene (see Figure 1.1). The double helix of DNA that forms a chromosome folds on itself at multiple scales, thus large distances along the sequence can translate into small distances in three-dimensional space, putting upstream transcription factors in the vicinity of a gene. Alternative splicing is itself regulated by specific proteins that bind to primary transcripts.

1.2   Genetic variations

The genome of a species is copied from individual to individual across multiple generations of a population, accumulating, like a chain letter, minimal differences at every step. The genomes of different individuals of the same species have typically the same number of chromosomes and most of the same bases in each chromosome: we can thus talk about the *consensus* or the *reference genome* of a species.

Variations from the consensus make every individual unique: most such variations are *single-nucleotide polymorphisms (SNPs)*, or equivalently variations of a single base. Insertions and deletions (in short, *indels*) are also common, with short indels being more frequent than large indels. Large-scale *structural variants*, like the *inversion* of entire subsequences of a chromosome, the *translocation* of a subsequence to a different location or to a completely different chromosome, the *fusion* of two chromosomes that get concatenated to one another, or the *fission* of a chromosome into two or more fragments, have the potential to induce *speciation events* that create new species, but they are typically *deleterious* and cause lethal anomalies.

How the genetic information transfers from one generation to the next differs according to the strategy adopted by a species to reproduce. *Sexual reproduction* is common in *eukaryotes* – complex organisms whose cells use internal membranes to store genetic material and other special-purpose structures. In sexual reproduction, every cell of an individual is endowed with two copies of each chromosome, one inherited from a "mother" and the other from a "father". Such copies are called *homologous chromosomes* or *haplotypes*: a human individual has 23 pairs of homologous chromosomes, one of which determines its sex. A cell with two copies of each chromosome is called *diploid*. A specific set of cells inside an individual, called *germ cells*, gives rise to another set of specialized cells, called *gametes*, that contain only one copy of each chromosome and are used by an individual to combine its chromosomes with those of another individual. Thus, mutations that occur in germ cells, called *germline mutations*, can be inherited by the offspring of an individual that reproduces sexually. *Somatic mutations* are instead those variations that do not affect germ cells, and thus cannot be inherited by the offspring of an individual. Somatic mutations that occur early in the development of an individual can, however, spread to a large fraction of its cells.

Before producing gametes, a germ cell randomly *recombines* the two homologous copies of every chromosome, so that the single copy given to each gamete consists of a random sequence of consecutive fragments, taken alternatively from one haplotype and from the other at corresponding positions, as simplified below:

```
ACGACGAtcggagcgatgAGTCGAGctagct
accacgaTCCCAGAGATGtgtccagCTACCT
```

In this example, lower-case letters denote one haplotype, and upper-case letters denote the other haplotype. Variants that occur in a single haplotype are called *heterozygous*, while variants that occur in both haplotypes are called *homozygous*. Every possible variant that can be found at a specific position of a chromosome is called an *allele*.

1.3 High-throughput sequencing

Sequencing is the process of deriving the sequence of bases that compose a DNA fragment. It is typically hard to sequence very long fragments, like entire chromosomes, from beginning to end. However, it is relatively easy to cut a long fragment into short pieces, create multiple copies of each piece (a process known as *amplification*), and sequence all the resulting pieces: this process is called *shotgun sequencing*, and the sequence of every piece is called a *read*. A read can originate from either of the two strands of a chromosome, and in the case of diploid cells from either of the two haplotypes. Sequencing loses the information about the position of a piece with respect to the original DNA fragment, therefore inferring the sequence of a chromosome from a corresponding set of reads is a nontrivial combinatorial puzzle, called *fragment assembly*, that we will solve in Chapter 13 by finding overlaps between pairs of reads. The term *high-throughput sequencing* refers to the fact that it is becoming increasingly cheaper to get a very large number of reads from a genome, or equivalently to cover every base of a genome with a large number of reads.

The technology to cut, amplify, and sequence DNA is under continuous development: rather than attempting a detailed description of the state of the art, we focus here on some fundamental concepts that most technologies have in common. By *short reads* we mean sequences whose length ranges from tens to hundreds of nucleotides: such reads typically contain just a few *sequencing errors*, most of which are single-base *substitutions* in which a base is mistaken for another base. By *long reads* we mean sequences that contain thousands of nucleotides: such reads are typically affected by a higher error rate.

For many species, accurate approximations of entire consensus genomes have been compiled by so-called *de novo sequencing* projects. Not surprisingly, the consensus human genome is quite accurate, and almost all of its roughly three gigabases are known. Common variants in the human population have also been mapped. As we will see in Chapter 14, it is relatively easy to detect SNP variants using a consensus genome and a set of reads from a specific individual, called a *donor*: indeed, it suffices to map every read to the position in the consensus genome that is "most similar" to it, and to study the resulting *pileup* of reads that cover a given position to detect whether the individual has a variation at that position. Collecting reads from a donor that belongs to an already sequenced species is called *whole genome resequencing*. In Chapter 10 we will study how reads can be mapped to their best matching locations using *sequence alignment* algorithms.

Figure 1.2 summarizes some key applications of high-throughput sequencing, in addition to the already mentioned de novo sequencing and whole genome resequencing. *RNA sequencing* allows one to sequence the complementary DNA *of RNA transcripts*: in this case, a read should be aligned to the reference genome allowing for one or more introns to split the read (see Section 10.6). In *targeted resequencing*, arbitrary areas of a genome are selected using specific marker sequences, called *primers*; such regions are then isolated, amplified, and sequenced, and the task is to detect their position in the reference genome using sequence alignment. We will mention an algorithm to compute primers in Section 11.2.3. In *chromatin immunoprecipitation sequencing* (often abbreviated to *ChIP-seq*) areas of the genome that contain the binding site of a specific set of transcription factors (or other DNA-binding proteins) are similarly isolated, amplified, and sequenced. Along the same lines, *bisulfite sequencing* can isolate and sequence areas of the genome with *methylation*. Methylation is a process in which cytosines (C) acquire a specific chemical compound, called a *methyl group*, and it is an example of an *epigenetic change*, or equivalently of a biochemical change that affects transcription *without altering the sequence of the genome*: indeed, methylation in promoter regions is known to affect the transcription rate of a gene. The set of all epigenetic changes of a genome, some of which are inheritable, is called the *epigenome*.

Finally, when a biological sample contains more than one species, the set of all the genetic material present in the sample is called the *metagenome*. As we will see

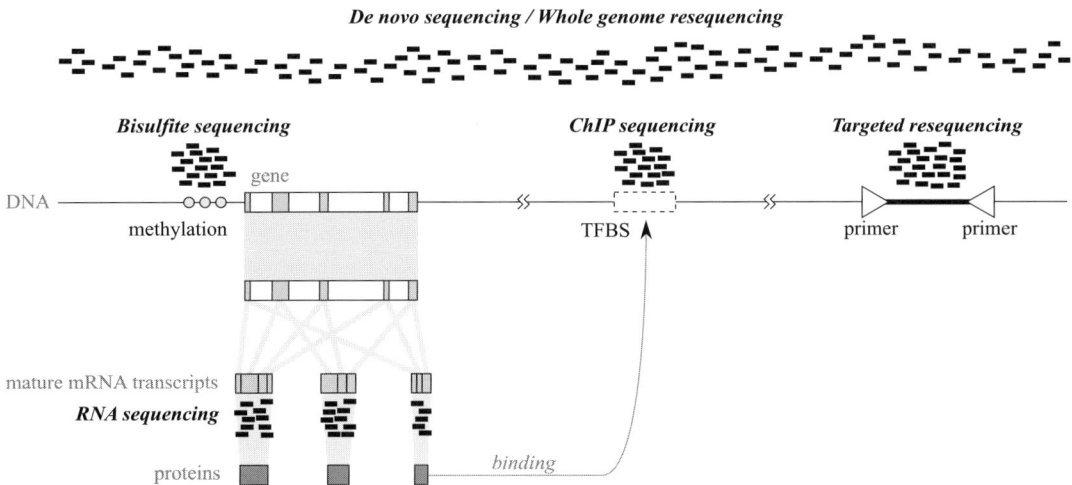

Figure 1.2 A schematic summary of high-throughput sequencing applications. Details are described in Section 1.3.

in Chapter 16, high-throughput sequencing can be used in this case to estimate the abundance of already sequenced species, and to reconstruct parts of the chromosomes of unknown species.

Exercises

1.1   Write a program that lists all the DNA sequences that encode a given protein sequence.

1.2   In a given organism some codons are used more frequently than others to encode the same amino acid. Given the observed frequency of every codon in a species, normalize it into probabilities and write a program that, given a protein sequence, samples a random DNA sequence that encodes that protein under such codon usage probabilities.

1.3   In a given organism, some *codon pairs* occur less frequently than others.

i.   Given the set of all exons of an organism, write a program that computes the ratio $z(XY)$ between the observed and the *expected* number of occurrences of every pair of consecutive codons XY. Note that the expected number of occurrences of pair XY depends on the frequency with which X and Y are used to encode their corresponding amino acids.

ii.   Given a DNA sequence S that encodes a protein P, let $f(S)$ be the average of $z(XY)$ over all consecutive pairs of codons XY in S. Write a program that computes, if it exists, a permutation S' of the codons of S, such that S' still encodes P, but $f(S') < f(S)$. Such an *artificially attenuated* version of S has been shown to decrease the rate of protein translation: for more details, see Coleman *et al.* (2008).

2 Algorithm design

Among the general algorithm design techniques, *dynamic programming* is the one most heavily used in this book. Since great parts of Chapters 4 and 6 are devoted to introducing this topic in a unified manner, we shall not introduce this technique here. However, a taste of this technique is already provided by the solution to Exercise 2.2 of this chapter, which asks for a simple one-dimensional instance.

We will now cover some basic primitives that are later implicitly assumed to be known.

2.1   Complexity analysis

The algorithms described in this book are typically analyzed for their *worst-case* running time and space complexity: complexity is expressed as a function of the input parameters on the worst possible input. For example, if the input is a string of length n from an alphabet of size σ, a *linear time* algorithm works in $O(n)$ time, where $O(\cdot)$ is the familiar big-O notation which hides constants. This means that the running time is upper bounded by cn elementary operations, where c is some constant.

We consider the alphabet size not to be a constant, that is, we use expressions of the form $O(n\sigma)$ and $O(n \log \sigma)$, which are not linear complexity bounds. For the space requirements, we frequently use notations such as $n \log \sigma (1 + o(1)) = n \log \sigma + o(n \log \sigma)$. Here $o(\cdot)$ denotes a function that grows asymptotically strictly slower than its argument. For example, $O(n \log \sigma / \log \log n)$ can be simplified to $o(n \log \sigma)$. Algorithms have an input and an output: by *working space* we mean the extra space required by an algorithm in addition to its input and output. Most of our algorithms are for processing DNA, where $\sigma = 4$, and it would seem that one could omit such a small constant in this context. However, we will later see a number of applications of these algorithms that are not on DNA, but on some derived string over an alphabet that can be as large as the input sequence length.

We often assume the alphabet of a string of length n to be $\Sigma = [1..\sigma]$, where $\sigma \leq n$. Observe that if a string is originally from an *ordered alphabet* $\Sigma \subseteq [1..u]$, with $\sigma = |\Sigma|$, it is easy to map the alphabet and the character occurrences in the string into alphabet $[1..\sigma]$ in $O(n \log \sigma)$ time using, for example, a *binary search tree*: see Section 3.1. Such mapping is not possible with an *unordered alphabet* Σ that supports only the operation $a = b$? for characters $a, b \in \Sigma$: comparison $a < b$? is not supported.

Yet still, the running time of some algorithms will be $O(dn^c)$, where c is a constant and d is some integer given in the input that can take arbitrarily large values, independently of the input size n. In this case we say that we have a *pseudo-polynomial* algorithm. More generally, a few algorithms will have running time $O(f(k)n^c)$, where c is a constant and $f(k)$ is an arbitrary function that does not depend on n but depends only on some other – typically smaller – parameter k of the input. In this case, we say that we have a *fixed-parameter tractable* algorithm, since, if k is small and $f(k)$ is sufficiently slow-growing, then this algorithm is efficient, thus the problem is *tractable*.

In addition to the notations $o(\cdot)$ and $O(\cdot)$, we will also use the notations $\Theta(\cdot)$, $\omega(\cdot)$, and $\Omega(\cdot)$ defined as follows. By $\Theta(\cdot)$ we denote a function that grows asymptotically as fast as its argument (up to a constant factor). Whenever we write $f(x) = \Theta(g(x))$, we mean that $f(x) = O(g(x))$ and $g(x) = O(f(x))$. The notions $\omega(\cdot)$ and $\Omega(\cdot)$ are the inverses of $o(\cdot)$ and $O(\cdot)$, respectively. Whenever we write $f(x) = \omega(g(x))$, we mean that $g(x) = o(f(x))$. Whenever we write $f(x) = \Omega(g(x))$, we mean that $g(x) = O(f(x))$.

None of the algorithms in this book will require any complex analysis technique in order to derive their time and space requirements. We mainly exploit a series of combinatorial properties, which we sometimes combine with a technique called *amortized analysis*. In some cases it is hard to bound the running time of a single step of the algorithm, but, by finer counting arguments, the total work can be shown to correspond to some other conceptual steps of the algorithm, whose total amount can be bounded. Stated differently, the time "costs" of the algorithm can all be "charged" to some other bounded resource, and hence "amortized". This technique deserves a concrete example, and one is given in Example 2.1.

Example 2.1 Amortized analysis – Knuth–Morris–Pratt

The *exact string matching* problem is that of locating the occurrences of a *pattern* string $P = p_1p_2 \cdots p_m$ inside a *text* string $T = t_1t_2 \cdots t_n$. The Morris–Pratt (MP) algorithm solves this problem in optimal linear time, and works as follows. Assume first that we already have available on P the values of a *failure function*, $\mathtt{fail}(i)$, defined as the length of the longest prefix of $p_1p_2 \cdots p_i$ that is also a suffix of $p_1p_2 \cdots p_i$. That is, $\mathtt{fail}(i) = i'$, where $i' \in [0 .. i - 1]$ is the largest such that $p_1p_2 \cdots p_{i'} = p_{i-i'+1}p_{i-i'+2} \cdots p_i$. The text T can be scanned as follows. Compare p_1 to t_1, p_2 to t_2, and so on, until $p_{i+1} \neq t_{j+1}$, where j is an index in T and equals i throughout the first iteration. Apply $i = \mathtt{fail}(i)$ recursively until finding $p_{i+1} = t_{j+1}$, or $i = 0$; denote this operation by $\mathtt{fail}^*(i, t_{j+1})$. Then, continue the comparisons, by again incrementing i and j, and again resetting $i = \mathtt{fail}^*(\cdot, \cdot)$ in case of mismatches. An occurrence of the pattern can be reported when $i = |P| + 1$.

To analyze the running time, observe first that each character t_j of T is checked for equality with some character of P exactly once. Thus, we need only estimate the time needed for the recursive calls $\mathtt{fail}^*(i, t_{j+1})$. However, without a deeper analysis, we can only conclude that each recursive call takes $O(m)$ time before one can increment j, which would give an $O(mn)$ bound for the time complexity of the algorithm. In other words, it is *not* enough to consider the worst-case running time of each step independently, and then just sum the results. However, we can show that the total number of recursive calls is at most the length of T, thus each recursive call can be "charged" to each element of T.

Observe that if $\mathtt{fail}^*(i, t_{j+1}) = i'$, then there are at most $i - i'$ recursive calls done for t_{j+1}. We "charge" these calls to the $i - i'$ characters $t_{j-i+1}t_{j-i+2}\cdots t_{j-i'}$ of T, characters which matched the prefix $p_1\cdots p_{i-i'}$. We need only show that each t_j gets charged by at most one recursive call, which will allow us to conclude that the total number of recursive calls is bounded by n. To see this, observe that after setting $i' = \mathtt{fail}^*(i, t_{j+1})$ we have $p_1\cdots p_{i'} = t_{j-i'+1}\cdots t_j$, hence new recursive calls get charged to positions from $j - i' + 1$ onwards, since this is where a prefix of P again starts matching.

We still need to consider how to compute the values of the $\mathtt{fail}(\cdot)$ function in $O(m)$ time; this is considered in Exercise 2.1.

The Knuth–Morris–Pratt (KMP) algorithm is a variant of the MP algorithm, where the $\mathtt{fail}(\cdot)$ function is optimized to avoid some unnecessary comparison: see Exercise 2.2. We will use the KMP algorithm and its combinatorial properties in Chapter 12. For those developments, the concept of a *border* is useful: the suffix $B = P_{m-i+1..m}$ of $P[1..m]$ is called a border if it is also a prefix of P, that is, if $B = P_{1..i}$. The length of the *longest border* is given by $i = \mathtt{fail}(m)$, and all other border lengths can be obtained by applying $\mathtt{fail}(\cdot)$ recursively, where $\mathtt{fail}(\cdot)$ is now referring to the MP algorithm.

The $\mathtt{fail}(\cdot)$ function is often referred to as a *(K)MP automaton*.

The KMP algorithm from the above example can be useful only if the text in which the pattern is being sought is relatively short. The search time will be huge if one has to search for a sequence inside a fairly large genome. Thus, KMP cannot be used for genome-scale algorithms.

In order to support fast searches, it is necessary to have an *index* of the text. This will be in contrast to the KMP algorithm which, through the $\mathtt{fail}(\cdot)$ function, indexed the pattern. In Chapter 8 we will study various indexing data structures, such as suffix trees and suffix arrays, which use $O(n)$ words of space for a text consisting of n characters, and which find occurrences of a pattern of length m in time $O(m)$. This is opposed to the KMP algorithm, which uses $O(m)$ words of space, but has a running time of $O(n)$. Still, for large genomes, another problem may arise, since the space usage $O(n)$ words might be too big to fit into the memory. In Chapter 9 we will introduce the concept of succinct text indexes, which can save an order of magnitude of space while still allowing fast queries. We introduce the concept of succinct data structures in the next section.

2.2   Data representations

We assume the reader is familiar with the elementary pointer-manipulation data structures such as *stacks* and *queues*, represented by doubly-linked lists, and with representations of *trees*. We also assume that the reader is familiar with bitvector manipulation routines (left/right shift, logical or/and) that give the possibility of allocating *arrays* with fixed-length *bit-fields*. If this is not the case, the assignments in the exercise section provide a crash course.

We emphasize that, throughout the book, we express all space requirements in *computer words*, except when we aim at optimizing the space needed by some large data structure, where we always add "bits of space" to make it clear. We use the *random access model* (RAM) to describe the content, with a model in which the computer word size w is considered to be $\Omega(\log U + \log n)$ bits, where U is the maximum input value and n the problem size. Basic arithmetic, logical, and bit-manipulation operations on words of length w bits are assumed to take constant time. For example, an array $A[1..U]$ with each $A[i] \in [1..U]$, can be represented in $U(\lfloor \log_2 U \rfloor + 1)$ bits (which we often simplify to $U \log U$), in the fixed-length bit-fields representation (see Exercise 2.7). A static tree of n nodes can be stored as a constant number of arrays of length n (Exercise 2.8), leading to $O(n \log n)$-bit representation. Even a dynamic tree can be maintained using $O(n \log n)$ bits, with tailor-made memory-allocation routines exploiting *dynamic tables*. We do not need to touch on this area in this book, since we have chosen to modify the algorithms to make use instead of *semi-dynamic structures*, in which elements are inserted at the beginning and updated later.

We often talk about succinct data structures. Such structures occupy space equal to the storage required for representing the input, plus a lower-order term. More formally, suppose that the input to the data structure is an element X taken from a set of elements $\mathcal{U}$. Then, a *succinct* data structure on X is required to occupy $\log |\mathcal{U}|(1 + o(1))$ bits of space. For example, a text T of length n from an alphabet Σ of size σ is an element of the set $U = \Sigma^n$ of all strings of length n over alphabet Σ. A data structure that takes as input T is succinct if it uses $n \log \sigma (1 + o(1))$ bits of space.

If a data structure built on T can be used to restore T after T itself has been deleted, we say that the data structure is a *representation* of T. If the same data structure occupies $n \log \sigma (1 + o(1))$ bits of space, we call it a *succinct representation* of T.

Observe that succinct space is the information-theoretic minimum, since otherwise two inputs would necessarily need to have the same representation. However, there could be *compressed* representations that occupy less than succinct space on *some inputs*. Many of the succinct data structures and representations covered in the book can be made compressed, but, for clarity, we do not explore this wide area of research systematically; this topic is touched on only in Section 10.7.1 and Chapter 12.

2.3 Reductions

One important algorithm design technique is to exploit connections between problems. There are some fundamental problems that have been studied for decades and sophisticated algorithms have already been developed for them. When formulating a new problem, for example, motivated by an analysis task in high-throughput sequencing, one should first look to see whether a solution to some widely studied problem could be used to solve the newly formulated problem. Sometimes one can exploit such a connection by modifying an algorithm for the old problem to solve the new problem, but, if the connection is close enough, one may even be able to *reduce* the new problem to the old problem.

This reduction means that any input to the new problem can be recast as an input of the old problem such that, on solving the old problem with some algorithm, one solves also the new problem. The running time of such an algorithm for the new problem is then the combined running time of recasting the input as an input to the old problem, the running time of the algorithm for the old problem, and then the time needed for recasting this output as an output for the original problem. Since we frequently exploit this kind of reduction in this book, we illustrate one such reduction in Example 2.2.

Example 2.2 Reduction to a tractable problem – longest increasing subsequence

Let the old problem be the one of finding the *longest common subsequence (LCS)* of two strings $A = a_1 a_2 \cdots a_m$ and $B = b_1 b_2 \cdots b_n$. For example, with

- $A = \text{ACAGTGA}$
- $B = \text{CCTGTAG}$

the answer is $C = \text{CGTA}$, since C is a subsequence of both strings, meaning that C can be obtained by deleting zero or more characters from A and from B, and there is no common subsequence of length 5 or more. We study this problem in Section 6.2, where we give an algorithm running in time $O(|M|\log \sigma)$, where M is the set of position pairs (i, j) of A and B with $a_i = b_j$, and σ is the alphabet size.

Let the new problem be the one of finding the *longest increasing subsequence (LIS)* in a sequence of numbers $S = s_1, s_2, \ldots, s_k$. For example, on $S = 1, 8, 4, 1, 10, 6, 4, 10, 8$, possible answers are $1, 4, 6, 8$ or $1, 4, 6, 10$. We can reduce LIS to LCS as follows. Sort the set of distinct numbers of S in $O(k \log k)$ time and replace each occurrence of a number in S with its *rank* in this sorted order: the smallest number gets rank 1, the second smallest gets rank 2, and so on. In our example, S is replaced by a sequence of ranks $R = 1, 4, 2, 1, 5, 3, 2, 5, 4$. Clearly, any longest increasing subsequence of ranks in R can be mapped to a longest increasing subsequence of S. Now $A = R$ (taken as a string) and $B = \min(R)(\min(R) + 1) \cdots \max(R) = 123 \cdots \max(R)$ form the input to the LCS. Observe that $\max(R) \leq k$. In our example,

- $A = 142153254$
- $B = 12345$.

Any longest common subsequence of A and B defined as above can be mapped to a longest subsequence of R, and hence of S. In our example, $C = 1234$ is one such longest common subsequence, and also the longest increasing subsequence of R, and hence of S.

The $O(|M|\log \sigma)$ algorithm for LCS simplifies to $O(k \log k)$, since $\sigma \leq k$, and each index i in A ($= R$) is paired with exactly one index j in B. The reduction itself took $O(k \log k)$ time to create, the solution to the reduced instance took the same

$O(k \log k)$ time, and casting the result of the LCS problem as a longest increasing subsequence in S can be done in time $O(k)$. Thus, the overall running time to solve the LIS problem is $O(k \log k)$.

Another kind of reduction will concern the *hardness* of some problems. There is a class of fundamental problems for which no efficient solutions are known. (By efficient, we mean an algorithm with *polynomial* running time with respect to the input size.) These hard problems are known only to have *exponential* time algorithms; such a running time can be achieved, for example, by enumerating through all possible answers, and checking whether one of the answers is a proper solution. A well-known example of a hard problem is the *Hamiltonian path* problem, in which one asks whether there is a route in a graph (for example, a railway network) that visits all the vertices (cities) exactly once. With n vertices, a trivial solution is to consider all $n! = \Omega(2^n)$ permutations of vertices, and to check for each one of them whether all consecutive vertices (cities) are connected by an edge (a railway line).

Assuming now that an old problem is hard, if we show that we can solve the old problem by recasting it as a new problem, by appropriately converting the input and output, then we can conclude that the new problem is at least as hard as the old problem. In this context, the running time of converting the input and output is just required to be polynomial in their sizes. Let us now describe these notions in slightly more formal terms.

Fix an alphabet Σ of size at least two. A *decision problem* is a subset of words over Σ^*. For example, if Σ is the ASCII alphabet, then under the standard convention for encoding graphs, an input to the Hamiltonian path problem is

$$w_G = (\{1, 2, 3, 4\}, \{(1, 2), (2, 3), (1, 3), (3, 4)\}).$$

If Π_H is the Hamiltonian path problem, $w_G \in \Pi_H$ holds. Having an algorithm for solving Π_H means that, for any input word $w \in \Sigma^*$, we can decide whether $w \in \Pi_H$.

Among all problems, one class of problems is of particular interest, called NP (nondeterministic polynomial time). A problem Π is in NP if each $w \in \Pi$ admits a "certificate" that can be checked in time polynomial in the length of w. For example, a certificate for Π_H is a permutation of the vertex set (the Hamiltonian path), which can be checked in polynomial time in the size of the vertex set to be a Hamiltonian path (see Exercises 2.4 and 2.5). A certificate for the particular instance w_G can be the sequence of vertices $w'_g = (1, 2, 3, 4)$ or the sequence $w''_g = (2, 1, 3, 4)$.

It is an open question whether every problem in NP can be also decided in time polynomial in the length of the input. To address this question, the following notion has been introduced: say that a problem Π is *NP-hard* if, for every problem $\Pi' \in$ NP, there is a polynomial-time algorithm that, given any $w \in \Sigma^*$, returns $f(w) \in \Sigma^*$ such that

$$w \in \Pi' \Leftrightarrow f(w) \in \Pi.$$

A problem is said to be *NP-complete* if it is NP-hard and it belongs to NP. Having a polynomial-time algorithm for an NP-complete problem would imply that all problems in NP can be solved in polynomial time. Many problems are NP-complete, including also the Hamiltonian path problem mentioned above. To show that some problem Π

is NP-hard, it suffices to exhibit a polynomial-time reduction to Π from *any* NP-hard problem. Example 2.3 provides one illustrative NP-hardness reduction.

Example 2.3   Reduction from an intractable problem – Vertex Cover Let the old NP-hard problem be the *Clique* problem: given an undirected graph G and an integer k, decide whether G has a clique of size k, namely a subset K of vertices such that every two vertices in K are connected by an edge. For example, in the graph G below, $K = \{a, b, c\}$ is a clique of size 3.

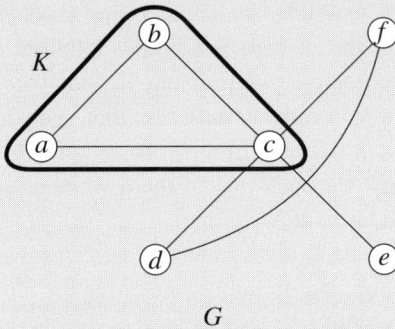

G

Let the new problem be the *Vertex Cover* problem: given an undirected graph G and an integer k, decide whether G has a vertex cover of size t, namely a subset T of vertices such that every edge of G has at least one extremity in T. For example, in the graph G above, $T = \{a, c, f\}$ is a vertex cover.

We now reduce Clique to Vertex Cover. Given a graph G, whose set of vertices is V, and an integer k, as input for the Clique problem, we construct the complement graph of G, denoted $\overline{G}$, which has the same set V of vertices, and in which an edge between two vertices is present if and only if it is absent in G. Observe that, if K is a clique in G, then $V \setminus K$ is a vertex cover in $\overline{G}$. Indeed, if there were an edge in $\overline{G}$ between u and v with both u and v outside $V \setminus K$ then in G there would be no edge between u and v, contradicting the fact that K is a clique. See the figure below for an example. Conversely, if T is a vertex cover in $\overline{G}$, then $V \setminus T$ is a clique in G.

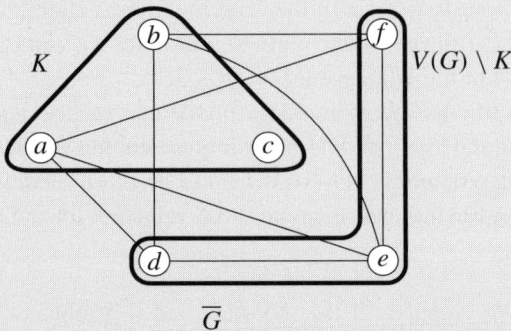

$\overline{G}$

Therefore, we can solve the Clique problem on G and k by solving the Vertex Cover problem on $\overline{G}$ and $t = |V| - k$. If the answer for the Vertex Cover problem is "yes", then G has a clique of size k; if the answer is "no", then G does not have a clique of size k.

We have shown that the Vertex Cover problem is at least as hard as the Clique problem (up to a polynomial-time reduction). Since the clique problem is NP-hard, we can conclude that also the Vertex Cover problem is NP-hard. Observe also that Vertex Cover too belongs NP, since given a subset T of vertices (the "certificate") we can check in polynomial time whether it is a vertex cover. Thus, Vertex Cover is NP-complete.

2.4   Literature

The Morris–Pratt algorithm appeared in Morris & Pratt (1970), and the Knuth–Morris–Pratt algorithm in Knuth *et al.* (1977). An exercise below hints how to extend this algorithm for multiple patterns, that is, to rediscover the algorithm in Aho & Corasick (1975).

The longest increasing subsequence (LIS) problem has a more direct $O(k \log k)$ solution (Fredman 1975) than the one we described here through the reduction to the longest common subsequence (LCS) problem. As we learn in Section 6.7, the LCS problem can also be solved in $O(|M| \log \log k)$ time, where k is the length of the shorter input string. That is, if a permutation is given as input to the LIS problem such that sorting can be avoided (the rank vector is the input), the reduction gives an $O(k \log \log k)$ time algorithm for this special case. For an extensive list of NP-complete problems and further notions, see Garey & Johnson (1979).

Exercises

2.1   Consider the $\mathtt{fail}(\cdot)$ function of the Morris–Pratt (MP) algorithm. We should devise a linear-time algorithm to compute it on the pattern to conclude the linear-time exact pattern-matching algorithm. Show that one can modify the same MP algorithm so that on inputs $P = p_1 p_2 \cdots p_m$ and $T = p_2 p_3 \cdots p_m \#^m$, where $\#^m$ denotes a string of m concatenated endmarkers #, the values $\mathtt{fail}(2), \mathtt{fail}(3), \ldots, \mathtt{fail}(m)$ can be stored on the fly before they need to be accessed.

2.2   The Knuth–Morris–Pratt (KMP) algorithm is a variant of the MP algorithm with optimized $\mathtt{fail}(\cdot)$ function: $\mathtt{fail}(i) = i'$, where i' is largest such that $p_1 p_2 \cdots p_{i'} = p_{i-i'+1} p_{i-i'+2} \cdots p_i$, $i' < i$, and $p_{i'+1} \neq p_{i+1}$. This last condition makes the difference from the original definition. Assume you have the $\mathtt{fail}(\cdot)$ function values computed with the original definition. Show how to update these values in linear time to satisfy the KMP optimization.

2.3   Generalize KMP for solving the *multiple pattern-matching* problem, where one is given a set of patterns rather than only one as in the exact string matching problem. The goal is to scan T in linear time so as to find exact occurrences of any pattern in the given set. *Hint.* Store the patterns in a tree structure, so that common prefixes of patterns share the same subpath. Extend $\mathtt{fail}(\cdot)$ to the positions of the paths in the tree. Observe that, unlike in KMP, the running time of the approach depends on the alphabet size σ. Can you obtain scanning time $O(n \log \sigma)$? Can you build the required tree data structure in $O(M \log \sigma)$ time, where M is the total length of the patterns? On top of the $O(n \log \sigma)$ time for scanning T, can you output all the occurrences of all patterns in linear time in the output size?

2.4   Show that a certificate for the Hamiltonian path problem can be checked in time $O(n)$ (where n is the number of vertices) assuming an adjacency representation of the graph that uses $O(n^2)$ bits. *Hint.* Use a table of n integers that counts the number of occurrences of the vertices in the given certificate.

2.5   Suppose that we can afford to use no more than $O(m)$ space to represent the adjacency list. Show that a certificate for the Hamiltonian path can now be checked in time $O(n \log n)$.

2.6   Find out how bit-manipulation routines are implemented in your favorite programming language. We visualize below binary representations of integers with the most-significant bit first. You might find useful the following examples of these operations:

- *left-shift*: 0000000000101001 << 2 = 0000000010100100,
- *right-shift*: 0000000010100100 >> 5 = 0000000000000101,
- *logical or*: 0000000000101001 | 1000001000001001 = 1000001000101001,
- *logical and*: 0000000000101001 & 1000001000001001 = 0000000000001001,
- *exclusive or*: 0000000000101001 ⊕ 1000001000001001 = 1000001000100000,
- *complement*: ~0000000000101001 = 1111111111010110,
- *addition*: 0000000000101001 + 0000000000100001 = 0000000001001010, and
- *subtraction*: 0000000000101001 − 0000000000100001 = 0000000000001000, 000000000001000 − 000000000000001 = 000000000000111.

These examples use 16-bit variables (note the overflow). Show two different ways to implement a function $\mathtt{mask}(B, d)$ that converts the d most significant bits of a variable to zero. For example, $\mathtt{mask}(1000001000001001, 7) = 0000000000001001$.

2.7   Implement with your favorite programming language a fixed-length bit-field array. For example, using C++ you can allocate with

```
A=new unsigned[(n*k)/w+1]
```

an array occupying roughly $n \cdot k$ bits, where w is the size of the computer word (unsigned variable) in bits and $k < w$. You should provide operations $\mathtt{setField(A,i,x)}$ and $\mathtt{x=getField(A,i)}$ to store and retrieve integer x from A, for x whose binary representation occupies at most k bits.

2.8   Implement using the above fixed-length bit-field array an $O(n \log n)$-bit representation of a node-labeled static tree, supporting navigation from the root to the children.

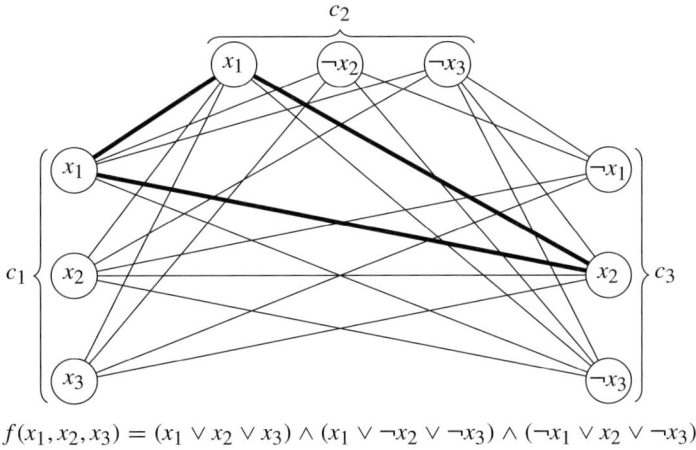

$$f(x_1, x_2, x_3) = (x_1 \lor x_2 \lor x_3) \land (x_1 \lor \neg x_2 \lor \neg x_3) \land (\neg x_1 \lor x_2 \lor \neg x_3)$$

Figure 2.1 A reduction of the 3-SAT problem to the Clique problem. A clique in G_f is highlighted; this induces either one of the truth assignments $(x_1, x_2, x_3) = (1, 1, 0)$ or $(x_1, x_2, x_3) = (1, 1, 1)$.

2.9 Recall how stack, queue, and *deque* work. Implement them using your favorite programming language using doubly-linked lists.

2.10 Given an undirected graph G, a subset $S \subseteq V(G)$ is called an *independent set* if no edge exists between the vertices of S. In the independent-set problem we are given an undirected graph G and an integer k and are asked whether G contains an independent set of size k. Show that the independent-set problem is NP-complete.

2.11 A Boolean formula $f(x_1, \ldots, x_n)$ is in *3-CNF form* if it can be written as

$$f(x_1, \ldots, x_n) = c_1 \land \cdots \land c_m,$$

where each c_i is $y_{i,1} \lor y_{i,2} \lor y_{i,3}$, and each $y_{i,j}$ equals x_k or $\neg x_k$, for some $k \in \{1, \ldots, n\}$ (with $y_{i,1}$, $y_{i,2}$, $y_{i,3}$ all distinct). The subformulas c_i are called *clauses*, and the subformulas $y_{i,j}$ are called *literals*. The following problem, called 3-SAT, is known to be NP-complete. Given a Boolean formula $f(x_1, \ldots, x_n)$ in 3-CNF form, decide whether there exist $\alpha_1, \ldots, \alpha_n \in \{0, 1\}$ such that $f(\alpha_1, \ldots, \alpha_n)$ is true (such values α_i are called a *satisfying truth assignment*).

Consider as a "new" problem the clique problem from Example 2.3. Show that clique is NP-complete by constructing a reduction from 3-SAT. *Hint.* Given a 3-CNF Boolean formula

$$f(x_1, \ldots, x_n) = (y_{1,1} \lor y_{1,2} \lor y_{1,3}) \land \cdots \land (y_{m,1} \lor y_{m,2} \lor y_{m,3}),$$

construct the graph G_f as follows (see Figure 2.1 for an example):

- for every $y_{i,j}$, $i \in \{1, \ldots, m\}$, $j \in \{1, 2, 3\}$, add a vertex $y_{i,j}$ to G_f; and
- for every $y_{i_1,j}$ and $y_{i_2,k}$ with $i_1 \neq i_2$ and $y_{i_1,j} \neq \neg y_{i_1,k}$, add the edge $(y_{i_1,j}, y_{i_2,k})$.

Show that f has a satisfying assignment if and only if G_f has a clique of size m.

3   Data structures

We will now describe a couple of basic data structures that will be used extensively throughout the book.

3.1   Dynamic range minimum queries

A *binary tree* is a tree with each internal node having at most two children. We denote by $\overleftarrow{v}$ and $\overrightarrow{v}$ the left-most and right-most leaves in the subtree rooted at v. For a tree node v, we denote by $key(v)$ the *search key* and by $value(v)$ the *value* associated with the key. A *binary search tree* is a binary tree with the following properties:

(a)   each leaf stores a pair $(key(v), value(v))$;
(b)   leaves are ordered left to right with increasing key; and
(c)   each internal node v stores $key(v)$ equalling $key(\overrightarrow{v})$.

The total order in (b) enforces the implicit assumption that all keys are unique. A *balanced binary search tree* is a binary search tree on n pairs $(key, value)$ such that each leaf can be reached with a path of length $O(\log n)$ from the root. Such a tree has $n - 1$ internal nodes. Observe that our definition follows the computational geometry literature and deviates slightly from the more common definition with the data stored in all nodes and not only in leaves.

   While there exist dynamic balanced binary search trees that allow elements to be inserted and deleted, a semi-dynamic version is sufficient for our purposes. That is, we can just construct a balanced search tree by sorting the set of keys, assigning them to the leaves of an empty tree in that order with values initialized to ∞, and building internal nodes level by level by pairing consecutive leaves (see Exercise 3.2). Such a fully balanced tree can be built in linear time after the keys have been sorted. Now we can consider how to update the values and how to query the content.

LEMMA 3.1   *The following two operations can be supported with a balanced binary search tree $\mathcal{T}$ in time $O(\log n)$, where n is the number of leaves in the tree.*

   update(k, val): *For the leaf w with* $key(w) = k$, *update* $value(w) = val$.
   RMQ(l, r): *Return* $\min_{w \,:\, l \leq key(w) \leq r} val(w)$ *(Range Minimum Query).*
   Moreover, the balanced binary search tree can be built in $O(n)$ time, given the n pairs
      $(key, value)$ *sorted by component* key.

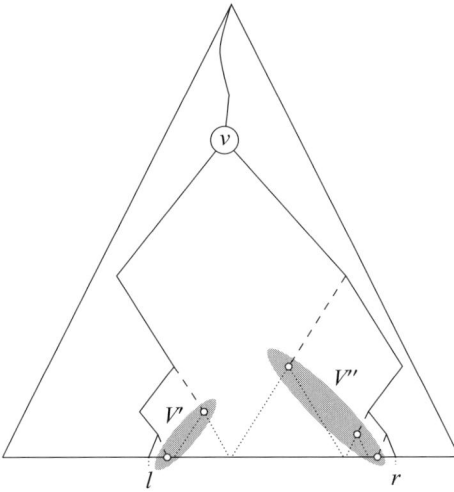

Figure 3.1 Answering a range minimum query with a binary search tree.

Proof Store for each internal node v the minimum among $\texttt{value}(i)$ associated with the leaves i under it. Analogously to the leaves, let us denote this minimum value by $\texttt{value}(v)$. These values can be easily updated after a call of an operation $\texttt{update}(k, \texttt{val})$; only the $O(\log n)$ values on the path from the updated leaf towards the root need to be modified.

It is hence sufficient to show that query $\mathsf{RMQ}(l, r)$ can be answered in $O(\log n)$ time. Find node v, where the search paths to keys l and r separate (it can be the root, or empty when there is at most one key in the query interval). Let $\texttt{path}(v, l)$ denote the set of nodes through which the path from v goes when searching for key l, excluding node v and leaf L where the search ends. Similarly, let $\texttt{path}(v, r)$ denote the set of nodes through which the path from v goes when searching for key r, excluding node v and leaf R where the search ends. Figure 3.1 illustrates these concepts.

Now, for all nodes in $\texttt{path}(v, l)$, where the path continues to the left, it holds that the keys k in the right subtree are at least l and at most r. Choose $vl = \min_{v' \in V'}(\texttt{value}(v'))$, where V' is the set of roots of their right subtrees. Similarly, for all nodes in $\texttt{path}(v, r)$, where the path continues to the right, it holds that the keys k in the left subtree are at most r and at least l. Choose $vr = \min_{v'' \in V''}(\texttt{value}(v''))$, where V'' is the set of roots of their left subtrees. If $L = l$ update $vl = \min(vl, \texttt{value}(L))$. If $R = r$ update $vr = \min(vr, \texttt{value}(R))$. The final result is $\min(vl, vr)$.

The correctness follows from the fact that the subtrees of nodes in $V' \cup V''$ contain all keys that belong to the interval $[l..r]$, and only them (excluding leaves L and R, which are taken into account separately). The running time is clearly $O(\log n)$. □

It is not difficult to modify the above lemma to support insertion and deletion operations as well: see Exercise 3.3.

3.2   Bitvector rank and select operations

A *bitvector* is an array $B[1..n]$ of Boolean values, that is, $B[i] \in \{0, 1\}$. An elementary operation on B is to compute

$$\text{rank}_c(B, i) = |\{i' \mid 1 \leq i' \leq i, B[i'] = c\}|, \qquad (3.1)$$

where $c \in \{0, 1\}$. Observe that $\text{rank}_0(B, i) = i - \text{rank}_1(B, i)$ for $B \in \{0, 1\}^*$. It is customary to use just $\text{rank}(B, i)$ to denote $\text{rank}_1(B, i)$.

Storing all values $\text{rank}(B, i)$ as an array would take $O(n \log n)$ bits, which is too much for our later purposes. Before entering into the details on how to solve rank queries more space-efficiently, let us consider a motivating example on a typical use of such structure. Consider a *sparse* table $A[1..n] \in \{0, 1, 2, \ldots, u\}^n$ having k non-zero entries, where k is much smaller than n. Let $B[1..n]$ have $B[i] = 1$ iff $A[i] \neq 0$ and let $A'[1..k]$ have $A'[\text{rank}(B, i)] = A[i]$ for $B[i] = 1$. We can now replace A that occupies $n \log u$ bits, with B and A' that occupy $n + k \log u$ bits plus the size of the data structure to support rank operations.

We shall soon see that $o(n)$ additional bits are enough to answer all rank queries in constant time. This solution is gradually developed in the following paragraphs. To simplify the formulas, we implicitly assume that all real numbers are rounded down to the nearest integer, unless explicitly indicated otherwise with, for example, the ceiling $\lceil \cdot \rceil$ operation.

Partial solution 1: Let us store each ℓth $\text{rank}(B, i)$ as is and scan the rest of the bits (at most $\ell - 1$), during the query. We then have an array $\text{first}[0..n/\ell]$, where $\text{first}[0] = 0$ and $\text{first}[i/\ell] = \text{rank}(B, i)$ when $i \bmod \ell = 0$ (/ is here integer division). If we choose $\ell = (\lceil (\log n)/2 \rceil)^2$, we need on the order of $n \log n/(\log^2 n) = n/(\log n)$ bits space for the array first. We can answer $\text{rank}(B, i)$ in $O(\log^2 n)$ time: $\text{rank}(B, i) = \text{first}[i/\ell] + \text{count}(B, [\ell \cdot (i/\ell) + 1..i])$, where $\text{count}(B, [i'..i])$ computes the amount of 1-bits in the range $B[i'..i]$.

Partial solution 2: Let us store more answers. We store inside each block of length ℓ induced by first answers for each kth position (how many 1-bits from the start of the block). We obtain an array $\text{second}[0..\ell/k]$, where $\text{second}[i/k] = \text{rank}(B, [\ell(i/\ell) + 1..i])$, when $i \bmod k = 0$. This uses overall space $O((n/k) \log \ell)$ bits. Choosing $k = \lceil (\log n)/2 \rceil$ gives $O(n \log \log n/(\log n))$ bits space usage. Now we can answer $\text{rank}(B, i)$ in $O(\log n)$ time, as $\text{rank}(B, i) = \text{first}[i/\ell] + \text{second}[i/k] + \text{count}(B, [k \cdot (i/k) + 1..i])$.

Final solution: We use the so-called *four Russians technique* to improve the $O(\log n)$ query time to constant. This is based on the observation that there are fewer than $\sqrt{n}$ possible bitvectors of length $k - 1 = \lceil (\log n)/2 \rceil - 1 < (\log n)/2$. We store a table $\text{third}[0..2^{k-1} - 1][0..k - 2]$, which stores the answers for rank queries for all possible $k - 1 < (\log n)/2$ positions on all possible $2^{k-1} \leq \sqrt{n}$ block configurations, where the answer is an integer of size $\log(k) \leq \log \log n$ bits. This takes in total $O(\sqrt{n} \log n \log \log n)$ bits. Let c_i be the bitvector $B[k \cdot (i/k) + 1..k \cdot (i/k + 1) - 1]$.

We obtain the final formula to compute $\mathrm{rank}(B, i)$:

$$\mathrm{rank}(B, i) = \mathtt{first}[i/\ell] + \mathtt{second}[i/k] + \mathtt{third}[c_i][(i \bmod k) - 1]. \qquad (3.2)$$

Integer c_i can be read in constant time from the bitvector B: see Exercise 2.7.
    Example 3.1 illustrates these computations.

Example 3.1   Simulate $\mathrm{rank}(B, 18)$ on $B = 0110101100011101010110\ldots$
assuming $\ell = 16$ and $k = 4$.

Solution

```
                                                  i = 18
  B        0 1 1 0 1 0 1 1 0 0 0 1 1 1 0 1 0 1 0 1 0 1 1 0...

first    0                                   9

second   0       2         5         6         0         11        13
```

third	0	1	2
000	0	0	0
001	0	0	1
010	0	1	1
011	0	1	2
100	1	0	0
101	1	1	2
110	1	2	2
111	1	2	3

$$\begin{aligned}
\mathrm{rank}(B, 18) &= \mathtt{first}[18/16] + \mathtt{second}[18/4] + \mathtt{third}[010][(18 \bmod 4) - 1] \\
&= \mathtt{first}[1] + \mathtt{second}[4] + \mathtt{third}[010][1] \\
&= 9 + 0 + 1 \\
&= 10
\end{aligned}$$

THEOREM 3.2   *There is a bitvector representation using $o(n)$ bits in addition to the bitvector $B[1..n]$ itself, such that operation $\mathrm{rank}(B, i)$ can be supported in constant time, assuming a RAM model with computer word size $w = \Omega(\log n)$. This representation can be built in $O(n)$ time and $n + o(n)$ space.*

We will also need the inverse operation $\mathtt{select}_1(B, j) = i$ that gives the maximal i such that $\mathrm{rank}_1(B, i) = j$, that is, the position i of the jth bit set in B. This operation can also be supported in constant time, after building a similar but a bit more involved data structure than for $\mathtt{rank}$: Exercise 3.7 asks to prove the following claim.

THEOREM 3.3   *There is a bitvector representation using $o(n)$ bits in addition to the bitvector $B[1..n]$ itself, such that operations $\mathtt{select}_1(B, j)$ and $\mathtt{select}_0(B, j)$ can* *be supported in constant time, assuming a RAM model with computer word size $w = \Omega(\log n)$. This representation can be built in $O(n)$ time and $n + o(n)$ space.*

3.3   Wavelet tree

The *wavelet tree* generalizes `rank` and `select` queries to sequences from any alphabet size. Let us denote by $\text{rank}_c(T, i)$ the count of characters c up to position i in T, and $\text{select}_c(T, j)$ the position of the jth occurrence of c in T, for $T \in \Sigma^*$. Note that $\text{rank}_c(T, \text{select}_c(T, j)) = j$.

Let us first consider a naive representation of a string $T[1..n] \in \Sigma^*$ as σ binary strings $B_c[1..n]$ for all $c \in \Sigma$, such that $B_c[i] = 1$ if $T[i] = c$, otherwise $B_c[i] = 0$. Now, $\text{rank}_c(T, i) = \text{rank}(B_c, i)$. After preprocessing binary strings B_c for `rank` queries, we can answer $\text{rank}_c(T, i)$ in constant time using $\sigma n(1 + o(1))$ bits of space.

We will next show a more space-efficient represenation that also reduces the problem of `rank` computation on general sequences to computation on binary sequences.

3.3.1   Balanced representation

Consider a perfectly balanced binary tree where each node corresponds to a subset of the alphabet $\Sigma = \{1, 2, \ldots, \sigma\}$. The children of each node partition the node subset into two. A bitvector B_v at node v indicates to which children each sequence position belongs. Each child then handles the subsequence of the parent's sequence corresponding to its alphabet subset. The root of the tree handles the sequence $T[1..n]$. The leaves of the tree handle single alphabet characters and require no space.

To answer query $\text{rank}_c(T, i)$, we first determine to which branch of the root c belongs. If it belongs to the left ($c \leq \sigma/2$), then we recursively continue at the left subtree with $i = \text{rank}_0(B_{\text{root}}, i)$. Otherwise we recursively continue at the right subtree with $i = \text{rank}_1(B_{\text{root}}, i)$. The value reached by i when we arrive at the leaf that corresponds to c is $\text{rank}_c(T, i)$.

The character t_i at position i is obtained similarly, this time going left or right depending on whether $B_v[i] = 0$ or 1 at each level, and finding out which leaf we arrived at. Denote this operation $\text{access}(T, i)$.

This hierarchical structure is called a *wavelet tree*. Example 3.2 illustrates `rank` and `access` operations. Note that string T can be replaced with its wavelet tree, and retain access to all its poistions.

> **Example 3.2** Visualize the wavelet tree of the string $T = $ AGTCGATTAC CGTGCGAGCTCTGA and solve $\text{rank}_C(T, 18)$ and $\text{access}(T, 18)$.
>
> **Solution**
> One can map the alphabet {A, C, G, T} into [1..4] by associating A = 1, C = 2, G = 3, and T = 4. Then [1..2] ({A, C}) go to the left and [3..4] ({G, T}) to the right, and so on. We obtain the following visualization, where the bitvectors and leaf labels form the wavelet tree.

$$r \qquad i = 18$$

$$
\begin{array}{c}
\text{A G T C G A T T A C C G T G C G A} \;\vert\; \text{G C T C T G A} \\
B_r \quad 0\ 1\ 1\ 0\ 1\ 0\ 1\ 1\ 0\ 0\ 0\ 1\ 1\ 1\ 0\ 1\ 0 \;\vert\; 1\ 0\ 1\ 0\ 1\ 1\ 0
\end{array}
$$

$$v \qquad i = 8 \qquad\qquad\qquad w \qquad i = 10$$

$$
\begin{array}{cc}
\begin{array}{c}
\text{A C A A C C C} \;\vert\; \text{A C C A} \\
B_v \quad 0\ 1\ 0\ 0\ 1\ 1\ 1 \;\vert\; 0\ 1\ 1\ 0
\end{array}
&
\begin{array}{c}
\text{G T G T T G T G G} \;\vert\; \text{G T T G} \\
B_w \quad 0\ 1\ 0\ 1\ 1\ 0\ 1\ 0\ 0 \;\vert\; 0\ 1\ 1\ 0
\end{array}
\end{array}
$$

$$i = 4$$

A   C   G   T

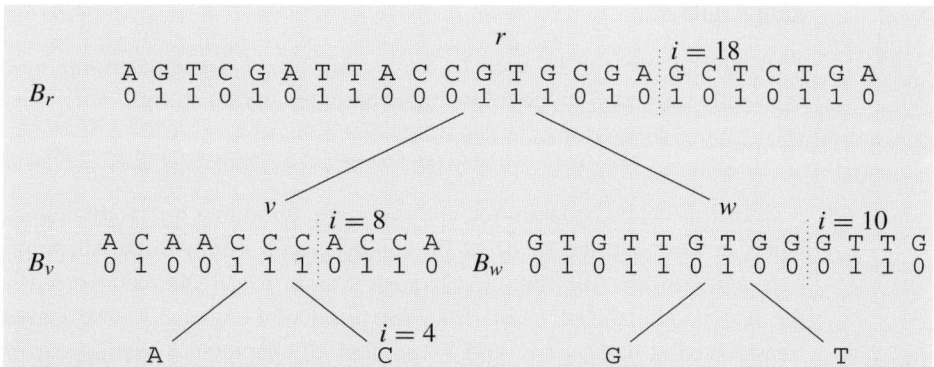

For solving $\mathrm{rank}_C(T, 18)$, we first compute $i = \mathrm{rank}_0(B_r, 18) = 8$ as C goes left from the root node r. At node v, C goes right, so we compute $i = \mathrm{rank}_1(B_v, 8) = 4$. We have reached the leaf, so this is our answer.

For solving $\mathrm{access}(T, 18)$, we first compute $i = \mathrm{rank}_{B_r[18]}(B_r, 18) = 10$ because $B_r[18] = 1$ indicates that the location is stored in the right subtree. Then we check that $B_w[10] = 0$, indicating that we should go left. To the left of w we have the leaf labeled G, so this is our answer.

One can construct the wavelet tree level by level, starting at the root. After marking $B_{\mathrm{root}}[i] = 0$ if $T[i] \le \sigma/2$ and $B_{\mathrm{root}}[i] = 1$ otherwise, one can extract subsequence T' of T marked by 0-bits and send it to the left child handling alphabet interval $[1..\sigma/2]$, and subsequence T'' of T marked by 1-bits and send it to the right child handling alphabet interval $[\sigma/2 + 1..\sigma]$. After this T can be deleted, and construction can continue recursively to the next levels, creating leaf nodes on encountering subsequences consisting of a single character. At any moment, the space is bounded by $3n \log \sigma + O(\sigma \log n)$ for representing the set of distinct subsequences of T at two consecutive levels, for the bitvectors forming the wavelet tree, and for the pointers recording the shape of the tree. Creating rank structures separately for each node and dropping the subsequences, the wavelet tree occupies $n \log \sigma (1 + o(1)) + O(\sigma \log n)$ bits.

To improve the space requirement further, one can concatenate all bitvectors level by level and build just one rank structure. The tree shape can then be implicitly reconstructed from the way the alphabet is partitioned, and from the content of the bitvectors. We leave the details of navigating this concatenated bitvector on rank and access queries for the reader: see Exercise 3.9.

THEOREM 3.4   *The wavelet tree representation of string $T[1..n] \in \{1, 2, \ldots, \sigma\}^*$ occupies $n \log \sigma (1 + o(1))$ bits and supports $\mathrm{rank}_c(T, i)$, $\mathrm{select}_c(T, i)$, and $\mathrm{access}(T, i)$ operations in $O(\log \sigma)$ time, assuming a RAM model with computer word size $w = \Omega(\log n)$. The structure can be built in $O(n \log \sigma)$ time and bits of space.*

Proof   See Exercise 3.10 for the solution for $\mathrm{select}_c(T, i)$. The other operations were considered above.   □

3.3.2 Range queries

The wavelet tree can be used for answering *two-dimensional range queries*. We are especially interested in the following counting query on string $T = t_1 \cdots t_n$:

$$\text{rangeCount}(T, i, j, l, r) = |\{k \mid i \leq k \leq j, l \leq t_k \leq r\}|. \qquad (3.3)$$

Recall the binary search tree and the range minimum query studied in Section 3.1. One can associate the shape of the binary search tree with the shape of the wavelet tree, and consider the same $[l..r]$ range search. Recall the node sets V' and V'' and the leaf nodes labeled L and R in the proof of Lemma 3.1. The leaves under subtrees rooted at nodes of V' and V'' contain all characters c that appear in T such that $l < c < r$.

Remember that the $\text{rank}_l(T, i)$ operation maintains on each node v encountered when searching l an updated value of i: denote this by i_v. Similarly, we obtain j_v when simulating $\text{rank}_l(T, j)$ operation, as well as i_w and j_w when simulating $\text{rank}_l(T, i)$ and $\text{rank}_r(T, j)$ on nodes w encountered when searching r. For each $v' \in V'$ and $v'' \in V''$ we have thus i_v, j_v, i_w, and j_w computed, where v is the parent of v' and w is the parent of v''. Then $i_{v'} = \text{rank}_1(B_v, i_v - 1)$, $j_{v'} = \text{rank}_1(B_v, j_v)$, $i_{v''} = \text{rank}_0(B_w, i_w - 1)$, and $j_{v''} = \text{rank}_0(B_w, j_w)$. We obtain an $O(\log \sigma)$ time algorithm to answer the range counting query, as

$$\begin{aligned}
\text{rangeCount}(T, i, j, l, r) = &\sum_{v' \in V'} j_{v'} - i_{v'} \\
&+ \sum_{v'' \in V''} j_{v''} - i_{v''} \\
&+ \text{lcount} + \text{rcount}, \qquad (3.4)
\end{aligned}$$

where $\text{lcount} = j_L - i_L$ if $L = l$, otherwise $\text{lcount} = 0$, and $\text{rcount} = j_R - i_R$ if $R = r$, otherwise $\text{rcount} = 0$.

One can also solve the $\text{rangeList}(T, i, j, l, r)$ operation that asks for a listing of all distinct characters inside $[l..r]$ appearing in $T[i..j]$ in lexicographic order. It is sufficient to continue from each $v' \in V'$ and $v'' \in V''$ down to the leaves, branching only to nodes w with $j_w - i_w > 0$. Labels of leaves reached occur at least once in $T[i..j]$. The path to each such reported leaf could be largely distinct from those for other reported leaves, so the running time is $O(d \log \sigma)$ for d being the size of the answer. An optimized search strategy yields $O(d \log(\sigma/d))$ time: see Exercise 3.11. For each such c, $l \leq c \leq r$, occurring at least once in $T[i..j]$, one can also compute the non-zero frequencies by

$$\text{freq}_c(T, i, j) = \text{rank}_c(T, j) - \text{rank}_c(T, i - 1). \qquad (3.5)$$

We will also need an operation $\text{isRangeUnary}(T, i, j)$ that returns true if substring $T[i..j]$ consists of a run of only one character c. To support this operation, we can use $c = \text{access}(T, i)$ and return true if $\text{freq}_c(T, i, j) = j - i + 1$, otherwise return false.

We will later see that these four operations play vital roles in navigating certain implicit search trees.

COROLLARY 3.5   *The wavelet tree of string $T[1..n] \in \{1, 2, \ldots, \sigma\}^*$ supports operations* rangeCount(T, i, j, l, r), freq$_c(T, i, j)$, *and* isRangeUnary(T, i, j) *in $O(\log \sigma)$ time, and operation* rangeList(T, i, j, l, r) *in $O(d \log(\sigma/d))$ time, where d is the size of the output.*

3.4   Literature

The range search with balanced binary search trees (Adelson-Velskii & Landis 1962; Bayer 1972) is typically presented as a listing problem in the computational geometry literature (de Berg *et al.* 2000, Chapter 5). We added here the aggregate operation of taking min over the elements in a range; one can verbatim change this to max or $\sum$, the latter leading to the *range counting* solution. For semi-infinite ranges one can obtain $O(\log \log n)$ time using *vEB trees* (van Emde Boas 1977) via a reduction (Gabow *et al.* 1984).

The $O(\log n)$ time solution we describe for dynamic range minimum queries is similar to the one presented by Arge *et al.* (2013). It is also possible to improve the time from $O(\log n)$ to $O(\log n / \log \log n)$ by using a balanced multiary tree in which every node has between $\log n / (4 \log \log n)$ and $\log n / (2 \log \log n)$ children instead of just 2. This will result in a tree of depth $O(\log n / \log \log n)$. In order to obtain the result, one needs to implement a local dynamic range minimum structure on $\log n / (2 \log \log n)$ elements in every node that supports updates and queries in constant time. This is possible using the Q-heap (Fredman & Willard 1994), which requires the use of $o(n)$ bits of space in a pre-computed table and allows predecessor queries in constant time and using the dynamic rank structure of Dietz (1989) to return the position of the predecessor in constant time too. This improved $O(\log n / \log \log n)$ time was only recently achieved (Brodal *et al.* 2011; Davoodi 2011) and is known to be the best possible by a lower bound proved in Alstrup *et al.* (1998).

Bitvector rank and select queries were first studied by Jacobson (1989) for the succinct representation of static trees supporting navigation. The constant time solutions in the RAM model described here are from Clark (1996) and Munro (1996), and the four Russians technique used inside the solutions is from Arlazarov *et al.* (1970). The construction time can be improved to $O(n / \log n)$ (Babenko *et al.* 2015). Wavelet trees were introduced by Grossi *et al.* (2003), although very similar concepts were implicitly known in computational geometry literature; the space-efficient improvement (Chazelle 1988) over *fractional cascading* (de Berg *et al.* 2000, Section 5.6) is almost identical to the two-dimensional range counting we covered (Mäkinen & Navarro 2007). Our description of rank, select, and wavelet trees follows closely the survey by Navarro & Mäkinen (2007). See Navarro (2012) for further extensions and applications of wavelet trees.

Exercises

3.1   Give an example of a perfectly balanced binary search tree storing eight (key, value) pairs in its leaves as described in Lemma 3.1. Give an example of a range minimum query for some non-empty interval.

3.2   Give a pseudocode for the algorithm to construct and initialize a balanced binary search tree given the sorted keys.

3.3   Recall how *red–black trees* work. Revisit the proof of Lemma 3.1 and consider how the tree can be maintained correctly updated for RMQ queries during the rebalancing operations needed if one adds the support for `Insert` and `Delete`.

3.4   Instead of taking the minimum among the values in Lemma 3.1 one could take a sum. If all leaves are initialized to value 1, what question does the operation analogous to RMQ answer?

3.5   Consider the sets V' and V'' in the proof of Lemma 3.1. The subtrees rooted at nodes in V' and V'' induce a *partitioning* of the set of characters appearing in the interval $[l..r]$ (see Figure 3.1). There are many other partitionings of $[l..r]$ induced by different subsets of nodes of the tree. Why is the one chosen in the proof the minimum size partitioning? Are there other partitionings that could give the same running time?

3.6   A *van Emde Boas tree* (vEB tree) supports in $O(\log\log n)$ time insertions, deletions, and *predecessor queries* for values in the interval $[1..n]$. A predecessor query returns the largest element i' stored in the vEB tree smaller than query element i. Show how the structure can be used instead of the balanced search tree of Lemma 3.1 to solve range minimum queries for semi-infinite intervals $(-\infty..i]$.

3.7   Prove Theorem 3.3. *Hint.* Start as in `rank` with $O(\log^2 n)$ size blocks, but this time on arguments of $select_1$. Call these *source blocks* and the areas they span in bitvector B *target blocks*. Define *long* target blocks so that there must be so few of those that you can afford to store all answers inside the corresponding source blocks. We are left with *short* target blocks. Apply the same idea recursively to these short blocks, adjusting the definition of a *long* target block in the second level of recursion. Then one should be left with short enough target blocks that the four Russians technique applies to compute answers in constant time. The solution to $select_0$ is symmetric.

3.8   Show how to reduce the preprocessing time in Theorem 3.2 and Theorem 3.3 from $O(n)$ to $O(n/\log n)$, by using the four Russians technique during the preprocessing.

3.9   Consider the wavelet tree in Example 3.2. Concatenate bitvectors B_r, B_v, and B_w. Give formulas to implement the example queries in Example 3.2 with just the concatenated bitvector. Derive the general formulas that work for any wavelet tree.

3.10   Consider the operation $select_c(A,j) = i$ that returns the position i of the jth occurrence of character c in string $A[1..n]$. Show that the wavelet tree with its bitvectors preprocessed for constant time $select_1(B,j)$ and $select_0(B,j)$ queries can answer $select_c(A,j)$ in $O(\log \sigma)$ time.

3.11   Show that the operation $rangeList(T,i,j,l,r)$ can be supported in $O(d\log(\sigma/d))$ time by optimizing the given search strategy. *Hint.* After finding the left-most element in the interval, go up until branching towards the second element occurs, and so on. Observe that the worst case is when the elements are equally distributed along the interval: see Section 8.1 for an analogous analysis.

3.12 Show how to efficiently implement the operation `rangeListExtended` (T, i, j, l, r) which returns not only the distinct characters from $[l..r]$ in $T[i..j]$, but also, for every such distinct character c, returns the pair $(\text{rank}_c(T, i-1), \text{rank}_c(T, j))$. *Hint.* Observe that the enumeration of the distinct characters uses rank queries on binary sequences that can also be used to compute the pairs of rank operations executed for the distinct characters.

4   Graphs

Graphs are a fundamental model for representing various relations between data. The aim of this chapter is to present some basic problems and techniques relating to graphs, mainly for finding particular paths in directed and undirected graphs. In the following chapters, we will deal with various problems in biological sequence analysis that can be reduced to one of these basic ones.

Unless stated otherwise, in this chapter we assume that the graphs have n vertices and m arcs.

4.1   Directed acyclic graphs (DAGs)

A directed graph is called *acyclic* if it does not contain a directed cycle; we use the shorthand DAG to denote a directed acyclic graph. DAGs are one of the most basic classes of graphs, being helpful also in many problems in bioinformatics. DAGs admit special properties that not all graphs enjoy, and some problems become simpler when restricted to DAGs.

4.1.1   Topological ordering

The acyclicity of the arc relation of a DAG allows us to build recursive definitions and algorithms on its vertices. In such a case, the value of a function in a vertex v of a DAG depends on the values of the function on the in-neighbors of v. In order to implement its computation, we need an ordering of the vertices of the DAG such that any vertex appears in this ordering after all its in-neighbors. Such an ordering is called a *topological ordering* of the DAG. See Figure 4.1 for an example.

Any DAG admits a topological ordering, but it need not be unique. Exercise 4.1 asks the reader to construct a DAG in which the number of topological orderings is exponential in the number of vertices. A consequence of the existence of a topological ordering is that every DAG has at least one source (the first node in a topological ordering) and at least one sink (the last node in a topological ordering).

A topological ordering can be found in time linear in the number of vertices and arcs of the DAG. This can be done by performing a depth-first search (DFS): when all the vertices reachable from a vertex v have been visited, v is added at the beginning of a queue. At the end, the queue traversed from the beginning to its last element gives a topological ordering.

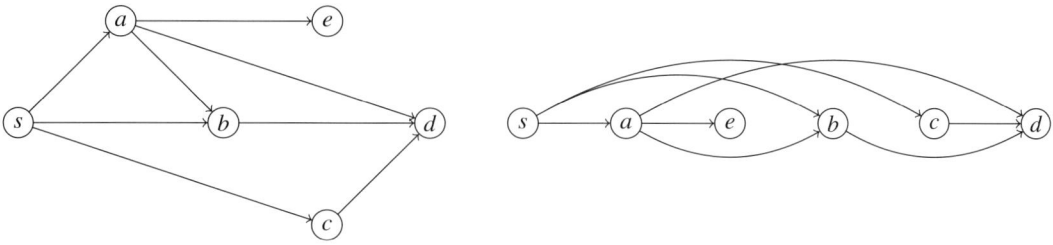

Figure 4.1 On the left, a DAG; on the right, a topological ordering of it.

Algorithm 4.1: Depth-first visit of a graph

Input: An undirected graph $G = (V, E)$.
Output: For every vertex $v \in V$, its value finishTime.

1 **foreach** $v \in V$ **do**
2 $\quad$ visited(v) $\leftarrow$ false;

3 time $\leftarrow 0$;
4 **def** *DFS-from-vertex(G, v)*:
5 $\quad$ visited(v) $\leftarrow$ true;
6 $\quad$ **foreach** $u \in N^+(v)$ **do**
7 $\quad\quad$ **if** *not* visited(u) **then**
8 $\quad\quad\quad$ *DFS-from-vertex(G, u)*;

9 $\quad$ finishTime(v) $\leftarrow$ time;
10 $\quad$ time $\leftarrow$ time $+ 1$;

11 **def** *DFS(G)*:
12 $\quad$ **foreach** $v \in V$ **do**
13 $\quad\quad$ **if** *not* visited(v) **then**
14 $\quad\quad\quad$ *DFS-from-vertex(G, v)*;

15 $\quad$ **return** finishTime;

THEOREM 4.1   *A topological ordering of a DAG G can be computed in time $O(n+m)$.*

Proof We perform a DFS as in Algorithm 4.1. Let $v_1, \ldots, v_n$ be the order of the vertices in decreasing order of finishTime. Since we record the finish time of v_i only after all out-neighbors of v_i have been visited, v_i appears in the ordering $v_1, \ldots, v_n$ after all its out-neighbors. $\qquad\square$

4.1.2   Shortest paths

One of the most basic problems on a graph is that of finding the shortest path between two vertices. Given a DAG G, we show that we can find in time $O(n + m)$ the shortest paths from a given vertex s to all other vertices of G. The acyclicity of G (in particular,

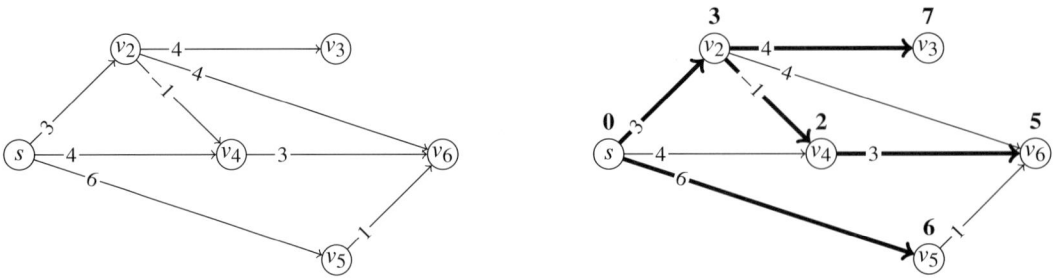

Figure 4.2 On the left, a DAG whose arcs are associated costs; on the right, each vertex v_i is labeled with $d(i)$; for every vertex we highlight an incoming arc minimizing (4.1).

the existence of a topological ordering) allows for such an algorithm even if the arcs have arbitrary positive or negative costs. In this case, the length (which we will also call the *weight*, or *cost*) of a path is the sum of the lengths (or weights, or costs) of its arcs.

Problem 4.1   Shortest path in a DAG

Given a DAG $G = (V, E)$, a vertex $s \in V$, and a cost function $c : E \to \mathbb{Q}$, find the paths of minimum cost from s to all other vertices $v \in V$.

THEOREM 4.2   *The shortest-path problem on a DAG can be solved in time $O(n + m)$.*

Proof   A shortest path starting at s and ending at v must arrive at v from one of the in-neighbors, say u, of v. If the shortest path from s to u is already known, then a shortest path from s to v coincides with it, and then proceeds to v.

Since G is acyclic, the optimal solution for all the in-neighbors of v can be computed without considering v (and all other vertices further reachable from v). Thus, we can decompose the problem along *any* topological ordering $v_1, \ldots, v_n$ of the vertices of G.

For every $i \in \{1, \ldots, n\}$, we define

$$d(i) = \text{the length of the shortest path from } s \text{ to } v_i,$$

where $d(i) = \infty$ if no path from s to v_i exists. We compute $d(i)$ as

$$d(i) = \min\{c(v_j, v_i) + d(j) \mid v_j \in N^-(v_i)\}, \tag{4.1}$$

by initializing $d(i) = 0$, if $v_i = s$, and $d(i) = \infty$ otherwise. (Recall that $N^-(v_i)$ denotes the set of in-neighbors of v_i.) Since $v_1, \ldots, v_n$ is a topological ordering of G, this recursive definition makes sense. In order to compute also one shortest path, we can also store, for every $i \in \{1, \ldots, n\} \setminus \{s\}$, one in-neighbor v_j of v_i minimizing (4.1).

Computing a topological ordering of G can be done in time $O(n + m)$, by Theorem 4.1. Every arc (v_j, v_i) of G is inspected only once in the computation of d, namely when computing $d(i)$. Thus the complexity of this algorithm is $O(n + m)$. $\qquad\square$

The solution of the shortest-path problem on a DAG is an elementary instance of *dynamic programming*, where one exploits the fact that the optimal solution to a larger instance (like the shortest path to the current vertex) can be constructed from the optimal solutions to smaller instances (the shortest paths to the in-neighbors of the current vertex). The *evaluation order* of a dynamic programming algorithm refers to the order in which the solutions to all required smaller instances become available, before computing the larger instance. In the case of DAGs, the topological ordering equals the evaluation order. We shall explore this connection in Chapter 6, but some of the exercises in this section already refer to dynamic programming.

4.2   Arbitrary directed graphs

4.2.1   Eulerian paths

A special kind of path (possibly with repeated vertices) in a connected directed graph is one using *every* arc *exactly* once, called an Eulerian path. An Eulerian path in which its first and last vertices coincide is called an *Eulerian cycle*. We say that a directed graph is *connected* if the undirected graph obtained from it by ignoring the orientation of its arcs is connected. A connected directed graph is said to be Eulerian if it has an Eulerian cycle.

Problem 4.2   Eulerian cycle

Given a connected directed graph G, check if G is Eulerian, and output an Eulerian cycle if this is the case.

THEOREM 4.3   *A connected directed graph G is Eulerian if and only if $|N^-(v)| = |N^+(v)|$ for every vertex v. When this is the case, an Eulerian cycle can be computed in time $O(m)$.*

Proof   If G is Eulerian, then $|N^-(v)| = |N^+(v)|$ must hold for every vertex v.

For the reverse implication, we proceed as follows. Start at an arbitrary vertex $v_0 \in V(G)$, and construct a cycle C (possibly with repeated vertices) as follows. Pick an arbitrary out-neighbor v_1 of v_0 and visit v_1 by removing the arc (v_0, v_1), and adding it to C. Since $|N^-(v_1)| = |N^+(v_1)|$ held before removing the arc (v_0, v_1), there exists an out-neighbor v_2 of v_1 in the graph. Remove again the arc (v_1, v_2), add it to C, and repeat this process from v_2.

Since the number of vertices of G is finite, this procedure stops on reaching v_0. If C already contains all the arcs of G, then we are done. Otherwise, thanks to the connectivity assumption (Exercise 4.7 asks the reader to detail how connectivity helps here), there must be a vertex w_0 on C for which $|N^-(w_0)| = |N^+(w_0)| \geq 1$. Repeat the above procedure starting at w_0 to obtain another cycle C'. Combine C and C' into one

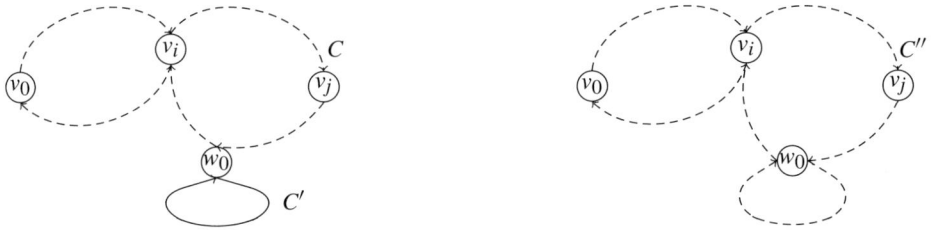

Figure 4.3 Combining two cycles C and C' into C''.

cycle C'' formed by taking the path from v_0 to w_0 on C, then the cycle C', then the path from w_0 to v_0 on C (see Figure 4.3).

We can iterate this process by checking again whether there is a vertex with at least one remaining out- and in-neighbor. This procedure terminates since the number of vertices of G is finite. At the end, we have constructed an Eulerian cycle of G.

Observe that every arc of G is visited exactly once. Exercise 4.8 ask the reader to show that, using doubly-linked lists, the above procedure can be implemented in time $O(m)$. □

A consequence of the above theorem is the following corollary about an Eulerian path with different endpoints.

COROLLARY 4.4   *A connected directed graph G has an Eulerian path with different endpoints if and only if there exist $s, t \in V(G)$ such that*

- $|N^-(s)| + 1 = |N^+(s)|$,
- $|N^-(t)| = |N^+(t)| + 1$, *and*
- $|N^-(v)| = |N^-(v)|$, *for all* $v \in V(G) \setminus \{s, t\}$.

In this case, we can find an Eulerian path in time $O(m)$.

Proof   If a connected directed graph G has an Eulerian path with different endpoints, then the above three conditions must hold.

For the reverse implication, add the arc (t, s) to G and find an Eulerian cycle in time $O(m)$ by Theorem 4.3. □

4.2.2   Shortest paths and the Bellman–Ford method

The algorithm given in Section 4.1.2 for the shortest-path problem does not work for arbitrary graphs, because one can no longer decompose the computation along a topological ordering. However, we can still decompose the problem by adding another parameter to a partial solution ending at a vertex v: the number of arcs of a path from s to v.

We restrict the problem to directed graphs without negative cost cycles.

Problem 4.3   Shortest path in a directed graph without negative cycles

Given a directed graph $G = (V, E)$, a vertex $s \in V$, and a cost function $c : E \to \mathbb{Q}$ such that G has no negative cost cycles, find the paths of minimum cost from s to every vertex $v \in V$.

THEOREM 4.5 (Bellman–Ford algorithm)   *The shortest-path problem on a directed graph G without negative cycles can be solved in time $O(nm)$.*

Proof   Let $v_1, \ldots, v_n$ be an arbitrary ordering of the vertices of G. To simplify notation in what follows we may assume $s = v_1$. For every $i \in \{1, \ldots, n\}$, we define

$$d(i, k) = \text{the length of a shortest path from } s \text{ to } v_i \text{ having at most } k \text{ arcs}, \qquad (4.2)$$

and $d(i, k) = \infty$ if no path having at most k arcs exists from s to v_i. Since G is assumed not to have negative cost cycles, any shortest path from s to any other vertex has at most $n - 1$ arcs. Thus, the solution is contained in $d(i, n - 1)$, for every $i \in \{1, \ldots, n\}$. See Figure 4.4 for an example.

For each $k \in \{1, \ldots, n - 1\}$ in increasing order, we compute $d(i, k)$ as

$$d(i, k) = \min \left(\min\{d(j, k - 1) + c(v_j, v_i) \mid v_j \in N^-(v_i)\}, \{d(i, k - 1)\} \right), \qquad (4.3)$$

where $d(1, 0) = 0$, and $d(i, 0) = \infty$ for every $i \in \{2, \ldots, n\}$. In other words, a shortest path from s to v_i having at most k arcs is either

- a shortest path from s to an in-neighbor v_j of v_i having at most $k - 1$ arcs, plus the arc (v_j, v_i), or
- a shortest path from s to v_i having at most $k - 1$ arcs.

Observe that this recursive definition makes sense because a solution for a subproblem parametrized by k depends only on the solutions to subproblems parametrized by $k - 1$, which have already been computed.

In order to compute also one shortest path, for every (i, k), $k > 1$, we store not only $d(i, k)$, but also the parent pair $p(i, k) = (j, k')$ minimizing (4.3).

For every $k \in \{1, \ldots, n - 1\}$, every arc (v_j, v_i) of G is inspected only once, when computing $d(i, k)$. Therefore, the complexity of the entire algorithm is $O(nm)$.   □

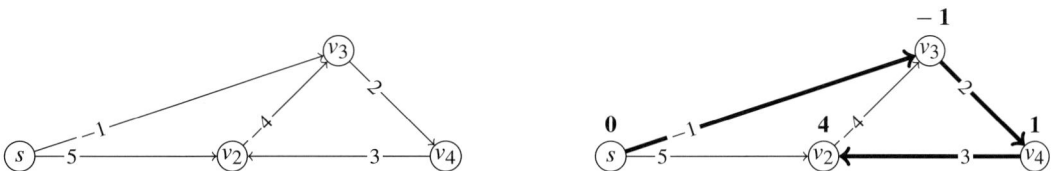

Figure 4.4 On the left, a directed graph whose arcs are associated costs; on the right, each vertex v_i is labeled with $d(i, n - 1)$; for every vertex we highlight an incoming arc minimizing (4.3) when $k = n - 1$.

The proof of Theorem 4.5 can also be interpreted as the operation of creating a DAG consisting of $n - 1$ copies of the vertices of G, one for each value of k, and making each arc (v_j, v_i) go from each v_j in one copy to the v_i in the next copy. More precisely, we can construct the graph G^* (see Figure 4.5):

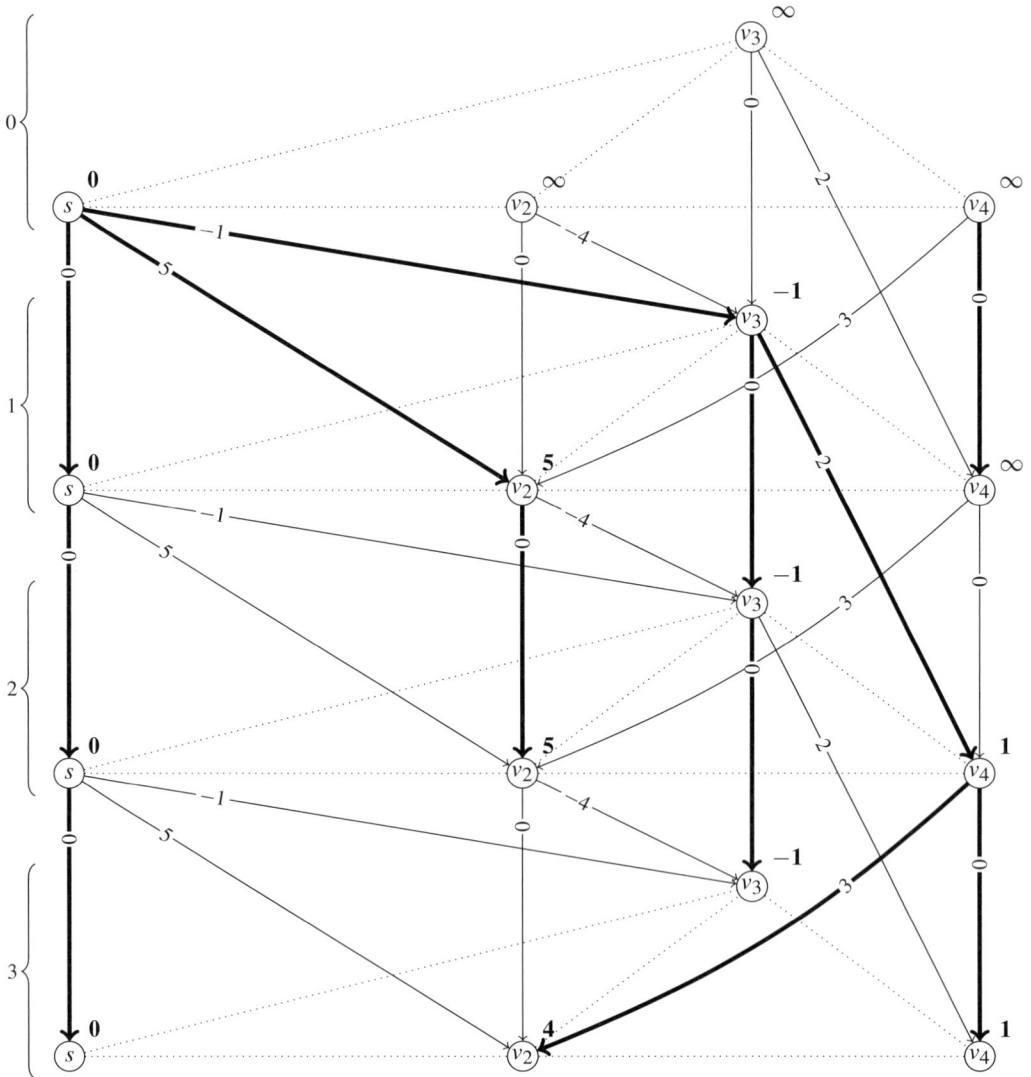

Figure 4.5 A reduction of the shortest-path problem in an arbitrary graph without negative cycles to the shortest-path problem in a DAG; each vertex (v_i, k) is labeled with $d(i, k)$; for every vertex we highlight an incoming arc minimizing (4.1).

- for every vertex $v_i \in V$, we add n vertices named $(v_i, 0), \ldots, (v_i, n-1)$;
- for every arc (v_j, v_i), we add the $n-1$ arcs $\big((v_j, 0), (v_i, 1)\big), \ldots, \big((v_j, n-2), (v_i, n-1)\big)$, each of cost $c(v_j, v_i)$; and
- for every vertex $v_i \in V$, we add $n-1$ arcs $\big((v_i, 0), (v_i, 1)\big), \ldots, \big((v_i, n-2), (v_i, n-1)\big)$, each of cost 0.

The shortest path from s to each of the vertices $(v_1, n-1), \ldots, (v_n, n-1)$ gives the shortest paths from s to each of $v_1, \ldots, v_n$, respectively. The DAG G^* has $n(n-1)$ vertices and $m(n-1)$ arcs. Assuming that G is connected (otherwise we can apply this reduction to each connected component of G), we have $m \geq n-1$. Thus, Theorem 4.5 also follows directly from Theorem 4.2.

The Bellman–Ford method also allows one to check whether a directed graph $G = (V, E)$, with a cost function $c : E \to \mathbb{Q}$, has a negative cost cycle.

Problem 4.4 Detection of a negative cycle

Given a directed graph G, and a cost function $c : E(G) \to \mathbb{Q}$, check whether G has a negative cost cycle, and output one if this is the case.

THEOREM 4.6 *Given a directed graph G, and a cost function $c : E(G) \to \mathbb{Q}$ the problem of detection of negative cycles on G can be solved in time $O(nm)$.*

Proof First, assume that the underlying graph of G is connected, otherwise we can work on each connected component of it. This assumption implies that $m \geq n-1$. We also assume that there is a vertex s from which all other vertices are reachable, otherwise we can add a dummy $(n+1)$th vertex s with arcs of cost 0 to all other vertices of G.

Let $v_1, \ldots, v_n$ be an arbitrary ordering of the vertices of G. For every $i \in \{1, \ldots, n\}$ and $k \in \{1, \ldots, n\}$ we compute $d(i, k)$ (lengths of shortest paths starting from s) as in the proof of Theorem 4.5, in time $O(nm)$.

If $d(i, n) < d(i, n-1)$ for some $i \in \{1, \ldots, n\}$, then, following the parent links from v_i, we can find a path P consisting of exactly n arcs, starting at s and ending at v_i. Since P has $n+1$ vertices, at least one vertex must appear twice on P. Thus, P contains a cycle C made up of the arcs of P between two consecutive occurrences of a vertex v such that no other vertex between these two occurrences is repeated. On removing C from P we obtain a path P' with at most n arcs.

For any path Q in G, let $c(Q)$ denote the sum of the costs of its arcs. Since $c(P') \geq d(i, n-1) > d(i, n) = c(P)$ and $c(P) = c(P') + c(C)$, we get $c(C) < 0$. Therefore, we can output the cycle C of negative cost.

Otherwise, $d(i, n) = d(i, n-1)$ holds for all $i \in \{1, \ldots, n\}$, and we must prove that G cannot have a negative cost cycle. Thanks to the Bellman–Ford recurrence (4.3), this implies that

$$d(j, n) + c(v_j, v_i) = d(j, n-1) + c(v_j, v_i) \geq d(i, n), \text{ for every arc } (v_j, v_i) \text{ of } G.$$

We argue by contradiction, and suppose that $v_{i_0}, v_{i_1}, \ldots, v_{i_\ell} = v_{i_0}$ is a cycle of negative cost $\sum_{t=1}^{\ell} c(v_{i_{t-1}}, v_{i_t}) < 0$. We sum up the above inequality for all vertices in the cycle and obtain

$$\sum_{t=1}^{\ell} d(i_t, n) \leq \sum_{t=1}^{\ell} \left(d(i_{t-1}, n) + c(v_{i_{t-1}}, v_{i_t}) \right) = \sum_{t=1}^{\ell} d(i_{t-1}, n) + \sum_{t=1}^{\ell} c(v_{i_{t-1}}, v_{i_t}).$$

Observe that $\sum_{t=1}^{\ell} d(i_t, n) = \sum_{t=1}^{\ell} d(i_{t-1}, n)$, which implies that $\sum_{t=1}^{\ell} c(v_{i_{t-1}}, v_{i_t}) \geq 0$, contradicting the initial assumption that the cycle $v_{i_0}, v_{i_1}, \ldots, v_{i_\ell} = v_{i_0}$ has negative cost. □

4.3   Literature

The reader can refer to the textbook *Introduction to Algorithms* (Cormen *et al.* 2009) for a gentle introduction to the algorithms presented here, or to the textbook *Introduction to Graph Theory* (West 2000) and the monograph *Combinatorial Optimization* (Schrijver 2003) for a more in-depth discussion. The Bellman–Ford method appeared in Shimbel (1955), Bellman (1958), Ford (1956), and Moore (1959).

Exercises

4.1 Find a family of DAGs in which the number of topological orderings is exponential in the number of vertices.

4.2 Find a family of DAGs with a number of distinct paths exponential in the number of vertices.

4.3 Show that a directed graph G is acyclic (does not contain a directed cycle) if and only if, for every subset of vertices $S \subseteq V(G)$, there exists a vertex $v \in S$ such that no out-neighbor of v belongs to S. Conclude that a DAG must have at least one sink.

4.4 Show that a directed graph G is acyclic if and only if it admits a topological ordering (the forward implication is Theorem 4.1). Conclude that we can check in $O(m)$ time whether a directed graph is acyclic.

4.5 Let G be a DAG with precisely one source s and one sink t. Show that for any $v \in V(G) \setminus \{s, t\}$ there exists a path from s to v and a path from v to t.

4.6 Let G be a DAG with a unique source. Prove that G is connected.

4.7 Explain in detail how the connectivity assumption is exploited in the proof of Theorem 4.3.

4.8 Show how the procedure in Theorem 4.3 can be implemented in time $O(m)$.

4.9 Consider a directed graph G in which for every pair of vertices $u, v \in V(G)$ exactly one of $(u, v) \in E(G)$ or $(v, u) \in E(G)$ holds (such a graph is called a *tournament*). Show that there exists a path in G passing through all vertices exactly once (such a path is called *Hamiltonian*).

4.10 Let G be an undirected graph. Show that either G or its complement $\overline{G}$ is connected.

4.11 Give an algorithm running in time $O(n + m)$ that checks whether an undirected graph G is connected, and if not, lists all of its connected components.

4.12 A connected undirected graph G is called *2-connected* (or *bi-connected*) if removing any vertex from G (along with its incident edges) results in a connected graph. Give an algorithm running in time $O(m)$ for checking whether a graph is 2-connected.

4.13 Refine the algorithm from Exercise 4.12 so that it also outputs the 2-connected components of G (that is, the maximal 2-connected subgraphs of G).

4.14 Given an undirected graph G, consider the graph $B(G)$ of 2-connected components of G: the vertices of $B(G)$ are the 2-connected components of G, and we add an edge between two vertices of $B(G)$ if the corresponding 2-connected components of G have a vertex in common. Show that $B(G)$ is a tree.

4.15 Consider a directed graph G, a vertex $s \in V(G)$, and $c : E(G) \to \mathbb{Q}$ so that no cycle of G has negative cost. Show that if G has only two 2-connected components C_1 and C_2, sharing a vertex v, such that $s \in V(C_1)$, then we can solve the shortest-path problem on G by solving it independently on C_1 and C_2, by appropriately initializing the dynamic programming algorithm on C_2 at vertex v.

4.16 Consider a DAG G, a vertex $s \in V(G)$, a cost function $c : V(G) \to \mathbb{Q}$, a partition $S = \{S_1, \ldots, S_k\}$ of $V(G)$, and an integer $t \le k$. Give a dynamic programming algorithm for computing a shortest path from s to any other vertex of G that passes through at most t sets of the partition S. What is the complexity of your algorithm? What if G is not acyclic?

4.17 Consider a DAG G, a partition $S = \{S_1, \ldots, S_k\}$ of $V(G)$, and an integer $t \le k$. We say that a path $P = u_1, u_2, \ldots, u_\ell$ in G is *t-restricted* if every maximal subpath of P of vertices from the same set of S has at least t vertices. (In other words, if P starts using vertices from a set $S_i \in S$, then it must do so for at least t vertices.) Give a dynamic programming algorithm that is additionally given a vertex $s \in V(G)$ and a cost function $c : V(G) \to \mathbb{Q}$, and finds a t-restricted shortest path from s to any other vertex of G. What is the complexity of your algorithm? What if G is not acyclic?

4.18 Given a directed graph $G = (V, E)$, $n = |V|$, $m = |E|$, and a cost function $c : E \to \mathbb{Q}$, we say that the *length* of a cycle $C = v_1, v_2, \ldots, v_t, v_{t+1} = v_1$ in G, denoted $l(C)$ is t, its *cost*, denoted $c(C)$, is $\sum_{i=1}^{t} c(v_i, v_{i+1})$, and its *mean cost* is $\mu(C) = c(C)/l(C)$. Denote by $\mu(G)$ the minimum mean cost of a cycle of G, namely

$$\mu(G) = \min_{C \text{ cycle of } G} \mu(C).$$

(a)   For each $v \in V(G)$, and each $k \in \{0, \ldots, n\}$, let $d(v, k)$ be the minimum cost of a path in G with exactly k edges, ending at v (where $d(v, 0) = 0$ by convention). Show that the bi-dimensional array d can be computed in time $O(nm)$.

(b)   Show that

$$\mu(G) = \min_{v \in V} \max_{0 \leq k \leq n-1} \frac{d(v, n) - d(v, k)}{n - k}, \tag{4.4}$$

by showing the following facts. Consider first the case $\mu(G) = 0$, and show that

- for any $v \in V$, there exists a $k \in \{0, \ldots, n-1\}$ such that $d(v, n) - d(v, k) \geq 0$, thus, the right-hand side of (4.4) is greater than or equal to 0;
- each cycle of G has non-negative cost: if C is a cycle of G, then there exists a vertex v on C such that, for every $k \in \{0, \ldots, n-1\}$, it holds that $d(v, n) - d(v, k) \leq c(C)$; and
- a cycle C of minimum mean cost $\mu(C) = 0$ also has $c(C) = 0$; use the above bullet to show that the right-hand side of (4.4) is equal to 0.

Conclude the proof of (4.4) by considering the case $\mu(G) \neq 0$. Show that

- if we transform the input (G, c) into (G, c') by subtracting $\mu(G)$ from the cost of every edge, then the minimum mean cost of a path of (G, c') is 0, and the paths of minimum mean cost of (G, c) are the same as those of (G, c'); and
- since relation (4.4) holds for (G, c'), it holds also for the original input (G, c).

(c)   Use (a) and the proof of (b) to conclude that a minimum mean cost cycle C in G can be found in time $O(nm)$.

5 Network flows

Many problems can be formalized as *optimization* problems: among all possible solutions of a problem, find one which minimizes or maximizes a certain cost. For example, in Chapter 4 we have seen algorithms that find the shortest path between two vertices of a graph, among all possible such paths. Another example is a *bipartite matching* problem, in which we have to match the elements of one set with the elements of another set, assuming that each allowed match has a known cost. Among all possible ways of matching the two sets of elements, one is interested in a matching with minimum total cost.

In this chapter we show that many optimization problems can be *reduced* (in the sense already explained in Section 2.3) to a *network flow* problem. This polynomially solvable problem is a powerful model, which has found a remarkable array of applications. Roughly stated, in a network flow problem one is given a transportation network, and is required to find the optimal way of sending some content through this network. Finding a shortest path is a very simple instance of a network flow problem, and, even though this is not immediately apparent, so is the bipartite matching problem.

One of the most well-known network flow problems is the *maximum flow* problem (whose definition we recall on page 46). In this chapter we focus instead on a more general version of it, called the *minimum-cost flow* problem. In Section 5.1 we give an intuition of flow, and we show that, in a DAG, a flow is just a collection of paths. This basic observation will allow us to reduce the assembly problems in Chapter 15 to minimum-cost flow problems. In Sections 5.3 and 5.4 we will discuss matching and covering problems solvable in terms of minimum-cost flow, which will in their turn find later applications in the book.

We must emphasize that algorithms tailored for a particular optimization problem can have a better complexity than the solution derived from a network flow algorithm. Nevertheless, a practitioner may prefer a quickly obtainable and implementable solution, given the existing body of efficient network flow solvers. Alternatively, a theoretician may first show that a problem is solvable in polynomial time by a reduction to a network flow problem, and only afterwards start the quest for more efficient solutions.

5.1 Flows and their decompositions

In this chapter we make the assumption that the directed graphs, be they acyclic or not, have exactly one source and one sink, which will be denoted s and t, respectively.

A path from s to t will be called an *s-t path*. Recall that all paths and cycles are directed.

For an intuition on what a minimum-cost flow problem is, consider a DAG with a cost associated with every arc. Suppose we want to find the optimal way of sending some content that continuously "flows" along the arcs of the DAG, from its source s to its sink t. If by optimal we mean that the cost of a possible solution is the sum of the costs of the arcs traversed by our content, then the optimal solution is just the shortest *s-t* path, in which costs are taken as lengths, the problem which we already solved in Section 4.1.2. Suppose, however, that each arc also has a *capacity*, that is, a limit on how much content it can be traversed by. This slight generalization renders it no longer solvable in terms of just one shortest-path problem, since our content must now flow along different *s-t* paths.

This brings us to the most basic meaning of *flow* in a DAG: a collection of *s-t* paths. Solving a minimum-cost flow problem amounts to solving simultaneously many shortest-path-like problems, tied together by different constraints, such as capacity constraints.

In this section we make the simplifying assumption that, for any two vertices x and y, if the arc (x, y) is present in the graph, then the reverse arc (y, x) is not present. Otherwise, we can simply *subdivide* (y, x), that its, we can remove (y, x), and then add a new vertex z together with the two arcs (y, z) and (z, x). Unless stated otherwise, graphs have n vertices and m arcs.

It will be convenient to regard a flow as a function assigning to every arc of the graph the amount of content it is being traversed by; we will call this content *the flow* of an arc. Clearly, a flow function arising from a collection of paths must satisfy the following condition:

> **Flow conservation.** The flow entering a vertex is equal to the flow exiting the vertex, with the exception of s and t.

Note that, if the flow conservation property holds, then the flow exiting s equals the flow entering t (Exercise 5.1). We can now give a formal definition of flow in a directed graph.

DEFINITION 5.1   *Given a directed graph G, a* flow *on G is a function $f : E(G) \to \mathbb{Q}_+$ satisfying the flow conservation property.*

Theorem 5.2 below shows that, in a DAG, collections of *s-t* paths and functions assigning flow values to the arcs of the DAG and satisfying the flow conservation property are equivalent notions. Namely, any flow in a DAG can be decomposed into a collection of weighted *s-t* paths, such that the flow on an arc equals the sum of the weights of the *s-t* paths using that arc (see Figure 5.1).

THEOREM 5.2   *Let f be a flow in DAG G. In time $O(nm)$ we can find $k \leq m$ s-t paths $P_1, \ldots, P_k$, each P_i having a weight $w_i > 0$, such that for every arc $(x, y) \in E(G)$ it holds that*

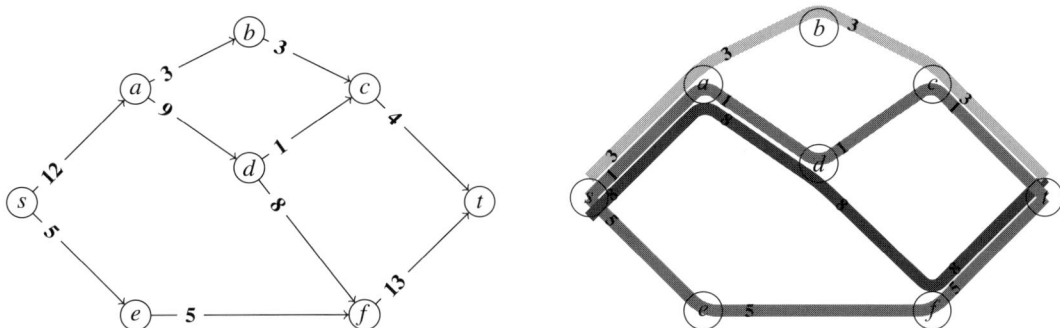

Figure 5.1 On the left, a flow f on a DAG G. On the right, the decomposition of f into four paths, (s,a,b,c,t), (s,a,d,c,t), (s,a,d,f,t), and (s,e,f,t), having weights 3, 1, 8, and 5, respectively.

$$f(x,y) = \sum_{P_i \,:\, (x,y) \in P_i} w_i.$$

Moreover, if f has only integer values, then the computed $w_1, \ldots, w_k$ are integers.

Proof At step i of the algorithm, we select an *s-t* path P_i and take its weight w_i to be the minimum flow value on its arcs (the value w_i is also called the *bottleneck* of P_i). We then remove P_i from G, in the sense that we subtract w_i from the flow value of all arcs of P_i, which results in a new flow function (that is, still satisfying the flow conservation property). If P_i is chosen such that its bottleneck is positive, then at least one more arc of G has flow value 0. Therefore, we can decompose any flow in G into $k \leq m$ paths.

In order to compute each P_i in $O(n)$ time, at the start of the algorithm we can remove the arcs of G with null flow value. As long as G has at least one arc, by the flow conservation property there exists an *s-t* path P, which we can find by a visit of G starting from s (since G is acyclic this visit must end in t). When we subtract w_i from the flow value of an arc on P_i, we also check whether its flow value becomes null, in which case we remove it from G. Since for every vertex in P_i we subtract w_i both from its in-coming flow and from its out-going flow, the result is still a flow function.

The fact that $w_1, \ldots, w_k$ are integers if f is integer-valued follows from the fact that at every step we are subtracting an integer value from an integer-valued flow. □

The proof of Theorem 5.2 can be extended to show that if a flow f is in a graph that is not necessarily acyclic, then f can be decomposed as a collection of at most m weighted paths or cycles (Exercise 5.3). See also Insight 5.1. In practice, we might be interested in decomposing a flow into the minimum number of components. However this is an NP-hard problem, as our next theorem shows.

Insight 5.1 Decomposing a flow in a DAG into a few paths in practice

One heuristic for decomposing a flow f in a DAG G into a few paths is to use the same strategy as in the proof of Theorem 5.2, but at each step to select the *s-t* path of maximum *bottleneck*, which path can be computed in time $O(m)$ (Exercise 5.5).

If an approximate solution is satisfactory, then one can decompose into a few paths only a fraction of the flow. This idea is based on the observation that, if f has value q, and the values of f on all arcs are multiples of an integer x, then f can be decomposed into exactly q/x paths each of weight x (Exercise 5.2). Therefore, one can round all values of a flow to the nearest multiple of an appropriately chosen integer, and then decompose the resulting flow. For example, through appropriate choices, it can be shown that, if the minimum number of paths into which a flow f of value q can be decomposed is k, then one can decompose $1/3$ of f into at most k paths in polynomial time; see the literature section.

THEOREM 5.3   *Given a flow f in a DAG G, the problem of decomposing f into the minimum number of weighted s-t paths is NP-hard.*

Proof   We reduce from the NP-hard subset sum problem, in which we are given a set $A = \{a_1, a_2, \ldots, a_n\}$ of positive integers, together with positive integers b and k, and we are asked whether there exists a subset $S \subseteq A$ such that $|S| \leq k$ and $\sum_{a \in S} a = b$. We denote by $\bar{b}$ the complementary value $(\sum_{a \in A} a) - b$.

For an instance (A, b, k) of the subset sum problem, we construct the DAG $G_{A,b}$, and the flow f in $G_{A,b}$, as follows (see also Figure 5.2):

- $V(G_{A,b}) = \{s, x_1, \ldots, x_n, y, z_1, z_2, t\}$;
- for every $i \in \{1, \ldots, n\}$, we add the arcs (s, x_i) and (x_i, y) to $G_{A,b}$, setting $f(s, x_i) = f(x_i, y) = a_i$;
- we add the arcs (y, z_1) and (z_1, t) to $G_{A,b}$, setting $f(y, z_1) = f(z_1, t) = b$; and
- we add the arcs (y, z_2) and (z_2, t) to $G_{A,b}$, setting $f(y, z_2) = f(z_2, t) = \bar{b}$.

We show that the subset sum problem admits a solution to an input (A, b, k) if and only if the flow f in $G_{A,b}$ can be decomposed into n weighted paths.

For the forward implication, let S be a solution to an input (A, b, k). For $i \in \{1, \ldots, n\}$, set

$$P_i = \begin{cases} s, x_i, y, z_1, t, & \text{if } a_i \in S, \\ s, x_i, y, z_2, t, & \text{if } a_i \notin S. \end{cases}$$

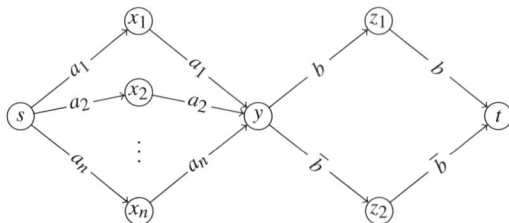

Figure 5.2 A reduction of the subset sum problem to the problem of splitting a flow in a DAG into the minimum number of weighted paths.

It is immediately seen that the flow f can be decomposed into the n paths $P_1, \ldots, P_n$, each P_i having weight $w_i = a_i$.

For the backward implication, let $P_1, \ldots, P_n$, each P_i having weight w_i, be a decomposition of the flow f into n paths. Observe first that, since each of the n arcs outgoing from s must be contained in at least one of the n paths, each such arc is contained in exactly one path in $\{P_1, \ldots, P_n\}$. Thus, $\{w_1, \ldots, w_n\} = \{a_1, \ldots, a_n\}$. Since the flow on the arc (y, z_1) is completely decomposed into a subset of the n paths, the solution S for the input (A, b, k) subset sum is given by the weights of the paths using the arc (y, z_1). Namely,

$$S = \{a_i \mid \text{the path } P \in \{P_1, \ldots, P_n\} \text{ having weight } a_i \text{ uses } (y, z_1)\}. \qquad \square$$

5.2   Minimum-cost flows and circulations

In this section we introduce the notion of a flow network, by enriching arcs with capacities. We make one further generalization by allowing the arcs to have also *demands*, namely, a lower bound on the amount of flow they must be traversed by. It is in fact the interplay among capacities, demands, and costs that renders network flows such a powerful model.

DEFINITION 5.4   *A flow network is a tuple $N = (G, \ell, u, c, q)$, where*

- *G is a directed graph (with the unique source s and the unique sink t);*
- *ℓ and u are functions assigning a non-negative demand and capacity, respectively, to every arc;*
- *c is a function assigning a cost per unit of flow to every arc;*
- *q is the required value of the flow.*

A flow over a flow network is a function satisfying the flow conservation property, the demand and capacity constraints, and having a given value.

DEFINITION 5.5   *Given a flow network $N = (G, \ell, u, c, q)$, a function $f : E(G) \rightarrow \mathbb{Q}_+$ is a flow over the network N if the following conditions hold.*

- ***Flow conservation***: *for every vertex $x \in V(G) \setminus \{s, t\}$, it holds that*

$$\sum_{y \in N^-(x)} f(y, x) = \sum_{y \in N^+(x)} f(x, y).$$

- ***Demand and capacity constraints***: *for every arc $(x, y) \in E(G)$, it holds that*

$$\ell(x, y) \leq f(x, y) \leq u(x, y).$$

- ***Required flow value q***:

$$\sum_{y \in N^+(s)} f(s, y) = \sum_{y \in N^-(t)} f(y, t) = q.$$

Even though this is not explicitly stated, some arcs can have unlimited capacity, in which case we say that their capacity is ∞, or that they do not have a capacity. Exercise 5.6 asks the reader to show that any such network can be reduced to one in which all arcs are assigned a finite capacity. Note also that some ill-constructed flow networks may not admit any flow defined as in Definition 5.5, because of ill-posed demand or capacity constraints, or because the required flow value is impossible to attain given the demands and capacities of the arcs. A flow satisfying the demand and capacity constraints will also be called *feasible*.

In the minimum-cost flow problem, given a flow network, over all its feasible flows, we have to find the one which minimizes the total cost of sending the required flow value q from s to t.

Problem 5.1   Minimum-cost flow

Given a flow network $N = (G, \ell, u, c, q)$, find a flow f over N that minimizes

$$\text{cost}(f) = \sum_{(x,y) \in E(G)} c(x, y) f(x, y).$$

In order to illustrate how a minimum-cost flow problem can be solved, we must introduce an apparent generalization, one that allows a concise statement of its solution. A *circulation* over a directed graph G is a function $f : E(G) \to \mathbb{Q}$ satisfying the flow conservation property *for every vertex*, and possibly taking negative values. Obviously, circulations must be defined only over directed graphs *without* sources or sinks. As such, a *circulation network* is a tuple $N = (G, \ell, u, c)$, where G is now a directed graph without sources or sinks and ℓ, u, and c have the same meanings as they do for flow networks. The *minimum-cost circulation* over N is a circulation $f : E(G) \to \mathbb{Q}$, satisfying the demand and capacity constraints of N (which can now take negative values), and minimizing $\text{cost}(f)$.

The minimum-cost flow problem over a flow network $N = (G, \ell, u, c, q)$ can be solved by a minimum-cost circulation problem by adding to G an arc from t to s with $\ell(t, s) = u(t, s) = q$ and $c(s, t) = 0$. In fact, these two problems are equivalent, since also a minimum-cost circulation problem can be solved in terms of a minimum-cost flow problem (Exercise 5.7).

Moreover, both the maximum flow problem and the shortest s-t path problem can be easily reduced to a minimum-cost circulation problem. In a *maximum flow problem*, one is given a flow network N with capacities only (that is, without demands and costs), and is required to find a flow of maximum value. This can be reduced to a minimum-cost circulation problem by adding, as above, the arc (t, s) with cost -1 and infinite capacity, and setting the cost of all other arcs to 0.

A shortest s-t path (recall Problem 4.1 and 4.3) can be obtained in a minimum-cost flow problem by setting the required flow value to 1, and keeping the same costs as in the original graph. Note that, in a case with multiple shortest s-t paths, one minimum-cost flow on this network may consist of multiple paths with sub-unitary weights which sum up to 1. However, we will later see in Corollary 5.10 that, whenever the demands, capacities, and flow value are integer-valued, an integer-valued minimum-cost flow can always be found.

5.2.1   The residual graph

Given a circulation f over a network $N = (G, \ell, u, c)$, we first show a condition guaranteeing that f can be transformed into another circulation over N of strictly smaller cost. This will be referred to as the *residual graph* $\mathsf{R}(f)$ of the circulation f, namely a directed graph having the same vertices as G, and in which, for every arc (x, y) of G, we do the following.

- We add the "increase" arc (x, y) to $\mathsf{R}(f)$, if the value of the circulation on (x, y) can be *increased* while satisfying the capacity of (x, y). This increase *adds* $c(x, y)$ to the cost of the resulting circulation.
- We add the "decrease" arc (y, x) to $\mathsf{R}(f)$, if the value of the circulation on (x, y) can be *decreased* while satisfying the demand of (x, y). This decrease *subtracts* $c(x, y)$ from the cost of the resulting circulation.

A circulation network is shown in Figure 5.3(a) and its residual graph in Figure 5.3(b). Formally, we have the following definition:

DEFINITION 5.6   *The residual graph $\mathsf{R}(f)$ of the circulation f is a directed graph with $V(\mathsf{R}(f)) = V(G)$, in which each arc is associated with the value by which the circulation can be increased or decreased, called its* residue, *and a cost. For every arc $(x, y) \in E(G)$:*

- *if $f(x, y) < u(x, y)$, the arc (x, y) is added to $\mathsf{R}(f)$ with residue $r(x, y) = u(x, y) - f(x, y)$ and cost $c(x, y)$; and*
- *if $\ell(x, y) < f(x, y)$, the reverse arc (y, x) is added to $\mathsf{R}(f)$ with residue $r(y, x) = f(x, y) - \ell(x, y)$ and cost $-c(x, y)$.*

The residual graph of a circulation f is a compact representation of the ways in which f can be altered, since each of its arcs encodes a possible increase or decrease of the value of f, with the corresponding cost adjustments. However, only proper cycles in $\mathsf{R}(f)$ (that is, cycles without repeated vertices apart from their first and last elements, and of length at least 3) lead to adjustments of f that satisfy the flow conservation property. A cycle C in $\mathsf{R}(f)$ gives a way of transforming f into another circulation f_C, as follows:

- let $b(C) > 0$ be the minimum residue of the arcs of C (the bottleneck of C);
- if C uses an "increase" arc (x, y), where $(x, y) \in E(G)$, then $f_C(x, y) = f(x, y) + b(C)$; and
- if C uses a "decrease" arc (y, x), where $(x, y) \in E(G)$, then $f_C(x, y) = f(x, y) - b(C)$.

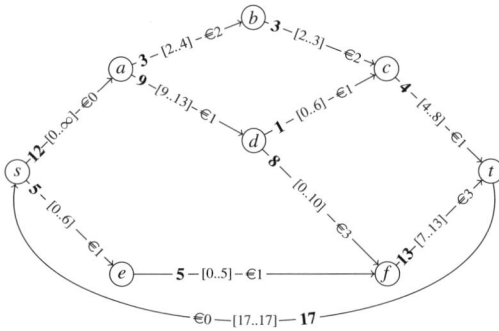

(a) A circulation network (G, ℓ, u, c) and a circulation f

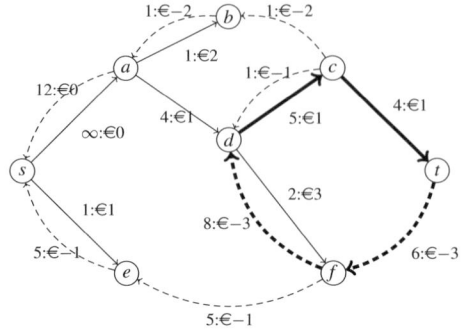

(b) The residual graph $\mathsf{R}(f)$ of f; cycle $C_1 = t, f, d, c, t$ has bottleneck $b(C_1) = 4$ and cost $c(C_1) = -4$

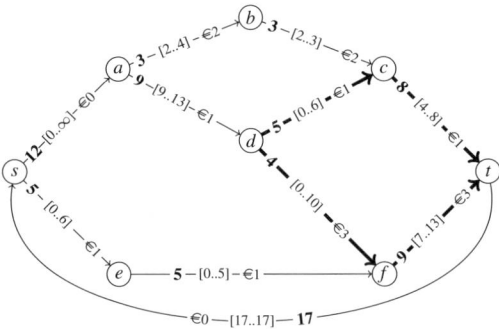

(c) The circulation $f_{C_1} = \mathsf{sw}(f, C_1)$; the arcs where it differs from f are drawn in bold

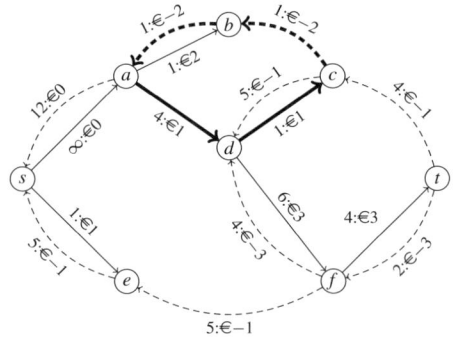

(d) The residual graph $\mathsf{R}(f_{C_1})$; cycle $C_2 = a, d, c, b, a$ has $b(C_2) = 1$, $c(C_2) = -2$

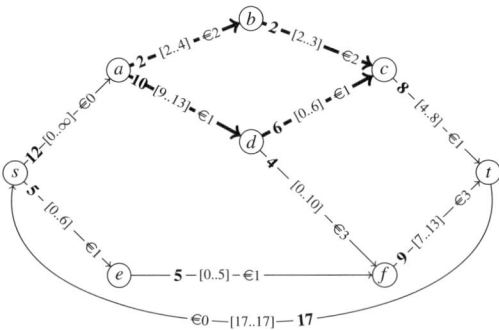

(e) The circulation $f_{C_2} = \mathsf{sw}(f_{C_1}, C_2)$; the arcs where it differs from f_{C_1} are drawn in bold

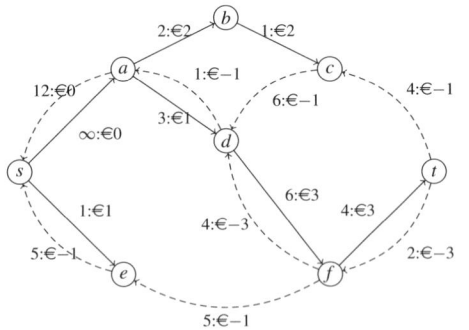

(f) The residual graph $\mathsf{R}(f_{C_2})$ has no negative-cost cycles

Figure 5.3 Circulations and their residual graphs. Arcs of circulation networks are labeled with f $[\ell..u]$ $\in c$, and arcs of residual graphs with r:$\in c$.

Onwards, we denote f_C also by $\mathsf{sw}(f, C)$, since it is obtained from f by "switching" the flow along C. Exercise 5.8 asks the reader to show that the function $\mathsf{sw}(f, C)$ is indeed a circulation over N. See Figure 5.3(c) for the circulation $\mathsf{sw}(f, C_1)$ obtained from f drawn in Figure 5.3(a), where C_1 is the cycle t, f, d, c, t of $\mathsf{R}(f)$.

We define the *cost* of a cycle C in $\mathsf{R}(f)$, and denote it $c(C)$, as the sum of the costs of the arcs of C. The following lemma gives a necessary condition for a circulation to be one of minimum cost.

LEMMA 5.7 *Let $N = (G, \ell, u, c)$ be a circulation network and let f be a circulation over N. If there is a cycle C in $\mathsf{R}(f)$ of negative cost, then the circulation $\mathsf{sw}(f, C)$ satisfies $\mathsf{cost}(\mathsf{sw}(f, C)) < \mathsf{cost}(f)$.*

Proof Let C be a cycle in $\mathsf{R}(f)$ of cost $c(C) < 0$. Circulations f and $\mathsf{sw}(f, C)$ differ only on the arcs of the cycle C, where their values differ by precisely $b(C) > 0$. By the definition of the cost function extended to $\mathsf{R}(f)$, it holds that $\mathsf{cost}(\mathsf{sw}(f, C)) = \mathsf{cost}(f) + b(C)c(C)$. Since $c(C) < 0$, the claim follows. □

For example, the circulation f_{C_1} in Figure 5.3(c) is not of minimum cost, since the cycle $C_2 = a, d, c, b, a$ in $\mathsf{R}(f_{C_1})$ has cost -2; the circulation $\mathsf{sw}(f_{C_1}, C_2)$ is shown in Figure 5.3(e).

The following result, showing that the above necessary condition is also a sufficient condition, is a fundamental result on minimum-cost circulations, one that will also give rise in the next section to an algorithm for solving a minimum-cost circulation problem. In particular, it implies that the circulation in Figure 5.3(e) is of minimum cost.

THEOREM 5.8 *Let $N = (G, \ell, u, c)$ be a circulation network. A circulation f over N is of minimum cost among all circulations over N if and only if $\mathsf{R}(f)$ has no cycles of negative cost.*

Proof The forward implication follows from Lemma 5.7. For the backward implication, let f' be an arbitrary circulation over N, and consider the function f_Δ over N defined as

$$f_\Delta(x, y) = f'(x, y) - f(x, y),$$

for all $(x, y) \in E(G)$. Clearly, f_Δ is a circulation over G, but it might not satisfy the arc constraints of N. Circulation f_Δ can instead be transformed into a circulation $\tilde{f}_\Delta$ over $\mathsf{R}(f)$ such that $\tilde{f}_\Delta(x, y) \geq 0$, for all $(x, y) \in E(\mathsf{R}(f))$, and $\mathsf{cost}(f_\Delta) = \mathsf{cost}(\tilde{f}_\Delta)$, as follows. For every $(x, y) \in E(G)$,

- if $f_\Delta(x, y) > 0$, then $f(x, y) < f'(x, y) \leq u(x, y)$ and thus the arc (x, y) is present in $\mathsf{R}(f)$; we set $\tilde{f}_\Delta(x, y) = f_\Delta(x, y)$; and
- if $f_\Delta(x, y) < 0$, then $\ell(x, y) \leq f'(x, y) < f(x, y)$ and thus the arc (y, x) is present in $\mathsf{R}(f)$; we set $\tilde{f}_\Delta(y, x) = -f_\Delta(x, y)$ (and $\tilde{f}_\Delta(x, y) = 0$ if the arc (x, y) is present in $\mathsf{R}(f)$).

See Figure 5.4 for an example. As in Theorem 5.2 and Exercise 5.3, circulation $\tilde{f}_\Delta$ can be decomposed into $k \leq |E(\mathsf{R}(f))|$ cycles of $\mathsf{R}(f)$, $C_1, \ldots, C_k$, each C_i having a weight $w_i > 0$, such that, for every arc $(x, y) \in E(\mathsf{R}(f))$, it holds that

$$\tilde{f}_\Delta(x, y) = \sum_{C_i \,:\, (x,y) \in C_i} w_i.$$

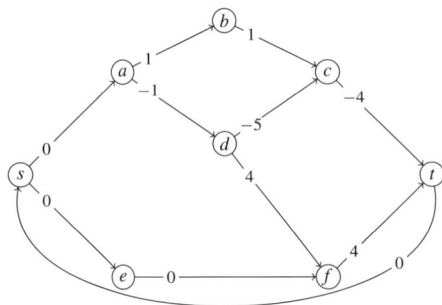

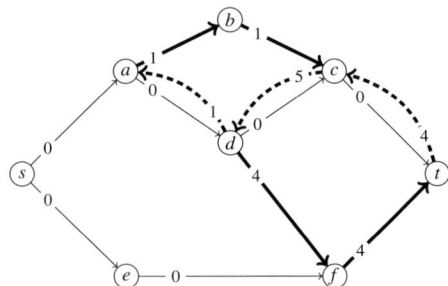

(a) The difference f_Δ between circulations f (Figure 5.3(a)) and f_{C_2} (Figure 5.3(e))

(b) The transformed circulation $\tilde{f}_\Delta$ over $\mathsf{R}(f_{C_2})$; we omitted costs and "decrease" arcs with null circulation. $\tilde{f}_\Delta$ can be decomposed into cycles a, b, c, d, a of weight 1 and cost 2, and t, c, d, f, t of weight 4 and cost 4

Figure 5.4 The difference between two circulations and its representation in the residual graph of the latter.

Therefore,

$$\mathsf{cost}(f') - \mathsf{cost}(f) = \mathsf{cost}(f_\Delta) = \mathsf{cost}(\tilde{f}_\Delta) = \sum_{i=1}^{k} w_i c(C_i).$$

Since no cycle in $\mathsf{R}(f)$ has negative cost, $\mathsf{cost}(f') - \mathsf{cost}(f) \geq 0$, implying that f is a minimum-cost circulation. □

5.2.2 A pseudo-polynomial algorithm

Theorem 5.8 shows, on the one hand, that the optimality of a given circulation f can be checked in polynomial time, by checking the existence of negative-cost cycles in $\mathsf{R}(f)$. This can be done, for example, in time $O(nm)$ with the Bellman–Ford method from Section 4.2.2. On the other hand, it leads to an algorithm for finding a minimum-cost circulation over a network N:

(1) start with any circulation f over N satisfying the demand and capacity constraints;
(2) as long as the residual graph $\mathsf{R}(f)$ of f has a cycle C of negative cost, replace f by $\mathsf{sw}(f, C)$.

COROLLARY 5.9   *If $N = (G, \ell, u, c)$ is a circulation network, with ℓ, u, and c integer-valued, then a minimum-cost circulation can be found in time $O(nm^2 CU)$, where $n = |V(G)|$, $m = |E(G)|$, C is the maximum absolute value of the costs, and U is the maximum capacity.*

Proof The problem of finding an initial feasible circulation over N satisfying the demand and capacity constraints can be reduced to finding a minimum-cost circulation in a network N' without demand constraints, a problem in which the null circulation is feasible. Moreover, N' can be constructed such that it has $O(n)$ vertices and $O(m)$ arcs, costs 0 or -1, and the maximum capacity is $O(U)$; thus a minimum-cost circulation on N can be found in time $O(nm^2U)$. Exercise 5.9 asks the reader to construct this reduction.

The cost of any feasible circulation is at most mCU, since, by definition, for each arc $(x, y) \in E(G)$, $-C \le c(x, y) \le C$ and $u(x, y) \le U$. Analogously, the cost of the optimal circulation is at least $-mCU$. At each iteration of the above algorithm, the cost of f is changed by $b(C)c(C) < 0$. If ℓ, u, and c are integer-valued, then $b(C) \ge 1$ and $|c(C)| \ge 1$, and thus the algorithm terminates after $O(mCU)$ iterations.

If C is found using the Bellman–Ford algorithm, we can conclude that a minimum-cost circulation over N can be found in time $O(nm^2CU)$. $\qquad\square$

Another corollary of Theorem 5.8 is the following.

COROLLARY 5.10 *If $N = (G, \ell, u, c)$ is a circulation network, with ℓ and u integer-valued, then the minimum-cost circulation problem always admits an integer-valued minimum-cost circulation.*

Proof In the proof of Corollary 5.9, we start either with a null circulation or with an optimal one in a circulation network with the same capacities, and integer costs and demands. At each iteration of the algorithm, each value of the circulation, if changed, is changed by $b(C)$. This is an integer, since ℓ and u are integer-valued. $\qquad\square$

The complexity of the above algorithm depends on the actual values of the input variables and can take exponential time on certain inputs. However, for adequate choices of a cycle C of negative cost in $\mathsf{R}(f)$, the number of iterations can be bounded only by a polynomial in n and m, independently of the integrality of ℓ, u, or c. For example, choosing C to be a cycle of minimum-mean cost (see Exercise 4.18 on how this can be found in $O(nm)$ time), the number of iterations can be shown to be at most $4nm^2\lceil \log n \rceil$, leading to an algorithm of complexity $O(n^2m^3\lceil \log n \rceil)$. Using more advanced ideas, more efficient algorithms for solving minimum-cost circulations can be devised; see the various complexity bounds and references cited in the literature section of this chapter.

See Insight 5.2 for an extension of the minimum-cost flow problem to flow networks with convex costs.

5.3 Bipartite matching problems

In the following sections we show how minimum-cost flows can solve various optimization problems, which often seem unrelated to flows at first. To ensure a polynomial-time solution, the size of the reduction must be polynomial in the original input size.

Insight 5.2 Convex cost functions

The minimum-cost flow problem discussed in this section has, for each arc, a cost function linear in the flow value. A more general problem is the one in which the cost function of each arc is convex in the value of the flow. We are given a flow network $N = (G, \ell, u, c, q)$, where $\ell, u : E(G) \to \mathbb{N}$, and $c : E(G) \times \mathbb{N} \to \mathbb{Q}$ is a convex function in the last argument, that is, for each arc (x, y), the function $c(x, y, \cdot)$ has a "bathtub" shape. The task is to find a flow $f : E(G) \to \mathbb{N}$ over N that minimizes

$$\text{cost}(f) = \sum_{(x,y) \in E(G)} c(x, y, f(x, y)).$$

Although efficient algorithms for this problem exist (see the references cited in the literature section), the following pseudo-polynomial reduction to a standard minimum-cost flow problem can turn out to be effective in practice. Construct the network N' from N, by replacing every arc $(x, y) \in E(G)$ by $u(x, y)$ parallel arcs between x and y, with demand 0 and capacity 1, such that the ith arc, $1 \le i \le u(x, y)$, has cost $c(x, y, i) - c(x, y, i - 1)$. Exercise 5.10 asks the reader to show that this reduction is correct, and that the cost of the optimal flow in N equals the cost of the optimal flow in N' plus $\sum_{(x,y) \in E(G)} c(x, y, 0)$. If an approximate solution is sufficient, then one can reduce the size of this reduction by introducing only $u(x, y)/t$ parallel arcs of capacity $t \ge 1$ such that the ith arc has cost $c(x, y, it) - c(x, y, (i - 1)t)$.

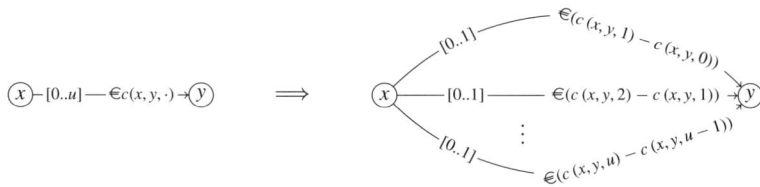

On the other hand, if the cost functions are concave, then this problem is generally NP-hard.

5.3.1   Perfect matching

In a bipartite matching problem one must match the objects of one set A with the elements on another set B, while satisfying some constraints and minimizing or maximizing a certain cost. Let us look at one of the simplest instances of such a problem, namely the minimum-cost perfect matching problem. Consider the problem in Example 5.2 below. We can model it as a *bipartite* graph, namely as an undirected graph $G = (V, E)$, where V is partitioned into two sets, A and B, such that there are no edges between vertices in A and there are no edges between vertices in B. When the two sets A and B are known in advance, we will write $G = (A \cup B, E)$.

Example 5.2   In a factory, n jobs need to be executed using one of the n available machines. We assume that each job takes one day to complete, and that each job can be performed by only a particular set of machines. Depending on the machine, each job can be performed with a given operating cost. Each machine cannot perform more than one job at the same time, and all jobs need to be finished on the same day. The task is to assign each job to a machine such that the total operating cost is minimized.

Set A to be the set of jobs, and set B to be the set of machines. Between every job x and every machine y that can solve job x we add the edge (x, y) to E. An assignment of the n jobs to the n machines is just a set M of edges of G with the property that no two edges of M share a vertex, and all vertices are touched by some edge of M.

More formally, given a set M of edges of G, and given $x \in V$, we denote by $d_M(x)$ the number of edges of M incident to x, namely

$$d_M(x) = |\{y \in V \mid (x, y) \in E\}|. \tag{5.1}$$

The set M is called a *matching* if, for all $x \in A \cup B$, $d_M(x) \leq 1$ holds. Solving the problem in Example 5.2 amounts to solving Problem 5.3 below (see Figure 5.5(a) for an example).

Problem 5.3 Minimum-cost perfect matching

Given a bipartite graph $G = (A \cup B, E)$, and a cost function $c : E \to \mathbb{Q}_+$, find a *perfect* matching M, in the sense that

- for every $x \in A \cup B$, $d_M(x) = 1$,

which minimizes $\sum_{(x,y) \in M} c(x, y)$, or report that no perfect matching exists.

Let us now see how the handles provided by a minimum-cost flow problem can be used to solve the minimum-cost perfect matching problem. We construct, as illustrated in Figure 5.5(b), the flow network $N = (G^*, \ell, u, c, q)$, where G^* is obtained from the input bipartite graph G, by orienting all edges from A toward B; they are assigned

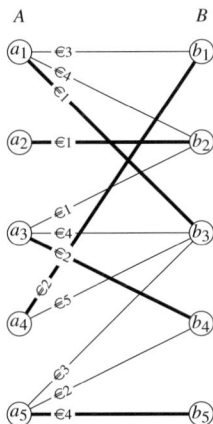

(a) A bipartite graph G whose edges are labeled with costs; the minimum-cost perfect matching is highlighted.

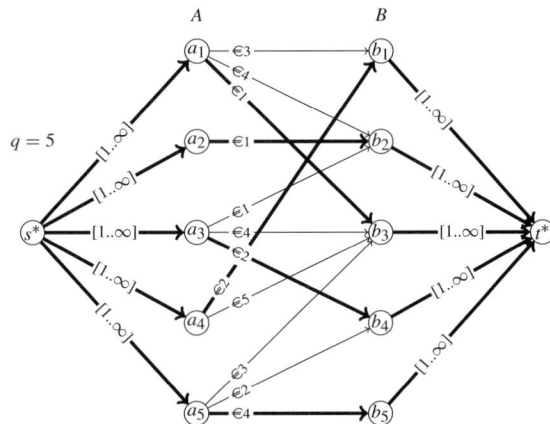

(b) The flow network N constructed from G; arcs incident to s^* or t^* have null cost; arcs between sets A and B have demand 0 and capacity 1; arcs with non-null value in the minimum-cost flow over N are highlighted.

Figure 5.5 Reducing minimum-cost perfect matching to minimum-cost flow.

demand 0 and cost as in the input to the minimum-cost perfect matching problem. A global source s^* is added, together with arcs from s^* to every vertex of A. Likewise, a global sink t^* is added, together with arcs from every vertex of B to t^*.

We set the flow value to be $q = |A|$, and set the demand of all arcs incident to s^* and t^* to 1, and their cost to 0. Observe that we have not assigned capacities to the arcs of N, since, if $q = |A|$ and all arcs incident to s^* have demand 1, then their flow values must be at most 1. Moreover, if a perfect matching exists, then $|A| = |B|$, and hence also all arcs incident to t^* have flow value at most 1.

By Corollary 5.10, network N admits an integer-valued minimum-cost flow, which *induces* a perfect matching M of G consisting of those edges between the sets A and B with flow value 1. Vice versa, every matching in G analogously *induces* a flow in N. A minimum-cost circulation over a network without capacity constraints, and whose demands are bounded by U, can be solved in time $O(n \log U(m + n \log n))$, by the algorithm of Gabow and Tarjan (see the literature section of this chapter). Therefore, since the flow network N constructed from G as described above has only demands, and they are all at most 1, we can further refine this reduction to obtain an algorithm of improved complexity.

THEOREM 5.11   *The minimum-cost perfect matching problem on a bipartite graph G with n vertices, m edges, and non-negative costs is solvable in time $O(nm + n^2 \log n)$.*

Proof   We transform the flow network N into a circulation network N' with demands only by adding the arc (t^*, s^*) with demand 0 and cost $c(t^*, s^*) = 1 + \sum_{(x,y) \in E(G)} c(x, y)$. Let f be an integer-valued minimum-cost circulation over N'.

Suppose that G admits a perfect matching M, thus also that $|A| = |B|$. Assume also for a contradiction that $f(t^*, s^*) > |A|$. The perfect matching M induces another circulation f', which satisfies $f'(t^*, s^*) = |A|$. By the choice of $c(t^*, s^*)$, we get $\mathsf{cost}(f') < \mathsf{cost}(f)$, which contradicts the minimality of f. Therefore, we have that indeed $f(t^*, s^*) = |A|$, and thus f' takes the value 1 on all arcs incident to s^* or to t^*. Thus, it induces a minimum-cost perfect matching in G.

If G does not admit a perfect matching, then some arc incident to s^* or t^* has a value of f' strictly greater than 1.   □

5.3.2   Matching with capacity constraints

In the next problem we are interested in possibly matching more elements of A to the same element of B (a *many-to-one matching*), but with two particular constraints on their number. Consider the following example.

Example 5.4   In another factory, n jobs need to be executed using one of the j available machines. In this problem, we assume that a machine can execute multiple jobs, but that there is a limit on the total number of jobs it can execute. As before, a job can be executed only by a particular set of machines, with a given operating cost. There is one additional constraint, namely that jobs are of two types, easy and hard, and that each machine also has a limit on the number of hard jobs it can execute.

The task is to assign each job to a machine such that the total operating cost is minimized, and that no machine executes more jobs than its total limit, and no more hard jobs than its limit for hard jobs.

The problem in Example 5.4 can be modeled as before, through a bipartite graph $G = (A \cup B, E)$ in which the set A of jobs is further partitioned into A_e, the set of easy jobs, and A_h, the set of hard jobs. Given a vertex $y \in B$ and a matching M, we denote

$$d_{M,A_h} = |\{x \in A_h \mid (x, y) \in M\}|. \qquad (5.2)$$

Solving the problem in Example 5.4 amounts to solving the following one (see Figure 5.6(a) for an example).

Problem 5.5   Minimum-cost many-to-one matching with constrained load factors

Given a bipartite graph $G = (A \cup B, E)$, with $A = A_e \cup A_h$, a cost function $c : E \to \mathbb{Q}_+$, and functions $\mathtt{ut}, \mathtt{uh} : B \to \mathbb{N}$, the capacities for the total number of jobs, and the total number of hard jobs, respectively, find a many-to-one matching M satisfying

- for every $x \in A$, $d_M(x) = 1$, and
- for every $y \in B$, $d_M(y) \le \mathtt{ut}(y)$, and $d_{M,A_h}(y) \le \mathtt{uh}(y)$,

and minimizing $\sum_{(x,y) \in M} c(x, y)$, or report that no such matching exists.

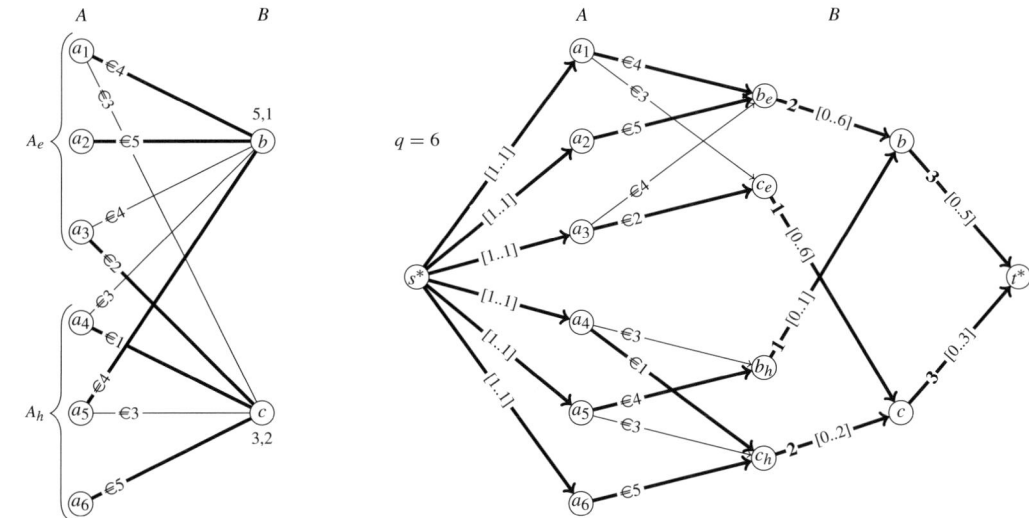

(a) A bipartite graph G and a minimum-cost many-to-one matching satisfying the load constraints of the vertices b and c, drawn as $\mathtt{ut}, \mathtt{uh}$

(b) The flow network N constructed from G; arcs incident to s^*, t^*, b, and c have null cost; arcs between sets A and B have demand 0 and capacity 1; arcs with non-null value in the minimum-cost flow over N are highlighted

Figure 5.6 Reducing minimum-cost many-to-one matching with constrained load factors to minimum-cost flow.

If the constraints on the hard jobs were missing, we could easily model this problem as done in the previous section, but removing the demand constraints of the arcs from each machine y to the global sink, and setting their capacity to the total limit of jobs each machine y can execute.

Therefore, for this particular many-to-one matching problem we construct the flow network $N = (G^*, \ell, u, c, q)$ as illustrated in Figure 5.6(b). For each vertex y of B we introduce three copies of it, y, y_e, and y_h. We add an arc from y_h to y with null demand and cost, and capacity $\mathrm{uh}(y)$, and an arc from y_e to y with null demand and cost, and infinite capacity. Moreover, we add an arc from y to the global sink t^* with null demand and cost, and capacity $\mathrm{ut}(y)$. For every job $x \in A_e$ that can be executed on machine y, we add the arc (x, y_e), and for every job $x \in A_h$ that can be executed on machine y, we add the arc (x, y_h). These arcs have demand 0, capacity 1, and cost as in the many-to-one matching problem instance. We also add arcs from the global source s^* to each vertex in A. The arcs incident to s^* have null cost, and demand and capacity 1. Finally, the flow is required to have value $q = |A|$.

An integer-valued flow in N induces a many-to-one matching M in G. Matching M satisfies the upper bound constraints for the total number of hard jobs of any machine $y \in B$, because of the capacity constraint of the arc (y_h, y). It also satisfies the upper bound constraint on the total number of jobs of machine y because of the capacity constraint of the arc (y, t^*). Conversely, a many-to-one matching M in G satisfying all load constraints induces a flow in N. Thus, a minimum-cost many-to-one matching with constrained load factors can be found from an integer-valued minimum-cost flow in N.

Graph G^* has $O(|V|)$ vertices and $O(|V| + |E|)$ arcs. The algorithms mentioned in the literature section of this chapter lead to various polynomial complexity bounds for this problem.

5.3.3   Matching with residual constraints

In the next many-to-one matching problem, we no longer have a rigid upper bound on the number of elements of A that can be matched with an element of B, but we are given ideal load factors for each element of B. Accordingly, we want to find a matching which better approximates these load factors. See Figure 5.7(a) for an example.

Example 5.6 In yet another factory, n jobs need to be executed using one of the j available machines. As before, a job can be executed only by a particular set of machines, with a given operating cost. For technical reasons, each machine must ideally execute a given number of jobs. If this number is not attained, the firm incurs an operating cost consisting of the absolute difference between the actual number of jobs the machine is assigned and its ideal load, multiplied by a fixed cost. The task is to assign each job to a machine such that the total operating cost is minimized.

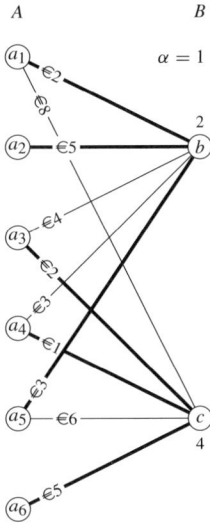

(a) A bipartite graph G and a many-to-one matching with optimal residual load factors; vertices b and c are labeled with their ideal load factors

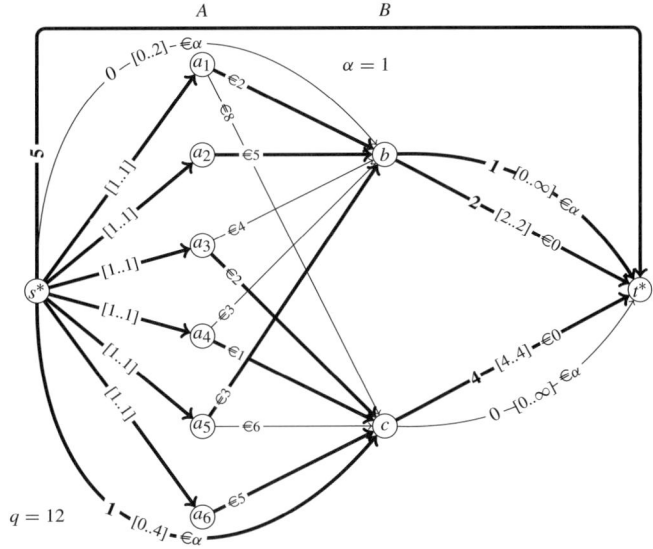

(b) The flow network N constructed from G; arcs from s^* to a vertex of A or to t^* have null cost; arcs between sets A and B have demand 0 and capacity 1; arcs with non-null value in the minimum-cost flow over N are highlighted

Figure 5.7 Reducing minimum-cost many-to-one matching with optimal residual load factors to minimum-cost flow.

Problem 5.7 Minimum-cost many-to-one matching with optimal residual load factors

Given a bipartite graph $G = (A \cup B, E)$, a cost function $c : E \to \mathbb{Q}_+$, constant $\alpha \in \mathbb{Q}_+$, and $\texttt{id} : B \to \mathbb{N}$, the optimal load factors for the vertices in B, find a many-to-one matching M satisfying

- for every $x \in A$, $d_M(x) = 1$,

and minimizing

$$\sum_{(x,y) \in M} c(x,y) + \alpha \sum_{y \in B} |\texttt{id}(y) - d_M(y)|.$$

We construct the flow network $N = (G^*, \ell, u, c, q)$ as in Figure 5.7(b). We start from G and orient all edges from A toward B, and set their demand to 0, their capacity to 1, and their cost as in the input many-to-one matching instance. We add a global source s^* with arcs toward all vertices in A, with demand and capacity 1, and null cost. Since we pay the fixed penalty α for every job below or above the ideal load factor of a machine y, we add the following arcs:

- an arc from y to a global sink t^* with $\ell(y, t^*) = u(y, t^*) = \mathtt{id}(y)$, and $c(y, t^*) = 0$;
- an arc from y to t^* with $\ell(y, t^*) = 0$, $u(y, t^*) = \infty$, and $c(y, t^*) = \alpha$; and
- an arc from s^* to y with $\ell(s^*, y) = 0$, $u(s^*, y) = \mathtt{id}(y)$, and $c(s^*, y) = \alpha$.

Finally, we set $q = |A| + \sum_{y \in B} \mathtt{id}(y)$, and also add the arc (s^*, t^*) with null demand and cost, and infinite capacity. The optimal matching for the many-to-one matching with optimal residual load factors consists of the edges of G whose corresponding arcs in G^* have non-null flow value in the integer-valued minimum-cost flow over N.

The correctness of this reduction can be seen as follows.

- As long as a machine y receives exactly $\mathtt{id}(y)$ jobs, no operating cost is incurred; this is represented by the arc (y, t^*) with demand and capacity $\mathtt{id}(y)$, and null cost.
- If y receives more jobs than $\mathtt{id}(y)$, then these additional jobs will flow along the parallel arc (y, t^*) of cost α.
- If y receives fewer jobs than $\mathtt{id}(y)$, then some compensatory flow comes from s^* through the arc (s^*, y), where these fictitious jobs again incur cost α. We set the capacity of (s^*, y) to $\mathtt{id}(s^*, y)$, since at most $\mathtt{id}(y)$ compensatory jobs are required in order to account for the lack of jobs for machine y.

Requiring flow value $q = |A| + \sum_{y \in B} \mathtt{id}(y)$ ensures that there is enough compensatory flow to satisfy the demands of the arcs of type (y, t^*). Having added the arc (s^*, t^*) with null cost ensures that there is a way for the compensatory flow which is not needed to satisfy any demand constraint to go to t^*, without incurring an additional cost.

Graph G^* has $O(|V|)$ vertices and $O(|V| + |E|)$ arcs. The algorithms mentioned in the literature section of this chapter give various polynomial-complexity bounds for this problem.

5.4   Covering problems

Generally speaking, in a covering problem one is given an arbitrary graph G and is required to cover the vertices of G with some subgraphs of G, under some constraints. For example, in a perfect matching problem one has to cover all the vertices of the graph with single edges that do not intersect. In this section we discuss two other covering problems, where the covering subgraphs are more complex.

5.4.1   Disjoint cycle cover

In the following problem, the covering subgraphs are disjoint cycles.

Example 5.8 A set of n players will take part in a competition. We know in advance which pairs of players are able to compete with one another, how much the spectators will enjoy a game between two players, and which of the two players is more likely to win. A set of games must be selected, such that

- each player plays exactly two games with other players;
- each player must be more likely to win in one of his games, and less likely to win in his other game; and
- the total enjoyment of the selected games be maximum.

We can model this problem as a directed graph $G = (V, E)$, where V is the set of players. A game in which player x is more likely to win than player y is represented by the arc (x, y). Observe that the set of games selected corresponds to a set of vertex-disjoint cycles of G covering all the vertices of G, since every vertex has precisely one in-neighbor and one out-neighbor in the set of games selected. Such a collection of cycles is called a *disjoint cycle cover*. Solving this problem amounts to solving the following one (see Figure 5.8(a) for an example).

Problem 5.9 Minimum-cost disjoint cycle cover

Given a directed graph $G = (V, E)$, and cost function $c : E \rightarrow \mathbb{Q}_+$, find a disjoint cycle cover $C_1, \ldots, C_k$ of G which minimizes

$$\sum_{i=1}^{k} \sum_{(x,y) \in C_i} c(x, y),$$

or report that no such cover exists.

Even though the problem in Example 5.8 is a maximization problem, and Problem 5.9 stated above is a minimization one, the positive costs accepted by Problem 5.9 can be suitably initialized as the sum of all possible spectators' enjoyment minus the benefit that the spectators receive if that game is played.

The minimum-cost disjoint cycle cover can be solved by a minimum-cost perfect matching problem in a bipartite graph, and thus by a minimum-cost flow problem. The idea of this reduction is that every vertex must choose its successor in some cycle of the optimal disjoint cycle cover. As such, we can construct a bipartite graph G^* to which we introduce, for every vertex $x \in V(G)$, two copies of it, x_{out} and x_{in}. For every arc (x, y) of G, we add an edge between x_{out} and y_{in} having cost $c(x, y)$. See Figure 5.8(b) for an example.

A disjoint cycle cover $C_1, \ldots, C_k$ in G naturally induces a matching M in G^* by adding to M the edge (x_{out}, y_{in}), for all arcs (x, y) in the cover. This is also a perfect matching since, in a disjoint cycle cover, every vertex is the start point of exactly one

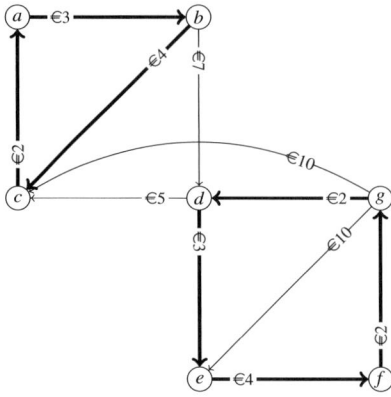

(a) A minimum-cost disjoint cycle cover in a directed graph G.

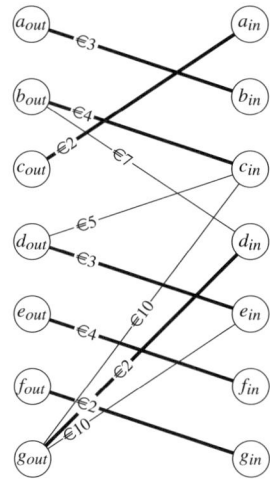

(b) A minimum-cost perfect matching in G^*.

Figure 5.8 Reducing minimum-cost disjoint cycle cover to minimum-cost perfect matching.

arc of the cover and the endpoint of exactly one arc of the cover. Conversely, every perfect matching M in G^* induces a disjoint cycle cover in the same way.

Observe that graph G^* has $O(|V|)$ vertices and $O(|E|)$ edges. From Theorem 5.11, we obtain the following corollary.

COROLLARY 5.12   *The minimum-cost disjoint cycle cover problem on a directed graph G with n vertices and m arcs and non-negative costs is solvable in time $O(nm+n^2 \log n)$.*

5.4.2   Minimum path cover in a DAG

In the next problem, the covering subgraphs are paths that may share vertices.

Example 5.10   A newspaper's office wants to cover a set of n events $e_1, \ldots, e_n$ taking place in a territory. Each event e_i starts at a known time $t(i)$ and lasts for a known duration $d(i)$. The newspaper is also aware of the time $t(i,j)$ and the cost $c(i,j)$ needed to travel from each event i to each other event j. The management would like to cover these events using the minimum number of reporters, and, among all such solutions, would like one that minimizes the total cost of traveling. Moreover, it is assumed that the reporters start and end their journey at the newspaper's premises.

In this example, the input graph $G = (V, E)$ is a DAG whose vertices are the events $e_1, \ldots, e_n$, and whose arcs are the possible connections between them: $(e_i, e_j) \in E$ if $t(i) + d(i) + t(i,j) \leq t(j)$. Each arc (e_i, e_j) has cost $c(i,j)$, the cost of traveling from event e_i to event e_j. We add two dummy events, e_0 and e_{n+1}, to model the requirement that the reporters start and end their journey at the newspaper's offices. For every $i \in \{1, \ldots, n\}$,

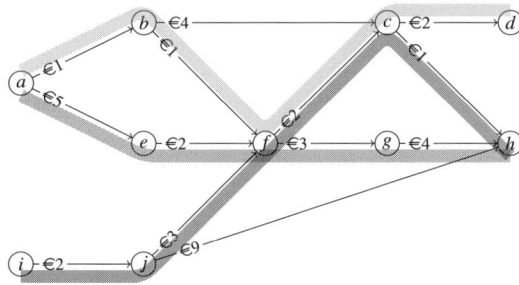

(a) A DAG G and three paths forming a minimum-cost minimum path cover. Path 1: a, b, f, c, d; path 2: a, e, f, g, h; path 3: i, j, f, c, h.

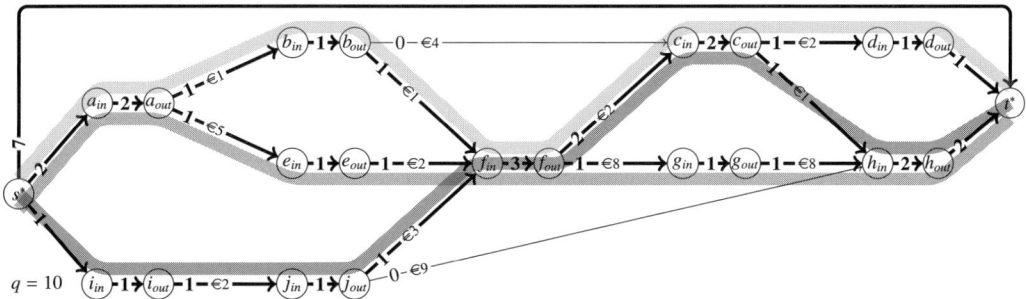

(b) The flow network N constructed from G; arcs of type (x_{in}, x_{out}) have demand 1 and cost 0, arcs of type (x_{out}, y_{in}) have demand 0 and cost as in G, arcs incident to s^* or t^* have demand 0; arcs incident to s^* have demand 0 and their cost equals the sum of all costs plus 1, arcs incident to t^* have demand 0 and cost 0; arcs with non-null value in the minimum-cost flow over N are highlighted, and a decomposition of the flow into three paths is shown

Figure 5.9 Reducing minimum-cost minimum path cover to minimum-cost flow.

we add the arcs (e_0, e_i) and (e_i, e_{n+1}) labeled with the cost of traveling between the newspaper's offices and event e_i. A path in G from the unique source e_0 to the unique sink e_{n+1} is one possible route of a reporter. Each vertex has to be covered by at least one path, since each event must be covered by a reporter. Such a collection of paths covering all the vertices of a directed graph is called a *path cover*. Therefore, solving the problem in Example 5.10 amounts to solving the following problem (see Figure 5.9(a) for an example).

Problem 5.11 Minimum-cost minimum path cover

Given a DAG $G = (V, E)$, and cost function $c : E \to \mathbb{Q}_+$, find the minimum number k of paths $P_1, \ldots, P_k$ such that

- every $v \in V$ belongs to some P_i (that is, $P_1, \ldots, P_k$ form a path cover of G),
- every P_i starts in a source of G and ends in a sink of G, and

among all path covers with k paths satisfying the above conditions, the function

$$\sum_{i=1}^{k} \sum_{(x,y)\in P_i} c(x,y)$$

is minimized.

We construct the flow network $N = (G^*, \ell, u, c, q)$ as shown in Figure 5.9(b). Namely,

- for every vertex x of G, we introduce two vertices x_{in} and x_{out} in G^*, and add the arc (x_{in}, x_{out}) to G^*, with null cost and demand 1;
- for every arc (y, x) incoming to x of G, we add the arc (y_{out}, x_{in}) to G^*, with cost $c(y,x)$ and demand 0; and
- for every arc (x, y) outgoing from x of G, we add the arc (x_{out}, y_{in}) to G^*, with cost $c(x,y)$ and demand 0.

We add a global source s^* with arcs to every vertex x_{in}, if x is a source in G, with demand 0. Likewise, we add a global sink t^* and arcs (x_{out}, t^*), for every sink x in G, with demand 0.

As opposed to the previous problems, here we also need to find a solution of minimum size. Accordingly, since the cost of any path from a source to a sink is at most the sum of all costs of G, we set the cost of each arc (s^*, x_{in}) as

$$c(s^*, x_{in}) = 1 + \sum_{(x,y)\in E(G)} c(x,y). \tag{5.3}$$

We can then set the cost of the arcs incident to t^* to 0. We set the value of the flow to $q = |V|$, since at most $|V|$ paths are needed in any path cover. To account for the supplementary flow, we also add the arc (s^*, t^*) with cost 0 and demand 0.

Having an integer-valued minimum-cost flow f in the network N, we show how f can be converted into a minimum path cover in G whose paths start in a source of G and end in a sink of G. Then we show that this path cover also has minimum cost among all such minimum path covers.

Even though decomposing an arbitrary flow into the minimum number of paths is an NP-hard problem (by Theorem 5.3), the following lemma shows that in our case this problem is simpler.

LEMMA 5.13   *If f is a minimum-cost flow over the acyclic flow network N constructed as above, then any decomposition of f, minus the arc (s^*, t^*), into integer-weighted paths has the same number of paths.*

Proof   Let $P_1, \ldots, P_k$ be a decomposition of f, minus the arc (s^*, t^*), into weighted paths, having integer weights $w_1, \ldots, w_k$, respectively. To prove our claim it suffices to show that $w_1 = \cdots = w_k = 1$.

Suppose for a contradiction that some path P_j has integer weight $w_j > 1$. Then, we can subtract 1 from the flow value of each of its arcs, and obtain another flow f', still feasible over N, of strictly smaller cost than f, which contradicts the minimality of f. This is true because at least the cost of the first arc of P_j outgoing from s^* is positive. $\square$

Hence f can induce a path cover in G in the following way. Since G^* is acyclic, we can consider a decomposition of f minus the arc (s^*, t^*) into the k paths $P_1, \ldots, P_k$ from s^* to t^* (all having weight 1, by the proof of Lemma 5.13). Removing the vertices s^* and t^* from each P_i, and contracting the arcs of the form (x_{in}, x_{out}) back into x, we obtain a collection $Q_1, \ldots, Q_k$ of paths in G.

We next show that k is also the minimum size of a path cover whose paths start in a source and end in a sink of G.

LEMMA 5.14   *The minimum number of paths starting in a source of G and ending in a sink of G and covering all its vertices is k, namely the number of paths that f is decomposable into.*

Proof   We argue by contradiction, and assume that there exists such a collection of paths $Q'_1, \ldots, Q'_j$ with $j < k$ paths. These paths in G induce a feasible flow f' over the network N. Since flow f has value k, and flow f' has value j, it holds that

$$\sum_{(s^*, x_{in}) \in E(G^*)} c(s^*, x_{in}) f(s^*, x_{in}) - \sum_{(s^*, x_{in}) \in E(G^*)} c(s^*, x_{in}) f'(s^*, x_{in})$$

$$\geq (k - j) \left(1 + \sum_{(x,y) \in E(G)} c(x, y) \right),$$

according to (5.3). As noted above, the cost of any path from s^* to t^* in N is at most $\sum_{(x,y) \in E(G)} c(x, y)$. Therefore, f' has strictly smaller cost than f, contradicting the minimality of f. $\square$

Therefore, $Q_1, \ldots, Q_k$ is a path cover of G with the minimum number of paths that start in a source of G and end in a sink of G. The fact that this path cover also has the minimum cost among all such path covers of size k follows from the optimality of f.

Graph G^* has $O(|V|)$ vertices and $O(|V| + |E|)$ arcs. Decomposing f into the k paths and constructing $Q_1, \ldots, Q_k$ can be done in time $O(|V| + |E|)$, by Theorem 5.2. Therefore, we obtain the following theorem.

THEOREM 5.15   *The minimum-cost minimum path cover problem on a DAG G with n vertices, m arcs, and non-negative costs is solvable in time $O(nm + n^2 \log n)$.*

Proof   A minimum-cost circulation over a network without capacity constraints and demands bounded by U can be solved in time $O(n \log U(m + n \log n))$ by the algorithm of Gabow and Tarjan (see the literature section of this chapter). Since the flow network N constructed from G as described above has only demands, and they are all at most 1, we can transform N into an equivalent circulation network N' with demands only, by adding the arc (t^*, s^*) with demand 0. $\square$

5.5   Literature

The seminal article on network flows is that by Ford & Fulkerson (1956). The first polynomial-time algorithm for the minimum-cost flow problem was given by Edmonds & Karp (1972), and the first strongly polynomial-time algorithm by Tardos (1985). Our presentation in this chapter is inspired by two well-known books on combinatorial optimization, *Network Flows* (Ahuja *et al.* 1993) and *Combinatorial Optimization* (Schrijver 2003).

The first proof of the fact that decomposing a flow into the minimum number of components is an NP-hard problem appeared in Vatinlen *et al.* (2008). The heuristics mentioned in Insight 5.1 are from Vatinlen *et al.* (2008) and Hartman *et al.* (2012).

Table 5.1, compiled from Schrijver (2003), lists some algorithms for minimum-cost circulation problems with integer-valued demands, capacities, and costs; the optimal solution returned is integer-valued.

One of the first algorithms for solving minimum-cost flows with convex cost functions was published by Hu (1966). See Weintraub (1974) and Minoux (1986) for polynomial-time algorithms for this problem, and Guisewite & Pardalos (1990) for minimum-cost flows with concave cost functions.

Algorithms for the maximum-cost bipartite matching problem include those published by Edmonds & Karp (1972), of complexity $O(n(m + n \log n))$, and by Gabow & Tarjan (1989), of complexity $O(\sqrt{n}m \log(nC))$, where C is the maximum absolute cost value. The minimum-cost perfect matching problem in an arbitrary graph can be solved in time $O(n(m + n \log n))$, by methods not based on network flows (Gabow 1990) (see Exercise 5.15 for using this algorithm in finding a minimum-cost maximum matching in an arbitrary graph).

The many-to-one matching with constrained load factors problem presented in Section 5.3.2 is adapted from Ahuja *et al.* (1993, Application 9.3). The many-to-one matching with optimal residual load factors problem presented in Section 5.3.3 is a simplified version of a problem proposed by Lo *et al.* (2013). The reduction of the disjoint cycle cover problem to a perfect matching problem used in Section 5.4.1 was first given by Tutte (1954).

Table 5.1 Some algorithms for minimum-cost circulations over integer-valued circulation networks. We denote by U the maximum value of a capacity (or demand), by U^+ the sum of all capacities, and by C the maximum absolute value of an arc cost.

$O(m \log U(m + n \log n))$	Capacity scaling (Edmonds & Karp 1972)
$O(n^{5/3} m^{2/3} \log(nC))$	Generalized cost scaling (Goldberg & Tarjan 1987)
$O(nm \log n \log(nC))$	Minimum-mean cost cycle canceling
	(Goldberg & Tarjan 1988, 1989)
$O(m \log n(m + n \log n))$	(Orlin 1988, 1993)
$O(nm \log(n^2/m) \log(nC))$	Generalized cost scaling (Goldberg & Tarjan 1990)
$O(nm \log \log U \log(nC))$	Double scaling (Ahuja *et al.* 1992)
$O((nm + U^+ \log U^+) \log(nC))$	(Gabow & Tarjan 1989)
$O(n \log U(m + n \log n))$	(Gabow & Tarjan 1989) – *for networks without capacities*

Our formulation of the minimum path cover problem from Section 5.4.2 included the constraint that the solution paths must start/end in a source/sink, respectively. This is usually absent from standard formulations of the problem; however, our minimum-cost flow reduction can accommodate any sets of starting and ending vertices of the solution paths.

The first result on path covers in DAGs is Dilworth's theorem (Dilworth 1950), which equates the minimum number of paths in a path cover to the maximum cardinality of an anti-chain. A constructive proof of this theorem (Fulkerson 1956) showed that the minimum path cover problem can be reduced to a maximum matching problem in a bipartite graph having $t(G)$ edges, where $m \leq t(G) \leq \binom{n}{2}$ is the number of arcs of the transitive closure of the input DAG G. Using the maximum matching algorithm of Hopcroft & Karp (1973), a minimum path cover can be computed in time $O(\sqrt{n}t(G))$. A minimum-cost maximum matching in the same graph, computable for example in time $O(n^2 \log n + nt(G))$ with the algorithm in Edmonds & Karp (1972), gives a minimum-cost minimum path cover. The connection between the minimum path cover problem and a network flow problem (namely, a minimum flow problem, see Exercise 5.12) appears for example in Ntafos & Hakimi (1979), Pijls & Potharst (2013), and Bang-Jensen & Gutin (2008). Note that the minimum path cover problem in arbitrary directed graphs is NP-complete, since a path cover with only one path is a Hamiltonian path. Example 5.10 is adapted from Bang-Jensen & Gutin (2008).

Exercises

5.1   Given an arbitrary graph G with a unique source s and a unique sink t, and given a function $f : E(G) \to \mathbb{Q}$ satisfying the flow conservation property on G, show that

$$\sum_{y \in N^+(s)} f(s, y) = \sum_{y \in N^-(t)} f(y, t).$$

5.2   Let G be a DAG with a unique source s and a unique sink t, and let $f : E(G) \to \mathbb{N}$ satisfy the flow conservation property on G, such that the flow exiting s equals q. Show that if there exists an integer x such that, for every $f(e)$, $e \in E(G)$ is a multiple of x, then f can be decomposed into q/x paths, each of weight x.

5.3   Show that a flow f on an arbitrary graph G with a unique source s and a unique sink t can be decomposed into at most $|E(G)|$ weighted paths or cycles. How fast can this be done?

5.4   In the 3-partition problem, we are given a set $A = \{a_1, \dots, a_{3q}\}$ of $3q$ positive integers, such that

- $\sum_{i=1}^{3q} a_i = qB$, where B is an integer, and
- for all $i \in \{1, \dots, 3q\}$ it holds that $B/4 < a_i < B/2$.

We are asked whether there exists a partition of A into q disjoint sets, such that the sum of the integers in each of these sets is B. This problem is known to be NP-hard, even when the values of the $3q$ integers in the set A are bounded by a certain value not depending on q (it is called *strongly* NP-hard). Use a similar reduction to that in the proof of Theorem 5.3 to show that the 3-partition problem can be reduced to the problem of decomposing a flow in a DAG into the minimum number of weighted paths.

5.5   Given a weighted DAG G, show that we can find an s-t path whose bottleneck is maximum among all s-t paths of G, in time $O(|E(G)|)$.

5.6   Suppose that in a flow network N with a unique source and a unique sink some arcs have infinite capacity. Show that there exists a flow f over N of minimum cost, with the additional property that the value of f on each arc without capacity is at most the sum of all arc capacities.

5.7   Show that the minimum-cost circulation problem can be reduced to the minimum-cost flow problem.

5.8   Given a circulation f over a circulation network N, show that, if C is a cycle in the residual graph $\mathsf{R}(f)$ of f, then the circulation f_C defined in Section 5.2 is a circulation over N.

5.9   Given a circulation network $N = (G, \ell, u, c)$, consider the problem of finding a feasible circulation over N, that is, a circulation satisfying the demand and capacity constraints of N. Construct the circulation network N' as follows:

- N' has the same vertices as N;
- for every arc (x, y) of N, add to N' two parallel arcs with demand 0: one with cost -1 and capacity $\ell(x, y)$, and one with cost 0 and capacity $u(x, y) - \ell(x, y)$.

Show that N admits a feasible circulation if and only if the minimum-cost circulation over N' has cost less than or equal to minus the sum of all demands of N. How can you obtain a feasible circulation f over N from a minimum-cost circulation f' over N'?

5.10   Show that the reduction of a minimum-cost flow problem with convex costs to a standard minimum-cost flow problem from Insight 5.2 is correct.

5.11   Given a DAG G with a unique source s and a unique sink t, show that we can find the maximum number of s-t paths without common vertices (apart from s and t) by a reduction to a maximum flow problem.

5.12   In the *minimum flow problem*, we are given a flow network N with arc demands only, and are asked to find a flow of minimum value over N satisfying all demand constraints. Show that the minimum flow problem can be solved by two applications of a maximum flow problem.

5.13   Show that finding a *maximum-cardinality matching* in a bipartite graph can be reduced to a minimum-cost flow problem.

5.14   Show that, among all maximum-cardinality matchings of a bipartite graph G, the problem of finding one of minimum cost can be reduced to a minimum-cost network flow problem.

5.15   Suppose that you have an algorithm for solving the minimum-cost perfect matching problem in an arbitrary graph (not necessarily bipartite). Show that, given any graph G with costs associated with its edges, you can use this algorithm, by appropriately transforming G, to find a *minimum-cost maximum matching* (that is, a matching of maximum cardinality in G, with the additional property that, among all matchings of maximum cardinality, it has minimum cost).

5.16   Consider Problem 5.7, in which one is required to minimize instead

$$\sum_{(x,y)\in M} c(x,y) + \alpha \sum_{y\in B} |\mathrm{id}(y) - d_M(y)|^2 .$$

How do you need to modify the reduction given for Problem 5.7 to solve this new problem? Does the resulting flow network still have size polynomial in the size of the input? *Hint.* Use the idea from Insight 5.2.

5.17   Show that the problem of finding a minimum-cost disjoint cycle cover in an undirected graph can be reduced to the problem of finding a minimum-cost disjoint cycle cover in a directed graph.

5.18   An *edge cover* of an undirected graph G is a set of edges covering all the vertices of G.

- Show that a minimum-cost edge cover in a bipartite graph can be solved by a minimum-cost flow problem.
- Show that a minimum-cost edge cover in an arbitrary graph G can be reduced to the problem of finding a maximum-weight matching in G.

Part II

Fundamentals of Biological Sequence Analysis

6   Alignments

An *alignment* of two sequences *A* and *B* aims to highlight how much in common the two sequences have. This concept arises naturally in settings where a sequence *A* changes over time into *B*, through some elementary *edit operations*, such as insertion or deletion of a character (operations called *indels*), or substitution/mutation of one character with/into another. An alignment of the characters which have survived over time could be defined informally as a list of pairs of indices (i, j), such that $A[i]$ is considered to *match B[j]*.

In a computational biology context, the sequences *A* and *B* could be short extracts from the genomes of two living species, fragments considered to have descended from an unknown ancestor sequence *C*. A biologically meaningful alignment of *A* and *B* should take into account the path in the evolutionary tree from *A* to *C* and from *C* to *B*. However, since in practice *C* and the evolutionary tree are unknown, then one could parsimoniously prefer an alignment with the minimum number of edit operations.

Taking into account some properties of biological sequences, one can assign different costs to the different edit operations. For example, in a coding region, the substitution of a base is usually less critical than an indel, which alters the synchronization between codons and the amino acid chain under translation. The former is likely to survive in an evolutionary selection, while the latter is not. Therefore, one could assign a high positive cost for indels, some low positive cost for substitutions, and cost 0 for an *identity*, that is, for two matched characters that are also equal. The biologically meaningful alignment could then be approximated by finding the alignment with minimum total cost. This cost of an optimal alignment is also called the *edit distance*.

In biological sequence analysis it has been customary to resort instead to a maximization framework, where in place of cost one assigns a *score* to each operation. The score of an alignment thus becomes a measure of the similarity between two sequences. Continuing our coding region example, identities are then given the highest score, the substitution of a base that does not affect the 3-base codon coding can be given a score almost as high as the one for identity. Indels are typically given negative scores. The correct alignment could then be approximated by finding the alignment with maximum total score.

Distances and similarities can typically be transformed from one to another through reductions, but each has some unique advantages. For example, with alignment similarity it is easier to give a probabilistic interpretation of an alignment, whereas a distance computation can more easily be related to tailored optimizations.

We start with a more generic distance interpretation of alignments that works for any two sequences, and then shift our focus to biological sequences.

6.1   Edit distance

Let us now formally define an alignment of two sequences. First, we say that a sequence $C = c_1 c_2 \cdots c_r$ is a *subsequence* of $A = a_1 a_2 \cdots a_m$, if C can be obtained by deleting zero or more characters from A.

DEFINITION 6.1   *An alignment of $A = a_1 \cdots a_m \in \Sigma^*$ and $B = b_1 \cdots b_n \in \Sigma^*$ is a pair of sequences $U = u_1 u_2 \cdots u_h$ and $L = l_1 l_2 \cdots l_h$ of length $h \in [\max\{m, n\}..n + m]$ such that A is a subsequence of U, B is a subsequence of L, U contains $h - m$ characters $-$, and L contains $h - n$ characters $-$, where $-$ is a special character not appearing in A or B.*

The following example illustrates this definition.

Example 6.1   Let us consider two expressions from Finnish and Estonian, having the same meaning (*welcome*), the Finnish `tervetuloa` and the Estonian `tere tulemast`. We set $A = $ `tervetuloa` and $B = $ `teretulemast` (by removing the empty space). Visualize a plausible alignment of A and B.

Solution
Two possible sequences U (for *upper*) and L (for *lower*) from Definition 6.1 for A and B are shown below:

```
tervetulo-a--
ter-etulemast
```

The *cost* of the alignment (U, L) of A and B is defined as

$$C(U, L) = \sum_{i=1}^{h} c(u_i, l_i),$$

where $c(a, b)$ is a non-negative cost of aligning any two characters a to b (the function $c(\cdot, \cdot)$ is also called a *cost model*). In the *unit* cost model, we set $c(a, -) = c(-, b) = 1$, $c(a, b) = 1$ for $a \neq b$, and $c(a, b) = 0$ for $a = b$. An alignment can be interpreted as a sequence of edits to convert A into B as shown in Example 6.2. We denote these edits as $u_i \to l_i$.

Example 6.2   Interpret the alignment from Example 6.1 as a sequence of edit operations applied *only* to A so that it becomes equal to B.

Solution
Sequence `ter` stays as it is, `v` is deleted, `etul` stays as it is, `o` is substituted by `e`, `m` is inserted, `a` stays as it is, `s` is inserted, and `t` is inserted.

Problem 6.3 Edit distance

Given two sequences $A = a_1 \cdots a_m \in \Sigma^*$ and $B = b_1 \cdots b_n \in \Sigma^*$, and a cost model $c(\cdot, \cdot)$, find an alignment of A and B having minimum *edit distance*

$$D(A, B) = \min_{(U,L) \in \mathcal{A}(A,B)} C(U, L),$$

where $\mathcal{A}(A, B)$ denotes the set of all valid alignments of A and B.

Fixing $c(\cdot, \cdot)$ as the unit cost model, we obtain the *unit cost* edit distance. This distance is also called the *Levenshtein distance*, and is denoted $D_L(A, B)$.

Setting the cost model to $c(a, -) = c(-, b) = \infty$, $c(a, b) = 1$ for $a \neq b$, and $c(a, b) = 0$ for $a = b$, the resulting distance is called the *Hamming distance*, and is denoted $D_H(A, B)$. This distance has a finite value only for sequences of the same length, in which case it equals the number of positions i in the two sequences which *mismatch*, that is, for which $a_i \neq b_i$.

6.1.1 Edit distance computation

Let us denote $d_{i,j} = D(a_1 \cdots a_i, b_1 \cdots b_j)$, for all $0 \leq i \leq m$ and $0 \leq j \leq n$. The following theorem leads to an efficient algorithm for computing the edit distance between the sequences A and B.

THEOREM 6.2   *For all $0 \leq i \leq m$, $0 \leq j \leq n$, with $i + j > 0$, the edit distances $d_{i,j}$, and in particular $d_{m,n} = D(A, B)$, can be computed using the recurrence*

$$d_{i,j} = \min\{d_{i-1,j-1} + c(a_i, b_j), d_{i-1,j} + c(a_i, -), d_{i,j-1} + c(-, b_j)\}, \qquad (6.1)$$

by initializing $d_{0,0} = 0$, and interpreting $d_{i,j} = \infty$ if $i < 0$ or $j < 0$.

Proof   The initialization is correct, since the cost of an empty alignment is zero. Fix i and j, and assume by induction that $d_{i',j'}$ is correctly computed for all $i' + j' < i + j$. Consider all the possible ways that the alignments of $a_1 a_2 \cdots a_i$ and $b_1 b_2 \cdots b_j$ can end. There are three cases.

(i) $a_i \rightarrow b_j$ (the alignment ends with substitution or identity). Such an alignment has the same cost as the best alignment of $a_1 a_2 \cdots a_{i-1}$ and $b_1 b_2 \cdots b_{j-1}$ plus the cost of a substitution or identity between a_i and b_j, that is, $d_{i-1,j-1} + c(a_i, b_j)$.

(ii) $a_i \rightarrow -$ (the alignment ends with the deletion of a_i). Such an alignment has the same cost as the best alignment of $a_1 a_2 \cdots a_{i-1}$ and $b_1 b_2 \cdots b_j$ plus the cost of the deletion of a_i, that is, $d_{i-1,j} + c(a_i, -)$.

(iii) $- \rightarrow b_j$ (the alignment ends with the insertion of b_j). Such an alignment has the same cost as the best alignment of $a_1 a_2 \cdots a_i$ and $b_1 b_2 \cdots b_{j-1}$ plus the cost of the insertion of b_j, that is, $d_{i,j-1} + c(-, b_j)$.

Since $c(-, -) = 0$, the case where a possible alignment ends with $- \rightarrow -$ does not need to be taken into account; such an alignment has equal cost to one ending with one of the cases above. Thus, the cost of the optimal alignment equals the minimum of the costs in the three cases (i), (ii), and (iii), and hence Equation (6.1) holds.   □

As is the case with any such recurrence relation, an algorithm to compute $d_{m,n} = D(A, B)$ follows directly by noticing that one can reduce this problem to a shortest-path problem on a DAG (recall Section 4.1). This DAG is constructed by adding, for every pair (i, j), with $0 \leq i \leq m$ and $0 \leq j \leq n$,

- a vertex $v_{i,j}$,
- the arc $(v_{i-1,j-1}, v_{i,j})$ with cost $c(a_i, b_j)$ (if $i, j > 0$),
- the arc $(v_{i-1,j}, v_{i,j})$ with cost $c(a_i, -)$ (if $i > 0$), and
- the arc $(v_{i,j-1}, v_{i,j})$ with cost $c(-, b_j)$ (if $j > 0$).

The cost of the shortest (that is, of minimum cost) path from $v_{0,0}$ to $v_{m,n}$ is then equal to $d_{m,n} = D(A, B)$. Using the algorithm given in Section 4.1.2, this path can be found in $O(mn)$ time, since the resulting DAG has $O(mn)$ vertices and $O(mn)$ arcs.

However, since the resulting DAG is a *grid*, a simpler tailored algorithm follows easily. Namely, values $d_{i,j}$, for all $0 \leq i \leq m$, $0 \leq j \leq n$, can be computed by *dynamic programming* (sometimes also called *tabulation*, since the algorithm proceeds by filling the cells of a table), as shown in Algorithm 6.1

Algorithm 6.1: Edit distance computation

Input: Sequences $A = a_1 a_2 \cdots a_m$ and $B = b_1 b_2 \cdots b_n$.
Output: Edit distance $D(A, B)$.

1 $d_{0,0} = 0$;
2 **for** $i \leftarrow 1$ *to* m **do**
3 $\quad$ $d_{i,0} \leftarrow d_{i-1,0} + c(a_i, -)$;
4 **for** $j \leftarrow 1$ *to* n **do**
5 $\quad$ $d_{0,j} \leftarrow d_{0,j-1} + c(-, b_j)$;
6 **for** $j \leftarrow 1$ *to* n **do**
7 $\quad$ **for** $i \leftarrow 1$ *to* m **do**
8 $\quad\quad$ $d_{i,j} \leftarrow \min\{d_{i-1,j-1} + c(a_i, b_j), d_{i-1,j} + c(a_i, -), d_{i,j-1} + c(-, b_j)\}$;
9 **return** $d_{m,n}$;

The time requirement of Algorithm 6.1 is $\Theta(mn)$. The computation is shown in Example 6.4; the illustration used there is called the *dynamic programming matrix*. Referring to this matrix, we often talk about its *column j* defined as

$$D_{*j} = \{d_{i,j} \mid 0 \leq i \leq m\}, \tag{6.2}$$

its *row i*, defined as

$$D_{i*} = \{d_{i,j} \mid 0 \leq j \leq n\}, \tag{6.3}$$

and its *diagonal k*, defined as

$$D^k = \{d_{i,j} \mid 0 \le i \le m, 0 \le j \le n, j - i = k\}. \tag{6.4}$$

Notice that, even though we defined D_{*j}, D_{i*}, and D^k as sets of numbers, we interpret them as sets of *cells* of the dynamic programming matrix.

The *evaluation order* for computing values $d_{i,j}$ can be any order such that the values on which each $d_{i,j}$ depends are computed before. Algorithm 6.1 uses the *column-by-column order*. By exchanging the two internal for-loops, one obtains the *row-by-row order*.

From Example 6.4 one observes that values $d_{i,j}$ at column j depend only on the values at the same column and at column $j - 1$. Hence, the space needed for the computation can be easily improved to $O(m)$ (see Exercise 6.2).

As is typical with dynamic programming, the steps (here, edit operations) chosen for the optimal solution (here, for $d_{m,n}$) can be traced back after the dynamic programming matrix has been filled in. From $d_{m,n}$ we can trace back towards $d_{0,0}$ and re-evaluate the decisions taken for computing each $d_{i,j}$. There is no need to store explicit back-pointers towards the cells minimizing the edit distance recurrence (6.1). The traceback is also visualized in Example 6.4. Notice that the whole matrix is required for the traceback, but there is an algorithm to do the traceback in optimal $O(m + n)$ space (see Exercise 6.16).

Example 6.4 Compute the unit cost edit distance for $A =$ tervetuloa and $B =$ teretulemast, and visualize the optimal alignments.

Solution
We can simulate Algorithm 6.1 as shown below. Notice that the convention is to represent the first sequence $A = a_1 a_2 \cdots a_m$ vertically (and we iterate the index i from top to bottom) and the second sequence $B = b_1 b_2 \cdots b_n$ horizontally (and we iterate the index j from left to right).

		t	e	r	e	t	u	l	e	m	a	s	t
	0	1	2	3	4	5	6	7	8	9	10	11	12
t	1	0	1	2	3	4	5	6	7	8	9	10	11
e	2	1	0	1	2	3	4	5	6	7	8	9	10
r	3	2	1	0	1	2	3	4	5	6	7	8	9
v	4	3	2	1	1	2	3	4	5	6	7	8	9
e	5	4	3	2	1	2	3	4	4	5	6	7	8
t	6	5	4	3	2	1	2	3	4	5	6	7	7
u	7	6	5	4	3	2	1	2	3	4	5	6	7
l	8	7	6	5	4	3	2	1	2	3	4	5	6
o	9	8	7	6	5	4	3	2	2	3	4	5	6
a	10	9	8	7	6	5	4	3	3	3	3	4	5

The final solution is in the bottom-right corner, $D_L(A, B) = 5$. The grayed area highlights the cells visited while tracing the optimal paths (alignments). There are two optimal alignments, shown below (the bottom-most path is on the left, and the top-most path is on the right):

```
tervetulo-a- -    tervetul-oa- -
ter-etulemast     ter-etulemast
```

6.1.2   Shortest detour

In the previous section we showed how the edit distance between the two sequences $A = a_1 a_2 \cdots a_m$ and $B = b_1 b_2 \cdots b_n$ (assume here that $m \leq n$) can be computed efficiently. We discovered a recurrence relation for $d_{m,n} = D(A, B)$, and gave an $\Theta(mn)$-time implementation. However, in practice we often expect to compute the edit distance between two strings that are similar, more precisely, for which $D(A, B)$ is fairly small compared with both n and m. In this section we give an algorithm exploiting this assumption, and running in time $O(dm)$, where $d = D(A, B)$.

Let an *optimal path* be a sequence of cells of the dynamic programming matrix (made up of some entries $d_{i,j}$). Consider a diagonal D^k of the dynamic programming matrix (recall its definition from (6.4)), and denote by $\Delta > 0$ the smallest cost of an insertion or deletion (which depends on the cost model chosen). The crucial observation here is that each time the optimal path changes diagonal, it incurs a cost of at least Δ. Hence, on an optimal path there can be at most $d_{m,n}/\Delta$ changes of diagonal when computing $d_{m,n}$. It is sufficient to limit the computation inside a *diagonal zone*, that includes diagonals $0, \ldots, n - m$ (since at least $n - m$ insertions into A are needed, or equivalently, $n - m$ deletions from B) and possibly some other diagonals. Since the value $d_{m,n}$ is not known beforehand, we actually cannot explicitly define the required diagonal zone beforehand.

Let us first consider the case where a threshold t is given for the distance and we have to answer the question of whether $d_{m,n} \leq t$, and, if so, what the actual value of $d_{m,n}$ is. First, if $t < (n - m)\Delta$, then the answer is "no". Otherwise, the solution for this problem limits the computation inside the diagonal zone $Z = [-x..n - m + x]$, where $x = \lceil t/(2\Delta) - (n - m)/2 \rceil$. Indeed, if the optimal path needs to go outside of this zone, then its cost is at least $((n - m) + 2(x + 1))\Delta > t$, and we can answer the question negatively. The running time of this algorithm is $O(((n - m) + 2x) - m) = O((t/\Delta)m)$, which translates to $O(tm)$ for the unit cost edit distance in which $\Delta = 1$. Thus, we obtained an algorithm whose time complexity depends on an assumption of the edit distance between A and B. See Example 6.5.

Exercise 6.3 asks the reader how to allocate space only for those cells that are to be computed, since otherwise the space allocation would dominate the $O(tm)$ running time of this algorithm.

Example 6.5   Given $A = $ `tervetuloa` and $B = $ `teretulemast`, compute the diagonal zone of the edit distance matrix to decide whether $D_L(A, B) \leq 5$.

Solution

The diagonal zone becomes $Z = [-2..4]$. The computation inside Z is shown below.

		t	e	r	e	t	u	l	e	m	a	s	t
	0	1	2	3	4								
t	1	0	1	2	3	4							
e	2	1	0	1	2	3	4						
r		2	1	0	1	2	3	4					
v			2	1	1	2	3	4	5				
e				2	1	2	3	4	4	5			
t					2	1	2	3	4	5	6		
u						2	1	2	3	4	5	6	
l							2	1	←2	3	4	5	6
o								2	2	←3	4	5	6
a									3	3	3	←4	←5

The final solution is in the bottom-right corner, $D_L(A, B) = 5$, so the answer is that the edit distance is within the threshold.

Consider now the case when we are not given an upper bound on the edit distance. Then we can run the above algorithm for increasing values of t (starting for example with $t = 1$), such that each value is double the previous one. If this algorithm gives the answer "yes", then we also have the correct edit distance, and we stop. Otherwise, we keep doubling t until we have reached the maximum possible value, which is n. This algorithm is called the *doubling technique*.

The fact that the overall complexity is the desired $O((d/\Delta)m)$, where $d = d_{m,n} = D(A, B)$, can be shown as follows. Let $t = 2^T$ be the *last* iteration of the algorithm, thus when we have that $2^{T-1} < d \leq 2^T$. The overall running time of this doubling technique algorithm is then

$$O\left(\frac{m}{\Delta}(2^0 + 2^1 + \cdots + 2^T)\right).$$

Exploiting the geometric sum formula $\sum_{i=0}^{T} 2^i = 2^{T+1} - 1$, and the fact that $2^{T+1} < 4d$ (since $2^{T-1} < d$), this running time is

$$O\left(\frac{m}{\Delta}(2^0 + 2^1 + \cdots + 2^T)\right) = O\left(\frac{m}{\Delta}4d\right) = O\left(\frac{d}{\Delta}m\right).$$

This implies that, for any cost model in which the smallest cost of an insertion or deletion is greater than a fixed *constant*, the corresponding edit distance d can be computed in time $O(dm)$. In particular, we have the following corollary.

COROLLARY 6.3   *Given sequences $A = a_1 \cdots a_m$ and $B = b_1 \cdots b_n$ with $m \le n$, if the unit cost edit (Levenshtein) distance $D_L(A,B)$ equals d, then $D_L(A,B)$ can be computed in time $O(dm)$.*

*6.1.3   Myers' bitparallel algorithm

The Levenshtein edit distance $D_L(A,B)$ has some properties that make it possible to derive an $O(\lceil m/w \rceil n)$-time algorithm, where w is the computer word size. Namely, the edit distance matrix $(d_{i,j})$ computed for $D_L(A,B)$ has the following properties:

The diagonal property: $d_{i,j} - d_{i-1,j-1} \in \{0,1\}$.
The adjacency property: $d_{i,j} - d_{i,j-1} \in \{-1,0,1\}$ and $d_{i,j} - d_{i-1,j} \in \{-1,0,1\}$.

For now, we assume $|A| < w$; the general case is considered later. Then one can encode the columns of the matrix with a couple of bitvectors. We will need R_j^+ and R_j^- coding the row-wise differences, C_j^+ and C_j^- the column-wise differences, and D_j the diagonal-wise differences, at column j:

$$
\begin{aligned}
R_j^+[i] &= 1 & \text{iff} && d_{i,j} - d_{i-1,j} &= 1, \\
R_j^-[i] &= 1 & \text{iff} && d_{i,j} - d_{i-1,j} &= -1, \\
C_j^+[i] &= 1 & \text{iff} && d_{i,j} - d_{i,j-1} &= 1, \\
C_j^-[i] &= 1 & \text{iff} && d_{i,j} - d_{i,j-1} &= -1, \\
D_j[i] &= 1 & \text{iff} && d_{i,j} - d_{i-1,j-1} &= 1.
\end{aligned}
$$

In the following, we assume that these vectors have already been computed for column $j-1$ and we want to derive the vectors of column j in *constant time*; this means that we need to find a way to use bitparallel operations (recall Exercise 2.6 from page 18). In addition to the vectors of the previous column, we need indicator bitvectors I^c having $I^c[i] = 1$ iff $A[i] = c$.

The update rules are shown in Algorithm 6.2. These should look magical until we go through the proof on why the values are computed correctly. To gather some intuition and familiarity with bitparallel notions, Example 6.6 simulates the algorithm.

Example 6.6 Simulate Algorithm 6.2 for computing the fourth column of the dynamic programming matrix in Example 6.4.

Solution
In the following visualization, the logical operators in column titles take one or two previous columns as inputs depending on the operator type and on the presence or absence of the bitvector name specified on the right-hand side. First, the part of the original matrix involved is shown and then the computation is depicted step by step. The top-most column titles mark the output bitvectors.

	e	r	e	R^+_{j-1}	R^-_{j-1}	I^e	$\&R^+_{j-1}$	$+R^+_{j-1}$	$\oplus R^+_{j-1}$	$\|I^e$	$\oplus R^-_{j-1}$	$\sim$	$\|1$
	2	3	4	0	0	0	0	0	0	0	0	1	1
t	1	2	3	0	1	0	0	0	0	0	1	0	0
e	0	1	2	0	1	1	0	0	0	1	1	0	0
r	1	0	1	0	1	0	0	0	0	0	1	0	0
v	2	1	1	1	0	0	0	1	0	0	0	1	1
e	3	2	1	1	0	1	1	0	1	1	1	0	0
t	4	3	2	1	0	0	0	0	1	1	1	0	0
u	5	4	3	1	0	0	0	0	1	1	1	0	0
l	6	5	4	1	0	0	0	0	1	1	1	0	0
o	7	6	5	1	0	0	0	0	1	1	1	0	0
a	8	7	6	1	0	0	0	0	1	1	1	0	0

Column groups: $j=4$; Inputs; Previous column input to next column; D_j.

Previous column input to next column		C^+_j			C^-_j
$\sim D_j$	$\|R^+_{j-1}$	$\sim$	$\|R^-_{j-1}$	$\sim D_j$	$\&R^+_{j-1}$
0	0	1	1	0	0
1	1	0	1	1	0
1	1	0	1	1	0
1	1	0	1	1	0
0	1	0	0	0	0
1	1	0	0	1	1
1	1	0	0	1	1
1	1	0	0	1	1
1	1	0	0	1	1
1	1	0	0	1	1
1	1	0	0	1	1

Previous column(s) input to next column					R^+_j			R^-_j
$\sim D_j$	$(C^+_j << 1)$	$\|$	$\sim$	$C^-_j << 1$	$\|$	$\sim D_j$	$(C^+_j << 1)$	$\&$
0	0	0	1	0	1	0	0	0
1	1	1	0	0	0	1	1	1
1	1	1	0	0	0	1	1	1
1	1	1	0	0	0	1	1	1
0	1	1	0	0	0	0	1	0
1	0	1	0	0	0	1	0	0
1	0	1	0	1	1	1	0	0
1	0	1	0	1	1	1	0	0
1	0	1	0	1	1	1	0	0
1	0	1	0	1	1	1	0	0
1	0	1	0	1	1	1	0	0

Algorithm 6.2: Myers' bitparallel algorithm for computing one column of edit distance computation on strings A and B, $|A| < w$

Input: Bitvectors $R^+_{j-1}, R^-_{j-1}, C^+_{j-1}, C^-_{j-1}, D_{j-1}, I^c$ for $c = B[j]$, $d = D(A, B_{1..j-1})$, and $m = |A|$.

Output: Edit distance $D(A, B_{1..j})$ and bitvectors encoding column j.

1 $D_j \leftarrow \sim(((I^c \,\&\, R^+_{j-1} + R^+_{j-1}) \oplus R^+_{j-1}) \mid I^c \mid R^-_{j-1}) \mid 1$;

2 $C^+_j \leftarrow R^-_{j-1} \mid \sim(\sim D_j \mid R^+_{j-1})$;

3 $C^-_j \leftarrow \sim D_j \,\&\, R^+_{j-1}$;

4 $R^+_j \leftarrow (C^-_j << 1) \mid \sim(\sim D_j \mid (C^+_j << 1))$;

5 $R^-_j \leftarrow \sim D_j \,\&\, (C^+_j << 1)$;

6 **if** $C^+[m] = 1$ **then**

7 $\quad\lfloor\ d \leftarrow d + 1$;

8 **if** $C^-[m] = 1$ **then**

9 $\quad\lfloor\ d \leftarrow d - 1$;

10 **return** d and the bitvectors $R^+_j, R^-_j, C^+_j, C^-_j, D_j$;

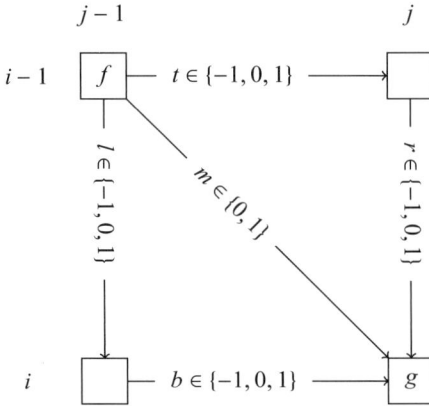

Figure 6.1 Difference DAG on four neighboring cells in a Levenshtein edit distance dynamic programming matrix.

In deriving Algorithm 6.2, we start with the four last update rules. For all of these, it suffices to consider how bits depend among cells $(i-1, j-1)$, $(i, j-1)$, $(i-1, j)$, and (i, j) and find bit-wise operation sets that implement the logical formulas simultaneously for all i. Instead of trying to mimic how these formulas were discovered in the first place, we follow a computerized approach. One could go over all total orders of the four variables for which we are trying to solve, through all logical formulas involving already computed variables, computing a truth table for each formula and the variable in question. Each combination of the input variables fixes some constraints on the edge weights in the *difference DAG* visualized in Figure 6.1.

As in Figure 6.1, we denote $f = d_{i-1,j-1}$, $l = d_{i,j-1} - d_{i-1,j-1}$, $t = d_{i-1,j} - d_{i-1,j-1}$, $m = d_{i,j} - d_{i-1,j-1}$, $b = d_{i,j} - d_{i,j-1}$, $r = d_{i,j} - d_{i-1,j}$, and $g = d_{i,j}$. We obtain equations $f + l + b = g$, $f + m = g$, and $f + t + r = g$ with the diagonal property and adjacency property as the initial constraints; each input value combination gives additional constraints, so that one can hope to solve for the variable in question. If the constraints define the output variable value on each input value combination and the logical formula in question gives the same answer, we have found a solution. Such a solution can be discovered in exponential time in the number of variables. With a constant number of variables, this is still feasible. Assuming the four update rules have been found in this way, it is sufficient to check the truth tables.

LEMMA 6.4　　$C_j^+ = R_{j-1}^- \mid {\sim}({\sim}D_j \mid R_{j-1}^+) \mid 1$

Proof　The satisfying truth table is shown below. Impossible input combinations, where $R_{j-1}^-[i]$ and $R_{j-1}^+[i]$ are 1 at the same time, are left out. The last column is derived by analyzing the constraints given by fixed input parameters. For example, the first row sets constraints $l \geq 0$, $f = g$, and $l \leq 0$ (see Figure 6.1). That is, one can solve b from $f + l + b = g$, giving $b = 0$ and thus $C_j^+[i] = 0$.

$x = R_{j-1}^-[i]$	$y = D_j[i]$	$z = R_{j-1}^+[i]$	$x \mid \sim(\sim y \mid z)$	$C_j^+[i]$
0	0	0	0	0
0	0	1	0	0
0	1	0	1	1
1	0	0	1	1
0	1	1	0	0
1	1	0	1	1

For $C_j^+[0]$ the formula is not defined, but the initialization $d_{0,j} = d_{0,j-1} + 1$ is enforced by $\mid 1$. $\qquad\square$

LEMMA 6.5 $C_j^- = \sim D_j \, \& \, R_{j-1}^+$

Proof   The satisfying truth table is shown below.

$x = D_j[i]$	$y = R_{j-1}^+[i]$	$\sim x \, \& \, y$	$C_j^-[i]$
0	0	0	0
0	1	1	1
1	0	0	0
1	1	0	0

For $C_j^-[0]$ the formula is not defined, but the initialization $d_{0,j} = d_{0,j-1}+1$ is guaranteed if, for example, $D_j[0]$ is initialized to 1. $\qquad\square$

LEMMA 6.6 $R_j^+ = (C_j^- << 1) \mid \sim(\sim D_j \mid (C_j^+ << 1))$

Proof   The satisfying truth table is shown below. Impossible input combinations, where $C_j^-[i-1]$ and $C_j^+[i-1]$ are 1 at the same time, are left out.

$x = C_j^-[i-1]$	$y = D_j[i]$	$z = C_j^+[i-1]$	$x \mid \sim(\sim y \mid z)$	$R_j^+[i]$
0	0	0	0	0
0	0	1	0	0
0	1	0	1	1
1	0	0	1	1
0	1	1	0	0
1	1	0	1	1

For $R_j^+[0]$ the formula is not defined, so we can leave the bit-wise formula to set that bit arbitrarily. The formula uses values shifted by one, and this is correctly executed by the bit-wise left-shifts. $\qquad\square$

LEMMA 6.7 $R_j^- = \sim D_j \, \& \, (C_j^+ << 1)$

Proof   The satisfying truth table is shown below.

$x = D_j[i]$	$y = C_j^+[i-1]$	$\sim\!x \,\&\, y$	$R_j^-[i]$
0	0	0	0
0	1	1	1
1	0	0	0
1	1	0	0

For $R_j^-[0]$ the formula is not defined, so we can leave the bit-wise formula to set that bit arbitrarily. The formula uses values shifted by one, and this is correctly executed by the bit-wise left-shift.   □

For the first update rule, we could follow the same procedure as above using local dependencies between cells, but that would give us a circular dependency. Instead, we need to use a more global property of the values to break this cycle.

LEMMA 6.8 $D_j = \sim\!(((I^c \,\&\, R_{j-1}^+ + R_{j-1}^+) \oplus R_{j-1}^+) \mid I^c \mid R_{j-1}^-) \mid 1$

Proof Recall Figure 6.1. For $I^c[i] = 1$ we obtain the equation $f + m = f$, so $m = D_j[i] = 0$. For $R_{j-1}^- = 1$ we obtain the equation $f - 1 + l = g$. Combined with $l \leq 1$, $f + m = g$, and $m \geq 0$, the only solution is $m = D_j[i] = 0$. For i such that $I^c[i] = 0$ and $R_{j-1}^-[i] = 0$ we claim that

$$d_{i,j} = d_{i-1,j-1} \text{ iff } \exists h \,:\, I^c[h] = 1 \text{ and } R_{j-1}^+[q] = 1, \text{ for } q = [h..i-1]. \qquad (6.5)$$

Assuming the right side holds, one can derive from $d_{h,j} = d_{h-1,j-1}$ and $d_{h,j-1} = d_{h-1,j-1} + 1$ that $d_{h,j} - d_{h,j-1} = -1$. On combining the latter with $d_{h+1,j} - d_{h,j} \leq 1$, one obtains $d_{h+1,j} = d_{h,j-1}$. Continuing in this way, one obtains $d_{i,j} = d_{i-1,j-1}$.

For the other direction, the assumptions $I^c[i] = 0$, $R_{j-1}^-[i] = 0$, and $d_{i,j} = d_{i-1,j-1}$ indicate that the optimal path to $d_{i,j}$ must come from $d_{i-1,j}$, and it follows that $d_{i-1,j} - d_{i-1,j-1} = -1$. Combining this with the diagonal and adjacency properties, we get that $d_{i-1,j-1} - d_{i-2,j-1} = 1$. If $I^c[i-1] = 0$, the optimal path to $d_{i-1,j}$ must come from $d_{i-2,j}$, and it follows that $d_{i-2,j} - d_{i-2,j-1} = -1$. Continuing in this way, there has to be h such that the optimal path to $d_{h,j}$ comes from $d_{h-1,j-1}$, that is, $I^c[h] = 1$: otherwise $d_{0,j} - d_{0,j-1} = -1$, which contradicts on how the matrix has been initialized.

Finally, the operation $I' = I^c \,\&\, R_{j-1}^+$ sets $I'[i] = 1$ iff $I^c[i] = 1$ and there is an overlapping run of 1-bits in R_{j-1}^+. The operation $P = I' + R_{j-1}^+$ propagates each $I'[i] = 1$ over the overlapping run of 1-bits in R_{j-1}^+, setting $P[i'] = 0$ for $i \leq i' \leq r$, and $P[r+1] = 1$, where r is the end of the run. Other parts of R_{j-1}^+ are copied to P as such. If there are multiple 1-bits in I^c overlapping the same run, all but the first location of those will be set to 1. The operation $P' = P \oplus C_{j-1}^+$ sets these back to 0, along with runs not overlapped with any $I^c[i] = 1$. In the end, exactly those indices i corresponding to Equation (6.5) have $P'[i] = 1$. With the complement operation, the claimed final formula follows, except for the $\mid 1$; this sets $D_j[0] = 1$, which is required in order to have $C_j^-[0]$ computed correctly (see the proof of Lemma 6.5).   □

It is straightforward to extend the algorithm for longer sequences by representing a column with arrays of bitvectors; one needs to additionally take care that the bits at the border are correctly updated.

THEOREM 6.9   *Given sequences $A = a_1 \cdots a_m$ and $B = b_1 \cdots b_n$, $m \leq n$, the edit distance $d = D_L(A, B)$ can be computed in time $O(\lceil m/w \rceil n)$ in the RAM model, where w is the computer word size.*

Adding traceback to Myers' algorithm for an optimal alignment is considered in Exercises 6.8 and 6.9.

6.2   Longest common subsequence

The edit distance in which the substitution operation is forbidden has a tight connection to the problem of finding a longest common subsequence of two sequences. Recall that a sequence $C = c_1 c_2 \cdots c_r$ is a *subsequence* of $A = a_1 a_2 \cdots a_m$, if C can be obtained by deleting zero or more characters from A.

Problem 6.7   Longest common subsequence (LCS)

Given two sequences A and B, find a sequence C that is a subsequence of both A and B, and is the longest one with this property. Any such sequence is called a *longest common subsequence* of A and B, and is denoted $\mathsf{LCS}(A, B)$.

Let $D_{\mathtt{id}}(A, B)$ be the (*indel*) edit distance with $c(a_i, b_j) = \infty$ if $a_i \neq b_j$, and $c(a_i, b_j) = 0$ if $a_i = b_j$, $c(a_i, -) = c(-, b_j) = 1$. Observe that, similarly to the general edit distance, also $D_{\mathtt{id}}(A, B)$ captures the minimum set of elementary operations needed to convert a sequence into any other sequence: any substitution can be simulated by a deletion followed by an insertion.

The connection between $D_{\mathtt{id}}(A, B)$ and $\mathsf{LCS}(A, B)$ is formalized in the theorem below, and is illustrated in Example 6.8.

THEOREM 6.10 $|LCS(A, B)| = (|A| + |B| - D_{\mathtt{id}}(A, B))/2.$

We leave the proof of this connection to the reader (Exercise 6.11).

Example 6.8   Simulate the computation of $D_{\mathtt{id}}(A, B)$ for $A = \mathtt{tervetuloa}$ and $B = \mathtt{teretulemast}$. Extract a longest common subsequence from an alignment of A and B.

Solution
The final solution is in the bottom-right corner, $D_{\mathtt{id}}(A, B) = 6$. Observe that $|\mathsf{LCS}(A, B)| = (|A| + |B| - D_{\mathtt{id}}(A, B))/2 = (10 + 12 - 6)/2 = 8$, as indicated by Theorem 6.10. The grayed area consists of the cells visited while tracing the optimal paths (alignments). One of the alignments is

		t	e	r	e	t	u	l	e	m	a	s	t
	0	1	2	3	4	5	6	7	8	9	10	11	12
t	1	0	1	2	3	4	5	6	7	8	9	10	11
e	2	1	0	1	2	3	4	5	6	7	8	9	10
r	3	2	1	0	1	2	3	4	5	6	7	8	9
v	4	3	2	1	2	3	4	5	6	7	8	9	10
e	5	4	3	2	1	2	3	4	5	6	7	8	9
t	6	5	4	3	2	1	2	3	4	5	6	7	8
u	7	6	5	4	3	2	1	2	3	4	5	6	7
l	8	7	6	5	4	3	2	1	2	3	4	5	6
o	9	8	7	6	5	4	3	2	3	4	5	6	7
a	10	9	8	7	6	5	4	3	4	5	4	5	6

```
tervetulo--a--
ter-etul-emast
```

Taking all the identities from the alignments, we have that

$$\mathsf{LCS}(\texttt{tervetuloa}, \texttt{teretulemast}) = \texttt{teretula}.$$

In fact, this is *the* longest common subsequence, as all optimal alignments give the same result.

In the next section we develop a sparse dynamic programming and invariant technique for computing the LCS. This method will also be deployed in Section 15.4 for aligning RNA transcripts to the genome.

6.2.1   Sparse dynamic programming

The goal in *sparse dynamic programming* is to compute only the cells of the dynamic programming matrix needed for obtaining the final value. Let $A = a_1 a_2 \cdots a_m$ and $B = b_1 b_2 \cdots b_n$ be two sequences, and denote by $M(A, B)$ the set of matching character pairs, that is, $M(A, B) = \{(i, j) \mid a_i = b_j\}$. We will here abbreviate $M(A, B)$ by M.

We will next show how to compute the edit distance $D_{\mathtt{id}}$ in time proportional to the size of M.

LEMMA 6.11   *Let $A = a_1 a_2 \cdots a_m$ and $B = b_1 b_2 \cdots b_n$ be two sequences, and let $M = M(A, B) = \{(i, j) \mid a_i = b_j\} \cup \{(0, 0)\}$. For all values $d_{i,j} = D_{\mathtt{id}}(A_{1..i}, B_{1..j})$, with $(i, j) \in M$, it holds that*

$$d_{i,j} = \min\{d_{i',j'} + i - i' + j - j' - 2 \mid i' < i, j' < j, (i', j') \in M\}, \qquad (6.6)$$

with the initialization $d_{0,0} = 0$. Moreover, we have that $D_{\mathtt{id}}(A, B) = d_{m+1,n+1}$.

Proof Consider two consecutive characters of A in $\mathsf{LCS}(A, B)$, say $a_{i'}$ and a_i. They have counterparts $b_{j'}$ and b_j in B. The corresponding optimal alignment $D_{\mathtt{id}}(A, B)$ contains only deletions and insertions between the two identity operations $a_{i'} \to b_{j'}$ and $a_i \to b_j$. The smallest number of deletions and insertions to convert $A_{i'+1..i-1}$ into $B_{j'+1..j-1}$ is $\big(i - 1 - (i'+1) + 1\big) + \big(j - 1 - (j'+1) + 1\big) = i - i' + j - j' - 2$. The recurrence considers, for all pairs (i, j) with $a_i = b_j$, all possible preceding pairs (i', j') with $a_{i'} = b_{j'}$. Among these, the pair that minimizes the overall cost of converting first $A_{1..i'}$ into $B_{1..j'}$ (cost $d_{i',j'}$) and then $A_{i'..i}$ into $B_{j'..j}$ (cost $i - i' + j - j' - 2$) is chosen. The initialization is correct: if $a_i \to b_j$ is the first identity operation, the cost of converting $A_{1..i-1}$ into $B_{1..j-1}$ is $i - 1 + j - 1 = d_{0,0} + i - 0 + j - 0 - 2 = d_{i,j}$.

If we assume by induction that all values $d_{i',j'}$, $i' < i, j' < j$, are computed correctly, then it follows that each $d_{i,j}$ receives the correct value. Finally, $d_{m+1,n+1} = D_{\mathtt{id}}(A, B)$, because, if (i', j') is the last identity in an optimal alignment, converting $A_{i'+1..m}$ into $B_{j'+1..n}$ costs $m - i' + n - j'$, which equals $d_{m+1,n+1} - d_{i',j'}$. □

A naive implementation of the recurrence (6.6) computes each $d_{i,j}$ in time $O(|M|)$, by just scanning M and selecting the optimal pair (i', j'). Let us now see how we can improve this time to $O(\log m)$. Observe that the terms i, j, and -2 in the minimization (6.6) do not depend on the pairs $(i', j') \in M$ with $i' < i, j' < j$. Therefore, we can rewrite (6.6) as

$$d_{i,j} = i + j - 2 + \min\{d_{i',j'} - i' - j' \mid i' < i, j' < j, (i', j') \in M\},$$

by bringing all such *invariant* values out from the minimization. Compute now all values $d_{i,j}$ in *reverse column-order*, namely in the order $<^{\mathtt{rc}}$ such that

$$(i', j') <^{\mathtt{rc}} (i, j) \text{ if and only if } j' < j \text{ or } (j' = j \text{ and } i' > i).$$

The crucial property of this order is that, if all the values $d_{i',j'}$, with $(i', j') <^{\mathtt{rc}} (i, j)$, *and only them*, are computed before computing $d_{i,j}$, then the condition $j' < j$ is automatically satisfied. Thus, among these values computed so far in reverse column-order, we need select only those pairs with $i' < i$. We can write the recurrence as

$$d_{i,j} = i + j - 2 + \min\{d_{i',j'} - i' - j' \mid i' < i, (i', j') <^{\mathtt{rc}} (i, j), (i', j') \in M\}. \quad (6.7)$$

We can then consider the cells $(i, j) \in M$ in the reverse column-order, and, after having computed $d_{i,j}$ we store the value $d_{i,j} - i - j$ in a range minimum query data structure $\mathcal{T}$ of Lemma 3.1 on page 20, with key i. Computing $d_{i,j}$ itself can then be done by retrieving the minimum value d from $\mathcal{T}$, corresponding to a pair (i', j') with $i' < i$. Then, $d_{i,j}$ is obtained as $d_{i,j} = i + j - 2 + d$, since we need to add to d the invariant terms $i + j - 2$. Figure 6.2 illustrates this algorithm, whose pseudocode is shown in Algorithm 6.3. The derivation used here is called the *invariant technique*.

THEOREM 6.12   *Given sequences $A = a_1 a_2 \cdots a_m$ and $B = b_1 b_2 \cdots b_n$ from an ordered alphabet, the edit distance $D_{\mathtt{id}}$ and $|\mathsf{LCS}(A, B)|$ can be computed using Algorithm 6.3 in time $O((m + n + |M|) \log m)$ and space $O(m + |M|)$, where $M = \{(i, j) \mid a_i = b_j\}$.*

Proof The reverse column-order guarantees that the call $\mathcal{T}.\mathsf{RMQ}(0, i-1)$ corresponds to taking the minimum in Equation (6.7). The only difference is that $\mathcal{T}$ contains only

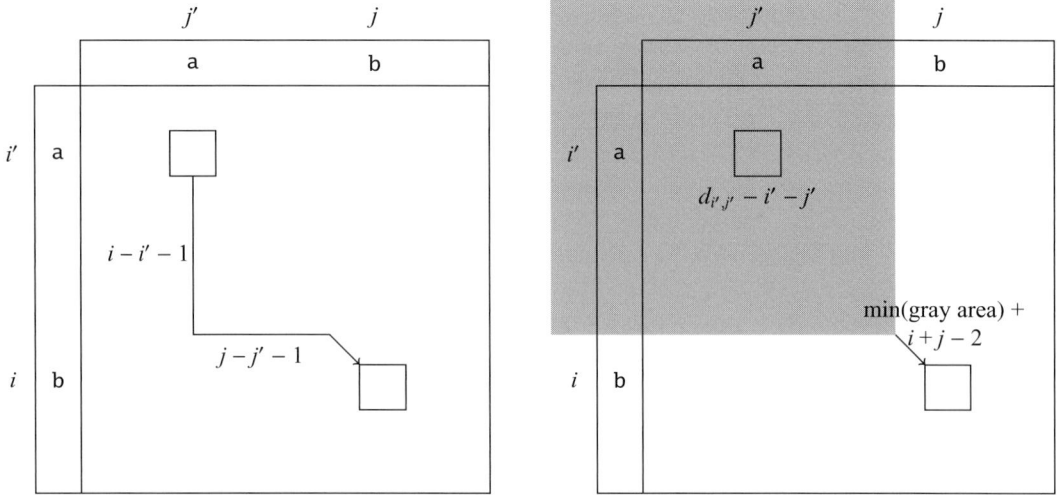

Figure 6.2 Geometric interpretation of Equation (6.7). On the left, we show a path from (i', j') to (i, j) such that in $\mathsf{LCS}(A, B)$ we have the consecutive identities $a_{i'} \to b_{j'}$ and $a_i \to b_j$ (with $a_{i'} = \mathsf{a}$ and $a_i = \mathsf{b}$). On the right, the invariant value is stored at (i', j') and the grayed area depicts from where the minimum value is taken for computing the cell (i, j). After taking the minimum, the actual minimum cost is computed by adding the invariant.

Algorithm 6.3: Sparse dynamic programming for edit distance $D_{\mathtt{id}}(A, B)$

Input: Sequences $A = a_1 a_2 \cdots a_m$ and $B = b_1 b_2 \cdots b_n$.

Output: Edit distance $D_{\mathtt{id}}(A, B)$.

1 Construct set $M = \{(i, j) \mid a_i = b_j\}$ and sort it into an array $M[1..|M|]$ in reverse column-order;

2 Initialize a search tree $\mathcal{T}$ with keys $0, 1, \ldots, m$;

3 $\mathcal{T}.\mathtt{update}(0, 0)$;

4 **for** $p \leftarrow 1$ **to** $|M|$ **do**

5     $(i, j) \leftarrow M[p]$;

6     $d \leftarrow i + j - 2 + \mathcal{T}.\mathsf{RMQ}(0, i - 1)$;

7     $\mathcal{T}.\mathtt{update}(i, d - i - j)$;

8 **return** $\mathcal{T}.\mathsf{RMQ}(0, m) + m + n$;

---

the smallest value at each row i', which does not affect the correctness. The algorithm calls $O(|M|)$ times the operations in $\mathcal{T}$. Each of these takes $O(\log m)$ time. Finally, set M can be easily constructed in $O(\sigma + m + n + |M|)$ time on an alphabet $[1..\sigma]$, or in $O((m + n) \log m + |M|)$ time on an ordered alphabet: see Exercise 6.12. $\qquad\square$

6.3   Approximate string matching

So far we have developed methods to compare two entire sequences. A more common scenario in biological sequence analysis is that one of the sequences represents the whole genome (as a concatenation of chromosome sequences) and the other sequence is a short extract from another genome. A simple example is a *homology search* in prokaryotes, where the short extract could be a contiguous gene from one species, and one wants to check whether the genome of another species contains a similar gene. This problem is captured as follows.

Let $S = s_1 s_2 \cdots s_n \in \Sigma^*$ be a *text* string and $P = p_1 p_2 \cdots p_m \in \Sigma^*$ a *pattern* string. Let k be a constant, $0 \le k \le m$. Consider the following search problems.

Problem 6.9   Approximate string matching under k mismatches

Search for all substrings X of S, with $|X| = |P|$, that differ from P in at most k positions, namely $D_H(P, X) \le k$.

Problem 6.10   Approximate string matching under k errors

Search for all the ending positions of substrings X of S for which $D_L(P, X) \le k$ holds.

In both problems, instead of fixing k, one could also take as input an error level α, and set k as αm. Recall Equation (6.1) for the edit distance. It turns out that only a minimal modification is required in order to solve Problem 6.10: initialize $d_{0,j} = 0$ for all $0 \le j \le n$, and consider the values $d_{i,j}$ otherwise computed as in Equation (6.1).

LEMMA 6.13   *Suppose that an optimal path ending at some $d_{m,j}$ (that is, $d_{m,j}$ equals the minimum value on row m) starts at some $d_{0,r} = 0$. Then $D_L(P, s_{r+1} s_{r+2} \cdots s_j) = d_{m,j}$ and $d_{m,j} = \min\{D_L(P, s_t s_{t+1} \cdots s_j) \mid t \le j\}$.*

Proof   Assume for a contradiction that there is $t \neq r+1$ for which $d = D_L(P, s_t \cdots s_j) < D_L(P, s_{r+1} \cdots s_j)$. It follows that the optimal path from $d_{0,t-1}$ to $d_{m,j}$ has cost $d < d_{m,j}$, which is a contradiction, since $d_{m,j}$ is assumed to be the cost of the optimal path.   □

The optimal path to $d_{m,j}$ is visualized in Figure 6.3.

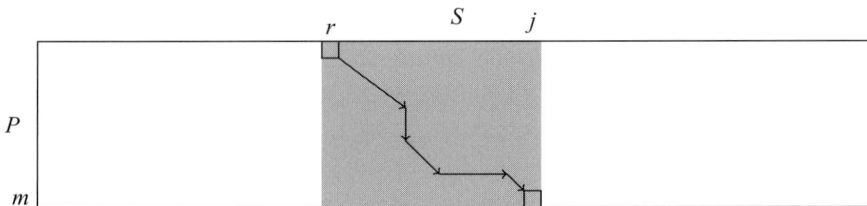

Figure 6.3 An optimal path in the dynamic programming matrix for solving approximate string matching under k errors (Problem 6.10).

THEOREM 6.14   *Approximate string matching under k errors (Problem 6.10) can be solved in $O(mn)$ time and $O(m)$ space, when the outputs are the occurrence end positions, and in space $O(m^2)$, if the start positions of the occurrences are also to be reported.*

Proof   Modify Algorithm 6.1 to take inputs P and S in place of A and B. Initialize instead $d_{0,j} = 0$, for all $0 \le j \le n$. After computing $d_{m,j}$ for column j, report j as the ending position of a k-errors match if $d_{m,j} \le k$. As in Exercise 6.2, the space of the algorithm can be improved to $O(m)$ by storing values only from two consecutive columns at a time.

To find an occurrence start position, one should keep the part of the matrix containing an optimal path ending at each column j with $d_{m,j} \le k$. Notice that such an optimal path must start at least at column $j - m - k$, so an $O(m^2)$ size portion of the matrix is required. When proceeding column-by-column from left to right, one can maintain the required part of the matrix in a circular buffer.   □

This simple modification of edit distance computation for the approximate matching solution is called the *first-row-to-zero* trick.

It is worth remarking that, with a different initialization, Myers' bitparallel algorithm can also be employed to solve the k-errors problem.

COROLLARY 6.15   *Approximate string matching under k errors (Problem 6.10) limited to the Levenshtein edit distance can be solved in $O(\lceil m/w \rceil n)$ time, when the outputs are the occurrence end positions, in the RAM model of computation, where w is the computer word size.*

The proof of this result is left as an exercise for the reader (Exercise 6.4). In Chapter 10 we will study more advanced approximate string matching techniques.

6.4   Biological sequence alignment

As discussed earlier, an alignment of two biological sequences is typically defined through a maximum score alignment, rather than a minimum-cost one, as in edit distance. Algorithm-wise, the differences are minimal, but the maximization framework allows a probabilistic interpretation for alignments. Let us define $s(a, b)$ to be the *score* of aligning character a with character b (that is, of substituting a with b). Let δ be the *penalty* of an *indel*, that is, the penalty for inserting or deleting a character. Then the weight of an alignment is defined by the sum of substitution scores minus the sum of the indel penalties in it.

For example, the following *substitution matrix* is often used for DNA sequence alignments:

$s(a,b)$	A	C	G	T
A	1	−1	−0.5	−1
C	−1	1	−1	−0.5
G	−0.5	−1	1	−1
T	−1	−0.5	−1	1

The reason for these numerical values is that so-called *transition* mutations (here with score -0.5) are twice as frequent as so-called *transversions* (here with score -1).

For example, if $\delta = 1$, then the weight, or total score, of the alignment

```
A  C  C  -  G  A  T  G
A  -  C  G  G  C  T  A
```

is $1 - 1 + 1 - 1 + 1 - 1 + 1 - 0.5 = 0.5$.

6.4.1 Global alignment

Global alignment is the dual of edit distance computation, with minimization replaced by maximization, defined as follows. The *weight*, or *score*, of an alignment U, L is

$$W(U, L) = \sum_{i=1}^{h} s(u_i, l_i), \tag{6.8}$$

where $s(a, -) = s(-, b) = -\delta$.

Problem 6.11   Global alignment

Given two sequences $A = a_1 \cdots a_m \in \Sigma^*$ and $B = b_1 \cdots b_n \in \Sigma^*$, find an alignment of A and B having the maximum score, that is, having the score

$$S(A, B) = \max_{(U,L) \in \mathcal{A}(A,B)} W(U, L),$$

where $\mathcal{A}(A, B)$ denotes the set of all valid alignments of A and B.

Let us denote $s_{i,j} = S(a_1 \cdots a_i, b_1 \cdots b_j)$, for all $0 \leq i \leq m, 0 \leq j \leq n$.

THEOREM 6.16   *The global alignment scores $s_{i,j}$, and, in particular, $s_{m,n} = S(A, B)$, can be computed using the recurrence*

$$s_{i,j} = \max\{s_{i-1,j-1} + s(a_i, b_j), s_{i-1,j} - \delta, s_{i,j-1} - \delta\}, \tag{6.9}$$

for all $1 \leq i \leq m$, $1 \leq j \leq n$, and using the initializations

$$s_{0,0} = 0,$$

$$s_{i,0} = -i\delta, \text{ for all } 1 \leq i \leq m, \text{ and}$$

$$s_{0,j} = -j\delta, \text{ for all } 1 \leq j \leq n.$$

The correctness proof is identical to the corresponding theorem on edit distance. For completeness, a pseudocode is given in Algorithm 6.4. The algorithm is also called *Needleman-Wunsch*, after its inventors.

Tracing back the optimal alignment(s) is identical to the corresponding procedure for edit distances.

Algorithm 6.4: Global alignment using dynamic programming

Input: Sequences $A = a_1 a_2 \cdots a_m$ and $B = b_1 b_2 \cdots b_n$.
Output: Global alignment score $S(A, B)$.

1 $s_{0,0} = 0$;
2 **for** $i \leftarrow 1$ **to** m **do**
3 $\quad \lfloor \; s_{i,0} \leftarrow s_{i-1,0} - \delta$;
4 **for** $j \leftarrow 1$ **to** n **do**
5 $\quad \lfloor \; s_{0,j} \leftarrow s_{0,j-1} - \delta$;
6 **for** $j \leftarrow 1$ **to** n **do**
7 $\quad$ **for** $i \leftarrow 1$ **to** m **do**
8 $\quad\quad \lfloor \; s_{i,j} \leftarrow \max\{s_{i-1,j-1} + s(a_i, b_j), s_{i-1,j} - \delta, s_{i,j-1} - \delta\}$;

9 **return** $s_{m,n}$;

6.4.2   Local alignment

An approximate pattern matching variant of global alignment is called *semi-local alignment*. This is the problem of finding the substring of a long sequence S having the highest scoring alignment with shorter sequence A. With the limitation of searching for only semi-local alignments with positive scores, one can use the first-row-to-zero trick from Section 6.3 in the global alignment computation. Then, the best-scoring semi-local alignments can be found analogously as explained in Section 6.3.

Continuing the example of a homology search in prokaryotes, we assume we have the full genomes of two different species. To identify potentially homologous genes, we should be able to identify a gene in one species in order to apply approximate string matching/semi-local alignment. However, there is a way to do the same thing without resorting to gene annotation, through *local alignment*, defined next.

Problem 6.12   Local alignment

Compute the maximum-scoring global alignment $L(A, B)$ over all substrings of $A = a_1 \cdots a_m$ and $B = b_1 \cdots b_n$, that is,

$$L(A, B) = \max \left\{ S(A_{i'..i}, B_{j'..j}) \mid \text{for all } i', i \in [1..m], j', j \in [1..n] \text{ with } i' \leq i, j' \leq j \right\}.$$

The pair $(A_{i'..i}, B_{j'..j})$ with $L(A, B) = S(A_{i'..i}, B_{j'..j})$ can be interpreted as a pair of "genes" with highest similarity in the homology search example. One can modify the problem statement into outputting all substring pairs (A', B') with high enough similarity, which constitute the sufficiently good candidates for homologies.

For an arbitrary scoring scheme, this problem can be solved by applying global alignment for all suffix pairs from A and B in $O((mn)^2)$ time. However, as we learn in

Insight 6.1, the scoring schemes can be designed so that local alignments with score less than zero are not statistically significant, and therefore one can limit the consideration to non-zero-scoring local alignments. To compute such non-zero-scoring alignments, one can use a version of the first-row-to-zero trick: use the global alignment recurrence, but add an option to start a new alignment at any suffix/suffix pair by assigning score 0 for an empty alignment. This observation is the *Smith–Waterman* algorithm for local alignment.

Insight 6.1 Derivation of substitution scores

Local alignments exploit some properties of substitution matrices, which we derive next. For clarity, we limit the discussion to the case of trivial alignments with the only allowed mutation being substitution. Consider now that two equal-length strings A and B are generated identically and independently by a *random* model R with probability q_a for each character $a \in \Sigma$. Then the joint probability under R for the two sequences A and B is

$$\mathbb{P}(A, B \mid R) = \prod_i q_{a_i} q_{b_i}. \qquad (6.10)$$

Now, consider the *match* model M, which assigns a joint probability p_{ab} for a substitution $a \to b$. This can be seen as the probability that a and b have a common ancestor in evolution. Then the probability of the trivial alignment is

$$\mathbb{P}(A, B \mid M) = \prod_i p_{a_i b_i}. \qquad (6.11)$$

The ratio $\mathbb{P}(A, B \mid M)/P(A, B \mid R)$ is known as the *odds ratio*. To obtain an additive scoring scheme, the product is transformed into summation by taking logarithm, and the result is known as the *log-odds ratio*:

$$S(A, B) = \sum_i s(a_i, b_i), \qquad (6.12)$$

where

$$s(a, b) = \log\left(\frac{p_{ab}}{q_a q_b}\right). \qquad (6.13)$$

This scoring matrix has the property that the expected score of a random match at any position is negative:

$$\sum_{a,b} q_a q_b s(a, b) < 0. \qquad (6.14)$$

This can be seen by noticing that

$$\sum_{a,b} q_a q_b s(a, b) = -\sum_{a,b} q_a q_b \log\left(\frac{q_a q_b}{p_{ab}}\right) = -\mathsf{H}(q^2 \mid p), \qquad (6.15)$$

where $\mathsf{H}(q^2 \mid p)$ is the *relative entropy* (or *Kullback–Leibler divergence*) of distribution $q^2 = q \times q$ with respect to distribution p. The value of $\mathsf{H}(q^2 \mid p)$ is always positive unless $q^2 = p$ (see Exercise 6.19).

A straightforward approach for estimating probabilities p_{ab} is to take a set of well-trusted alignments and just count the frequencies of substitutions $a \to b$. One can find in the literature more advanced methods that take into account different evolutionary divergence times, that is, the fact that alignment of a to b could happen through a sequence of substitutions $a \to c \to \cdots \to b$.

Let us denote $l_{i,j} = \max\{S(a_{i'} \cdots a_i, b_{j'} \cdots b_j) \mid i' \le i, j' \le j\}$, for all $0 \le i \le m$, $0 \le j \le n$.

THEOREM 6.17 *The non-negative local alignment scores $l_{i,j}$ can be computed using the recurrence*

$$l_{i,j} = \max\{0, l_{i-1,j-1} + s(a_i, b_j), l_{i-1,j} - \delta, l_{i,j-1} - \delta\}, \tag{6.16}$$

for all $1 \le i \le m$, $1 \le j \le n$, and using the initializations

$$l_{0,0} = 0,$$
$$l_{i,0} = 0, \text{ for all } 1 \le i \le m, \text{ and}$$
$$l_{0,j} = 0, \text{ for all } 1 \le j \le n.$$

It is easy to modify Algorithm 6.4 for computing the values $l_{i,j}$. After the computation, one can either locate max $l_{i,j}$ or maintain the maximum value encountered during the computation. In the latter case, $O(m)$ space is sufficient for finding the maximum (computing the recurrence column-by-column). To trace back the alignment, one cannot usually afford $O(mn)$ space to store the matrix. Instead, one can easily modify the dynamic programming algorithm to maintain, in addition to the maximum score, also the left-most positions in A and B, say i' and j', respectively, where a maximum-scoring alignment ending at $l_{i,j}$ starts. Then it is easy to recompute a matrix of size $(i - i' + 1) \times (j - j' + 1)$ to output (all) local alignment(s) with maximum score ending at $l_{i,j}$. The space is hence quadratic in the length of the longest local alignment with maximum score. Exercise 6.15 asks you to develop the details of this idea and Exercise 6.16 asks for an optimal space solution.

Figure 6.4 illustrates the differences of global, local, semi-local, and overlap alignment, the last of which is our next topic.

6.4.3   Overlap alignment

An *overlap alignment* between two strings ignores the prefix of the first string and the suffix of the second one, as long as the remaining two (suffix–prefix overlapping) substrings have a good alignment. Computing overlap alignments is important for *fragment assembly*, a problem in which a genomic sequence is reconstructed, or *assembled*, from only a random subset of its substrings. Many assembly methods are based on a so-called

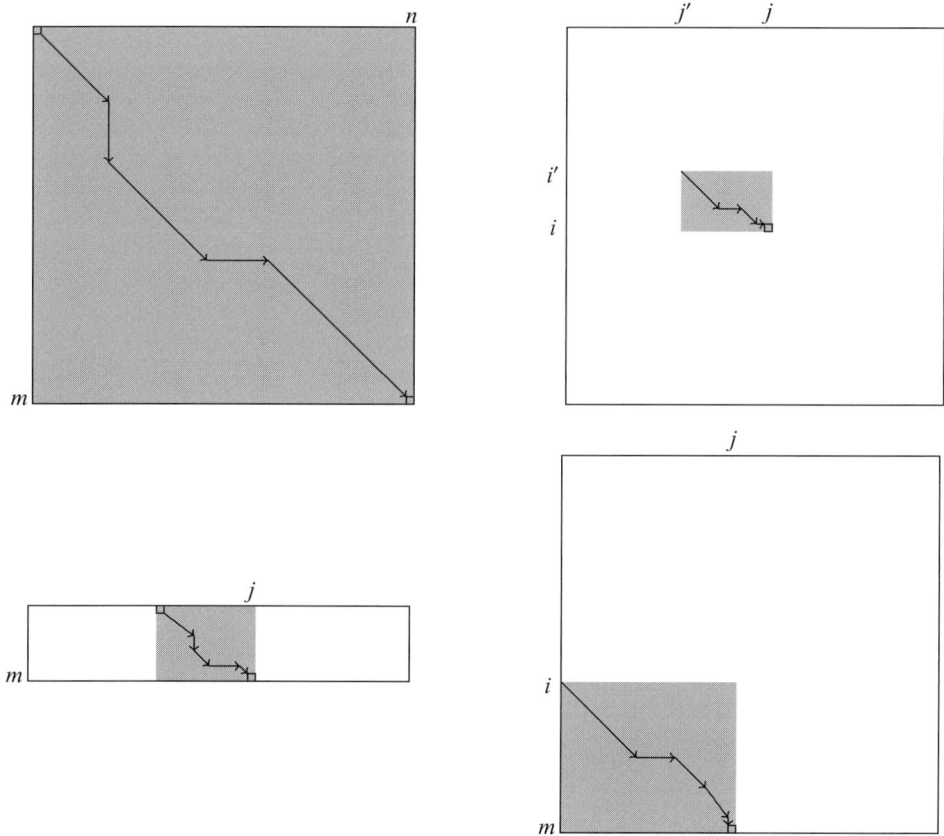

Figure 6.4 Summary of interpretations of global (top-left), local (top-right), semi-local (bottom-left), and overlap alignment (bottom-right).

overlap graph, in which these substrings are vertices, and in which two substrings are connected by an arc if they have an overlap alignment with a high enough score. Paths in this overlap graph represent possible partial assemblies of the originating sequence. In Chapter 13 we will study the fragment assembly problem in greater detail.

Problem 6.13   Overlap alignment

Compute the maximum-scoring global alignment $O(A, B)$ between the suffixes of A and the prefixes of B, that is,

$$O(A, B) = \max \left\{ S(A_{i..m}, B_{1..j}) \mid \text{for all } i \in [1..m], j \in [1..n] \right\}.$$

Like in the local alignment problem, we are interested only in alignments with score greater than zero. It is easy to see that a variant of the first-row-to-zero trick works again, as follows.

Let us denote $o_{i,j} = \max\{S(a_{i'} \cdots a_i, b_1 \cdots b_j) \mid i' \leq i\}$, for all $0 \leq i \leq m$, $0 \leq j \leq n$.

THEOREM 6.18   *The non-negative overlap alignment scores $o_{i,j}$ can be computed using the recurrence*

$$o_{i,j} = \max\{o_{i-1,j-1} + s(a_i, b_j), o_{i-1,j} - \delta, o_{i,j-1} - \delta\}, \qquad (6.17)$$

for all $1 \leq i \leq m$, $1 \leq j \leq n$, and using the initializations

$$o_{0,0} = 0,$$
$$o_{i,0} = 0, \text{ for all } 1 \leq i \leq m, \text{ and}$$
$$o_{0,j} = -\delta j, \text{ for all } 1 \leq j \leq n.$$

Again, the modifications to Algorithm 6.4 are minimal. Figure 6.4 illustrates the idea behind the computation. We shall study overlap alignments further in Section 10.4 and in Chapter 13.

6.4.4   Affine gap scores

In biological sequence evolution, indels often occur in blocks, and the simple *linear gap model* scoring scheme studied so far, with a penalty δ for each inserted or deleted base, is not well grounded. We cover here a more realistic scheme, called the *affine gap model*, which also admits efficient algorithms.

Recall the definition of alignment through sequences U and L containing A and B as subsequences, respectively, together with some gap characters $-$. We say that $U[l..r], L[l..r]$ is a *run of gaps* in the alignment U, L if

$$U[i] = -, \text{ or } L[i] = -, \text{ for all } i \in [l..r].$$

Symmetrically, we say that $U[l..r], L[l..r]$ is a *run of matches* in the alignment U, L if

$$U[i] \neq - \text{ and } L[i] \neq - \text{ for all } i \in [l..r].$$

Notice that a "match" refers here both to an identity and to a substitution, differing from the meaning we attributed to it in the context of longest common subsequences.

Any alignment can be partitioned into a sequence of runs of gaps and runs of matches. Let $U = U^1 U^2 \cdots U^k$ and $L = L^1 L^2 \cdots L^k$ denote such a partitioning, where, for all $i \in [1..k]$, we have $|U^i| = |L^i|$, and U^i, L^i is either a run of gaps or a run of matches. Let $\mathcal{P}(U, L)$ denote the set of all possible such partitions. Let (see Example 6.14)

$$W_G(U^i, L^i) = \begin{cases} W(U^i, L^i) & \text{if } U^i, L^i \text{ is a run of matches,} \\ -\alpha - \beta(|U^i| - 1) & \text{if } U^i, L^i \text{ is a run of gaps,} \end{cases}$$

where we recall that $W(U^i, L^i)$ is the global alignment score defined in (6.8). In the above definition, α is the penalty for opening a gap, and β is the penalty for extending a gap. In practice, α is chosen to be greater than β, the idea being that starting a gap always involves significant costs, but making it longer is cheap.

Example 6.14 Consider the following alignment U, L of $A = \texttt{ACTCGATC}$ and $B = \texttt{ACGTGAATGAT}$, where we have marked a partition of U into $U^1 U^2 U^3 U^4$ and of L into $L^1 L^2 L^3 L^4$:

$$
\begin{array}{c}
\quad\;\; U^1 \quad U^2 \quad U^3\; U^4 \\
U = \boxed{\texttt{ACT}\;\texttt{---CC}\;\texttt{ATC}\;\texttt{--}} \\
L = \boxed{\texttt{ACG}\;\texttt{TGA--}\;\texttt{ATC}\;\texttt{AT}} \\
\quad\;\; L^1 \quad\; L^2 \quad\;\; L^3\; L^4
\end{array}
$$

We have

- $W_G(U^1, L^1) = s(\texttt{T}, \texttt{G})$,
- $W_G(U^2, L^2) = -\alpha - 4\beta$,
- $W_G(U^3, L^3) = 0$,
- $W_G(U^4, L^4) = -\alpha - \beta$.

The following problem asks for a best global alignment under affine gap scores.

Problem 6.15 Global alignment under affine gap scores

Given two sequences $A = a_1 \cdots a_m \in \Sigma^*$ and $B = b_1 \cdots b_n \in \Sigma^*$, find an alignment of A and B having maximum score under the affine gap model, that is, having score

$$
S_G(A, B) = \max_{(U,L) \in \mathcal{A}(A,B)} \max \left\{ \sum_{i=1}^{k} W_G(U^i, L^i) \mid (U^1 \cdots U^k, L^1 \cdots L^k) \in \mathcal{P}(U, L) \right\},
$$

where $\mathcal{A}(A, B)$ denotes the set of all valid alignments of A and B.

Let us denote $\mathsf{sm}_{i,j} = S_G(a_1 \cdots a_{i-1}, b_1 \cdots b_{j-1}) + s(a_i, b_j)$ for all $0 \leq i \leq m$, $0 \leq j \leq n$, where we interpret $s(a_0, b_0) = s(a_{m+1}, s_{n+1}) = 0$. In particular, $S_G(A, B) = \mathsf{sm}_{m+1,n+1}$. That is, we aim to store the optimal alignment score for each prefix pair over all alignments ending with a match.

THEOREM 6.19 *The scores $\mathsf{sm}_{i,j}$, and in particular the global alignment under affine gap model $S_G(A, B) = \mathsf{sm}_{m+1,n+1}$, can be computed using the recurrence*

$$
\mathsf{sm}_{i,j} = \max \Big(\mathsf{sm}_{i-1,j-1} + s(a_i, b_j),
$$

$$
\max_{i' < i,\, j' < j,\, (i',j') \neq (i-1,j-1)} \{ \mathsf{sm}_{i',j'} - \alpha - \beta(j - j' + i - i' - 3) + s(a_i, b_j) \} \Big),
$$

$$
(6.18)
$$

for all $1 \leq i \leq m$, $1 \leq j \leq n$, and using the initializations

$$sm_{0,0} = 0,$$
$$sm_{i,0} = -\infty, \text{ for all } 1 \leq i \leq m, \text{ and}$$
$$sm_{0,j} = -\infty, \text{ for all } 1 \leq j \leq n.$$

Proof First, note that the initializations are correct, since no alignment with an empty string can end with a match. By definition, the alignment score for two empty strings is 0.

Assume by induction that the recurrence is correct for all i', j' such that $i' < i, j' < j$. In any alignment of $a_1 \cdots a_i$ and $b_1 \cdots b_j$ ending with the match $a_i \rightarrow b_j$, there must be some previous match $a_{i'} \rightarrow b_{j'}$ (possibly the conceptual initialization $a_0 \rightarrow b_0$). The run of gaps between the matches $(a_{i'}, b_{j'})$ and (a_i, b_j) has score $-\alpha - \beta(j - j' + i - i' - 3)$, since the gap length is $j - 1 - (j' + 1) + 1 + i - 1 - (i' + 1) + 1 = j - j' + i - i' - 2$. An exception is the gap of length 0 that is taken into account separately by the term $sm_{i-1,j-1} + s(a_i, b_j)$. We take the maximum over all such possible shorter alignments ending with a match, adding the score of the gap, or the score $s(a_i, b_j)$. ☐

The running time for computing $S_G(A, B)$ with the recurrence in Equation (6.18) is $O((mn)^2)$.

Notice that this recurrence is general, in the sense that $-\alpha - \beta(j - j' + i - i' - 3)$ can be replaced with any other function on the gap length. For example, it is easy to modify the recurrence to cope with various alternative definitions of a "run of gaps". One could separately define a run of insertions and a run of deletions. In fact, this definition is more relevant in the context of biological sequence evolution, since insertions and deletions are clearly separate events. For this alternative definition, the recurrence changes so that one takes separately $\max_{i' < i} sa_{i',j} - \alpha - \beta(i - i' - 1)$ and $\max_{j' < j} sa_{i,j'} - \alpha - \beta(j - j' - 1)$, where $sa_{i,j}$ stores the optimal alignment score over all alignments (not just those ending with a match). The full details are left to the reader. The running time then reduces to $O(mn(m + n))$.

In what follows, we will continue with the general definition, and ask the reader to consider how to modify the algorithms for separate runs of insertions and deletions.

There are several ways to speed up the affine gap model recurrences. The most common one is *Gotoh's algorithm*, which fills two or more matrices simultaneously, one for alignments ending at a match state and the others for alignments ending inside a run of gaps. Let

$$S_G(A, B \mid \text{match}) = \max_{(U,L) \in \mathcal{A}_{\mathcal{M}}(A,B)} \max \left\{ \sum_{i=1}^{k} W_G(U^i, L^i) \mid (U^1 \cdots U^k, L^1 \cdots L^k) \in \mathcal{P}(U, L) \right\},$$

where $\mathcal{A}_{\mathcal{M}}(A, B)$ denotes the set of all valid alignments of A and B ending with a match. Let

$$S_G(A, B \mid \text{gap}) = \max_{(U,L) \in \mathcal{A}_{\mathcal{G}}(A,B)} \max \left\{ \sum_{i=1}^{k} W_G(U^i, L^i) \mid (U^1 \cdots U^k, L^1 \cdots L^k) \in \mathcal{P}(U, L) \right\},$$

where $\mathcal{A}_{\mathcal{G}}(A, B)$ denotes the set of all valid alignments of A and B ending with a gap.

Then the value $\text{sm}_{i,j}$ we defined earlier equals $S_G(A_{1..i}, B_{1..j} \mid \text{match})$, and we also let $\text{sg}_{i,j}$ denote $S_G(A_{1..i}, B_{1..j} \mid \text{gap})$. It holds that $S_G(A_{1..i}, B_{1..j}) = \max(\text{sm}_{i,j}, \text{sg}_{i,j})$.

THEOREM 6.20 *The scores* $\text{sm}_{i,j}$ *and* $\text{sg}_{i,j}$, *and in particular the global alignment under affine gap model* $S_G(A, B) = \text{sm}_{m+1,n+1} = \max(\text{sm}_{m,n}, \text{sg}_{m,n})$, *can be computed using the recurrences*

$$
\begin{aligned}
\text{sm}_{i,j} &= \max\{\text{sm}_{i-1,j-1} + s(a_i, b_j), \text{sg}_{i-1,j-1} + s(a_i, b_j)\}, \\
\text{sg}_{i,j} &= \max\{\text{sm}_{i-1,j} - \alpha, \text{sg}_{i-1,j} - \beta, \text{sm}_{i,j-1} - \alpha, \text{sg}_{i,j-1} - \beta\},
\end{aligned}
\tag{6.19}
$$

where $1 \le i \le m$, $1 \le j \le n$, *and using the initializations*

$$
\begin{aligned}
\text{sm}_{0,0} &= 0, \\
\text{sm}_{i,0} &= -\infty, \text{ for all } 1 \le i \le m, \\
\text{sm}_{0,j} &= -\infty, \text{ for all } 1 \le j \le n, \\
\text{sg}_{0,0} &= 0, \\
\text{sg}_{i,0} &= -\alpha - \beta(i-1), \text{ for all } 1 \le i \le m, \text{ and} \\
\text{sg}_{0,j} &= -\alpha - \beta(j-1), \text{ for all } 1 \le j \le n.
\end{aligned}
$$

Proof The initialization works correctly, since in the match matrix the first row and the first column do not correspond to any valid alignment, and in the gap matrix the cases correspond to alignments starting with a gap.

The correctness of the computation of $\text{sm}_{i,j}$ follows easily by induction, since take to next line $\max\{\text{sm}_{i-1,j-1}, \text{sg}_{i-1,j-1}\}$ is the maximum score of all alignments where a match can be appended. The correctness of the computation of $\text{sg}_{i,j}$ can be seen as follows. An alignment ending in a gap is either (1) opening a new gap or (2) extending an existing one. In case (1) it is sufficient to take the maximum over the scores for aligning $A_{1..i-1}$ with $B_{1..j}$ and $A_{1..i}$ with $B_{1..j-1}$ so that the previous alignment ends with a match, and then add the cost for opening a gap. These maxima are given (by the induction assumption) by $\text{sm}_{i-1,j}$ and $\text{sm}_{i,j-1}$. In case (2) it is sufficient to take the maximum over the scores for aligning $A_{1..i-1}$ with $B_{1..j}$ and $A_{1..i}$ with $B_{1..j-1}$ so that the previous alignment ends with a gap, and then add the cost for extending a gap. These maxima are given (by the induction assumption) by $\text{sg}_{i-1,j}$ and $\text{sg}_{i,j-1}$. □

The running time is now reduced to $O(mn)$. Similar recurrences can be derived for local alignment and overlap alignment under affine gap score.

6.4.5 The invariant technique

It is instructive to study an alternative $O(mn)$ time algorithm for alignment under affine gap scores. This algorithm uses the *invariant technique*, in a similar manner to that in Section 6.2.1 for the longest common subsequence problem.

Consider the formula $\text{sm}_{i',j'} - \alpha - \beta(j - j' + i - i' - 3) + s(a_i, b_j)$ inside the maximization in Equation (6.18). This can be equivalently written as $-\alpha - \beta(j + i - 3) + s(a_i, b_j) + \text{sm}_{i',j'} + \beta(j' + i')$. The maximization goes over all valid values of i', j'

and the first part of the formula, namely $-\alpha - \beta(j + i - 3) + s(a_i, b_j)$, is not affected. Recall the search tree $\mathcal{T}$ of Lemma 3.1 on page 20. With a symmetric change it can be transformed to support range maximum queries instead of range minimum queries; let $\mathcal{T}.\mathsf{RMaxQ}(c, d)$ return the *max* of values associated with keys $[c..d]$. Then, by storing values $\mathrm{sm}_{i,j} + \beta(j + i)$ together with key i to the search tree, we can query the maximum value and add the invariant. Using the reverse column-order as in Algorithm 6.3, we obtain Algorithm 6.5, which is almost identical to the sparse dynamic programming solution for the longest common subsequence problem; the main difference is that here we need to fill the whole matrix. We can thus obtain an $O(mn \log m)$ time algorithm for computing the global alignment under affine gap scores.

Algorithm 6.5: Affine gap score $S_G(A, B)$ using range maximum queries

Input: Sequences $A = a_1 a_2 \cdots a_m$ and $B = b_1 b_2 \cdots b_n$.
Output: Global alignment under affine gap scores $S_G(A, B)$.
1 Initialize the search tree $\mathcal{T}$ of Lemma 3.1 with keys $0, 1, \ldots, m$;
2 $\mathcal{T}.\mathrm{update}(0, 0)$;
3 $\mathrm{sm}_{0,0} \leftarrow 0$;
4 **for** $j \leftarrow 1$ **to** n **do**
5   **for** $i \leftarrow m$ **down to** 1 **do**
6          $\mathrm{sm}_{i,j} \leftarrow \max\{\mathrm{sm}_{i-1,j-1} + s(a_i, b_j),$
            $\mathcal{T}.\mathsf{RMaxQ}(0, i-1) - \alpha - \beta(i+j-3) + s(a_i, b_j)\}$;
7          $\mathcal{T}.\mathrm{update}(i, \mathrm{sm}_{i,j} + \beta(i+j))$;

8 $\mathrm{sm}_{m+1,n+1} \leftarrow \max\{\mathrm{sm}_{m,n}, \mathcal{T}.\mathsf{RMaxQ}(0, m) - \alpha - \beta(m+n-1)\}$;
9 **return** $\mathrm{sm}_{m+1,n+1}$;

However, the use of a data structure in Algorithm 6.5 is in fact unnecessary. It is sufficient to maintain the maximum value on each row, and then maintain on each column the maximum of the row maxima; this is the same as the incremental computation of maxima for semi-infinite rectangles in a grid as visualized in Figure 6.5. The modified $O(mn)$ time algorithm is given as Algorithm 6.6.

6.5   Gene alignment

Consider the problem of having a protein sequence (produced by an unknown gene) and looking for this unknown gene producing it in the DNA. With prokaryotes, the task, which we call *prokaryote gene alignment*, is relatively easy: just modify the semi-local alignment (global alignment with the first row initialized to zero) so that $s_{i-1,j-1} + s(a_i, b_j)$ is replaced by $s_{i-1,j-3} + s(a_i, \mathrm{aa}[B[j-2..j]])$, where the first sequence A is the protein sequence, the second sequence B is the DNA sequence, and $\mathrm{aa}[xyz]$ is the amino acid encoded by codon xyz (recall Table 1.1). For symmetry, also $s_{i,j-1} - \delta$ should be replaced by $s_{i,j-3} - \delta$, and the initializations modified accordingly. An alternative and more rigorous approach using the DAG-path alignment of Section 6.6.5 is considered in Exercise 6.27.

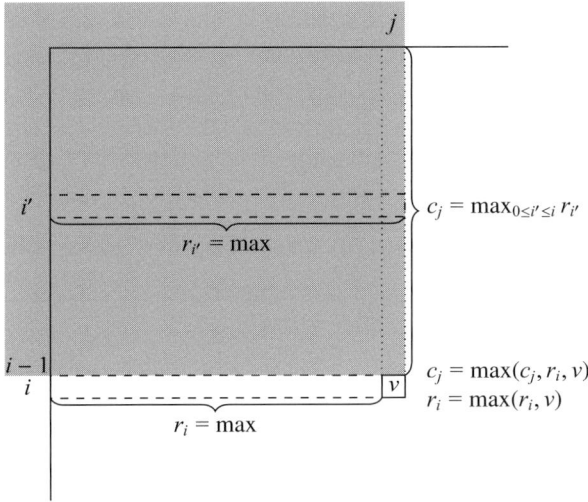

Figure 6.5 Maintaining sliding maxima in semi-infinite rectangles $[-\infty..i] \times [-\infty..j]$ over a grid of values. Here r_i is the maximum at row i computed with column-by-column evaluation order before reaching cell (i, j). After the updates visualized, c_j keeps the maximum over values in $[-\infty..i] \times [-\infty..j]$.

With eukaryotes, introns need to be taken into account. The most common way to do this is to use affine gap penalties and assign a large gap opening cost and a very small gap extension cost in the DNA side, and use a linear gap cost on the protein side. Call this approach *eukaryote gene alignment with affine gaps*. We leave it as an exercise for the reader to formulate an $O(mn)$ algorithm for solving this variant. However, this approach is clearly problematic because intron lengths can vary from tens to tens of thousands of nucleotides: there is no good choice for the gap extension cost.

Moreover, combining scores from two different phenomena into one objective function is like comparing apples and oranges. In what follows, we systematically avoid such formulations, and instead choose to optimize only one feature at a time, while limiting the others with thresholds.

For simplicity of exposition, we consider in what follows that the pattern $P = A$ is a cDNA sequence (the complement of an RNA sequence) instead of a protein sequence and denote the genomic DNA sequence $T = B$; the protein-to-DNA versions follow using the idea explained above.

Let us now develop a better approach for gene alignment in eukaryotes exploiting the fact that the number of introns in genes is usually quite small. For example, in the human genome, the average number of introns is 7.8, the maximum being 148. We can fix the maximum number of introns allowed in an alignment as follows.

Problem 6.16 Eukaryote gene alignment with a limited number of introns

For all $i \in [1..m]$ and $j' \in [1..j]$, let $e_{i,j,k}$ be the maximum-scoring semi-local alignment under linear gap scores of $P_{1..i}$ and $T_{j^s..j}$, using k *maximal free runs* of deletions in

$T_{j^s..j-1}$. Here by *free* we mean that each deletion in such a run incurs no cost to the overall score and by *maximal* that a free run cannot be succeeded by another free run. Compute the value $\max_{1 \le j \le n, 0 \le k \le \texttt{maxin}} e_{m,j,k}$, where $\texttt{maxin}$ is a given limit for the allowed number of introns.

Algorithm 6.6: Affine gap score $S_G(A, B)$ maintaining maxima of semi-infinite rectangles

---

Input: Sequences $A = a_1 a_2 \cdots a_m$ and $B = b_1 b_2 \cdots b_n$.
Output: Global alignment under affine gap score $S_G(A, B)$.

1 $\text{sm}_{0,0} \leftarrow 0$;
2 **for** $i \leftarrow 1$ *to* m **do**
3 $\quad$ $\text{sm}_{i,0} \leftarrow -\infty$;
4 $\quad$ $r_i \leftarrow -\infty$;
5 **for** $j \leftarrow 1$ *to* n **do**
6 $\quad$ $\text{sg}_{0,j} \leftarrow -\infty$;
7 $\quad$ $c_i \leftarrow -\infty$;
8 **for** $j \leftarrow 1$ *to* n **do**
9 $\quad$ **for** $i \leftarrow 1$ *to* m **do**
10 $\quad\quad$ $\text{sm}_{i,j} \leftarrow \max\{\text{sm}_{i-1,j-1} + s(a_i, b_j),$
$\quad\quad\quad\quad\quad\quad \max(c_{j-1}, r_{i-1}) - \alpha - \beta(i+j-3) + s(a_i, b_j)\}$;
11 $\quad\quad$ $c_{j-1} \leftarrow \max(c_{j-1}, r_{i-1})$;
12 $\quad$ **for** $i \leftarrow 1$ *to* m **do**
13 $\quad\quad$ $r_i \leftarrow \max(r_i, \text{sm}_{i,j} + \beta(i+j))$;
14 **for** $i \leftarrow 1$ *to* m **do**
15 $\quad$ $c_n \leftarrow \max(c_n, r_{i-1})$;
16 $\text{sm}_{m+1,m+1} \leftarrow \max\{\text{sm}_{m,n}, \max(c_n, r_m) - \alpha - \beta(m+n-1)\}$;
17 **return** $\text{sm}_{m+1,n+1}$;

This problem can be easily solved by extending the semi-local alignment recurrences with the fact that each j can be the start of an exon, and hence position j can be preceded by any $j' < j - 1$, where j' is the end of a previous exon. In such cases, one needs to spend one more free run of deletions in the genomic DNA to allow an intron. That is, one needs to look at alignment scores with $k - 1$ runs of deletions, stored in $e_{*,*,k-1}$, when computing the best alignment with k runs of deletions. The solution is given below.

THEOREM 6.21 *The eukaryote gene alignment with a limited number of introns problem can be solved by computing all the scores $e_{i,j,k}$, using the recurrence*

$$e_{i,j,k} = \max \left\{ e_{i-1,j-1,k} + s(p_i, t_j), e_{i-1,j,k} - \delta, e_{i,j-1,k} - \delta, \right.$$

$$\left. \max_{j' < j-1} \{e_{i-1,j',k-1} + s(p_i, t_j), e_{i,j',k-1} - \delta\} \right\} \qquad (6.20)$$

where $1 \leq i \leq m$, $1 \leq j \leq n$, *and using the initializations*

$$e_{0,0,0} = 0,$$
$$e_{i,0,0} = -\delta i, \text{ for all } 1 \leq i \leq m,$$
$$e_{0,j,0} = 0, \text{ for all } 1 \leq j \leq n, \text{ and}$$
$$e_{i,j,k} = -\infty \text{ otherwise.}$$

The running time is $O(mn^2 \cdot \texttt{maxin})$, but this can be improved to $O(mn \cdot \texttt{maxin})$ by replacing each $\max_{j' < j-1} e_{x,j',k-1}$ with $m_{x,k-1}$ such that $m_{*,k}$ maintains the row maximum for each k during the computation of the values $e_{i,j,k}$. That is, $m_{i,k} = \max(m_{i,k}, e_{i,j,k})$. It is left as an exercise for the reader to find a correct evaluation order for the values $e_{i,j,k}$ and $m_{i,k}$ so that they get computed correctly.

Obviously, the values j giving the maximum for $\max_{1 \leq j \leq n, 0 \leq k \leq \texttt{maxin}} e_{m,j,k}$ are plausible ending positions for the alignment. The alignment(s) can be traced back with the standard routine.

6.6 Multiple alignment

Multiple alignment is the generalization of pair-wise alignment to more than two sequences. Before giving a formal definition, see below one possible alignment of three sequences:

```
A G A C - G A T T A
A C - C A G C T T A
A - A C - G G T T -
```

What is no longer obvious is how to score the alignments and how to compute them efficiently.

6.6.1 Scoring schemes

A multiple alignment of d sequences $T^1, T^2, \ldots, T^d \in \Sigma^*$ is a $d \times n$ matrix M. We denote by M_{i*} the ith row in the alignment, and by M_{*j} the jth column in the alignment. The alignment is required that, for all $i \in [1..d]$, T^i must be a subsequence of M_{i*}, and the remaining $n - |T^i|$ characters of M_{i*} must be $-$, where $- \notin \Sigma$.

The *sum-of-pairs (SP) score* for M is defined as

$$\sum_{j=1}^{n} \sum_{\substack{i',i\in[1..d] \\ i'<i}} s(M[i',j], M[i,j]),$$

where $s(a, b)$ is the same score as that used for pair-wise alignments. The sum-of-pairs score can be identically written row-by-row as $\sum_{i',i\in[1..d],\, i'<i} W(M_{i'*}, M_{i*})$, using our earlier notion of pair-wise alignment scores.

Although this is a natural generalization from pair-wise alignments, this score has the drawback of not being well motivated from an evolutionary point of view: the same substitutions are possibly counted for many pairs of sequences, although they could in some cases be explained by just one substitution in the evolutionary tree of the biological sequences. An obvious improvement over the SP score for biological sequences is to find the most parsimonious evolutionary tree for the sequences to be aligned, and to count the sum of scores of substitutions and indels from that tree. This will be called the *parsimony score* of the multiple alignment.

To define this measure more formally, let us first assume that an evolutionary tree (V, E) with d leaves is given for us, where V denotes the set of nodes and E the set of edges. We will compute a score for every column M_{*j} of a multiple alignment. The final score of the alignment can be taken as the sum of the column scores.

Thus, for every column M_{*j} of a multiple alignment, we label each leaf i of the tree with $\ell(i) = M[i, j]$. The *small parsimony* problem is to label also the internal nodes $v \in V$ with some $\ell(v) \in \Sigma \cup \{-\}$ such that the sum of the scores $s(\ell(v), \ell(w))$ over all edges $(v, w) \in E$ is maximized.

For example, a solution to the small parsimony problem for a column ACAT of a multiple alignment could be the following:

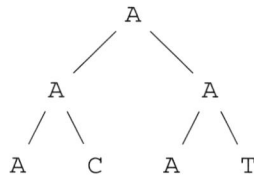

With this example, the sum of scores would be $4s(\mathrm{A, A}) + s(\mathrm{A, C}) + s(\mathrm{A, T})$.

To find this optimal assignment for a fixed column M_{*j}, a simple dynamic programming algorithm works. Let $s_v(a)$ denote the optimal score of the subtree rooted at v, when v is labeled with the character $a \in \Sigma \cup \{-\}$. It is easy to see that the following recurrence holds:

$$s_v(a) = \sum_{w\,:\,(v,w)\in E} \left(\max_{b\in\Sigma\cup\{-\}} s_w(b) + s(a, b) \right). \tag{6.21}$$

One can compute $s_v(a)$, for all $v \in V$ and $a \in \Sigma \cup \{-\}$, in a bottom-up manner, from leaves to the root. The optimal score is $\max_{a\in\Sigma\cup\{-\}} s_{\mathrm{root}}(a)$. The optimal labeling can then be computed top-down from the root to the leaves, by selecting the labels leading to the maximum-score solution.

Recall our original goal of defining a parsimony score for a given multiple alignment. Assuming that an evolutionary tree is given for us, we can now compute the score in optimal time on constant alphabets. Without this assumption, we should enumerate through all possible trees, and choose the one that gives the maximum score summed over all columns of the multiple alignment. There are $\Theta(4^{2d}/d^{3/2})$ binary evolutionary trees with d leaves, so the computation of the parsimony score of a given alignment is exponential in d.

6.6.2   Dynamic programming

Having described different scoring schemes for a multiple alignment, let us now see how to actually compute an optimal multiple alignment, for such scoring schemes.

Let $s_{i_1,i_2,\ldots,i_d}$ denote the optimal scoring alignment of $T^1_{1..i_1}, T^2_{1..i_2}, \ldots, T^d_{1..i_d}$, under some fixed score. To compute $s_{i_1,i_2,\ldots,i_d}$, one needs to consider all the $2^d - 1$ ways the last column of the alignment can look (any combinations of j sequences that can have a gap, for all $0 \leq j < d$). For example, in the three-dimensional case we have

$$
s_{i_1,i_2,i_3} = \max \begin{cases}
s_{i_1-1,i_2,i_3} + s(t^1_{i_1}, -, -) \\
s_{i_1,i_2-1,i_3} + s(-, t^2_{i_2}, -) \\
s_{i_1,i_2,i_3-1} + s(-, -, t^3_{i_3}) \\
s_{i_1-1,i_2-1,i_3} + s(t^1_{i_1}, t^2_{i_2}, -) \\
s_{i_1-1,i_2,i_3-1} + s(t^1_{i_1}, -, t^3_{i_3}) \\
s_{i_1,i_2-1,i_3-1} + s(-, t^2_{i_2}, t^3_{i_3}) \\
s_{i_1-1,i_2-1,i_3-1} + s(t^1_{i_1}, t^2_{i_2}, t^3_{i_3})
\end{cases} \tag{6.22}
$$

where $s(x, y, z)$ denotes the score for a column. The running time of the d-dimensional case is $O(n^d(d^2 + 2^d))$ in the case of an SP score, $O(n^d(d + 2^d))$ in the case of a parsimony score with a fixed evolutionary tree, and $O(n^d(d + 2^d)4^{2d})$ in the case of a general parsimony score (enumerating over all trees).

6.6.3   Hardness

We will now show that it is unlikely that one will be able to find much faster algorithms (namely, polynomial) for multiple alignment. For such a proof it suffices to consider the simplest scoring scheme.

Problem 6.17   Longest common subsequence (LCS) on multiple sequences

Find a longest sequence C that is a subsequence of all the given sequences $T^1, T^2, \ldots, T^d$.

This problem is identical to finding a maximum-scoring multiple alignment with the scoring scheme $s(t^1_{i_1}, t^2_{i_2}, \ldots, t^d_{i_d}) = 1$ if and only if $t^1_{i_1} = t^2_{i_2} = \cdots = t^d_{i_d}$, and 0, otherwise.

THEOREM 6.22   *The longest common subsequence problem on multiple sequences is NP-hard.*

Proof   We reduce from the independent set problem by showing that, given a graph $G = (V = \{1, \ldots, n\}, E)$, we can construct n strings $S^1, \ldots, S^n$, over the alphabet V, such that G has an independent set of size k if and only if $S^1, \ldots, S^n$ have a common subsequence of length k.

For each vertex i, let $j_1 < j_2 < \cdots < j_{t(i)}$ denote all the vertices of G which are not adjacent to i and are smaller than i. Then the ith string is

$$S^i = j_1 \cdot j_2 \cdot \cdots \cdot j_{t(i)} \cdot i \cdot 1 \cdot 2 \cdots (i-1) \cdot \cdots (i+1) \cdot (i+2) \cdot \cdots \cdot n,$$

where, in particular, $S^1 = 1 \cdot 2 \cdot \cdots \cdot n$.

If $\{i_1, \ldots, i_k\}$, with $i_1 < \cdots < i_k$, is an independent set of size k of G, then $i_1, \ldots, i_k$ is a common subsequence of $S^1, \ldots, S^k$. Indeed, if $j \notin \{i_1, \ldots, i_k\}$, by construction $i_1, \ldots, i_k$ appear in this order in S^j. If $j \in \{i_1, \ldots, i_k\}$ then the elements of $\{i_1, \ldots, i_k\} \setminus \{j\}$ smaller than j have been added, by construction, also to the left of j in S^j.

Conversely, assuming that $i_1, \ldots, i_k$ is a common subsequence of $S^1, \ldots, S^n$, it is also a subsequence of S_1, and thus $i_1 < \cdots < i_k$. We argue that $\{i_1, \ldots, i_k\}$ is an independent set of size k of G. Assume for a contradiction that there is an edge between $i_\ell, i_m \in \{i_1, \ldots, i_k\}$, where, say, $i_\ell < i_m$. Then in S^m, by construction, i_ℓ appears only to the right of i_m, which contradicts the fact that $i_1, \ldots, i_k$ is a subsequence also of S^m. $\square$

6.6.4   Progressive multiple alignment

Owing to the unfeasibility of multiple alignment with many sequences, most bioinformatics tools for multiple alignments resort to ad-hoc processes, like building the multiple alignment incrementally: first do pair-wise alignments and then combine these into larger and larger multiple alignments. The process of combining alignments can be guided by an evolutionary tree for the sequences.

Assume we have constructed such a *guide tree*. We can construct a multiple alignment bottom-up by doing first pair-wise alignment combining neighboring leaves and then pair-wise alignments of two multiple alignments (of size $1, 2, \ldots$) until we reach the root. The only difficulty is how to do pair-wise alignment of two multiple alignments (called *profiles* in this context). The trick is to simply consider each profile as a sequence of columns and to apply normal pair-wise alignment when defining $s(M_{*j}, M'_{*j})$, for example as an SP score, where M and M' are the two profiles in question. This ad-hoc process is called the *once a gap, always a gap* principle, since the gaps already in M and M' are not affected by the alignment; new gaps are formed one complete column at a time. Example 6.18 clarifies the process.

Example 6.18   Construct a multiple alignment of sequences $A = \text{AAGACAC}$, $B = \text{AGAACC}$, $C = \text{AGACAC}$, and $D = \text{AGAGCAC}$ using progressive alignment.

Solution
Assume the guide tree has an internal node v with A and B as children, an internal node w with C and D as children, and a root with v and w as children. To form the profile for node v, we need to compute the optimal pair-wise alignment of A and B, which could be

$$\begin{array}{ccccccc} \text{A} & \text{A} & \text{G} & \text{A} & \text{C} & \text{A} & \text{C} \\ \text{A} & \text{G} & \text{A} & \text{A} & \text{C} & \text{-} & \text{C} \end{array}$$

with some proper choice of scores. Likewise, the profile for w could be

```
A G A - C A C
A G A G C A C
```

Finally, for the root, we take the profiles of v and w as sequences of columns, and obtain the following multiple alignment as the result:

```
A A G A C A C
A G A A C - C          A A G A C A C
                       A G A A C - C
                       A G A - C A C
A G A - C A C          A G A G C A C
A G A G C A C
```

6.6.5   DAG alignment

The alignment of profiles raises the question of whether there is some approach to avoid adding a full column of gaps at a time. We will next explore an approach based on labeled DAGs (recall Chapter 4) that tackles this problem.

A multiple alignment M can be converted to a labeled DAG as follows. On each column j, create a vertex for each character c appearing in it. Let such a vertex be $v_{j,c}$. Add an arc from $v_{j',c'}$ to $v_{j,c}$ if there is an i such that $M[i,j'] = c'$, $M[i,j] = c$, and $M[i,j''] = -$ for all j'' such that $j' < j'' < j$. Add a global source by connecting it to all previous sources of the DAG. Likewise, add a global sink and arcs from all previous sinks to it. Example 6.20 on page 106 illustrates the construction (but not showing the added global source and global sink).

An interesting question is as follows: given two labeled DAGs as above, how can one align them? The alignment of DAGs could be defined in many ways, but we resort to the following definition that gives a feasible problem to be solved.

Problem 6.19   DAG-path alignment

Given two labeled DAGs, $A = (V^A, E^A)$ and $B = (V^B, E^B)$, both with a unique source and sink, find a source-to-sink path P^A in A and a source-to-sink path P^B in B, such that the global alignment score $S(\ell(P^A), \ell(P^B))$ is maximum over all such pairs of paths, where $\ell(p)$ denotes the concatenation of the labels of the vertices of P.

Surprisingly, this problem is easy to solve by extending the global alignment dynamic programming formulation. Consider the recurrence

$$s_{i,j} = \max \left\{ \max_{i' \in N_A^-(i), j' \in N_B^-(j)} s_{i',j'} + s(\ell^A(i), \ell^B(j)), \ \max_{i' \in N_A^-(i)} s_{i',j} - \delta, \ \max_{j' \in N_B^-(j)} s_{i,j'} - \delta \right\},$$

(6.23)

where i and j denote the vertices of A and B with their topological ordering number, respectively, $N_A^-(\cdot)$ and $N_B^-(\cdot)$ are the functions giving the set of in-neighbors of a given vertex, and $\ell^A(\cdot)$ and $\ell_B(\cdot)$ are the functions giving the corresponding vertex labels. By initializing $s_{0,0} = 0$ and evaluating the values in a suitable evaluation order (for example, $i = 1$ to $|V^A|$ and $j = 1$ to $|V^B|$), one can see that $s_{|V^A|,|V^B|}$ evaluates to the solution of the DAG-path alignment of Problem 6.19: an optimal alignment can be extracted with a traceback.

THEOREM 6.23   *Given two labeled DAGs $A = (V^A, E^A)$ and $B = (V^B, E^B)$, the DAG-path alignment problem (Problem 6.19) can be solved in $O(|E^A||E^B|)$ time.*

Proof   The correctness can be seen by induction. Assume $s_{i',j'}$ gives the optimal alignment score between any two paths from the source of A to i' and from the source of B to j', and this value is computed correctly for all $i' \in N^-(i)$ and $j' \in N^-(j)$ for some fixed (i,j). Any alignment of a pair of paths to i and j ends by $\ell^A(i) \rightarrow \ell^B(j)$, by $\ell^A(i) \rightarrow -$, or by $- \rightarrow \ell^B(j)$. Equation (6.23) chooses among all the possible shorter alignments the one that gives the maximum score for $s_{i,j}$.

Each pair of arcs of A and B is involved in exactly one cross-product in Equation (6.23), which gives the claimed running time.   □

The idea of using the DAG alignment in progressive multiple alignment is to convert the pair-wise alignments at the bottom level to DAGs, align the DAGs at each internal node bottom-up so that one always extracts the optimal paths into a pair-wise alignment, and converts it back to a DAG. One can modify the substitution scores to take into account how many bottom-level sequences are represented by the corresponding vertex in the DAG. This ad-hoc process is illustrated in Example 6.20.

Example 6.20   Construct a multiple alignment of the sequences $A = $ AAGACAC, $B = $ AGAACC, $C = $ AGACAC, and $D = $ AGAGCAC using a DAG-based progressive alignment.

Solution
Assume the same guide tree as in Example 6.18. That pair-wise alignment of A and B can be converted to the following DAG:

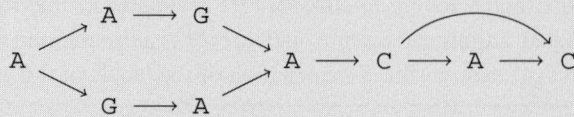

Likewise, that pair-wise alignment of C and D can be converted to the following DAG:

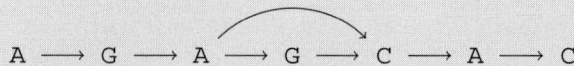

For the root, we find the pair of paths from the two DAGs that have the best alignment. These paths form the following alignment:

```
A A G A - C A C
A - G A G C A C
```

We project the final aligment to the four input sequences using the columns of the root profile as anchors, and partially re-align the areas between consecutive projected anchors. This can be done recursively using the same approach. We obtain the following multiple alignment, where ⋆ indicates the anchors projected to all four sequences and ⁒ indicates the anchors projected to some of the sequences:

```
⋆ ⁒ ⁒   ⋆   ⋆ ⁒ ⋆
A A G - A - C A C
A - G A A - C - C
A - G - A - C A C
A - G - A G C A C
```

The re-alignment between the two first ⋆ anchors was made between the corresponding part of the root DAG and the corresponding part of all sequences causing the partial anchors ⁒. In this case, sequence B alone caused the partial anchors ⁒. That is, we aligned GA with the DAG:

$$A \longrightarrow G$$

Similarly, sequence B causes the last partial anchor ⁒, and there the only possible alignment is adding a gap in B.

6.6.6   Jumping alignment

Consider a database of multiple alignments. For example, protein sequences are often clustered into *protein families* sharing some common functionality. Each such family can be represented as a multiple alignment. With a new protein whose function is unclear, one can try to align the new sequence to all families in the database, and link that sequence to the family with which it best aligns.

Jumping alignment can be used to model this alignment problem. The idea is to limit the amount of times the sequence to be aligned can change row inside the multiple alignment (profile). This can be accomplished for example by assigning a large penalty for changing row, or by fixing some upper bound on the number of allowed jumps.

Let us consider the case of a limited number of jumps. Let $M[1..d, 1..n]$ denote the profile and $A = a_1 \cdots a_m$ the sequence to be aligned to M. Let $s^l_{i,j,k}$ denote the score of the best global alignment of $a_1 \cdots a_i$ against $M[k_0, 1..j_0]M[k_1, j_0 + 1..j_1] \cdots M[k_l, j_l + 1..j]$ with $k_l = k$. The recurrence for computing the values $s^l_{i,j,k}$ is

$$s^l_{i,j,k} = \max_{k'} (\, s^l_{i-1,j-1,k} + s(a_i, M[k,j]), s^l_{i,j-1,k} - \delta, s^l_{i-1,j,k} - \delta,$$

$$s^{l-1}_{i-1,j-1,k'} + s(a_i, M[k',j]), s^{l-1}_{i-1,j,k'} - \delta, s^{l-1}_{i,j-1,k'} - \delta).$$

With a proper initialization and a suitable evaluation order, values $s^l_{i,j,k}$ can be computed in time $O(mndL)$, where L is the limit for the number of jumps.

This formulation has the shortcoming that existing gaps in M are aligned to A. Exercise 6.27 asks for a better formulation, whose idea is presented in Example 6.21.

Example 6.21 Find an optimal alignment of $A =$ AAGAAGCC to the profile

$$M = \begin{matrix} A & A & G & - & A & - & C & A & C \\ A & - & G & A & A & - & C & - & C \\ A & - & G & - & A & - & C & A & C \\ A & - & G & - & A & G & C & A & C \end{matrix}$$

with at most $L = 3$ jumps.

Solution

Using the better formulation asked in Exercise 6.27, there is a solution with exact match, visualized below with lower-case letters. In this formulation, the gaps in M can be bypassed for free:

$$\begin{matrix} a & a & g & - & A & - & C & A & C \\ A & - & G & a & a & - & c & - & c \\ A & - & G & - & A & - & C & A & C \\ A & - & G & - & A & g & C & A & C \end{matrix}$$

6.7   Literature

Edit distance computation (Levenshtein 1966; Wagner & Fischer 1974) is a widely studied stringology problem that has been explored thoroughly in many textbooks. We covered here one general speedup technique, namely the shortest detour (Ukkonen 1985), and one more specific technique, namely the bitparallel algorithm tailored for the Levenshtein edit distance and approximate string matching (Exercise 6.4) (Myers 1999). For the latter, we followed essentially the version described by Hyyrö (2001). Interestingly, two other kinds of technique also lead to sub-quadratic running times (Masek & Paterson 1980; Crochemore *et al.* 2002): we develop some ideas related to the former in Exercise 6.10. Space-efficient traceback (see the exercises) is from Hirschberg (1975). The approximate string matching algorithms under Levenshtein and Hamming distances can be sped up to $O(kn)$ time (Landau & Vishkin 1988) using the index structures we study in Chapter 8. Many practical techniques for obtaining efficient exact and approximate string matching algorithms are covered in Navarro & Raffinot (2002).

The first sparse dynamic programming algorithm for edit distances is the Hunt–Szymanski algorithm for the LCS problem (Hunt & Szymanski 1977). We explained another algorithm in order to demonstrate the invariant technique and the connection to range search data structures (Mäkinen *et al.* 2003). Observe that, using a more efficient range search structure for semi-infinite ranges as in Exercise 3.6, the $O(|M|\log m)$ time algorithm can be improved to $O(|M|\log \log m)$.

There also exist sparse dynamic programming algorithms for other variants of edit distances (Eppstein *et al.* 1992a,b).

The global and local alignment algorithms are from Needleman & Wunsch (1970) and Smith & Waterman (1981), respectively. For affine gap scores, the first algorithm is from Gotoh (1982) and the second one (invariant technique, range maximum) was developed for this book. The distribution on the number of introns is taken from Sakharkar *et al.* (2004). The derivation of substitution scores in Insight 6.1 follows the presentation of Durbin *et al.* (1998).

The small parsimony problem we studied in the context of multiple alignment scores is from Sankoff (1975). The NP-hardness of the longest common subsequence problem on multiple sequences was first examined in Maier (1978); that reduction (from vertex cover) is stronger and more complex than the one we described here (from an independent set), insofar as it shows that the problem is NP-hard even on binary sequences. The basic version (on a large alphabet) of that reduction is, however, not much more difficult than ours (see Exercise 6.24).

The literature on multiple alignments is vast; we refer the reader to another textbook (Durbin *et al.* 1998) for further references. The idea of using DAGs to replace profiles appears in Lee *et al.* (2002) and in Löytynoja *et al.* (2012); we loosely followed the latter.

Jumping alignments are from Spang *et al.* (2002); we followed the same idea here, but gave a quite different solution.

Exercises

6.1 Show that the edit distance is a *metric*, namely that it satisfies

i. $D(A, B) \geq 0$,

ii. $D(A, B) = 0$, if and only if $A = B$,

iii. $D(A, B) = D(B, A)$,

iv. $D(A, B) \leq D(A, C) + D(C, B)$.

6.2 Modify Algorithm 6.1 on page 74 to use only one vector of length $m + 1$ in the computation.

6.3 The shortest-detour algorithm is able to compute the dynamic programming matrix only partially. However, just allocating space for the rectangular matrix of size $m \times n$ requires $O(mn)$ time. Explain how one can allocate just the cells that are to be computed. Can you do the evaluation in $O(d)$ space, where $d = D(A, B)$?

6.4 Modify Myers' bitparallel algorithm to solve the k-errors problem on short patterns.

6.5 Optimize Myers' bitparallel algorithm to use fewer basic operations.

6.6 Write a program to automatically verify each row in the truth tables in the lemmas proving the correctness of Myers' bitparallel algorithm.

6.7   Write a program to automatically discover the four "easy" operations (those of four the first lemmas) in Myers' bitparallel algorithm.

6.8   Describe how to trace back an optimal alignment with Myers' bitparallel algorithm.

6.9   Consider Exercise 6.16. Show that the same technique can be combined with Myers' bitparallel algorithm to obtain linear space traceback together with the bitparallel speedup. *Hint.* Compute the algorithm also for the reverse of the strings. Then you can deduce j_{mid} from forward and reverse computations. See Section 7.4 for an analogous computation.

6.10   The four Russians technique we studied in Section 3.2 leads to an alternative way to speed up edit distance computation. The idea is to encode each column as in Myers' bitparallel algorithm with two bitvectors R_j^+ and R_j^- and partition these bitvectors to blocks of length b. Each such pair of bitvector blocks representing an *output encoding* of a block $d_{i-b+1..i,j}$ in the original dynamic programming matrix is uniquely defined by the *input parameters* $A_{i-b+1..i}$, b_j, $d_{i-b,j-1}$, and $d_{i-b,j}$ and the bitvector blocks representing the block $d_{i-b+1..i,j-1}$. Develop the details and choose b so that one can afford to precompute a table storing, for all possible input parameters, the output encoding of a block. Show that this leads to an $O(mn/\log_\sigma n)$ time algorithm for computing the Levenshtein edit distance.

6.11   Prove Theorem 6.10.

6.12   Show that the set $M = M(A, B) = \{(i,j) \mid a_i = b_j\}$ which is needed for sparse dynamic programming LCS computation, sorted in reverse column-order, can be constructed in $O(\sigma + |A| + |B| + |M|)$ time on a constant alphabet and in $O((|A| + |B|)\log|A| + |M|)$ time on an ordered alphabet. Observe also that this construction can be run in parallel with Algorithm 6.3 to improve the space requirement of Theorem 6.12 to $O(m)$.

6.13   The sparse dynamic algorithm for distance $D_{\text{id}}(A, B)$ can be simplified significantly if derived directly for computing $|\text{LCS}(A, B)|$. Derive this algorithm. *Hint.* The search tree can be modified to answer range maximum queries instead of range minimum queries (see Section 6.4.5).

6.14   Give a pseudocode for local alignment using space $O(m)$.

6.15   Give a pseudocode for tracing an optimal path for the maximum-scoring local alignment, using space quadratic in the alignment length.

6.16   Develop an algorithm for tracing an optimal path for the maximum-scoring local alignment, using space linear in the alignment length. *Hint.* Let $[i'..i] \times [j'..j]$ define the rectangle containing a local alignment. Assume you know j_{mid} for row $(i - i')/2$ where the optimal alignment goes through. Then you can independently recursively consider rectangles defined by $[i'..(i - i')/2] \times [j'..j_{\text{mid}}]$ and $[(i - i')/2] \times [j_{\text{mid}}..j]$. To find j_{mid} you may, for example, store, along with the optimum value, the index j where that path crosses the middle row.

6.17 Prove the correctness of Algorithms 6.5 and 6.6 on pages 98 and 100.

6.18 The sequencing technology SOLiD^{TM} from Applied Biosystems produces short reads of DNA in *color-space* with a two-base encoding defined by the following matrix (row = first base, column = second base):

```
  A C G T
A 0 1 2 3
C 1 0 3 2
G 2 3 0 1
T 3 2 1 0
```

For example, T012023211202102 equals TTGAAGCTGTCCTGGA (the first base is always given). Modify the overlap alignment algorithm to work properly in the case where one of the sequences is a SOLiD read and the other is a normal sequence.

6.19 Show that the relative entropy used in Equation (6.15) on page 91 is non-negative.

6.20 Give a pseudocode for prokaryote gene alignment.

6.21 Give a pseudocode for eukaryote gene alignment with affine gaps.

6.22 Give a pseudocode for the $O(mn \cdot \texttt{maxin})$ time algorithm for eukaryote gene alignment with limited introns.

6.23 Modify the recurrence for gene alignment with limited introns so that the alignment must start with a start codon, end with a stop codon, and contain GT and AG dinucleotides at intron boundaries. Can you still solve the problem in $O(mn \cdot \texttt{maxin})$ time? Since there are rare exceptions when dinucleotides at intron boundaries are something else, how can you make the requirement softer?

6.24 Give an alternative proof for the NP-hardness of the longest common subsequence problem on multiple sequences, by using a reduction from the vertex cover problem defined on page 16. *Hint.* From a graph $G = (V = \{1, \dots, |V|\}, E)$ and integer k, construct $|E| + 1$ sequences having LCS of length $|V| - k$ if and only if there is vertex cover of size k in G. Recall that vertex cover V' is a subset of V such that all edges in E are incident to at least one vertex in V.

6.25 Develop a sparse dynamic programming solution for computing the longest common subsequence of multiple sequences. What is the expected size of the match set on random sequences from an alphabet of size σ? *Hint.* The search tree can be extended to higher-dimensional range queries using recursion (see the range trees from the computational geometry literature).

6.26 Reformulate the DAG-path alignment problem (Problem 6.19 on page 105) as a local alignment problem. Show that the Smith–Waterman approach extends to this, but is there a polynomial algorithm if one wishes to report alignments with negative score? Can the affine gap cost alignment be extended to the DAG-path alignment? Can one obtain a polynomial-time algorithm for that case?

6.27 Show how to use DAG-path alignment to align a protein sequence to DNA. *Hint.* Represent the protein sequence as a *codon-DAG*, replacing each amino acid by a sub-DAG representing its codons.

6.28 Consider a DAG with predicted exons as vertices and arcs formed by pairs of exons predicted to appear inside a transcript: this is the *splicing graph* considered in Chapter 15. Detail how DAG-path alignment on the splicing graph and the codon-DAG of the previous assignment can be used to solve the eukaryote gene alignment problem.

6.29 Reformulate jumping alignment so that A is just aligned to the sequences forming the multiple alignment and not to the existing gaps inside the multiple alignment, as in Example 6.21. Show that the reformulated problem can be solved as fast as the original one.

7   Hidden Markov models (HMMs)

In the previous chapter we discussed how to analyze a sequence with respect to one or more other sequences, through various alignment techniques. In this chapter we introduce a different model, one that can abstract our biological insight about the available sequences, and can provide findings about a new sequence S. For example, given the simple assumption that in prokaryote genomes the coding regions have more C and G nucleotides than A and T nucleotides, we would like to find a *segmentation* of S into coding and non-coding regions. We will develop this toy problem in Example 7.1.

The model introduced by this chapter is called a *hidden Markov model (HMM)*. Even though this is a probabilistic one, which thus copes with the often empirical nature of our biological understanding, we should stress that the focus of our presentation is mainly algorithmic. We will illustrate how simple techniques such as dynamic programming, and in particular the Bellman–Ford method from Section 4.2.2, can provide elegant solutions in this probabilistic context. The exercises given at the end of this chapter derive some other probabilistic models that connect back to the problem of aligning biological sequences from the previous chapter.

Let us start by recalling that a *Markov chain* is a random process generating a sequence $S = s_1 s_2 \cdots s_n$ such that the probability of *generating*, or *emitting*, each s_i is fixed beforehand and depends only on the previously generated symbols $s_1 \cdots s_{i-1}$; we will denote this probability by $\mathbb{P}(s_i \mid s_1 \cdots s_{i-1})$. An important subclass of Markov chains consists in those of *finite history*, namely those in which the probability of generating any symbol s_i depends only on the last k previous symbols, for some constant k; that is, $\mathbb{P}(s_i \mid s_1 \cdots s_{i-1}) = \mathbb{P}(s_i \mid s_{i-k} \cdots s_{i-1})$ (k is called the *order* of the Markov chain).

Consider now for concreteness a Markov chain model of order 1, that is, $\mathbb{P}(s_i \mid s_1 \cdots s_{i-1}) = \mathbb{P}(s_i \mid s_{i-1})$. In such a Markov chain, the probability of emitting s_i depends only on the previous symbol s_{i-1}. In an HMM, however, the model can have different states, in which the probability of emitting s_i depends instead *only* on the current state, which is unknown, or *hidden*, from an observer. Moreover, at each step, the model can also probabilistically choose to transition to another hidden state. In other words, if we observe S, we cannot specify the state of the HMM in which each s_i was generated. Assuming that we know the HMM which generated S, that is, we know the possible transitions and their probabilities between the hidden states, and the emission probabilities in each state, we can instead try to find the *most probable* sequence of states of the HMM which generated S. We will solve this problem in Section 7.2, with the Viterbi algorithm, which is in close kinship with the Bellman–Ford algorithm for finding a shortest path in a graph. Another interesting question is that of how to compute the probability that the HMM generated S, which we solve in Section 7.3 by adapting the Viterbi algorithm into the forward and backward algorithms. Finally, in Section 7.4 we discuss how we get to know the parameters of an HMM in the first place, by showing a technique for learning the transition and emission probabilities from data.

7.1   Definition and basic problems

Before giving the formal definition of an HMM, let us give a concrete example for the segmentation problem mentioned above.

Example 7.1   The segmentation problem

In prokaryote genomes, the coding and non-coding regions have some easily recognizable statistical properties on their nucleotide content. The simplest statistical property to consider is the *GC content*, since coding regions typically have more G and C nucleotides than A and T. The segmentation problem is to partition a genome into coding and non-coding regions.

We can model this problem with an HMM that has two hidden states: the coding state, in which it emits G and C nucleotides with a higher probability than it emits A and T, and the non-coding state, in which it emits any one of all four nucleotides with the same probability. The system can remain in each of these two states with a large probability, but also transition between them with a smaller one. We assume that HMMs always have a single *start* state, the initial state of the model, in which no symbol is emitted. Analogously, there is a unique *end* state, the final state of the model. We draw such an HMM below:

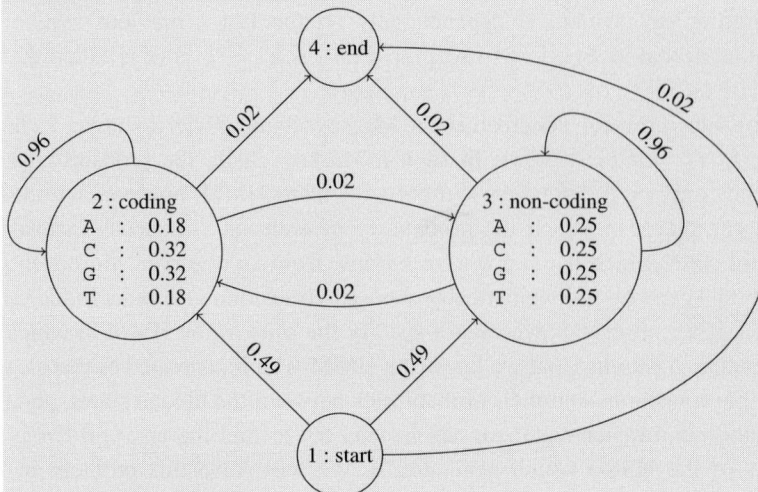

In this HMM, the probability of transitioning between the coding and non-coding states is 0.02, and the probability of remaining in the same coding or non-coding state is 0.96. In the coding state, the probability of emitting a C or a G is 0.32, and the probability of emitting an A or a T is 0.18. In the non-coding state, all nucleotides are emitted with the same probability.

We should underline that these probabilities have to be derived in the first place from a set of training data; we present in Section 7.4 a technique for obtaining these probabilities.

For a simple example emission sequence, suppose that from the *start* state the HMM chooses, with probability 0.49, to transition to the non-coding state. There, it emits an A, and chooses, with probability 0.96, to remain in the same state. It then emits a T, and chooses with probability 0.02 to transition to the coding state. Once there, it emits a G, and onwards, twice it chooses to remain in the same state, from where it emits C and then G. Finally, it transitions back to the non-coding state, from where it emits A, and then it transitions to the *end* state, with probability 0.02. What we observe in the end is just the sequence of nucleotides $S = $ ATGCGA, but we do not know the sequence of hidden states from which each nucleotide was generated. If we manage to find the *most probable* sequence of coding and non-coding states from which S is more likely to have been generated, then we can find a segmentation of S and solve the gene prediction problem.

Observe that an HMM can also be seen as an arc-labeled directed graph, with emission probabilities associated with its vertices, thus we will also refer to a sequence of states as a *path* in the HMM.

Formally, a *hidden Markov model (HMM)* is a tuple $(H, \Sigma, T, E, \mathbb{P})$, where $H = \{1, \ldots, |H|\}$ is the set of hidden states, Σ is the set of symbols, $T \subseteq H \times H$ is the set of transitions, $E \subseteq H \times \Sigma$ is the set of emissions, and $\mathbb{P}$ is the probability function for elements of T and E, satisfying the following conditions.

- There is a single *start state* $h_{\text{start}} \in H$ with no transitions $(h, h_{\text{start}}) \in T$, and no emissions (for this reason, h_{start} is also called a *silent state*).
- There is a single *end state* $h_{\text{end}} \in H$ with no transitions $(h_{\text{end}}, h) \in T$, and no emissions (also h_{end} is a *silent state*).
- Let $\mathbb{P}(h \mid h')$ denote the probability for the transition $(h', h) \in T$, and let $\mathbb{P}(c \mid h)$ denote the probability of an emission $(h, c) \in E$, for $h', h \in H$ and $c \in \Sigma$. It must hold that

$$\sum_{h \in H} \mathbb{P}(h \mid h') = 1, \text{for all } h' \in H \setminus \{h_{\text{end}}\},$$

and

$$\sum_{c \in \Sigma} \mathbb{P}(c \mid h) = 1, \text{for all } h \in H \setminus \{h_{\text{start}}, h_{\text{end}}\}.$$

Observe that we denote the probability of a transition *from h' to h* by $\mathbb{P}(h \mid h')$, rather than with a notation like $p(h', h)$ (observe the reverse order of the symbols h' and h), contrary to what the reader might have been accustomed to after Chapter 4.

A *path through an HMM* is a sequence P of hidden states $P = p_0 p_1 p_2 \cdots p_n p_{n+1}$, where $(p_i, p_{i+1}) \in T$, for each $i \in \{0, \ldots, n\}$. The *joint probability* of P and a sequence $S = s_1 s_2 \cdots s_n$, with each $s_i \in \Sigma$, is

$$\mathbb{P}(P, S) = \prod_{i=0}^{n} \mathbb{P}(p_{i+1} \mid p_i) \prod_{i=1}^{n} \mathbb{P}(s_i \mid p_i). \tag{7.1}$$

We will be mainly interested in the set $\mathcal{P}(n)$ of all paths $p_0 p_1 \cdots p_{n+1}$ through the HMM, of length $n + 2$, such that $p_0 = h_{\texttt{start}}$, and $p_{n+1} = h_{\texttt{end}}$.

> **Example 7.2** In the HMM in Example 7.1, we have $\Sigma = \{\texttt{A, C, G, T}\}$ and $H = \{1, 2, 3, 4\}$, where we have used the names 1, start; 2, coding; 3, non-coding; and 4, end. We took $h_{\texttt{start}} = $ start and $h_{\texttt{end}} = $ end. The sets T and E, and their corresponding probabilities are shown in the tables below.
>
> | h' | h | $\mathbb{P}(h \mid h')$ |
> |------|-----|------------------------|
> | start | coding | 0.49 |
> | start | non-coding | 0.49 |
> | start | end | 0.02 |
> | coding | coding | 0.96 |
> | coding | non-coding | 0.02 |
> | coding | end | 0.02 |
> | non-coding | non-coding | 0.96 |
> | non-coding | coding | 0.02 |
> | non-coding | end | 0.02 |
>
> (a) The set T of transitions, and their corresponding probabilities
>
> | h | c | $\mathbb{P}(c \mid h)$ |
> |-----|-----|------------------------|
> | coding | A | 0.18 |
> | coding | C | 0.32 |
> | coding | G | 0.32 |
> | coding | T | 0.18 |
> | non-coding | A | 0.25 |
> | non-coding | C | 0.25 |
> | non-coding | G | 0.25 |
> | non-coding | T | 0.25 |
>
> (b) The set E of emissions, and their corresponding probabilities
>
> Given the path $P \in \mathcal{P}(6)$ and the string S,
>
> $$\begin{array}{ccccccccc} P = & 1 & 3 & 3 & 2 & 2 & 2 & 3 & 4 \\ S = & & \texttt{A} & \texttt{T} & \texttt{G} & \texttt{C} & \texttt{G} & \texttt{A} & \end{array}$$
>
> the joint probability of P and S is
>
> $$\begin{aligned} \mathbb{P}(P, S) = \; & (0.49 \cdot 0.96 \cdot 0.02 \cdot 0.96 \cdot 0.96 \cdot 0.02 \cdot 0.02) \cdot \\ & (0.25 \cdot 0.25 \cdot 0.32 \cdot 0.32 \cdot 0.32 \cdot 0.25). \end{aligned}$$

In the next two sections we solve the following two basic problems on HMMs, whose variations find many applications in biological sequence analysis. In the first one, Problem 7.3 below, we are given a sequence S, and we want to find the most probable sequence of states of the HMM that generated S.

Problem 7.3   Most probable path of a sequence in an HMM

Given an HMM M over an alphabet Σ, and a sequence $S = s_1 s_2 \cdots s_n$, with each $s_i \in \Sigma$, find the path P^* in M having the highest probability of generating S, namely

$$P^* = \arg\max_{P \in \mathcal{P}(n)} \mathbb{P}(P, S) = \arg\max_{P \in \mathcal{P}(n)} \prod_{i=0}^{n} \mathbb{P}(p_{i+1} \mid p_i) \prod_{i=1}^{n} \mathbb{P}(s_i \mid p_i). \qquad (7.2)$$

In the second one, Problem 7.4 below, we are given an HMM and we need only to compute the probability of it generating S.

Problem 7.4   Generation probability of a sequence by an HMM

Given an HMM M over an alphabet Σ, and a sequence $S = s_1 s_2 \cdots s_n$, with each $s_i \in \Sigma$, compute the probability

$$\mathbb{P}(S) = \sum_{P \in \mathcal{P}(n)} \mathbb{P}(P, S) = \sum_{P \in \mathcal{P}(n)} \prod_{i=0}^{n} \mathbb{P}(p_{i+1} \mid p_i) \prod_{i=1}^{n} \mathbb{P}(s_i \mid p_i). \qquad (7.3)$$

Their solutions are based on dynamic programming. The first one can be solved using the *Viterbi* algorithm. The second one is solved using the so-called *forward* algorithm, obtained from the Viterbi algorithm just by replacing the maximum operation with summation. For some variants of these problems (to be covered later), also the reverse of the forward algorithm, called the *backward* algorithm, is useful.

For ease of notation in these algorithms, we also give the following two analogs of $\mathbb{P}(P, S)$, in which we ignore the probabilities of the first and last, respectively, transitions of a path through the HMM. For a path $P = p_0 p_1 p_2 \cdots p_n$ through the HMM, we define

$$\mathbb{P}_{\texttt{prefix}}(P, S) = \prod_{i=0}^{n-1} \mathbb{P}(p_{i+1} \mid p_i) \prod_{i=1}^{n} \mathbb{P}(s_i \mid p_i).$$

Given a path $P = p_1 p_2 \cdots p_n p_{n+1}$ through the HMM, we define

$$\mathbb{P}_{\texttt{suffix}}(P, S) = \prod_{i=1}^{n} \mathbb{P}(p_{i+1} \mid p_i) \prod_{i=1}^{n} \mathbb{P}(s_i \mid p_i).$$

Finally, we must draw the attention of the reader to the numerical errors arising from the repeated multiplications and additions needed in implementing these algorithms. We touch on this issue in Exercises 7.5 and 7.6. However, more advanced techniques for addressing this problem are beyond the scope of this book, and we assume here that all arithmetic operations are exact, and take $O(1)$ time.

7.2   The Viterbi algorithm

The Viterbi algorithm solves Problem 7.3 and is very similar to the Bellman–Ford algorithm, which was studied in Section 4.2.2. Recall that in the Bellman–Ford algorithm we dealt with the possible cycles of the graph by adding another parameter to the recurrence, namely the number of arcs in a solution path. The additional recursion parameter exploited by the Viterbi algorithm for "breaking" the possible cycles on the HMM is analogous, namely the length of the prefix of the sequence S generated so far.

Accordingly, for every $i \in \{1, \ldots, n\}$ and every $h \in \{1, \ldots, |H|\}$, define

$$v(i, h) = \max \left\{ \mathbb{P}_{\mathrm{prefix}}(P, s_1 \cdots s_i) \mid P = h_{\mathrm{start}} p_1 \cdots p_{i-1} h \right\}$$

as the largest probability of a path starting in state h_{start} and ending in state h, given that the HMM generated the prefix $s_1 \cdots s_i$ of S (symbol s_i being emitted by state h). Notice the similarity with the dynamic programming table (4.2) in the Bellman–Ford algorithm.

We can easily derive the following recurrence relation for $v(i, h)$:

$$v(i, h) = \max\{\mathbb{P}_{\mathrm{prefix}}(h_{\mathrm{start}} p_1 \cdots p_{i-1} h', s_1 \cdots s_{i-1}) \mathbb{P}(h \mid h') \mathbb{P}(s_i \mid h) \mid (h', h) \in T\}$$
$$= \mathbb{P}(s_i \mid h) \max\{v(i - 1, h') \mathbb{P}(h \mid h') \mid (h', h) \in T\}, \tag{7.4}$$

where we take by convention $v(0, h_{\mathrm{start}}) = 1$ and $v(0, h) = 0$ for all $h \neq h_{\mathrm{start}}$. Indeed, $v(i, h)$ equals the largest probability of getting to a predecessor h' of h, having generated the prefix sequence $s_1 \cdots s_{i-1}$, multiplied by the probability of the transition (h', h) and by $\mathbb{P}(s_i \mid h)$. Observe that $\mathbb{P}(s_i \mid h)$ depends only on h and s_i, and not on the predecessors of h; this is the reason why we have written $\mathbb{P}(s_i \mid h)$ outside of the max notation in (7.4).

The largest probability of a path for the entire string S (that is, the value maximizing (7.2)) is the largest probability of getting to a predecessor h' of h_{end}, having generated the entire sequence $S = s_1 \cdots s_n$ (symbol s_n being emitted by state h'), multiplied by the probability of the final transition (h', h_{end}). Expressed in terms of v,

$$\max_{P \in \mathcal{P}(n)} \mathbb{P}(P, S) = \max \left\{ v(n, h') \mathbb{P}(h_{\mathrm{end}} \mid h') \mid (h', h_{\mathrm{end}}) \in T \right\}. \tag{7.5}$$

The values $v(\cdot, \cdot)$ can be computed by filling a table $V[0..n, 1..|H|]$ row-by-row in $O(n|T|)$ time. The most probable path can be traced back in the standard dynamic programming manner, by checking which predecessor h' of h_{end} maximizes (7.5) and then, iteratively, which predecessor h' of the current state h maximizes (7.4). Figure 7.1 illustrates the dynamic programming recurrence.

7.3   The forward and backward algorithms

The *forward* algorithm solves Problem 7.4 and is very similar to the Viterbi algorithm. The only difference is the fact that the maximum operation is replaced by

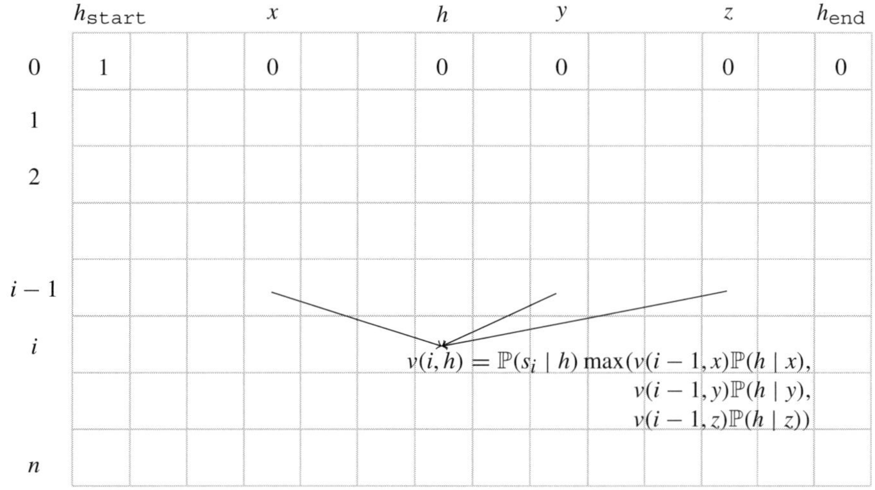

	h_{start}	x	h	y	z	h_{end}
0	1		0		0	0	0
1						
2						
$i-1$						
i						
n						

$$v(i,h) = \mathbb{P}(s_i \mid h) \max(v(i-1,x)\mathbb{P}(h \mid x),$$
$$v(i-1,y)\mathbb{P}(h \mid y),$$
$$v(i-1,z)\mathbb{P}(h \mid z))$$

Figure 7.1 The idea behind the Viterbi algorithm, assuming that the predecessors of state h are the states x, y, and z. Notice the similarity between this row-by-row computation of $V_{0..n,1..|H|}$ and the interpretation of the Bellman–Ford algorithm as finding the shortest path in a DAG made up of layered copies of the input graph, illustrated in Figure 4.5.

summation, since we now have to compute $\mathbb{P}(S) = \sum_{P \in \mathcal{P}(n)} \mathbb{P}(P,S)$, instead of $\arg\max_{P \in \mathcal{P}(n)} \mathbb{P}(P,S)$.

More precisely, for every $i \in \{1, \ldots, n\}$ and every $h \in \{1, \ldots, |H|\}$, we define

$$f(i,h) = \sum_{P = h_{\text{start}} p_1 \cdots p_{i-1} h} \mathbb{P}_{\texttt{prefix}}(P, s_1 \cdots s_i)$$

as the probability that the HMM, having started in state h_{start}, is generating symbol s_i of S while in state h. We compute each $f(i,h)$ as

$$f(i,h) = \mathbb{P}(s_i \mid h) \sum_{(h',h) \in T} f(i-1,h')\mathbb{P}(h \mid h'), \qquad (7.6)$$

where we take by convention $f(0, h_{\text{start}}) = 1$ and $f(0,h) = 0$ for all $h \neq h_{\text{start}}$. As before, the values $f(\cdot, \cdot)$ can be computed by filling in a table $F[0..n, 1..|H|]$ row-by-row in total time $O(n|T|)$. The solution of Problem 7.4 is obtained as

$$\mathbb{P}(S) = \sum_{(h', h_{\text{end}}) \in T} f(n,h')\mathbb{P}(h_{\text{end}} \mid h'). \qquad (7.7)$$

We can ask the converse question, what is the probability that the HMM generates symbol s_i of S in state h, and it finally goes to state h_{end} after generating the symbols $s_{i+1} \cdots s_n$ of S? Let us call this probability $b(i,h)$, where b stands for *backward*:

$$b(i,h) = \sum_{P=hp_{i+1}\cdots p_n h_{\mathrm{end}}} \mathbb{P}_{\mathtt{suffix}}(P, s_i s_{i+1} \cdots s_n).$$

Analogously, for every $i \in \{1, \ldots, n\}$ and every $h \in \{1, \ldots, |H|\}$, we can compute each $b(i,h)$ as

$$b(i,h) = \mathbb{P}(s_i \mid h) \sum_{(h,h')\in T} b(i+1, h')\mathbb{P}(h' \mid h), \qquad (7.8)$$

where we take by convention $b(n+1, h_{\mathrm{end}}) = 1$ and $b(n+1, h) = 0$ for all $h \neq h_{\mathrm{end}}$. The values $b(\cdot, \cdot)$ can also be computed by filling in a table $B[1..n+1, 1..|H|]$ row-by-row in time $O(n|T|)$. This can be done row-by-row, backwards from the nth row up to the first row. Observe that this *backward* algorithm also solves Problem 7.4, since it holds that

$$\mathbb{P}(S) = \sum_{(h_{\mathrm{start}}, h)\in T} b(1, h)\mathbb{P}(h \mid h_{\mathrm{start}}). \qquad (7.9)$$

Moreover, observe that the backward algorithm can be seen as identical to the forward algorithm applied on the reverse sequence, and on the HMM in which all transitions have been reversed, and h_{start} and h_{end} have exchanged meaning.

Having computed $f(i,h)$ and $b(i,h)$ for every $i \in \{1, \ldots, n\}$ and $h \in \{1, \ldots, |H|\}$, we can answer more elaborate questions; for example, what is the probability that the HMM emits the ith symbol of the given string S while being in a given state h? Using the tables f and b, this probability can be obtained as

$$\sum_{P=p_0 p_1 \cdots p_n p_{n+1}\in \mathcal{P}(n)\,:\,p_i=h} \mathbb{P}(P \mid S) = \sum_{(h',h)\in T} f(i-1, h')\mathbb{P}(h \mid h')b(i,h). \qquad (7.10)$$

In the next section we will see another example where both tables $F[0..n, 1..|H|]$ and $B[1..n+1, 1..|H|]$ are useful.

7.4   Estimating HMM parameters

Where do we get the transition and emission probabilities from in the first place? Let us come back to our example of segmenting a DNA sequence into coding and non-coding regions based on the GC content.

Assume first that we have *labeled training data*, that is, a collection of DNA sequences in which each position has been labeled as coding or non-coding. In this case, it suffices to count how many times each transition and emission is used in the training data. Let $\mathtt{TC}(h', h)$ denote the number of times transition $(h', h) \in T$ takes place (*transition count*), and let $\mathtt{EC}(h, c)$ denote the number of times emission $(h, c) \in E$ takes place (*emission count*) in the training data. Then, we can set

$$\mathbb{P}(h \mid h') = \frac{\mathtt{TC}(h', h)}{\sum_{h''\in H} \mathtt{TC}(h', h'')} \quad \text{and} \quad \mathbb{P}(c \mid h) = \frac{\mathtt{EC}(h, c)}{\sum_{c'\in \Sigma} \mathtt{EC}(h, c')}. \qquad (7.11)$$

Insight 7.1 HMM parameters and pseudocounts

Consider the HMM in Example 7.2, but with unknown transition and emission probabilities, and the following labeled sequence as training data:

labels   NNNNNNNNNNNCCCCCCCCCCCCCCNNNNNNNNNNNNNNCCCCCCCCCCC

S ATTACATCTAGCGCGAGCGCGCGTATATCTATTATCGATGCGCGCG

where the N label indicates a non-coding symbol of S, and the C label indicates a coding symbol of S. Using relations (7.11), we get for example $\mathbb{P}(\text{A} \mid \text{non-coding}) = 8/23$, and $\mathbb{P}(\text{coding} \mid \text{non-coding}) = 2/22$.

Moreover, we have $\mathbb{P}(\text{G} \mid \text{non-coding}) = 0/23 = 0$, which prevents *any* emission of G in the non-coding state. This situation is due to there being too few training samples, since we do expect to find symbol G in non-coding regions. To avoid such null probabilities, a common simple solution is to use so-called *pseudocounts*, by adding a constant b to every counter $\text{TC}(h', h)$ and $\text{EC}(h, c)$. For example, using $b = 1$, we obtain $\mathbb{P}(\text{coding} \mid \text{non-coding}) = (2 + 1)/(22 + 3)$, $\mathbb{P}(\text{A} \mid \text{non-coding}) = (8 + 1)/(23 + 4)$, and $\mathbb{P}(\text{G} \mid \text{non-coding}) = 1/(23 + 4)$.

When the sequences in the training dataset are not labeled (*unlabeled training data*), we cannot distinguish which paths the training sequences correspond to. One way to proceed is to use the following iterative approach, called *Viterbi training*. Assume there are some prior estimates for the transition and emission probabilities. Use the Viterbi algorithm to find the most likely paths for the training sequences, and then use these paths as new labeled training data to improve the prior estimates. Stop when the parameters of the HMM no longer change significantly. See Insight 7.1.

A less greedy approach, called *Baum–Welch training*, is to iteratively estimate each $\text{TC}(h', h)$ and $\text{EC}(h, c)$ by taking into account all the paths through the HMM. For instance, we want to estimate $\text{EC}(h, c)$ as the expected value of emitting a symbol $c \in \Sigma$ while in a state $h \in H$, over all paths in the HMM that emit the training string $S = s_1 \cdots s_n$.

Recall from the previous section that relation (7.10) allows us to compute, using the forward and backward algorithms, the probability

$$\sum_{P = p_0 p_1 \cdots p_n p_{n+1} \in \mathcal{P}(n) \,:\, p_i = h} \mathbb{P}(P \mid S)$$

that the ith symbol of S is generated by the HMM while in state h. Moreover, the sequence S itself has a probability $\mathbb{P}(S)$ of being generated by the HMM, where $\mathbb{P}(S)$ is the solution to Problem 7.4, computable with either the forward or the backward algorithm (recall relations (7.7) and (7.9)).

Therefore, given the sequence S, the expected value of being in a state $h \in H$, over all paths in the HMM that emit the training string $S = s_1 \cdots s_n$, is

$$\frac{\sum_{P = p_0 p_1 \cdots p_n p_{n+1} \in \mathcal{P}(n) \,:\, p_i = h} \mathbb{P}(P \mid S)}{\mathbb{P}(S)}. \tag{7.12}$$

Finally, we can obtain $\text{EC}(h, c)$ by summing the above probability (7.12) over positions $1 \le i \le n$ of S such that $s_i = c$:

$$
\begin{aligned}
\text{EC}(h, c) &= \sum_{i \,:\, s_i = c} \frac{\displaystyle\sum_{P = p_0 p_1 \cdots p_n p_{n+1} \in \mathcal{P}(n) \,:\, p_i = h} \mathbb{P}(P \mid S)}{\mathbb{P}(S)} \\
&= \sum_{i \,:\, s_i = c} \frac{\sum_{(h', h) \in T} f(i - 1, h') \mathbb{P}(h \mid h') b(i, h)}{\sum_{(h', h_{\text{end}}) \in T} f(n, h') \mathbb{P}(h_{\text{end}} \mid h')}.
\end{aligned}
\tag{7.13}
$$

One can derive an expected case formula for $\text{TC}(h', h)$ analogously.

7.5   Literature

The Viterbi algorithm (Viterbi 1967) and its multiple variations have been vital in the development of genome analysis. Gene discovery was used here as a toy example, but also in practice very similar, yet more tailored, HMM techniques are in use (Mathé *et al.* 2002). Another very popular application of HMMs is the representation of protein families with *profile HMMs* (Krogh *et al.* 1994; Eddy 2011): aligning a new sequence against all such profiles in a database gives a way to categorize the new sequence to the closest family. We will see in the exercises that the Viterbi recurrences are very easy to generalize to such profile HMMs. A more theoretical application of HMMs is to extend them for modeling pair-wise alignments. Such HMMs have match states emitting a symbol from two input sequences simultaneously, and symmetric insertion and deletion states emitting only from one input sequence. These *pair HMMs* (Durbin *et al.* 1998, Chapter 4) offer a fully probabilistic interpretation of alignments, and give a rigorous way to do probabilistic sampling of alignments and to define suboptimal alignments. We will elaborate this connection in one of the assignments below.

We covered only the fundamental algorithmic properties of HMMs. As was already noticeable in Section 7.4 when deriving the Baum–Welch formulas (Baum 1972) for the case of unlabeled training data, the derivations pertaining to HMMs are tightly connected to the probabilistic interpretation, and iterative machine-learning approaches are required for further development of the models. These methods are beyond the scope of this book, and we recommend the excellent textbook by Durbin *et al.* (1998) for an in-depth understanding; our treatment here is evidently highly influenced by this classic book. For a more mathematical introduction to HMMs, we also recommend the book on pattern discovery by Parida (2007).

Finally, we should mention that, if the sequence is compressible, then it is possible to achieve faster implementations of the Viterbi, forward-backward, and Baum–Welch algorithms (Lifshits *et al.* 2009).

Exercises

7.1 Argue that $O(|H|)$ working space suffices in Problem 7.3 to compute the final numerical value (not the actual solution path) with the Viterbi algorithm, and in Problem 7.4 with the forward and backward algorithms.

7.2 To trace back the solution of Viterbi algorithm, the naive solution requires $O(n|H|)$ working space. Show that this can be improved to $O(\sqrt{n}|H|)$ by sampling every $\sqrt{n}$th position in S, without asymptotically affecting the running time of the traceback.

7.3 The standard algorithm for the multiplication of two matrices of m rows by m columns takes time $O(m^3)$. More sophisticated approaches achieve time $O(m^\omega)$ with $\omega < 3$ (the current record is $\omega \leq 2.38$ and $\omega = 2$ is the best one can hope for). Suppose that we have to compute many instances of the forward and backward algorithms on different sequences, but on the same HMM. Show how to use fast matrix multiplication to improve the running time of such a computation.

7.4 Given an HMM and a sequence $S = s_n \cdots s_n$, derive an $O(n|H|)$ time algorithm to compute, for all $h \in H$,

$$\operatorname*{argmax}_i \{\mathbb{P}(P, s_1 \cdots s_n) \mid P = p_0 p_1 \cdots p_n p_{n+1} \in \mathcal{P}(n) \text{ and } p_i = h\}. \tag{7.14}$$

7.5 Multiplication is the source of numerical errors in HMM algorithms, when the numbers become too large to fit into a computer word. Show how the Viterbi algorithm can be implemented using a sum of logarithms to avoid these numerical problems.

7.6 For the forward and backward algorithms the sum of logarithms conversion is not enough for numerical stability. Browse the literature to find a solution for this problem.

7.7 The flexibility of choosing the states, transitions, emissions, and their probabilities makes HMMs a powerful modeling device. So far we have used a *zeroth-order Markov model* for emission probabilities (probabilities depended only on the state, not on the sequence context). We could just as well use *first-order Markov chains* or, more generally, *kth-order Markov chains*, in which the probability depends on the state and on the last k symbols preceding the current one: $\mathbb{P}(s_i \mid s_{i-k} \cdots s_{i-1}) = \mathbb{P}(s_i \mid s_1 \cdots s_{i-1})$.

Notice that the states of the HMM are independent, in the sense that each state can choose a Markov chain of a different order from that of the Markov chain for its emission probabilities. In addition to the use of different-order Markov chains, we could adjust how many symbols are emitted in each state. Use these considerations to design a realistic HMM for *eukaryote gene prediction*. Try to take into account intron/exon boundary dinucleotides, codon adaptation, and other known features of eukaryote genes. Consider also how you can train the HMM.

7.8 *Profile HMMs* are an extension of HMMs to the problem of aligning a sequence with an existing multiple alignment (profile). Consider for example a multiple alignment of a protein family:

```
AVLSLSKTTNNVSPA
AV-SLSK-TANVSPA
A-LSLSK-TANV-PA
A-LSSSK-TNNV-PA
AS-SSSK-TNNV-PA
AVLSLSKTTANV-PA
```

We considered the problem of aligning a sequence A against a profile in the context of progressive multiple alignment in Section 6.6.4, and the idea was to consider the multiple alignment as a sequence of columns and apply normal pair-wise alignment with proper extensions of substitution and indel scores. Consider $A = $ AVTLSLSTAANVSPA aligned to our example profile above, for example, as follows:

```
AVTLSLS--TAANVSPA
AV-LSLSKTTN-NVSPA
AV--SLSK-TA-NVSPA
A--LSLSK-TA-NV-PA
A--LSSSK-TN-NV-PA
AS--SSSK-TN-NV-PA
AV-LSLSKTTA-NV-PA
```

Here we have added two gaps to the sequence and two gap columns to the profile following the "once a gap, always a gap" principle.

Profile HMMs are created using *inhomogeneous* Markov chains, such that each of the columns will form separate match, insertion, and deletion states, and transitions go from left to right, as illustrated in Figure 7.2. Match and deletion states emit the columns of the profile, so they do not contain self-loops. Insertion states emit symbols from the input sequence, so they contain self-loops to allow any number of symbols emitted between states that emit also columns of the profile.

Since the resulting HMM is reading only one sequence, the Viterbi, forward, and backward algorithms are almost identical to the ones we have studied so far. The only

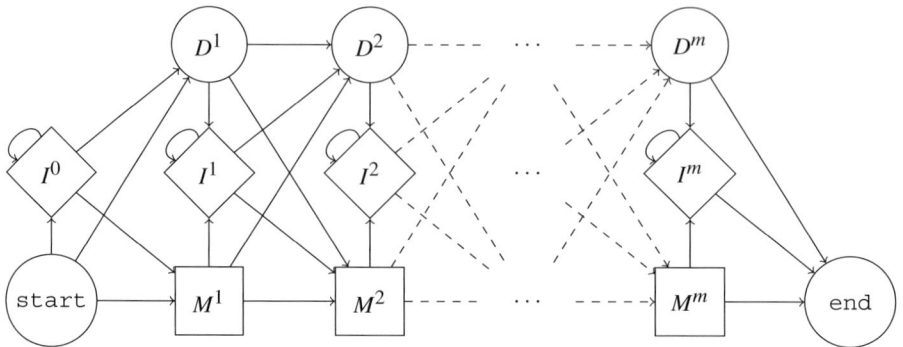

Figure 7.2 Profile HMM illustration without showing the transition and emission probabilities.

difference is that deletion states are *silent* with respect to the input string, since they do not emit any symbol.

(a)   Modify the Viterbi recurrences to handle both emitting and silent states.

(b)   Derive the Viterbi recurrences specific to profile HMMs.

7.9   Derive a local alignment version of profile HMMs.

7.10   *Pair HMMs* are a variant of HMMs emitting two sequences, such that a path through the HMM can be interpreted as an alignment of the input sequences. Such pair HMMs have a *match* state emitting a symbol from both sequences simultaneously, and symmetric *insertion* and *deletion* states to emit only from one input sequence.

(a)   Fix a definition for pair HMMs and derive the corresponding Viterbi, forward, and backward recurrences. *Hint.* The result should look very similar to Gotoh's algorithm for global alignment with affine gap costs from Section 6.4.4.

(b)   Apply a derivation for pair HMMs similar to the ones we used in obtaining relation (7.13), in order to define the probability of a_i aligning to b_j over all alignments of $A = a_1 \cdots a_m$ and $B = b_1 \cdots b_n$.

(c)   Let p_{ij} denote the probability derived above to align a_i to b_j. We say that the *most robust alignment* of A and B is the alignment maximizing the sum of values p_{ij} over i,j such that the $a_i \rightarrow b_j$ substitution is part of the alignment. Derive a dynamic programming algorithm to compute this most robust alignment.

Part III

Genome-Scale Index Structures

8 Classical indexes

A *full-text index* for a string $T = t_1 t_2 \cdots t_n$ is a data structure that is built once, and that is kept in memory for answering an arbitrarily large number of queries on the position and frequency of *substrings* of T, with a time complexity that depends sublinearly on n. Consider for example a set $\mathcal{R} = \{R^1, R^2, \ldots, R^d\}$ of reads generated by a high-throughput sequencing machine from a genome G: assuming that each R^i is an exact copy of a substring of G, a routine question in read analysis is finding the position of each R^i in G. The lens of full-text indexes can also be used to detect common features of a set of strings, like their *longest common substring*: such analyses arise frequently in evolutionary studies, where substrings that occur exactly in a set of biological sequences might point to a common ancestral relationship.

The problem of matching and counting *exact substrings* might look artificial at a first glance: for example, reads in $\mathcal{R}$ are not exact substrings of G in practice. In the forthcoming chapters we will present more realistic queries and analyses, which will nonetheless build upon the combinatorial properties and construction algorithms detailed here. Although this chapter focuses on the fundamental data structures and design techniques of classical indexes, the presentation will be complemented with a number of applications to immediately convey the flexibility and power of these data structures.

8.1 *k*-mer index

The oldest and most popular index in *information retrieval* for natural-language texts is the so called *inverted file*. The idea is to sort in lexicographic order all the distinct words in a text T, and to precompute the set of occurrences of each word in T. A query on any word can then be answered by a binary search over the sorted list. Since biological sequences have no clear delimitation in words, we might use all the distinct substrings of T of a fixed length k. Such substrings are called *k-mers*.

DEFINITION 8.1 *The k-mer index (also known as the q-*gram index*) of a text $T = t_1 t_2 \cdots t_n$ is a data structure that represents* occurrence lists $\mathcal{L}(W)$ *for each k-mer W such that $j \in \mathcal{L}(W)$ iff $T_{j..j+k-1} = W$.*

For example, the *k*-mer index of string $T = \text{AGAGCGAGAGCGCGC\#}$ for $k = 2$ is shown in Figure 8.1(a).

(a)

AG →	1	3	7	9
GA →	2	6	8	
GC →	4	10	12	14
C# →	15			
CG →	5	11	13	

(b)

1	16	#
2	1	AGAGCGAGAGCGCGC#
3	7	AGAGCGCGC#
4	3	AGCGAGAGCGCGC#
5	9	AGCGCGC#
6	15	C#
7	5	CGAGAGCGCGC#
8	13	CGC#
9	11	CGCGC#
10	6	GAGAGCGCGC#
11	2	GAGCGAGAGCGCGC#
12	8	GAGCGCGC#
13	14	GC#
14	4	GCGAGAGCGCGC#
15	12	GCGC#
16	10	GCGCGC#

(c)

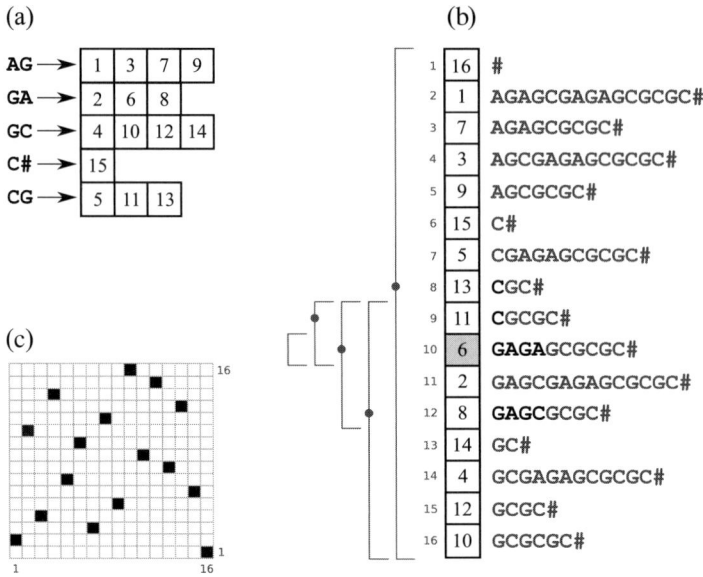

Figure 8.1 The 2-mer index (a) and the suffix array (b) of string $T = $ AGAGCGAGAGCGCGC#. Suffixes in (b) are shown just for clarity: they are not part of the suffix array. Intervals in panel (b) show the iterations of a binary search for the starting position of the interval of suffixes matching pattern $P = $ GAG. Note that P cannot be easily searched for in (a). The matrix in panel (c) visualizes the fact that SA_T is a permutation of $[1..|T|]$: cell (i, j) is black iff $\mathsf{SA}_T[i] = j$. Reading the matrix by row gives SA_T, and reading it by column gives the inverse suffix array of T.

Note that any substring P of length at most k can be searched for in the list of k-mers in $O(|P| \log n)$ time, and that its list of occurrences can be reconstructed by taking the union of the lists of all k-mers prefixed by P. However, it is not possible to do this efficiently for strings longer than k: we will waive this constraint in Section 8.2 using *suffix arrays*.

If each position is represented in $\log_2 n$ bits, the occurrence lists of all k-mers can take up exactly $n \log_2 n$ bits of space in total. Since the *difference* between consecutive numbers in the sorted occurrence lists is small, one could encode each difference using fewer than $\log_2 n$ bits. Concatenating such *variable-length* bit-fields into one long stream of bits makes decoding impossible, unless one encodes the lengths of the fields separately. Another option is to use *self-delimiting codes* that encode the length of each code internally. One well-known family of such codes is called *Elias codes*. Let $x^{\texttt{bin}}$ denote the binary representation of an integer x. The Elias *gamma code* of $x > 0$ is defined as

$$\gamma(x) = \underbrace{0 \cdots\cdots 0}_{|x^{\texttt{bin}}| - 1 \text{ times}} \quad x^{\texttt{bin}}$$

That is, the first run of zeros encodes the number of bits to represent $x^{\texttt{bin}}$; with $x > 0$ the binary code $x^{\texttt{bin}}$ starts always with 1 (if one needs to encode 0, one can use systematically $\gamma(x + 1)$). For example, $\gamma(5) = 00\,101$, since $5^{\texttt{bin}} = 101$. Note that a sequence of gamma-coded non-negative integers can be decoded in a unique way, and that $\gamma(x)$ takes exactly $2\lceil\log_2 x\rceil - 1$ bits. The idea of gamma codes can be applied recursively, to obtain the asymptotically smaller *delta codes* for $x > 0$:

$$\delta(x) = \underbrace{0 \cdots\cdots\cdots 0}_{|(|x^{\mathrm{bin}}|)^{\mathrm{bin}}|-1 \text{ times}} (|x^{\mathrm{bin}}|)^{\mathrm{bin}} x^{\mathrm{bin}}.$$

For example, $\delta(5) = 0^1 \, 3^{\mathrm{bin}} \, 5^{\mathrm{bin}} = 0\,11\,101$.

Assume now that the sorted occurrence list $\mathcal{L}(W)$ for k-mer W starts with 0 (which is not encoded explicitly), and define $\mathtt{diff}_W[i] = \mathcal{L}(W)[i] - \mathcal{L}(W)[i-1]$ for $1 \le i \le |\mathcal{L}(W)|$. Then, the totality of gamma-coded lists takes the following number of bits:

$$\sum_{W\in\Sigma^k} \sum_{i=1}^{|\mathcal{L}(W)|} |\gamma(\mathtt{diff}_W[i])| = \sum_{W\in\Sigma^k} \sum_{i=1}^{|\mathcal{L}(W)|} (2\lceil\log(\mathtt{diff}_W[i])\rceil - 1)$$

$$\le \sum_{W\in\Sigma^k} \sum_{i=1}^{|\mathcal{L}(W)|} (2\log(\mathtt{diff}_W[i]) + 1)$$

$$\le n + 2 \sum_{W\in\Sigma^k} |\mathcal{L}(W)|\log\left(\frac{n}{|\mathcal{L}(W)|}\right) \tag{8.1}$$

$$\le n + 2n\log\ell, \tag{8.2}$$

where ℓ is the number of non-empty lists, that is, the number of distinct k-mers in T. Inequality (8.1) derives from the facts that $\sum_{i=1}^{|\mathcal{L}(W)|} \mathtt{diff}_W[i] \le n$ for each W and $\sum_{W\in\Sigma^k} |\mathcal{L}(W)| = n$, and application of Jensen's inequality

$$\frac{\sum_i a_i f(x_i)}{\sum_i a_i} \le f\left(\frac{\sum_i a_i x_i}{\sum_i a_i}\right), \tag{8.3}$$

which holds for any concave function f. Here one should set $a_i = 1$ for all i, $x_i = \mathtt{diff}_W[i]$, and $f = \log$. Inequality (8.2) derives from the same facts, setting $a_i = |\mathcal{L}(W)|$, $x_i = n/|\mathcal{L}(W)|$, and $f = \log$ in Jensen's inequality.

THEOREM 8.2 *The k-mer index of a text $T = t_1 t_2 \cdots t_n \in \Sigma^*$ can be represented in $|\Sigma|^k(1+o(1)) + n(1+o(1)) + 2n\log\ell + \ell\log(n+2n\log\ell)$ bits of space such that each occurrence list $\mathcal{L}(W)$ can be decoded in $O(k+|\mathcal{L}(W)|)$ time, where ℓ is the number of non-empty occurrence lists.*

Proof See Exercise 8.1 on how to access the list corresponding to W in $O(k)$ time and Exercise 8.2 on how to encode the pointers to the non-empty elements using $|\Sigma|^k(1+o(1)) + \ell\log(n+2n\log\ell)$ bits. It remains to show that each γ-coded value can be decoded in constant time. We use the same four Russians technique as in Chapter 3 to build a table of $\sqrt{n}$ entries storing for each bitvector of length $(\log n)/2$ the location of the first bit set. At most two accesses to this table are required for decoding a γ-coded value. $\qquad\square$

See Exercises 8.3 and 8.4 for alternative ways to link k-mers to their occurrence lists.

8.2   Suffix array

The obvious problem with the k-mer index is the limitation to searching for patterns of fixed maximum length. This can be alleviated as follows. Consider all suffixes of string T that start with k-mer W. If one replaces W with those suffixes, and places the suffixes in lexicographic order, a binary search for the pattern can continue all the way to the end of each suffix if necessary. Listing all suffixes like this takes quadratic space, but it is also not required, since it suffices to store the starting positions of suffixes: see Figure 8.1(b).

DEFINITION 8.3   *The suffix array* $\mathsf{SA}_T[1..n]$ *of a text* $T = t_1 t_2 \cdots t_n$ *is the permutation of* $[1..n]$ *such that* $\mathsf{SA}_T[i] = j$ *iff suffix* $T_{j..n}$ *has position i in the list of all suffixes of T taken in lexicographic order.*

We say that suffix $T_{j..n}$ has *lexicographic rank i* iff $\mathsf{SA}_T[i] = j$. To ensure that suffixes can be sorted, we assume that $t_n = \#$ is an artificial character smaller than any other character in Σ, or in other words that $\mathsf{SA}_T[1] = n$. For example, the suffix array of $T = \mathtt{AGAGCGAGAGCGCGC\#}$ is shown in Figure 8.1(b). We will frequently use the shorthand SA to mean SA_T when T is clear from the context, and we will call *inverse suffix array* an array $\mathsf{ISA}[1..n]$ such that $\mathsf{ISA}[\mathsf{SA}[j]] = j$ for all $j \in [1..n]$ (Figure 8.1(c)).

Let us now consider the binary search for the pattern in more detail. First observe that a substring P of T corresponds to a unique, contiguous *interval* $\overrightarrow{P}$ in SA_T: the interval that contains precisely the suffixes of T that are prefixed by P. Knowing the interval $\overrightarrow{P}$ allows one to derive in constant time the number of occurrences of P in T, and to output the occurrences themselves in time linear in the size of the output. Searching for P in SA_T corresponds thus to finding the starting and ending position of its interval, and this naturally maps to an $O(|P|\log n)$-time binary search over SA_T as illustrated in Example 8.1 and Figure 8.1(b).

Example 8.1   Let $P = \mathtt{GAG}$ and $T = \mathtt{AGAGCGAGAGCGCGC\#}$, and assume that we want to determine the starting position $\overset{\bullet}{\overrightarrow{P}}$ of $\overrightarrow{P}$ in SA_T (see Figure 8.1(b)). We begin by considering the whole suffix array as a candidate region containing $\overrightarrow{P}$; that is, we start from the interval $[1..16]$ in SA_T, and we compare $P\#_1$ with the suffix in the middle of the current search interval, that is, suffix $\mathsf{SA}[1 + \lfloor(16 - 1)/2\rfloor] = \mathsf{SA}[8] = 13$. Note that, since we are searching for the starting position of P, we append the artificial delimiter $\#_1$ to the end of P, where $\#_1$ is lexicographically smaller than any other character in $\Sigma \cup \{\#\}$. Symmetrically, searching for $\overrightarrow{P}$ requires appending \$ to the end of P, where \$ is lexicographically larger than any other character in $\Sigma \cup \{\#_1, \#\}$. Since $P\#_1 = \mathtt{GAG}\#_1$ is lexicographically larger than suffix $T_{13..16} = \mathtt{CGC\#}$, we update the search interval to $[9..16]$ and we compare $P\#_1$ with suffix $\mathsf{SA}[9 + \lfloor(16 - 9)/2\rfloor] = \mathsf{SA}[12] = 8$ in the middle of the new search interval. $P\#_1 = \mathtt{GAG}\#_1$ is lexicographically smaller than suffix $T_{8..16} = \mathtt{GAGCGCGC\#}$, thus we update the search interval to $[9..12]$, and we compare $P\#_1$ with suffix $\mathsf{SA}[9 + \lfloor(12 - 9)/2\rfloor] = \mathsf{SA}[10] = 6$. Now $P\#_1 = \mathtt{GAG}\#_1$ is lexicographically smaller than suffix $T_{6..16} = \mathtt{GAGAGCGCGC\#}$, thus we update the search interval to

[9..10], and we compare $P\#_1$ with suffix $\mathsf{SA}[9 + \lfloor (10-9)/2 \rfloor] = \mathsf{SA}[9] = 11$. Since $P\#_1 = \mathrm{GAG}\#_1$ is lexicographically larger than suffix $T_{11..16} = \mathrm{CGCGC}\#$, we finally update the search interval to $[10..10]$ and we stop, since the interval has reached size one. A symmetric binary search on $P\$ = \mathrm{GAG}\$$ results in $\overrightarrow{P} = 12$. That is, $\overrightarrow{\mathrm{GAG}} = [10..12]$.

THEOREM 8.4   *Given the suffix array $\mathsf{SA}_T[1..n]$ of a text $T = t_1 t_2 \cdots t_n$, and a pattern $P = p_1 p_2 \cdots p_m$, one can find in $O(m \log n)$ time the maximal interval $\overrightarrow{P}$ in SA_T such that $T_{\mathsf{SA}[i]..\mathsf{SA}[i]+m-1} = P$ for all $i \in \overrightarrow{P}$.*

The list of all suffixes of a string T in lexicographic order enjoys a remarkably regular structure. We will now detail some of its combinatorial properties and show how they can be leveraged for building SA_T from T.

8.2.1   Suffix and string sorting

It is not surprising that algorithms for building suffix arrays have deep connections to algorithms for sorting integers. We will first consider sorting fixed-length strings (all of length k) as an extension of sorting integers, then proceed to sorting variable-length strings, and finally show how these can be used as subroutines to sort the suffixes of a text.

Sorting a set of strings of fixed length

Let $L[1..n]$ be a sequence of integers $L[i] \in [1..\sigma]$ for $1 \leq i \leq n$. The folklore *counting sort* algorithm to sort L into L' such that $L'[i] \leq L'[i+1]$, for $1 \leq i \leq n-1$, works as follows. Build a vector $\mathrm{count}[1..\sigma]$ that stores in $\mathrm{count}[c]$ the number of entries i with value $L[i] = c$, with a linear scan over L. These counts can be turned into cumulative counts in another *block pointer* vector $B[1..\sigma]$ by setting $B[k] = B[k-1] + \mathrm{count}[k-1]$ for $k \in [2..\sigma]$, after initializing $B[1] = 1$. Finally, setting $L'[B[L[i]]] = L[i]$ and $B[L[i]] = B[L[i]] + 1$, for $i \in [1..n]$, produces the required output. Observe that before the iteration $B[c]$ points to the beginning of the block of characters c in the final L', and after the iteration $B[c]$ points to the beginning of the next block. For this analogy, let us call this step *block pointer iteration* of counting sort.

Assume now that L is a list of *pairs* with the ith pair denoted $(L[i].p, L[i].v)$, where p stands for *primary key* and v for *value*. One can verbatim modify the counting sort to produce L' such that $L'[i].p \leq L'[i+1].p$ for $1 \leq i \leq n-1$. For cases $L'[i].p = L[i+1].p$ one often wants to use the original entry as an *implicit secondary key*. This is accomplished automatically if i is processed left to right in the block pointer iteration of counting sort by setting $L'[B[L[i].p]] = L[i]$ and $B[L[i].p] = B[L[i].p] + 1$. Such sorted order is called *stable*.

Continuing the induction towards fixed-length strings, consider now that L is a list of *triplets* with the ith triplet denoted $(L[i].p, L[i].s, L[i].v)$, where we have added an *explicit secondary key* $L[i].s$. Obviously, we wish to stable sort L into L' first by the primary key and then by the secondary key, and this is easy to achieve by extending the counting sort into *radix sort*, as follows.

LEMMA 8.5   *Let $L[1..n]$ be a sequence of (primary key $L[i].p$, secondary key $L[i].s$, value $L[i].v$) triplets such that $L[i].p, L[i].s \in [1..\sigma]$ for $i \in [1..n]$. There is an algorithm to stable sort L into L' such that either $L'[i].p < L'[i+1].p$, or $L'[i].p = L'[i+1].p$ and $L'[i].s \leq L'[i+1].s$, for $i \in [1, n-1]$, in $O(\sigma + n)$ time and space.*

Proof   Sort L independently into sequences L^p and L^s using stable counting sort with $L[i].p$ and $L[i].s$ as the sort keys, respectively. Let $B[1..\sigma]$ store the cumulative counts *before* the execution of the block pointer iteration of counting sort, when producing L^p. That is, $B[c]$ points to the first entry i of the final solution L' with primary key $L'[i].p = c$. The block pointer iteration is replaced by the setting of $L'[B[L^s[k].p]] = L^s[k]$ and $B[L^s[k].p] = B[L^s[k].p] + 1$ iterating k from 1 to n. Observe that here the explicit secondary keys replace the role of implicit secondary keys in the underlying stable counting sort.   □

Consider now a sequence of strings $\mathcal{S} = W^1, \ldots, W^n \mid W^i \in [1..\sigma]^*$, and assume that we want to compute $\mathcal{S}^*$, the permutation of $\mathcal{S}$ in lexicographic order. Assuming the strings are all of some fixed length, the radix sort approach can be applied iteratively.

COROLLARY 8.6   *Let $\mathcal{S} = W^1, \ldots, W^n \mid W^i \in [1..\sigma]^*$ be a sequence of strings of fixed length k, that is, $k = |W^i|$ for all i. There is an algorithm to sort $\mathcal{S}$ into lexicographic order in $O(k(\sigma + n))$ time.*

Proof   Apply Lemma 8.5 on $L = (w_1^1, w_2^1, 1), (w_1^2, w_2^2, 2), \ldots, (w_1^n, w_2^n, n)$ to produce L' such that $L'[j].v = i$ identifies the string W^i in $\mathcal{S}$ that has order j when considering only prefixes of length 2 in stable sorting. Let $R[1..n]$ store the *lexicographic ranks* of $L'[1..n]$ with $R[j]$ defined as the number of *distinct* primary and secondary element pairs occurring in prefix $L'[1..j]$. This array is computed by initializing $R[1] = 1$ and then setting $R[j] = R[j-1]$ if $L'[j].p = L'[j-1].p$ and $L'[j].s = L'[j-1].s$, otherwise $R[j] = R[j-1] + 1$, iterating j from 2 to n.

Now, form $L = (R[1], w_3^{L'[1].v}, L'[1].v), (R[2], w_3^{L'[2].v}, L'[2].v), \ldots, (R[n], w_3^{L'[n].v}, L'[n].v)$ and apply again Lemma 8.5 to produce L' such that $L'[j].v = i$ identifies the string W^i in $\mathcal{S}$ which has order j when considering only prefixes of length 3 in stable sorting. Then recompute lexicographic ranks $R[1..n]$ as above. Continuing this way until finally applying Lemma 8.5 on $L = (R[1], w_{k-1}^{L'[1].v}, L'[1].v), (R[2], w_k^{L'[2].v}, L'[2].v), \ldots, (R[n], w_k^{L'[n].v}, L'[n].v)$, one obtains L' such that $L'[j].v = i$ identifies the string W^i in $\mathcal{S}$ which has the order j when considering the full strings in stable sorting. The requested permutation of $\mathcal{S}$ in lexicographic order is obtained by taking $\mathcal{S}^*[j] = W^{L'[j].v}$.   □

*Sorting a set of strings of variable length

Now let $\mathcal{S} = W^1, \ldots, W^n \mid W^i \in [1..\sigma]^*$ be a sequence of strings of *variable* length. We follow basically the same scheme as for strings of fixed length, except that we need to take into account the fact that, on moving from prefixes of length p to prefixes of length $p + 1$, some strings can end. Filling tables of size $O(n)$ at each level p is not sufficient, because the running time should be proportional to the sparse set of strings that are still alive at that level. We shall now proceed to the details on how to achieve the optimal running time by maintaining the sparse set of strings at each level.

Let m be max$\{|W^k| : k \in [1..n]\}$, and for a given integer p consider the set of distinct prefixes $\mathcal{S}^p = \{W^k_{1..p} \mid k \in [1..n], p \in [1..m]\}$, where we assume that $W^k_{1..p} = W^k$ if $p > |W^k|$. We could sort $\mathcal{S}$ by induction over position $p \in [1..m]$: in particular, assuming that $\mathcal{S}$ is already sorted by $\mathcal{S}^{p-1}$ at iteration p, we could derive the lexicographic order induced by $\mathcal{S}^p$ from the lexicographic order of the characters that occur at position p (see Figure 8.2(a)). More formally, every string $W \in \mathcal{S}^p$ maps to a contiguous interval $\overleftrightarrow{W} = [\overleftarrow{W}..\overrightarrow{W}]$ in $\mathcal{S}^*$ that contains all strings in $\mathcal{S}$ that start with W, where $\overleftarrow{W}$ is the first position and $\overrightarrow{W}$ is the last position of the interval. Moreover, $\mathcal{I}^p = \{\overleftrightarrow{W} \mid W \in \mathcal{S}^p\}$ is a *partition* of $\mathcal{S}^*$, and partition $\mathcal{I}^p$ is a *refinement* of partition $\mathcal{I}^{p-1}$, in the sense that $\overleftrightarrow{Wa} \subseteq \overleftrightarrow{W}$ for any $a \in \Sigma$. We can thus transform $\mathcal{S}$ into $\mathcal{S}^*$ by iteratively assigning to each string W^k value $\overleftarrow{W^k_{1..p}}$ for increasing p, until all strings have distinct $\overleftarrow{W_{1..p}}$. The following lemma, illustrated in Figure 8.2, shows how to implement this idea using any algorithm for sorting triplets as a black box, as well as data structures analogous to those of counting sort.

LEMMA 8.7   *A sequence of strings $\mathcal{S} = W^1, \ldots, W^n \mid W^i \in [1..\sigma]^*$ can be sorted in lexicographic order in $O(\sigma + N)$ time, where $N = \sum_{k\in[1..n]} |W^k|$.*

Proof   Consider the list L of all possible triplets (pos, char, str) for str $\in [1..n]$, pos $\in [1..m]$, and char $= W^{\text{str}}[\text{pos}]$. Sort L by its primary and secondary keys with Lemma 8.5. For clarity, denote by L^p the contiguous interval of the sorted L whose triplets have primary key p. Vectors L^p represent permutations of $\mathcal{S}$ sorted only by characters that occur at position p, and are thus homologous to the primary- and secondary-key vectors L^p and L^s used in Lemma 8.5. Note that each L^p is partitioned into contiguous intervals, each corresponding to triplets with the same character. In what follows, we will keep refining the sorted order from shorter prefixes to longer prefixes as in the case of fixed-length strings by iterating $p \in [1..m]$, but we will engineer each refinement to take $O(|L^p|)$ rather than $O(n)$ time.

Set $p = 1$, and assign to each triplet in L^1 the starting position of its corresponding interval (block) in L^1, that is, initialize a block pointer vector $B[1..n]$ so that $B[k] = |\{h : w^h_1 < w^k_1\}| + 1$ as in counting sort. In the generic iteration we are considering a specific value of $p > 1$, and we assume that $B[k]$ contains the starting position $\overleftarrow{W^k[1..p-1]}$ of the interval of string $W^k[1..p-1]$ in $\mathcal{S}^*$. Recall that every interval of $\mathcal{I}_p$ is completely contained inside an interval of $\mathcal{I}^{p-1}$, and that the existence of an interval $\overleftrightarrow{T} \in \mathcal{I}^{p-1}$ implies the existence of an interval $\overleftrightarrow{Ta} \in \mathcal{I}^p$ with $\overleftarrow{Ta} = \overleftarrow{T}$, where a is the smallest character at position p preceded by string T. Let $Q[1..n]$, $S[1..n]$, and $C[1..n]$ be arrays of temporary variables with the following meaning: Q records the current position inside an interval, $Q[\overleftrightarrow{T}] = \overleftarrow{Ta}$, S stores the current size of each interval, $S[\overleftrightarrow{T}] = |\overleftrightarrow{T}|$, and C stores the current character corresponding to each interval, $C[\overleftrightarrow{T}] = a$, for every interval $\overleftrightarrow{T} \in \mathcal{I}^{p-1}$ and for a *temporary* interval $\overleftrightarrow{Ta} \in \mathcal{I}^p$. Initialize Q in $O(|L^p|)$ time using two scans over L^p, as follows. In the first scan, consider the string $Ta = W^{L^p[k].\text{str}}[1..p]$

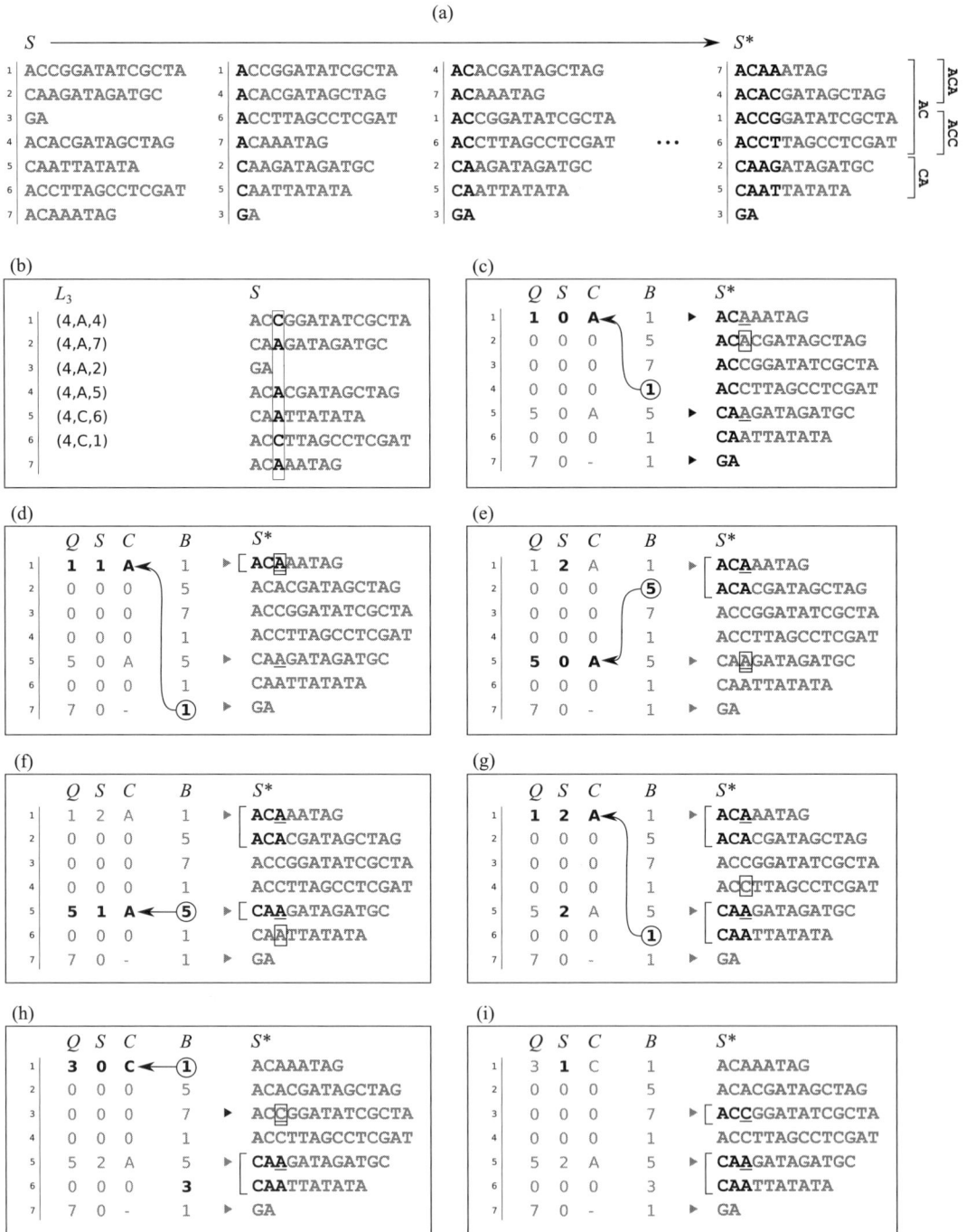

Figure 8.2 A sample run of the algorithm in Lemma 8.7. (a) The high-level strategy of the algorithm, with the intervals in $\mathcal{I}^2 \cup \mathcal{I}^3$ of size greater than one. (b) The initial sequence of strings $\mathcal{S}$ and the vector of triples L^3. (c–h) The data structures before processing $L^3[1], \ldots, L^3[6]$. (i) The data structures after processing $L^3[6]$. Triangles, square brackets, and underlined characters in $\mathcal{S}^*$ illustrate the meanings of Q, S, and C, respectively.

pointed by $L^p[k]$ for each $k \in [1..|L_p|]$, determine the starting position $\overset{\bullet}{T}$ of the interval of T in $\mathcal{S}^*$ by setting $\overset{\bullet}{T} = R[L^p[k].\mathtt{str}]$, and set $Q[\overset{\bullet}{T}]$ to zero. In the second scan repeat the same steps for every $k \in [1..|L_p|]$, but if no Q value has yet been assigned to position $\overset{\bullet}{T}$, that is, if $Q[\overset{\bullet}{T}] = 0$, then set $Q[\overset{\bullet}{T}]$ to $\overset{\bullet}{T}$ itself, and initialize $C[\overset{\bullet}{T}]$ to character $a = L_p[k].\mathtt{char}$. Note that a is the lexicographically smallest character that follows any occurrence of T as a prefix of a string of $\mathcal{S}$.

To compute in $O(|L^p|)$ time a new version of block pointer vector B that stores at $B[k]$ the starting position $W^k_{1..p}$ of the interval of string $W^k_{1..p}$ in $\mathcal{S}^*$, proceed as follows. Again, scan L^p sequentially, consider the string $Ta = W^{L^p[k].\mathtt{str}}_{1..p}$ pointed at by $L^p[k]$ for each $k \in [1..|L^p|]$, and compute $\overset{\bullet}{T}$ by setting $\overset{\bullet}{T} = B[L^p[k].\mathtt{str}]$. Recall that the values of $Q[\overset{\bullet}{T}]$, $S[\overset{\bullet}{T}]$, and $C[\overset{\bullet}{T}]$ encode the starting position, size, and character of a sub-interval $\overset{\bullet}{Tb}$ of $\overset{\bullet}{T}$ for some $b \in \Sigma$. If character a equals b, that is, if $C[\overset{\bullet}{T}] = L^p[k].\mathtt{char}$, string Ta belongs to the current sub-interval of $\overset{\bullet}{T}$, so $B[L^p[k].\mathtt{str}]$ is assigned to $Q[\overset{\bullet}{T}]$ and $S[\overset{\bullet}{T}]$ is incremented by one. Otherwise, Ta does not belong to the current sub-interval of $\overset{\bullet}{T}$, so we encode in $Q[\overset{\bullet}{T}]$, $S[\overset{\bullet}{T}]$ and $C[\overset{\bullet}{T}]$ the new sub-interval $\overset{\bullet}{Ta}$ by setting $Q[\overset{\bullet}{T}] = Q[\overset{\bullet}{T}] + S[\overset{\bullet}{T}]$, $S[\overset{\bullet}{T}] = 0$ and $C[\overset{\bullet}{T}] = L^p[k].\mathtt{char}$. Then we set $B[L^p[k].\mathtt{str}] = Q[\overset{\bullet}{T}]$. □

Suffix sorting by prefix doubling

Lemma 8.7 immediately suggests a way to build SA_T from T in $O(\sigma + |T|^2)$ time: consider the suffixes of T just as a set of strings, disregarding the fact that such strings are precisely the suffixes of T. But how does the set of suffixes of the same string differ from an arbitrary set of strings?

Denote again by $\mathcal{S}$ the set of suffixes of T, by $\mathcal{S}^*$ the permutation of $\mathcal{S}$ in lexicographic order, and by $\mathcal{S}_p$ the set of all distinct prefixes of length p of strings in $\mathcal{S}$. It is by now clear that every string $W \in \mathcal{S}_2$ maps to a contiguous interval $\overset{\bullet}{W}$ of $\mathcal{S}^*$, and thus of SA_T. Assume again that we have an approximation of SA_T, in which such intervals are in lexicographic order, but suffixes inside the same interval are not necessarily in lexicographic order. Denote this permutation by SA^2. Clearly, SA^2 contains enough information to build a refined permutation SA^4 in which suffixes in the same interval of SA^2 are sorted according to their prefix of length four. Indeed, if $T_{i..n}$ and $T_{j..n}$ are such that $T_{i..i+1} = T_{j..j+1}$, then the lexicographic order of substrings $T_{i..i+3}$ and $T_{j..j+3}$ is determined by the order of $T_{i+2..i+3}$ and $T_{j+2..j+3}$ in SA^2. The idea of the *prefix-doubling technique* for suffix array construction consists in iterating this inference $O(\log|T|)$ times, progressively refining SA^2 into SA by considering prefixes of exponentially increasing length of all suffixes of T.

LEMMA 8.8   *There is a $O((\sigma + n) \log n)$-time algorithm to construct SA_T from T.*

Proof   Sort the list of triples $L = (t_1, t_2, 1), (t_2, t_3, 2), \ldots, (t_n, t_{n+1}, n)$ into a new list L' in $O(\sigma + n)$ time, and assign to each suffix $T_{k..n+1}$ the starting position $\overset{\bullet}{T_{k..k+1}}$ of the interval of its length-2 prefix $T_{k..k+1}$ in SA^2, as follows. Initialize an array $R[1..2n]$ with ones. Scan L' and set $R[L'[k].v] = R[L'[k-1].v]$ if $L'[k].p = L'[k-1].p$ and $L'[k].s = L'[k-1].s$, otherwise set $R[L'[k].v] = R[L'[k-1].v] + 1$. Values in $R[n+1..2n]$ are left as one, and will not be altered by the algorithm. At this point, refining SA^2 into SA^4

requires just sorting a new list $L = (R[1], R[3], 1), (R[2], R[4], 2), \ldots, (R[n], R[n+2],$ $n)$, that is, sorting suffixes using as primary key their prefix of length two and as secondary key the substring of length two that immediately follows their length-two prefix. Having all values in $R[n+1..2n]$ constantly set to one guarantees correctness. Updating R and repeating this process $O(\log n)$ times produces SA. Indeed, the pth step of the algorithm sorts a list $L = (R[1], R[1+2^p], 1), (R[2], R[2+2^p], 2), \ldots, (R[n], R[n+2^p], n)$, where $R[k]$ contains the starting position $\overset{\bullet}{T[k..k+2^p]}$ of the interval of string $T[k..k+2^p]$ in SA. Thus, after the pth step, $R[k]$ is updated to the starting position $\overset{\bullet}{T[k..k+2^{p+1}]}$ of the interval of string $T[k..k+2^{p+1}]$ in SA. Clearly, when $2^p \geq n$ every cell of R contains a distinct value, the algorithm can stop, and $\mathsf{SA}[R[k]] = k$. $\square$

Suffix sorting in linear time

An alternative strategy for building SA_T could be to iteratively fill a sparse version $\widetilde{\mathsf{SA}}_T$ initialized with just some seed values (see Figure 8.3). More specifically, let $\mathcal{I} = \{i_1, i_2, \ldots, i_m\} \subseteq [1..n]$ be a set of positions in T, and assume that $\widetilde{\mathsf{SA}}[k] = \mathsf{SA}[k]$ for $\mathsf{SA}[k] \in \mathcal{I}$, and that $\widetilde{\mathsf{SA}}[k] = 0$ otherwise. Again, store in vector $Q[c]$ the position of the smallest zero in $\widetilde{\mathsf{SA}}[c..c]$ for every $c \in [1..\sigma]$. Assume that we are at position k in $\widetilde{\mathsf{SA}}$, and that we have computed the value of $\widetilde{\mathsf{SA}}[h]$ for all $h \in [1..k]$. Consider the *previous suffix* $T[\widetilde{\mathsf{SA}}[k] - 1..n]$ and set $c = T[\widetilde{\mathsf{SA}}[k]]$. If $T[\widetilde{\mathsf{SA}}[k] - 1..n]$ is lexicographically

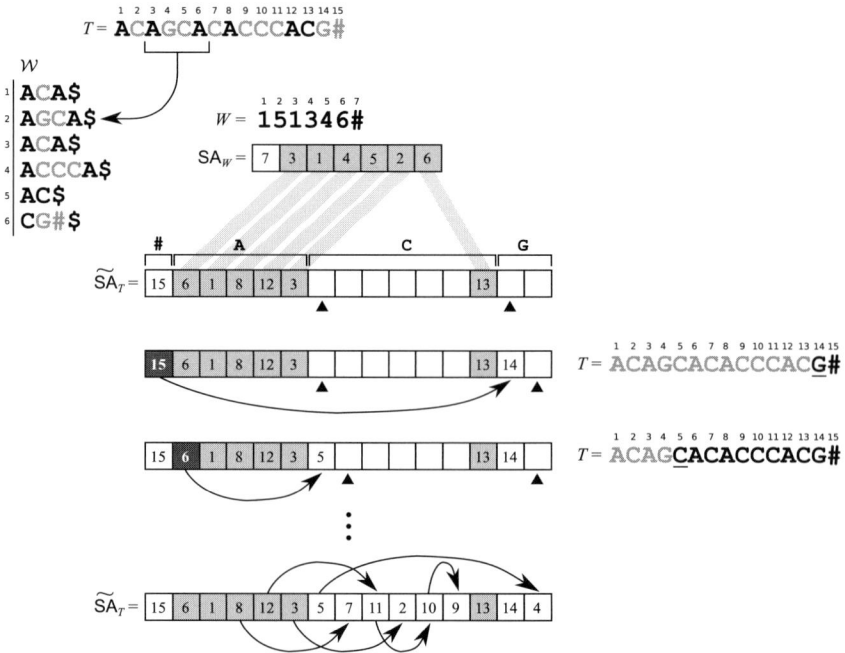

Figure 8.3 A strategy for the linear-time construction of SA_T. Set $\mathcal{I}$ of small suffixes is marked by black letters in the top-most occurrence of T, and by gray cells in $\widetilde{\mathsf{SA}}_T$. Triangles signal the current content of array Q. Note that substring ACA$ occurs twice in $\mathcal{W}$, and that substring AC$ occurs once.

smaller than the current suffix $T[\widetilde{\mathsf{SA}}[k]..n]$, then it already appears at an index smaller than k in $\widetilde{\mathsf{SA}}$. Otherwise, we can safely append suffix $T[\widetilde{\mathsf{SA}}[k]-1..n]$ to the current list of suffixes that start with c, that is, we can set $\widetilde{\mathsf{SA}}[Q[c]]$ to $\widetilde{\mathsf{SA}}[k]-1$ and we can increment pointer $Q[c]$ by one. Clearly this $O(n)$ scan of $\widetilde{\mathsf{SA}}$ works only if $\widetilde{\mathsf{SA}}[k] \neq 0$ at the current position k, and this holds either because $\mathsf{SA}[k] \in \mathcal{I}$ or because suffix $T[\mathsf{SA}[k]+1..n]$ has already been processed by the algorithm; that is, because the current suffix $T[\mathsf{SA}[k]..n]$ is *lexicographically larger* than suffix $T[\mathsf{SA}[k]+1..n]$. This motivates the definition of *small* and *large suffixes*.

DEFINITION 8.9 *A suffix $T_{i..n}$ of a string T is* large *(respectively,* small*) if $T_{i..n}$ is lexicographically larger (respectively, smaller) than $T_{i+1..n}$.*

It follows that, if $\mathcal{I}$ is the set of starting positions of all *small* suffixes of T, and if $\widetilde{\mathsf{SA}}$ is initialized according to $\mathcal{I}$, we can fill all the missing values of $\widetilde{\mathsf{SA}}$ in a single scan, obtaining SA: see Exercise 8.12.

Small and large suffixes enjoy a number of combinatorial properties: we focus on small suffixes here, with the proviso that everything we describe for small suffixes can easily be mirrored for large suffixes. Note that, in order to initialize $\widetilde{\mathsf{SA}}$, we need at least a Boolean vector $I[1..n]$ such that $I[k] = 1$ iff suffix $T_{i..n}$ is small. This vector can easily be computed in $O(n)$ time: see Exercise 8.10. Assume thus that $\mathcal{I} = \{i_1, i_2, \ldots, i_m\}$ is known through I, that $\mathcal{S}$ is the sequence of suffixes $T[i_1..n], T[i_2..n], \ldots, T[i_m..n]$, and that $\mathcal{S}^*$ is the permutation of $\mathcal{S}$ in lexicographic order. In particular, let vector R store in $R[k]$ the position of suffix $T[i_k..n]$ in $\mathcal{S}^*$ and assume $R[m+1] = \#$. Note that we can initialize the entries of $\widetilde{\mathsf{SA}}$ that correspond to all small suffixes of T by scanning the suffix array SA_R *of array R considered as a string*, in overall $O(n)$ time: see Exercise 8.11. The bottleneck to building SA_T becomes thus computing SA_R.

Consider again set $\mathcal{I} = \{i_1, i_2, \ldots, i_m\}$: intuitively, most of the information in R is already determined by the substrings $T[i_1..i_2], T[i_2..i_3], \ldots, T[i_m..n]$. We might thus try to compute SA_R by sorting such substrings and by extrapolating SA_R from the sorted list. Specifically, consider the sequence of substrings $\mathcal{W} = T[i_1..i_2]\$, T[i_2..i_3]\$, \ldots, T[i_m..n]\$$ of total length $O(n)$, where the artificial separator $\$$ is lexicographically larger than σ, and denote by $\mathcal{W}^*$ the permutation of $\mathcal{W}$ in lexicographic order. Recall that we can build $\mathcal{W}^*$ from $\mathcal{W}$ in $O(\sigma + n)$ time using Lemma 8.7. In particular, the lemma returns a vector W such that $W[k]$ is the position of string $T[i_k..i_{k+1}]\$$ in $\mathcal{W}^*$ (see Figure 8.3), where again we assume $W[m+1] = \#$. Our plan is to infer SA_R from SA_W by exploiting another property of small suffixes.

LEMMA 8.10 *Let $\overleftrightarrow{c}$ be the interval of the single-character string $c \in [1..\sigma]$ in SA_T. In $\overleftrightarrow{c}$, large suffixes occur before small suffixes, for any $c \in [1..\sigma]$.*

Proof Let $T_{i..n}$ and $T_{j..n}$ be two suffixes of T such that $t_i = t_j = c$, and assume that $T_{i..n}$ is a small suffix and $T_{j..n}$ is a large suffix. Then $T_{i..n} = c^p dv\#$ with $p \geq 1$, $d \in [c+1..\sigma]$, and $v \in \Sigma^*$, and $T_{j..n} = c^q bw\#$ with $q \geq 1$, $b < c$, and $w \in \Sigma^*$. Thus, $T_{j..n}$ is lexicographically smaller than $T_{i..n}$. □

Note in particular that interval $\tilde{c}$ with $c = 1$ contains only small suffixes, and that interval $\tilde{c}$ with $c = \sigma$ contains only large suffixes. As anticipated, Lemma 8.10 is the bridge between SA_W and SA_R:

LEMMA 8.11 $\mathsf{SA}_R = \mathsf{SA}_W$.

Proof Consider two suffixes $T[i_k..n]$ and $T[i_h..n]$ with i_k and i_h in $\mathcal{I}$: we prove that, if i_k follows i_h in SA_W, then i_k follows i_h also in SA_R. Assume that i_k follows i_h in SA_W and substring $S\$ = T[i_k..i_{k+1}]\$$ is different from substring $S'\$ = T[i_h..i_{h+1}]\$$. This could happen because a prefix $S_{1..p}$ with $p \le |S|$ is not a prefix of S': in this case, it is clear that $T[i_k..n]$ follows $T[i_h..n]$ in SA_R as well. Alternatively, it might be that S is a prefix of S', but $S\$$ is not (for example, for strings $S = \text{AC}$ and $S' = \text{ACA}$ in Figure 8.3). In this case, character $S'_{|S|}$ equals the last character of S: let c be this character, and let $S = Vc$ for $V \in \Sigma^*$. Then, suffix $T[i_k..n]$ equals $V \cdot T[i_{k+1}..n]$ and suffix $T[i_h..n]$ equals $V \cdot T[i_{h+|S|-1}..n]$. Note that both suffix $T[i_{k+1}..n]$ and suffix $T[i_{h+|S|-1}..n]$ start with character c, but $i_{h+|S|-1}$ does not belong to $\mathcal{I}$, thus, by Lemma 8.10, suffix $T[i_{h+|S|-1}..n]$ is lexicographically smaller than suffix $T[i_{k+1}..n]$. It follows that suffix $T[i_h..n]$ is lexicographically smaller than suffix $T[i_k..n]$ as well.

Assume now that i_k follows i_h in SA_W, but $S\$ = T[i_k..i_{k+1}]\$$ equals $S'\$ = T[i_h..i_{h+1}]\$$, i.e. $W[k] = W[h]$ (for example, for $S = S' = \text{ACA}$ in Figure 8.3). Thus, the sequence of numbers $W[k+1..m]\#$ follows the sequence of numbers $W[h+1..m]\#$ in lexicographic order, i.e. $T[i_{k+x}..i_{k+x+1}]$ follows $T[i_{h+x}..i_{h+x+1}]$ in lexicographic order for some $x \ge 1$, and $T[i_{k+y}..i_{k+y+1}]$ equals $T[i_{h+y}..i_{h+y+1}]$ for all $y < x$. This implies that suffix $T[i_k..n]$ follows suffix $T[i_h..n]$ in lexicographic order.   □

It thus remains to compute SA_W. If $|\mathcal{I}| \le n/2$, i.e. if T has at most $n/2$ small suffixes, we could apply the same sequence of steps recursively *to* W, i.e. we could detect the small suffixes of W, split W into substrings, sort such substrings, infer the lexicographic order of all small suffixes of W, and then extrapolate the lexicographic order of all suffixes of W. If T has more than $n/2$ small suffixes, we could recurse on large suffixes instead, with symmetrical arguments. This process takes time:

$$f(n, \sigma) = an + b\sigma + f(n/2, n/2)$$
$$= an + b\sigma + an/2 + bn/2 + an/4 + bn/4 + \cdots + a + b$$
$$\in O(\sigma + n),$$

where a and b are constants. We are thus able to prove the main theorem of this section, as follows.

THEOREM 8.12   *The suffix array SA_T of a string T of length n can be built in $O(\sigma + n)$ time.*

8.3   Suffix tree

A *trie* is a labeled tree $\mathcal{T}$ that represents a set of strings. Specifically, given a set of variable-length strings $\mathcal{S} = \{S^1, \ldots, S^n\}$ over alphabet $\Sigma = [1..\sigma]$, the *trie of $\mathcal{S}$* is a tree

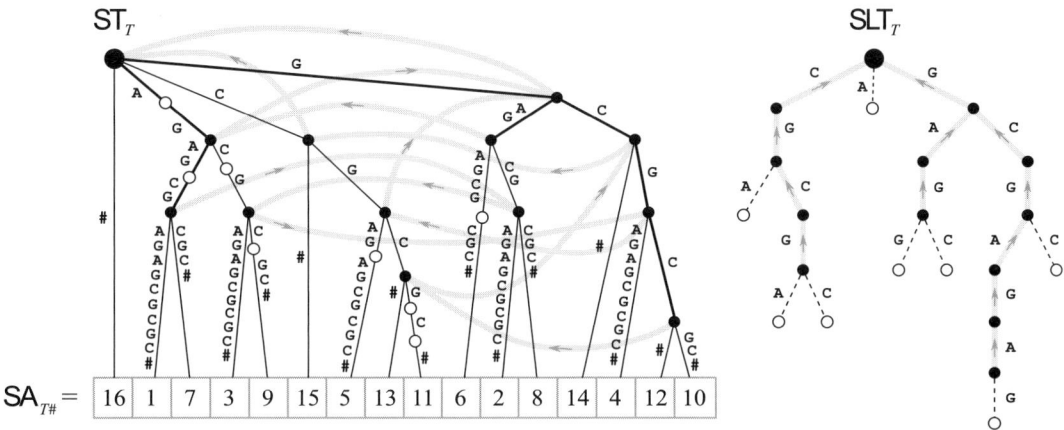

Figure 8.4 Relationships among the suffix tree, the suffix-link tree, and the suffix array of string $T = $ AGAGCGAGAGCGCGC. Thin black lines: edges of ST_T; thick gray lines: suffix links; thin dashed lines: implicit Weiner links; thick black lines: the subtree of ST_T induced by maximal repeats. Black dots: nodes of ST_T; large black dot: r; white dots: destinations of implicit Weiner links; squares: leaves of ST_T; numbers: starting position of each suffix in T. For clarity, implicit Weiner links are not overlaid to ST_T, and suffix links from the leaves of ST_T are not drawn.

$\mathcal{T}$ with at most n leaves and exactly n marked nodes, such that (1) every edge (v_i, v_j) is labeled by a character $\ell(v_i, v_j) \in \Sigma$; (2) the edges that connect a node to its children have distinct labels; (3) every leaf is marked; and (4) every path from the root to a marked node spells a distinct string of $\mathcal{S}$. A *compact trie*, or *Patricia trie*, is a tree $\mathcal{T}'$ whose edges are labeled by *strings* in Σ^+, rather than by single characters, and it can be thought of as obtained from a trie $\mathcal{T}$ as follows: for every path $v_1, v_2, \ldots, v_k$ such that v_i has only one child for every $i \in [1..k-1]$, remove nodes $v_2, v_3, \ldots, v_{k-1}$ and the incoming edges to $v_2, v_3, \ldots, v_k$, and add edge (v_1, v_k) with label $\ell(v_1, v_2) \cdot \ell(v_2, v_3) \cdot \cdots \cdot \ell(v_{k-1}, v_k)$. Clearly $\mathcal{T}'$ has at most $n - 1$ internal nodes, where n is the number of leaves in $\mathcal{T}$.

The *suffix tree* is one of the oldest and most powerful full-text indexes, and it can be thought of as a compact trie built on the set of all *suffixes* of a string, or equivalently as a search tree built on top of the suffix array, as visualized in Figure 8.4.

DEFINITION 8.13   *The suffix tree* $\mathsf{ST}_T = (V, E)$ *of a string* $T \in [1..\sigma]^n$ *is a tree rooted in* $r \in V$. *The edges are labeled by substrings of* $T\#$. *The leaves of the tree correspond to suffixes of* $T\#$, *and the path from the root to a leaf spells the corresponding suffix. The label of each edge is of maximal length, so that the tree branches at each internal node, and the labels that connect a node to its children start with distinct characters.*

We assume that the children of a node v are ordered lexicographically according to the labels of the corresponding edges. We denote by $\ell(e)$ the label of an edge $e \in E$ and by $\ell(v)$ the string $\ell(r, v_1) \cdot \ell(v_1, v_2) \cdot \cdots \cdot \ell(v_{k-1}, v)$, where $r, v_1, v_2, \ldots, v_{k-1}, v$ is the path of v in ST_T. We say that node v has *string depth* $|\ell(v)|$. Since the tree has exactly $n + 1$ leaves, and each internal node is branching, there are at most n internal nodes. Observe that the edge labels can be stored not as strings, but as pointers to the corresponding substring of $T\#$: see Figure 8.4. Hence, each node and edge will store a constant number of values or pointers, and each such value or pointer can be encoded using $\log n$ bits. A careful implementation requires $(3|V|+2n)\log n$ bits, without pointers from child to parent and other auxiliary information.

Searching T for a pattern $P = p_1 p_2 \ldots p_k$ clearly amounts to following the path labeled by P from r in ST_T and performing a binary search at each internal node. Indeed, if P is a substring of T, then it is the prefix of some suffix of T; thus it is a prefix of a label $\ell(v)$ for some node $v \in V$. For example, in Figure 8.4 $P = \mathsf{AGC}$ is a prefix of suffixes starting AGCG. We call node v the *proper locus* of P if the search for P ends at an edge (w, v). In our example, the search for $P = \mathsf{AGC}$ ends at an edge (w, v) such that $\ell(w) = \mathsf{AG}$ and $\ell(v) = \mathsf{AGCG}$. Then, the leaves in the subtree rooted at v are exactly the starting positions of P in T. In our example, leaves under v point to the suffixes $T_{3..n}$ and $T_{9..n}$, which are the only suffixes prefixed by $P = \mathsf{AGC}$. This search takes $O(|P|\log\sigma + |\mathcal{L}_T(P)|)$ time, where $\mathcal{L}_T(P)$ is the set of all starting positions of P in T.

8.3.1   Properties of the suffix tree

In the forthcoming chapters we shall be heavily exploiting some key features of suffix trees. This is a good point to introduce those notions, although we are not using all of these features yet; the reader might opt to come back to some of these notions as required when they arise.

Building and using ST_T often benefits from an additional set of edges, called *suffix links*. Let the label $\ell(v)$ of the path from the root to a node $v \in V$ be aX, with $a \in \Sigma$. Since string X occurs at all places where aX occurs, there must be a node $w \in V$ with $\ell(w) = X$; otherwise v would not be a node in V. We say that there is a *suffix link from v to w labeled by a*, and we write $sl(v) = w$. More generally, we say that the set of labels $\{\ell(v) \mid v \in V\}$ enjoys the *suffix closure* property, in the sense that, if a string W belongs to this set, so does every one of its suffixes. Note that, if v is a leaf, then $sl(v)$ is either a leaf or r. We denote the set of all suffix links of ST_T by $L = \{(v, sl(v), a) \mid v \in V, sl(v) \in V, \ell(v) = a\ell(sl(v)), a \in \Sigma\}$. It is easy to see that the pair (V, L) is a trie rooted at r: we call such a trie the *suffix-link tree* SLT_T of string T: see Figure 8.4.

Inverting the direction of all suffix links yields the so-called *explicit Weiner links*, denoted by $\underline{L} = \{(sl(v), v, a) \mid v \in V, sl(v) \in V, \ell(v) = a\ell(sl(v)), a \in \Sigma\}$. Clearly a node in V might have more than one outgoing Weiner link, and all such Weiner links have different labels. Given a node v and a character $a \in \Sigma$, it might happen that string $a\ell(v)$ does occur in T, but that it does not label any node in V: we call all such extensions of nodes in V *implicit Weiner links*, and we denote the set of all implicit Weiner links by $\underline{L}'$: see Figure 8.4. Note that Weiner links from leaves are explicit.

We summarize the following characteristics of suffix trees.

OBSERVATION 8.14   *The numbers of suffix links, explicit Weiner links, and implicit Weiner links in the suffix tree ST_T of string $T = t_1 t_2 \cdots t_{n-1}$ are upper bounded by $2n - 2$, $2n - 2$, and $4n - 3$, respectively.*

Proof   Each of the at most $2n-2$ nodes of a suffix tree (other than the root) has a suffix link. Each explicit Weiner link is an inverse of a suffix link, so their total number is also at most $2n-2$.

Consider a node v with only one implicit Weiner link $e = (\ell(v), a\ell(v))$ in ST_T. The number of such nodes, and thus the number of such implicit Weiner links, is bounded by $2n-1$. Call these the implicit Weiner links of class I, and the remaining ones the implicit Weiner links of class II. Consider then a node v with two or more (class II) implicit Weiner links forming set $W_v = \{c \mid e_c = (\ell(v), c\ell(v))$ is a Weiner link in $\mathsf{ST}_T\}$. This implies the existence of a node w in the suffix tree of the *reverse* of T, $\mathsf{ST}_{\underline{T}}$, labeled by the reverse of $\ell(v)$, denoted by $\underline{\ell(v)}$; each $c \in W_v$ can be mapped to a distinct edge of $\mathsf{ST}_{\underline{T}}$ connecting w to one of its children. This is an injective mapping from class II implicit Weiner links to the at most $2n-2$ edges of the suffix tree of $\underline{T}$. The sum of class I and II, that is, all implicit Weiner links, is hence bounded by $4n-3$.   □

The bound on the number of implicit Weiner links can easily be improved to $2n-2$: see Exercise 8.18. With more involved arguments, the bound can be further improved to n: see Exercise 8.19.

8.3.2   Construction of the suffix tree

Despite its invention being anterior to the suffix array, the suffix tree can be seen as an extension of it: building ST_T from SA_T clarifies the deep connections between these two data structures. Throughout this section we assume that $T = t_1 t_2 \cdots t_n$ with $t_n = \#$, and we assume that we have access both to $\mathsf{SA}_T[1..n]$ and to an additional *longest common prefix array* $\mathsf{LCP}[2..n]$ such that $\mathsf{LCP}[i]$ stores the length of the longest prefix that is common to suffixes $T[\mathsf{SA}[i-1]..n]$ and $T[\mathsf{SA}[i]..n]$. Building LCP_T in $O(n)$ time from SA_T is left to Exercise 8.14. To build ST, we first insert the suffixes of T as leaves in lexicographic order, i.e. we assign suffix $T[\mathsf{SA}[i]..n]$ to leaf i. Then, we build the internal nodes by iterating over LCP_T from left to right, as follows.

Consider a temporary approximation $\widetilde{\mathsf{ST}}^{i-1}$ of ST, which is built using only suffixes $T[\mathsf{SA}[1]..n]$, $T[\mathsf{SA}[2]..n]$, ..., $T[\mathsf{SA}[i-1]..n]$. In order to update this tree to contain also suffix $T[\mathsf{SA}[i]..n]$, we can traverse $\widetilde{\mathsf{ST}}^{i-1}$ bottom-up starting from its right-most leaf, until we find a node v such that $|\ell(v)| \leq \mathsf{LCP}[i]$. If $|\ell(v)| = \mathsf{LCP}[i]$ we insert a new edge from v to leaf i, otherwise we split the right-most edge from v, say (v, w), at depth $\mathsf{LCP}[i] - |\ell(v)|$, and create a new internal node with one child pointing to node w and another child pointing to leaf i (Figures 8.5(b) and (c)). Clearly, after scanning the whole LCP we have $\widetilde{\mathsf{ST}}^n = \mathsf{ST}$. Observe that, if we traverse from v to its parent during the traversal, we never visit v again. Each traversal to add a new suffix ends with a constant number of operations to modify the tree. Thus the total traversal and update time can be amortized to the size of ST, and we obtain the following result.

THEOREM 8.15   *The suffix tree ST_T of a string $T = t_1 t_2 \cdots t_{n-1} \in [1..\sigma]^{n-1}\#$ can be built in time $O(\sigma + n)$ so that the leaves of ST_T are sorted in lexicographic order.*

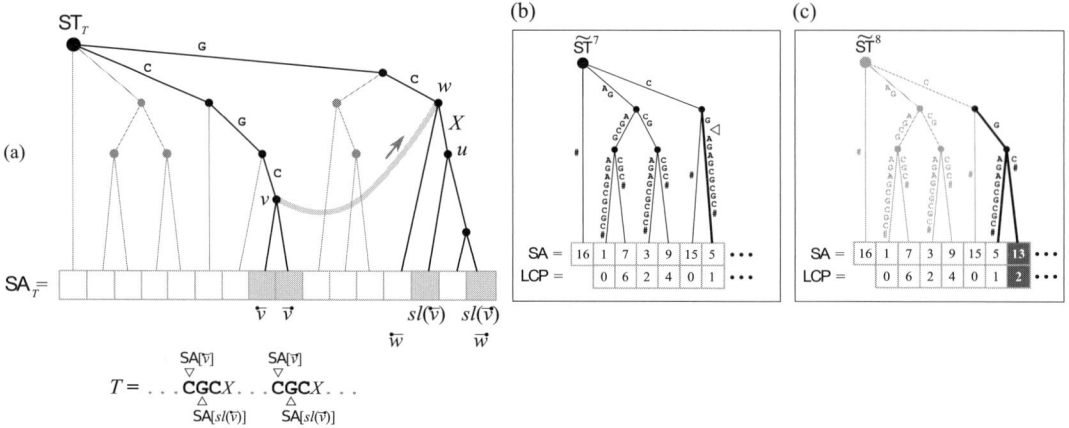

Figure 8.5 (a) Visualizing Lemma 8.16. (b,c) Building the suffix tree from the suffix array and the LCP array: the current approximation $\widetilde{\mathsf{ST}}^7$ (b) and the new approximation $\widetilde{\mathsf{ST}}^8$ (c).

Adding suffix links to the leaves of ST_T using the suffix array and the inverse suffix array of T is easy: see Exercise 8.16. But how can we augment ST_T with the suffix links of internal nodes?

Let $Y \subset V$ be the set of leaves of ST_T. Note that every internal node of ST_T corresponds to an interval $\overleftarrow{v}$ in SA_T, where $\overleftarrow{v}$ and $\overrightarrow{v}$ are the leaves with lexicographically smallest and largest labels in the subtree of ST_T rooted at v. Note also that, if node v is an ancestor of node w in ST_T, then $\overline{w}$ is contained in $\overline{v}$. Assign thus to each node $v \in V \setminus Y$ values $\overleftarrow{v}$ and $\overrightarrow{v}$, and consider set $N = \{\overline{v} \mid v \in V \setminus Y\}$ of *node intervals* and set $Q = \{[sl(\overleftarrow{v})..sl(\overrightarrow{v})] : v \in V \setminus Y\}$ of *query intervals*.

LEMMA 8.16   *Let v and w be two nodes in V such that $sl(v) = w$. Then, $\overline{w}$ is the smallest interval in N that contains $[sl(\overleftarrow{v})..sl(\overrightarrow{v})]$. In other words, w is the lowest common ancestor (LCA) of leaves $sl(\overleftarrow{v})$ and $sl(\overrightarrow{v})$ in ST_T.*

Proof   Assume that node $u \in V \setminus Y$ is such that interval $\overline{u}$ includes $[sl(\overleftarrow{v})..sl(\overrightarrow{v})]$ and $\overline{u}$ is smaller than $\overline{w}$. Then u must be a descendant of w, so $\ell(u) = \ell(w) \cdot X$, where $\ell(v) = a \cdot \ell(w)$, $a \in \Sigma$, and $|X| > 0$. It follows that $T[\mathsf{SA}[sl(\overleftarrow{v})]..\mathsf{SA}[sl(\overleftarrow{v})] + |\ell(w)| + |X| - 1] = T[\mathsf{SA}[sl(\overrightarrow{v})]..\mathsf{SA}[sl(\overrightarrow{v})] + |\ell(w)| + |X| - 1] = \ell(w)X$, and thus $T[\mathsf{SA}[\overleftarrow{v}]..\mathsf{SA}[\overleftarrow{v}] + |\ell(w)| + |X|] = T[\mathsf{SA}[\overrightarrow{v}]..\mathsf{SA}[\overrightarrow{v}] + |\ell(w)| + |X|] = a\ell(w)X$. Hence v would have a single outgoing edge prefixed by X, which amounts to a contradiction. $\quad\square$

We thus need to find, for each interval in Q, the smallest interval in N that includes it. For this we can use a general result on *batched lowest common ancestor* queries.

LEMMA 8.17   *Let $\mathcal{T} = (V, E)$ be a tree on n leaves numbered $1, 2, \ldots, n$. Given a set of query intervals $Q = \{[i..j] \mid 1 \le i \le j \le n\}$ of size $O(n)$, one can link each $q \in Q$ to its lowest common ancestor $v \in V$ in $O(n)$ time.*

Proof   During a left-to-right depth-first traversal of $\mathcal{T}$ we open an annotated parenthesis "$(_v$" when we first visit a node v, and we close an annotated parenthesis "$)_v$" when we last visit a node v. This produces a valid parenthesization (see Figure 8.6, top).

Assume that, when we visit a leaf i, we first print "$[_q$" for each query $q \in Q$ such that $\overleftarrow{q} = i$, and then we print "$]_q$" for each $q \in Q$ such that $\overrightarrow{q} = i$. Assume that the resulting parenthesization is stored in a doubly-linked list, with additional links from each "$)_v$" to the corresponding "$(_v$", from each "$]_q$" to the corresponding "$[_q$", and from each "$[_q$" to the closest parenthesis of type "(" on its left. Use call prev[q] to denote this link, which is initially set to the immediate predecessor of "$[_q$" in the linked list if it is of type "(", and otherwise remains undefined. Then, scan the parenthesization from left to right. If we meet a closing parenthesis "$)_v$" we delete it, we reach the corresponding "$(_v$", and we delete it as well. If we meet a closing parenthesis "$]_q$", we delete it, we reach the corresponding "$[_q$", we delete it as well, and we start a right-to-left iteration by following links in prev to locate the closest opening parenthesis of type "(" to the left of "$[_q$". If prev[q'] points to a "$(_w$" for some "$[_{q'}$" encountered during this iteration, we stop and we set prev[q''] to point to "$(_w$" as well for every "$[_{q''}$" that has been traversed so far. Note that the list before "$[_q$" contains only opening parentheses "$(_w$" for which no corresponding closing parenthesis "$)_w$" has yet been found, therefore $\overrightarrow{w}$ is necessarily the smallest interval that contains q and it is safe to create the link q to w. Clearly, at the end of the process, all links to lowest common ancestors from query intervals are created. The size of the parenthesization is $O(n)$, and processing each parenthesis takes *amortized* constant time: indeed, after following k links while searching for the closest parenthesis of type "(", we assign $k - 1$ prev values, and each such prev value is assigned exactly once. □

Figure 8.6 (bottom) shows an instantiation of Lemma 8.17 on a suffix tree and on query intervals $Q = \{[sl(\overleftarrow{v})..sl(\overrightarrow{v})] : v \in V \setminus Y\}$. The following theorem immediately follows.

THEOREM 8.18   *Assume we are given* ST_T *and the suffix links from all its leaves. Then, the suffix links from all internal nodes of* ST_T *can be built in* $O(n)$ *time.*

8.4   Applications of the suffix tree

In this section we sketch some prototypical applications of suffix trees. The purpose is to show the versatility and power of this fundamental data structure. The exercises of this chapter illustrate some additional uses, and later chapters revisit the same problems with more space-efficient solutions.

In what follows, we often consider the suffix tree of a set of strings instead of just one string. Then, a property of a set $\mathcal{S} = \{S^1, S^2, \ldots, S^d\}$ should be interpreted as a property of the concatenation $C = S^1 \$_1 S^2 \$_2 \cdots \$_{d-1} S^d \#$, where characters $\$_i$ and $\#$ are all distinct and do not appear in the strings of $\mathcal{S}$.

8.4.1   Maximal repeats

A *repeat* in a string $T = t_1 t_2 \cdots t_n$ is a substring X that occurs more than once in T. A repeat X is *right-maximal* (respectively, *left-maximal*) if it cannot be extended to the

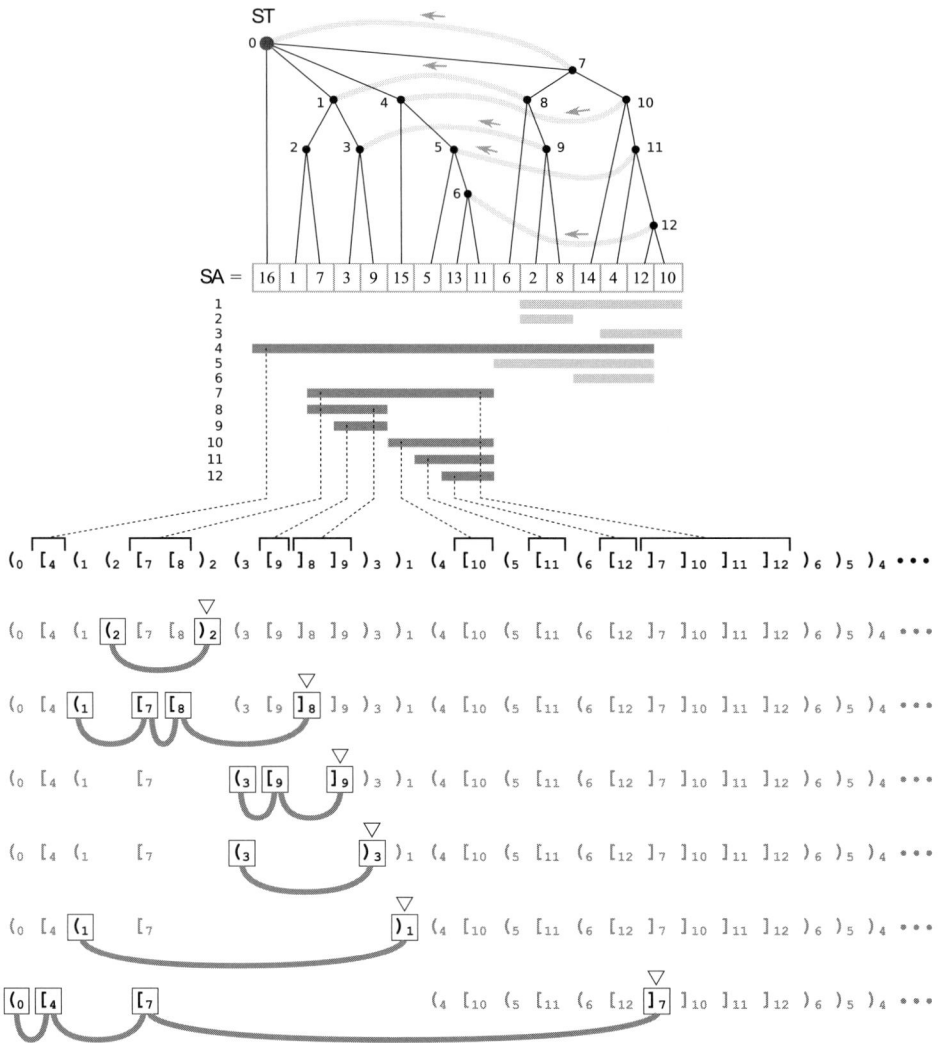

Figure 8.6 Building the suffix links of internal nodes from the suffix links of leaves. (Top) The intervals $[sl(i_v)..sl(j_v)]$ for every internal node v (gray bars), where i_v and j_v are the first and the last position of the interval of v in SA, and the parenthesization produced by a depth-first traversal of ST. For clarity, character labels are omitted and nodes are assigned numerical identifiers. (Bottom) Left-to-right scan of the parenthesization, represented as a doubly-linked list with shortcuts connecting open and closed parentheses. Triangles show the current position of the pointer.

right (respectively, to the left), by even a single character, without losing at least one of its occurrences. A repeat is *maximal* if it is both left- and right-maximal. For example, the maximal repeats of $T = \text{ACAGCAGT}$ are A and CAG.

Maximal repeat detection (Problem 8.2) is tightly connected to the properties of the suffix tree of T.

Problem 8.2   Maximal repeats

Find all maximal repeats in a string T, that is, all substrings of T that are both left- and right-maximal.

It suffices to build $\mathsf{ST}_T = (V, E)$ and to traverse V. Indeed, note that the right-maximal substrings of T are exactly the labels of the internal nodes of ST_T, thus the maximal repeats of T correspond to a subset of $V \setminus Y$, where Y is the set of leaves of ST_T. To test the left-maximality of an internal node, it suffices to perform a linear-time post-order traversal of ST_T, keeping an integer $\texttt{previous}[u] \in \Sigma \cup \{\#\}$ at each node u. Assume that all children of an internal node u have already been explored, and that each such child v has $\texttt{previous}[v] = \#$ if V is the leaf that corresponds to suffix $T[1..n]$, or if there are at least two leaves in the subtree of v whose corresponding suffixes $T[i..n]$ and $T[j..n]$ in T are such that $T[i-1] \neq T[j-1]$. Otherwise, $\texttt{previous}[v] = c \in \Sigma$, where c is the character that precedes all suffixes of T that correspond to leaves in the subtree of v. Then, we set $\texttt{previous}[u] = \#$ if there is a child v of u with $\texttt{previous}[v] = \#$, or if there are at least two children v and w of u such that $\texttt{previous}[v] \neq \texttt{previous}[w]$. Otherwise, we set $\texttt{previous}[u] = \texttt{previous}[v]$ for some child v of u. The maximal repeats of T are then the labels of all internal nodes u with $\texttt{previous}[u] = \#$, and their occurrences in T are stored in the leaves that belong to the subtree rooted at u. By Theorem 8.15 we thus have the following fact.

THEOREM 8.19   *Problem 8.2 of listing all maximal repeats of a string $T = t_1 t_2 \cdots t_n \in [1..\sigma]^*$ and their occurrences can be solved in $O(\sigma + n)$ time.*

Maximal repeats can also be characterized through the suffix-link tree SLT_T: recall Figure 8.4. Note that every leaf of SLT_T either has more than one Weiner link or corresponds to $T\#$. The set of all leaves of SLT_T (excluding $T\#$) and of all its internal nodes with at least two (implicit or explicit) Weiner links coincides with the set of all substrings of T that are maximal repeats. The set of all left-maximal substrings of T enjoys the *prefix closure* property, in the sense that, if a string is left-maximal, so is any of its prefixes. It follows that the maximal repeats of T form a subtree of the suffix tree of T rooted at r (thick black lines in Figure 8.4).

Figure 8.7 visualizes concepts that are derived from maximal repeats: a *supermaximal repeat* is a maximal repeat that is not a substring of another maximal repeat, and a *near-supermaximal repeat* is a maximal repeat that has at least one occurrence that is not contained inside an occurrence of another maximal repeat. Maximal exact matches will be defined in Section 11.1.3 and maximal unique matches will be our next topic.

8.4.2   Maximal unique matches

A *maximal unique match (MUM)* of two string S and T is a substring X that occurs exactly once in both S and T and that cannot be extended to the left or to the right without losing one of the occurrences. That is, for a MUM X it holds that $X = s_i s_{i+1} \cdots s_{i+k-1}$,

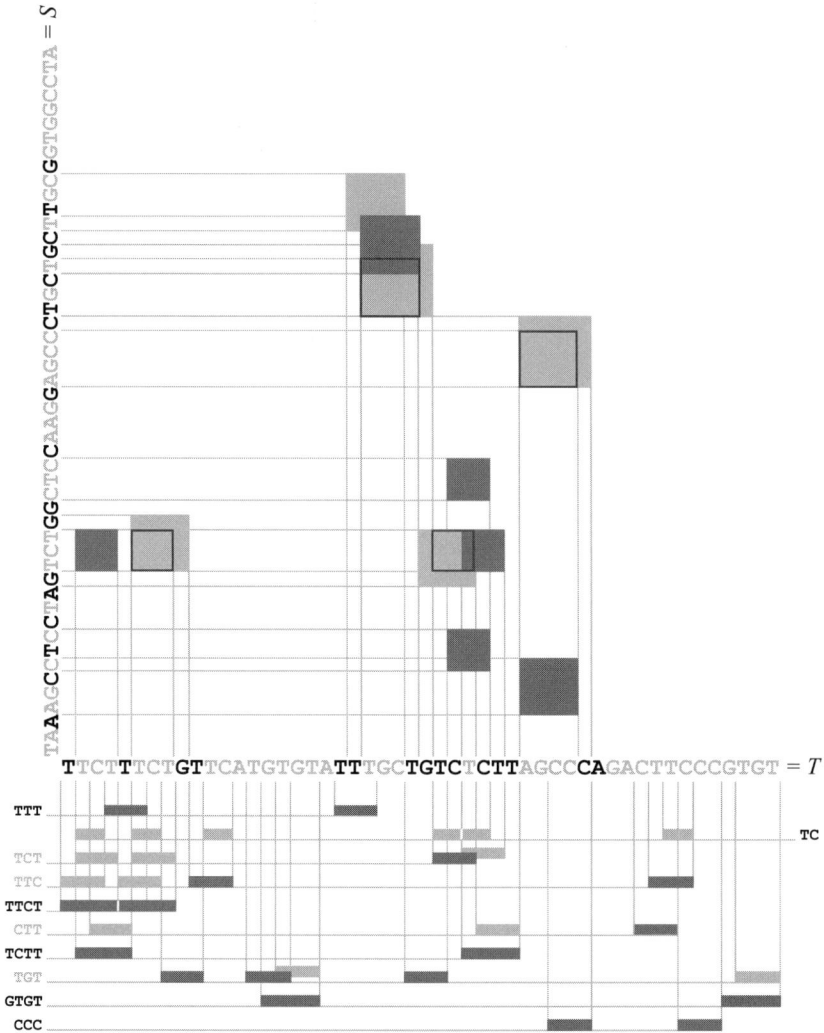

Figure 8.7 (Top) Examples of maximal exact matches (dark-gray rectangles) and maximal unique matches (light-gray rectangles) between string S and string T. Empty rectangles with black border show pairs of occurrences of AGCC, TGCT, and TCT that are not maximal exact matches, because at least one of their flanking characters matches in S and T. (Bottom) Examples of supermaximal repeats (black strings) and of near-supermaximal repeats (gray strings) of string T, and their occurrences (rectangles). The occurrences of a repeat that are not covered by the occurrences of any other repeat of T are highlighted in dark gray. Note that maximal repeat TC is not near-supermaximal, since all its occurrences are covered by occurrences of TTC or TCT.

$X = t_j t_{j+1} \cdots t_{j+k-1}$, $s_{i-1} \neq t_{j-1}$, and $s_{i+k} \neq t_{j+k}$ (for clarity, we omit the boundary cases here). To detect all the MUMs of two strings S and T, it suffices again to build the suffix tree $\mathsf{ST}_C = (V, E)$ of the concatenation $C = S\$T\#$ and to traverse V. Indeed, MUMs are right-maximal substrings of C, so, like maximal repeats, they are a subset of $V \setminus Y$. Moreover, only internal nodes $v \in V$ with exactly two leaves as children can be MUMs. Let the two leaves of a node v be associated with suffixes $C[i..|C|]$ and

$C[j..|C|]$, respectively. Then, i and j must be such that $i \leq |S|$ and $j > |S + 1|$, and the left-maximality of v can be checked by accessing $S[i - 1]$ and $T[j - 1]$ in constant time.

The notion of maximal unique matches extends naturally to a set of strings, as stated in Problem 8.3.

Problem 8.3 Maximal unique matches

Given a set of strings $\mathcal{S} = \{S^1, S^2, \ldots, S^d\}$ of total length $n = \sum_{i=1}^{d} |S^i|$, report all maximal repeats X of $\mathcal{S}$ that are *maximal unique matches*, that is, X appears exactly once in each string of $\mathcal{S}$.

To solve Problem 8.3, we can build again the generalized suffix tree $\mathsf{ST}_C = (V, E)$ of the concatenation $C = S^1 \$_1 S^2 \$_2 \cdots \$_{d-1} S^d \#$. We can determine left-maximal nodes with the same post-order traversal of ST_C we used for maximal repeats, and we can detect nodes $v \in V$ with exactly d leaves in their subtree. However, we need also to check that no two leaves in the subtree of v come from the same string of $\mathcal{S}$. This can be done by filling a bitvector $I[1..d]$ (initialized to zeros), setting at each leaf i under v value $I[k] = 1$ if suffix $C_{\mathsf{SA}[i]..|C|}$ starts inside string S^k. Value k can be stored at leaf i during the construction of the suffix tree. After filling I under node v, if all bits are set to 1, each of the d suffixes under v must originate from a different string, and hence maximal node v defines a maximal unique match. This process can be repeated for each maximal node v having exactly d leaves under it. It follows that the checking of d bits at node v can be amortized to the leaves under it, since each leaf corresponds to only one v to which this process is applied. By Theorem 8.15 we thus have the following result.

THEOREM 8.20 *Problem 8.3 of finding all the maximal unique matches (MUMs) of a set of strings of total length n can be solved in $O(\sigma + n)$ time.*

8.4.3 Document counting

A generalization of the MUM problem is to report in how many strings a (right-)maximal repeat appears, as stated in Problem 8.4.

Problem 8.4 Document counting

Given a set of strings $\mathcal{S} = \{S^1, S^2, \ldots, S^d\}$ of total length $n = \sum_{i=1}^{d} |S^i|$, compute for each (right-)maximal repeat X of $\mathcal{S}$ the number of strings in $\mathcal{S}$ that contain X as a substring.

To solve Problem 8.4, we can start exactly as in the solution for MUMs by building the generalized suffix tree $\mathsf{ST}_C = (V, E)$ of the concatenation $C = S^1 \$_1 S^2 \$_2 \cdots \$_{d-1} S^d \#$. For a node $v \in V$, let $n(v, k)$ be the number of leaves in the subtree rooted at v that correspond to suffixes of C that start inside string S^k. Recall the algorithm for building the suffix tree of a string T from its suffix array and LCP array (Figures 8.5(b) and (c)). We now follow the same algorithm, but add some auxiliary computations to its execution. Assign to each internal node v the temporary counter `leafCount[v]`, storing the total number of leaves in the subtree rooted at v, and the temporary counter `duplicateCount[v]` $= \sum_{i=1}^{d} \max\{0, n(v,k) - 1\}$. At the end of the process, we make sure that these temporary counters obtain their final values. The solution to Problem 8.4 is given by `leafCount[v]` $-$ `duplicateCount[v]` and available for each internal node v, corresponding thus to a right-maximal repeat. Observe that the solution to the MUMs computation is a special case with `leafCount[v]` $= d$ and `duplicateCount[v]` $= 0$ for nodes v that also correspond to left-maximal repeats.

To start the update process, counters `leafCount[v]` and `duplicateCount[v]` are initially set to zero for each v. At each step of the process, `leafCount[v]` stores the number of leaves in the current subtree rooted at any node v, except for nodes in the *right-most path* of the current suffix tree, for which `leafCount[v]` stores the number of leaves in the current subtree of v, minus the number of leaves in the subtree rooted at the *right-most child* of v. Let us call this property the *leaf count invariant*. When we insert a new leaf below an internal node w, it suffices to increment `leafCount[w]` by one, and when we visit a node v while traversing the right-most path bottom-up, it suffices to increment `leafCount[u]` by `leafCount[v]`, where u is the parent of v. These updates guarantee that the leaf count invariant will hold.

We can compute `duplicateCount[v]` by reusing the batched lowest common ancestor computation from Lemma 8.17. Assume that the set of query intervals Q is empty at the beginning. Let `document[1..|C|]` be a vector that stores at `document[i]` value k if $\mathsf{SA}_C[i]$ starts inside string S^k in C. Building this array is left to Exercise 8.20. Assume that we are scanning SA_C from left to right, and that we are currently at position i. Our strategy consists in incrementing `duplicateCount[v]` for the node v that is the lowest common ancestor of two closest leaves i and j of ST_C such that `document[i]` $=$ `document[j]`, that is, `document[k]` $\neq$ `document[i]` for all $k \in [i+1..j-1]$. Observe that this process is like *dueling* i and j, since one of them is announced as a duplicate at v. Specifically, assume that we are currently processing leaf i. We maintain for each document identifier $k \in [1..d]$ a variable `previous[k]` that stores the largest $j \in [1..i-1]$ such that `document[j]` $= k$. Then we can access $j =$ `previous[document[i]]` at leaf i. Once we have updated the current suffix tree using suffix $\mathsf{SA}_C[i]$, we set `previous[document[i]]` $= i$ and we insert $[j..i]$ into the set of query intervals Q. After processing all leaves, we apply Lemma 8.17 on ST_C and set Q. As a result, we obtain for each query interval $q \in Q$ the lowest common ancestor v in ST_C. For each pair (q, v) we increment `duplicateCount[v]` by one.

Finally, we *propagate* the values of `duplicateCount` (increment the parent counter by the child counters) from such lowest common ancestors to all nodes of ST_C by a linear-time traversal of the tree. This guarantees that all counter values will be correctly computed since a duplicate leaf under node v is a duplicate leaf under each ancestor of v. By Theorem 8.15 we thus have the following result.

THEOREM 8.21   *Problem 8.4 of document counting on a set of strings of total length n can be solved in $O(\sigma + n)$ time.*

8.4.4   Suffix–prefix overlaps

In later chapters we will present variants of the *fragment assembly* problem, which consists in connecting together substrings of an unknown string (typically short DNA fragments, known as *short reads*) on the basis of their shared prefixes and suffixes. As a preview, we consider here the following problem.

Problem 8.5   All-against-all suffix–prefix overlaps

Given a set of strings $\mathcal{R} = \{R^1, R^2, \dots, R^d\}$ of total length n and a threshold τ, output all pairs (R^i, R^j) such that $R^i = XY$ and $R^j = YZ$ for some X, Y, Z in Σ^*, where $|Y| \geq \tau$.

Problem 8.5 can be solved again with a suitable depth-first traversal of a suffix tree. Indeed, build the generalized suffix tree ST_T of $T = R^1 \$_1 R^2 \$_2 \cdots \$_{d-1} R^d \$_d \#$, and imagine that an empty stack is allocated to every string in $\mathcal{R}$. Assume that, during a depth-first traversal of ST_T, we encounter for the first time a node v with string depth at least τ: we push its string depth to the stacks of all R^i such that there is an edge from v starting with character $\$_i$. Symmetrically, when v is visited for the last time, we pop from all the corresponding stacks. Clearly, when we reach a leaf node corresponding to suffix $R^i \$_i \cdots$ of T, the non-empty stacks in memory are exactly those that correspond to strings R^j having a suffix of length at least τ that equals a prefix of R^i, and the top of the non-empty stack j contains the length of the longest suffix of R^j that equals a prefix of R^i.

We should keep in memory only the non-empty stacks at each step of the traversal. This can be done using a doubly-linked list, to which we append in constant time a new stack, and we delete in constant time an empty stack. An auxiliary table $P[1..d]$ points to the (variable) position in the linked list of the stack associated with each string in $\mathcal{R}$. The following result is thus immediately obtained.

THEOREM 8.22   *Problem 8.5 of all-against-all suffix–prefix overlaps can be solved in $O(\sigma + n + \texttt{pairs})$ time, where* `pairs` *is the number of pairs in the given set of string $\mathcal{R}$ that overlap by at least τ characters.*

8.5   Literature

We just scratched the surface of k-mer indexes, giving almost folklore solutions. Elias coding is from Elias (1975). The literature on inverted files (see for example Baeza-Yates & Ribeiro-Neto (2011)) contains more advanced techniques that can also be adapted to k-mer indexes. Suffix arrays were first introduced in Gonnet *et al.* (1992) and Manber & Myers (1993), and our doubling algorithm for their construction mimics the one in Manber & Myers (1993). The linear-time construction algorithm presented here is from Ko & Aluru (2005). Simultaneously and independently, two other linear-time construction algorithms were invented (Kärkkäinen *et al.* 2006, Kim *et al.* 2005): all of these follow similar divide-and-conquer recursive strategies, but differ considerably in technical details. We chose to present Ko's & Aluru's solution because it contains as a sub-result a useful technique for sorting sets of strings. By encapsulating this sub-result in a conceptual black box, the rest of the algorithm can be presented in relatively self-contained modules. Nevertheless, as with other linear-time suffix array construction algorithms, the very interplay among all steps is key to achieving linear time: such deep relationships make these algorithms one of the most fascinating achievements in the field of string processing. For readers wishing to obtain a more detailed exposition on the different suffix array construction algorithms, there is an excellent survey available by Puglisi *et al.* (2007).

Suffix trees are an earlier invention: the 24th Annual Symposium on Combinatorial Pattern Matching (CPM 2013) celebrated the 40th anniversary of Weiner's foundational paper (Weiner 1973). Construction algorithms with different properties have been considered (McCreight 1976; Ukkonen 1995; Farach 1997). We propose an exercise that sketches the online construction (Ukkonen 1995). Many textbooks describe this algorithm in detail, because it is the simplest to understand and very likely the most educative. We chose to give the readers of this book the chance to "rediscover" some of its parts. The first linear-time suffix tree construction algorithm on large alphabets was given in Farach (1997): this algorithm can indirectly construct suffix arrays in linear time, and it already uses a divide-and-conquer recursion strategy similar to the later direct suffix array constructions mentioned above. Owing to the existence of efficient and practical suffix array construction algorithms, we chose to give the details of just the folklore construction of the suffix tree from the suffix array and LCP array. The surprisingly simple construction of the LCP array (left as an exercise) is from Kasai *et al.* (2001).

For learning more advanced properties of suffix trees and their variants, such as suffix automata and compact suffix automata, we refer the interested reader to stringology textbooks like Crochemore & Rytter (2002). We mentioned just a few illustrative applications of suffix trees to biological sequence analysis, whose connections to the rest of the book will become clear in the forthcoming chapters: the maximal unique matches computation is folklore and hinted to us by Gonzalo Navarro, document counting is from Hui (1992), and the overlap computation is from Gusfield *et al.* (1992). More related applications are described by Gusfield (1997), whose presentation we partly followed to introduce the repeat finding concepts. We made a significant conceptual simplification by introducing *batched lowest common ancestor queries*, which we also use for the construction of suffix links. Usually these queries are solved with rather advanced data structures for *range minimum queries* (Bender & Farach-Colton 2000; Harel & Tarjan 1984; Fischer & Heun 2011), but we showed that one can collect all such queries and then use a simple scan to collect all answers at once.

Other theoretically important text indexing data structures we did not discuss are the *directed acyclic word graph* (DAWG) and its compacted version, the CDAWG: these are covered by Exercises 8.24 and 8.25. The DAWG was introduced in Blumer *et al.* (1985), and the CDAWG in Crochemore & Vérin (1997b). Efficient construction algorithms are presented in Apostolico & Lonardi (2002) and Crochemore & Vérin (1997a). The relationship between CDAWG and maximal repeats described in Exercise 8.25 was studied in Raffinot (2001).

Exercises

8.1 Consider a k-mer index in which the pointers to the lists of occurrences of k-mers are encoded using a table of size σ^k in which the ith element is a pointer to the list of the occurrence positions of k-mer $P = p_1 p_2 \cdots p_k$ in $T = t_1 t_2 \cdots t_n$ for $i = p_1 \cdot \sigma^{k-1} + p_2 \cdot \sigma^{k-2} + \cdots + p_k$, where each $p_j \in \{0, 1, \ldots, \sigma - 1\}$. Show that both this table and the gamma-encoded lists can be constructed in $O(\sigma^k + n)$ time by two scans from left to right over T. *Hint.* Use a first scan to determine the size of the lists and a second scan to fill both the lists and the table.

8.2 The table of the previous assignment is sparse if k is large. Show that by using techniques from Section 3.2 one can represent the table in $\sigma^k(1 + o(1)) + \ell \log(n + 2n \log \ell)$ bits, where ℓ is the number of non-empty lists, while still retaining constant-time access to the list. *Hint.* Use a bitvector marking the non-empty elements and supporting rank queries. Concatenate all encodings of the occurrence lists and store a pointer to the starting position of each list inside the concatenation.

8.3 Show how to represent the pointer table of the previous assignment in $\sigma^k(1 + o(1)) + (\ell + n(1 + 2 \log \ell))(1 + o(1))$ bits. *Hint.* Use a bitvector supporting rank queries and another supporting select queries.

8.4 Derive an analog of Theorem 8.2 replacing γ-coding with δ-coding. Do you obtain a better bound?

8.5 The *order-k de Bruijn graph* on text $T = t_1 t_2 \cdots t_n$ is a graph whose vertices are $(k-1)$-mers occurring in T and whose arcs are k-mers connecting $(k-1)$-mers consecutive in T. More precisely, let $\mathtt{label}(v) = \alpha_1 \alpha_2 \cdots \alpha_{k-1}$ denote the $(k-1)$-mer of vertex v and $\mathtt{label}(w) = \beta_1 \beta_2 \cdots \beta_{k-1}$ the $(k-1)$-mer of vertex w. There is an arc e from vertex v to w with $\mathtt{label}(e) = \alpha_1 \alpha_2 \cdots \alpha_{k-1} \beta_{k-1}$ iff $\alpha_2 \alpha_3 \cdots \alpha_{k-1} = \beta_1 \beta_2 \cdots \beta_{k-2}$ and $\mathtt{label}(e)$ is a substring of T. See Section 9.7 for more details on de Bruijn graphs and on their efficient representation. Modify the solution to Exercise 8.1 to construct the order-k de Bruijn graph.

8.6 The table of size σ^k on large k makes the above approach infeasible in practice. Show how hashing can be used for lowering the space requirement.

8.7 Another way to avoid the need for a table of size σ^k is to use a suffix array (enhanced with an LCP array) or a suffix tree. In the suffix tree of T, consider all paths from the root that spell strings of length k; these are the distinct k-mers occurring in T. Consider one such path labeled by X ending at an edge leading to node v. The leaves in the subtree of v contain all the suffixes prefixed by k-mer X in lexicographic order. Show that one can sort all these lexicographically sorted lists of occurrences of k-mers in $O(n)$ time, to form the k-mer index.

8.8 Modify the approach in the previous assignment to construct the de Bruijn graph in linear time.

8.9 Visualize the suffix array of the string ACGACTGACT# and simulate a binary search on the pattern ACT.

8.10 Recall the indicator vector $I[1..n]$ for string $T = t_1 t_2 \cdots t_n$ with $I[i] = 1$ if suffix $T_{i..n}$ is small, that is, $T_{i..n} < T_{i+1..n}$, and $I[i] = 0$ otherwise. Show that it can be filled with one $O(n)$ time scan from right to left.

8.11 Recall R and SA_R from the linear-time suffix array construction. Give pseudocode for mapping SA_R to suffix array SA_T of T so that $\mathsf{SA}_T[i]$ is correctly computed for small suffixes.

8.12 Give the pseudocode for the linear-time algorithm sketched in the main text to fill suffix array entries for large suffixes, assuming that the entries for small suffixes have already been computed.

8.13 We assumed there are fewer small suffixes than large suffixes. Consider the other case, especially, how does the string sorting procedure need to be changed in order to work correctly for substrings induced by large suffixes? Solve also the previous assignments switching the roles of small and large suffixes.

8.14 Show how to compute the LCP$[2..n]$ array in linear time given the suffix array. *Hint.* With the help of the inverse of the suffix array, you can scan the text $T = t_1 t_2 \cdots t_n$ from left to right, comparing suffix $t_1 t_2 \cdots$ with its predecessor in suffix array order, suffix $t_2 \cdots$ with its predecessor in suffix array order, and so on. Observe that the common prefix length from the previous step can be used in the next step.

8.15 Visualize the suffix tree of a string ACGACTGACT and mark the nodes corresponding to maximal repeats. Visualize also Weiner links.

8.16 Consider the suffix tree built from a suffix array. Give a linear-time algorithm to compute suffix links for its leaves.

8.17 Visualize the suffix tree on a concatenation of two strings of your choice and mark the nodes corresponding to maximal unique matches. Choose an example with multiple nodes with two leaf children, of which some correspond to maximal unique matches, and some do not.

8.18 Show that the number of implicit Weiner links can be bounded by $2n - 2$. *Hint.* Show that all implicit Weiner links can be associated with unique edges of the suffix tree of the reverse.

8.19 Show that the number of implicit Weiner links can be bounded by n. *Hint.* Characterize the edges where implicit Weiner links are associated in the reverse. Observe that these edges form a subset of a cut of the tree.

8.20 Recall the indicator vector $I[1..|C|]$ in the computation of MUMs on multiple sequences. Show how it can be filled in linear time.

8.21 The *suffix trie* is a variant of suffix tree, where the suffix tree edges are replaced by unary paths with a single character labeling each edge: the concatenation of unary path labels equals the corresponding suffix tree edge label. The suffix trie can hence be quadratic in the size of the text (which is a reason for the use of suffix trees instead). Now consider the suffix link definition for the nodes of the suffix trie that do *not* correspond to nodes of the suffix tree. We call such nodes *implicit nodes* and such suffix links *implicit suffix links* of the suffix tree. Show how to *simulate* an implicit suffix link $sl(v,k) = (w,k')$, where $k > 0$ ($k' \geq 0$) is the position inside the edge from the parent of v to v (respectively, parent of w to w), where $v = (v,0)$ and $w = (w,0)$ are *explicit* nodes of the suffix tree and (v,k) (possibly also (w,k')) is an implicit node of the suffix tree. In the simulation, you may use parent pointers and explicit suffix links. What is the worst-case computation time for the simulation?

8.22 Consider the following *descending suffix walk* by string $S = s_1 s_2 \cdots s_m$ on the suffix *trie* of $T = t_1 t_2 \cdots t_n$. Walk down the suffix tree of T as deep as the prefix of S still matches the path followed. When you cannot walk any more, say at s_i, follow a suffix link, and continue walking down as deep as the prefix of $s_i s_{i+1} \cdots$ still matches the path followed, and follow another suffix link when needed. Continue this until you reach the end of S. Observe that the node v visited during the walk whose string depth is greatest defines the *longest common substring* of S and T: the string X labeling the path from root to v is such a substring. Now, consider the same algorithm on the suffix *tree* of T. Use the simulation of implicit suffix links from the previous assignment. Show that the running time is $O(\sigma + |T| + |S| \log \sigma)$, i.e. the time spent in total for simulating implicit suffix links can be *amortized* on scanning S.

8.23 The suffix tree can also be constructed directly without requiring the suffix array. *Online* suffix tree construction works as follows. Assume you have the suffix tree $\mathsf{ST}(i)$ of $t_1 t_2 \cdots t_i$, then update it into the suffix tree $\mathsf{ST}(i+1)$ of $t_1 t_2 \cdots t_{i+1}$ by adding t_{i+1} at the end of each suffix. Let $(v_1, k_1), (v_2, k_2), \ldots, (v_i, k_i)$ be the (implicit) nodes of $\mathsf{ST}(i)$ corresponding to paths labeled $t_1 t_2 \cdots t_i$, $t_2 t_3 \cdots t_i$, $\ldots$, t_i, respectively (see the assigments above for the notion of implicit and explicit nodes). Notice that $sl(v_j, k_j) = (v_{j+1}, k_{j+1})$.

(a) Show that $(v_1, k_1), (v_2, k_2), \ldots, (v_i, k_i)$ can be split into lists $(v_1, 0), \ldots, (v_l, 0)$, $(v_{l+1}, k_{l+1}), \ldots, (v_a, k_a)$, and $(v_{a+1}, k_{a+1}), \ldots, (v_i, k_i)$ such that nodes in the first list are leaves before and after appending t_{i+1}, nodes in the second list do not yet have a branch (or a next character along the edge) starting with character t_{i+1}, and nodes in the third list already have a branch (or a next character along the edge) starting with character t_{i+1}.

(b)   From the above it follows that the status of nodes $(v_1, 0), \ldots, (v_l, 0)$ remains the same and no updates to their content are required (the start position of the suffix is a sufficient identifier). For $(v_{l+1}, k_{l+1}), \ldots, (v_a, k_a)$ a new leaf for t_{i+1} needs to be created. In the case of explicit node $(v, 0)$, this new leaf is inserted under node v. In the case of implicit node (v, k), the edge from the parent of v to v is split after position k, and a new internal node is created, inheriting v as child and the old parent of v as parent, and t_{i+1} as a new leaf child. The suffix links from the newly created nodes need to be created during the traversal of the list $(v_{l+1}, k_{l+1}), \ldots, (v_{a+1}, k_{a+1})$. Once one has detected (v_{a+1}, k_{a+1}) being already followed by the required character and created the suffix link from the last created node to the correct successor of (v_{a+1}, k_{a+1}), say (w, k'), the list does not need to be scanned further. To insert t_{i+2}, one can start from the new *active point* (w, k') identically as above, following suffix links until one can walk down with t_{i+2}. Notice the analogy to a descending suffix walk, and use a similar analysis to show that the simulation of implicit suffix links amortizes and that the online construction requires $O(n)$ steps (the actual running time is multiplied by an alphabet factor depending on how the branching is implemented).

8.24   The *directed acyclic word graph* (DAWG) is an automaton obtained by minimizing a suffix trie. Suppose that we build the suffix trie of a string $T\#$. Then two nodes v and u of the suffix trie will be merged if the substring that labels the path to v is a suffix of the substring that labels the path to u and the frequencies of the two substrings in $T\#$ are the same. Show that the number of vertices in the DAWG is linear in the length of T, by showing that the following statements hold.

(a)   There is a bijection between the nodes of the suffix-link tree of $\underline{T}\#$ and the vertices of the DAWG labeled by strings whose frequency is at least two in $T\#$.

(b)   There is a bijection between all prefixes of $T\#$ that have exactly one occurrence in $T\#$ and all the vertices of the DAWG labeled by strings whose frequency is exactly one in $T\#$.

Then show that the number of edges is also linear, by showing a partition of the edges into three categories:

(a)   edges that are in bijection with the edges of the suffix-link tree of $\underline{T}\#$,

(b)   edges that are in bijection with a subset of the edges of the suffix tree of $T\#$, and

(c)   edges that connect two nodes that are in bijection with two prefixes of $T\#$ of lengths differing by one.

8.25   The *compacted DAWG* (CDAWG) of a string $T\#$ is obtained by merging every vertex of the DAWG of $T\#$ that has just one outgoing arc with the destination of that arc. Equivalently, the CDAWG can be obtained by minimizing the suffix tree of T in the same way as the suffix trie is minimized to obtain the DAWG. Note that the resulting automaton contains exactly one vertex with no outgoing arc, called the *sink*. Show that there is a bijection between the set of maximal repeated substrings of $T\#$ and the set of vertices of the CDAWG, excluding the root and the sink.

9 Burrows–Wheeler indexes

Consider the largest sequenced genome to date, the approximately $25.5 \cdot 10^9$ base-pair-long genome of the white spruce *Picea glauca*. The suffix array of this genome would take approximately $1630 \cdot 10^9$ bits of space, or approximately 204 gigabytes, if we represent it with 64-bit integers, and it would take approximately 110 gigabytes if we represent it with fixed-length bit-field arrays. The suffix tree of the same genome would occupy from three to five times more space. However, the genome itself can be represented in just $51 \cdot 10^9$ bits, or approximately 6.4 gigabytes, assuming an alphabet of size $\sigma = 4$: this is equivalent to $n \log \sigma$ bits, where n is the length of the string. The gaps between the sizes of the suffix tree, of the suffix array, and of the original string are likely to grow as more species are sequenced: for example, the unsequenced genome of the flowering plant *Paris japonica* is estimated to contain $150 \cdot 10^9$ base pairs. The size of metagenomic samples is increasing at an even faster pace, with files containing approximately $160 \cdot 10^9$ base pairs already in public datasets.

Burrows–Wheeler indexes are a space-efficient variant of suffix arrays and suffix trees that take $n \log \sigma (1 + o(1))$ bits for a genome of length n on an alphabet of size σ (or just about 10 gigabytes for *Picea glauca*), and that support a number of high-throughput sequencing analyses approximately as fast as suffix arrays and suffix trees. Recall from Section 2.2 that we use the term *succinct* for data structures that, like Burrows–Wheeler indexes, occupy $n \log \sigma (1 + o(1))$ bits. This chapter walks the reader through the key algorithmic concepts behind Burrows–Wheeler indexes, leaving applications to Part IV and Part V.

After describing the Burrows–Wheeler transform of a string and related data structures for counting and locating all the occurrences of a pattern, the focus shifts to the *bidirectional* Burrows–Wheeler index, a powerful data structure that allows one to enumerate all internal nodes of the suffix tree of a string in a small amount of space. The bidirectional index will be used heavily in the following chapters, but for concreteness a sampler of its flexibility will be provided in Section 9.7.3. The chapter covers also advanced topics, like the space-efficient construction of the Burrows–Wheeler transform of a string, and concludes with a unified description of Burrows–Wheeler indexes for labeled trees, directed acyclic graphs, and de Bruijn graphs, which will be the substrate of a number of applications in Part IV and V.

9.1   Burrows–Wheeler transform (BWT)

Given a string T, its Burrows–Wheeler transform can be seen as the permutation of T induced by scanning its suffix array sequentially, and by printing the character that precedes each position in the suffix array. More formally, we have the following definition.

DEFINITION 9.1   *Let $T = t_1 t_2 \cdots t_n$ be a string such that $t_i \in \Sigma = [1..\sigma]$ for $i \in [1..n-1]$ and $t_n = \#$, where $\# = 0$ is a character that does not belong to alphabet Σ. The* Burrows–Wheeler transform *of T, denoted by BWT_T, is the permutation $L[1..n]$ of T defined as follows:*

$$L[i] = \begin{cases} T[\mathsf{SA}_T[i] - 1] & \textit{if } \mathsf{SA}[i] > 1, \\ t_n = \# & \textit{if } \mathsf{SA}[i] = 1. \end{cases}$$

See Figure 9.1 for an illustration of the transform.

Alternatively, BWT_T can be defined in terms of cyclic shifts of T: see Insight 9.1. In what follows, we repeatedly use the following function on two integers i and n:

$$i \;(\mathrm{mod}_1\; n) = \begin{cases} n & \text{if } i = 0, \\ i & \text{if } i \in [1..n], \\ 1 & \text{if } i = n+1. \end{cases}$$

	SA$_T$			BWT$_T$	
1	16		1	C	#
2	1		2	#	AGAGCGAGAGCGCGC# $= T$
3	7		3	G	AGAGCGCGC#
4	3		4	G	AGCGAGAGCGCGC#
5	9		5	G	AGCGCGC#
6	15		6	G	C#
7	5		7	G	CGAGAGCGCGC#
8	13		8	G	CGC#
9	11		9	G	CGCGC#
10	6		10	C	GAGAGCGCGC#
11	2		11	A	GAGCGAGAGCGCGC#
12	8		12	A	GAGCGCGC#
13	14		13	C	GC#
14	4		14	A	GCGAGAGCGCGC#
15	12		15	C	GCGC#
16	10		16	A	GCGCGC#

Figure 9.1 The Burrows–Wheeler transform of string $T = \mathsf{AGAGCGAGAGCGCGC\#}$, shown for clarity with the suffix array of T and with the corresponding set of lexicographically sorted suffixes of T

Insight 9.1 Burrows–Wheeler transform and cyclic shifts

The Burrows–Wheeler transform was originally defined in terms of the *cyclic shifts* of a string $T = t_1 \cdots t_n \in [1..\sigma]^+$. Indeed, assume that we build the n cyclic shifts of T, or equivalently strings $t_1 t_2 \cdots t_n, t_2 t_3 \cdots t_n t_1, \ldots, t_n t_1 \cdots t_{n-1}$, and assume that we sort them into lexicographic order. For concreteness, let $T = $ mississippi. After sorting the cyclic shifts of T, we get the following matrix of characters M:

i	M
1	imississip**p**
2	ippimissis**s**
3	issippimis**s**
4	ississippi**m**
5	mississipp**i**
6	pimississi**p**
7	ppimississ**i**
8	sippimissi**s**
9	sissippimi**s**
10	ssippimiss**i**
11	ssissippim**i**

The *first column F* of M contains the characters of T in sorted order, and the *last column L* of M (highlighted with bold letters) contains the BWT of T, that is, the string pssmipissii. The suffix of T in each cyclic permutation is underlined. The original string mississippi is row $x_1 = 5$ of matrix M, and its right-shift imississipp is row $x_2 = 1$. Note that shifting mississippi to the right moves its last character $L[x_1] = $ i to the beginning of the next cyclic permutation imississipp, that is, $L[x_1] = F[x_2]$. It follows that row x_2 must belong to the interval of all sorted cyclic shifts that satisfy $F[x_2] = $ i: see Section 9.1 for an algorithm that computes x_2 using just x_1 and L. Once we know how to move from the row x_i of a cyclic permutation to the row x_{i+1} of its right shift, we can reconstruct T *from right to left*, by printing the sequence $L[x_n] \cdots L[x_2]L[x_1]$.

Note that, since the string of this example is not periodic, no two cyclic shifts are identical, therefore the lexicographic order between every pair of cyclic shifts is well defined. In general, appending an artificial character $\# = 0 \notin [1..\sigma]$ at the end of T ensures that all cyclic shifts of $T\#$ are distinct for any T, and moreover that sorting the cyclic shifts of $T\#$ coincides with sorting the suffixes of $T\#$.

A key feature of BWT_T is that it is *reversible*: given $\mathsf{BWT}_T = L$, one can reconstruct the unique T of which L is the Burrows–Wheeler transform. Indeed, let V and W be two suffixes of T such that V is lexicographically smaller than W, and assume that both V and W are preceded by character a in T. It follows that suffix aV is lexicographically smaller than suffix aW, thus there is a bijection between suffixes *preceded by* a and suffixes that *start with* a that preserves the relative order of such suffixes. Consider thus suffix $T[i..n]$, and assume that it corresponds to position p_i in SA_T. If $L[p_i] = a$ is the

kth occurrence of a in $L[1..p_i]$, then the suffix $T[i-1 \pmod_1 n)..n]$ that starts at the *previous position* in T must be the kth suffix that starts with a in SA_T, and its position p_{i-1} in SA_T must belong to the compact interval $\vec{a}$ that contains all suffixes that start with a. For historical reasons, the function that projects the position p_i in SA_T of a suffix $T[i..n]$ to the position p_{i-1} in SA_T of the *previous suffix in circular string order* $T[i-1 \pmod_1 n)..n]$ is called LF-*mapping* (see Insight 9.1) and it is defined as

$$\mathsf{LF}(i) = j, \text{ where } \mathsf{SA}[j] = \mathsf{SA}[i] - 1 \pmod_1 n). \tag{9.1}$$

In what follows we will use $\mathsf{LF}^{-1}(j) = i$ to denote the *inverse* of function LF, or equivalently the function that projects position j in SA onto the position i in SA such that $\mathsf{SA}[i] = \mathsf{SA}[j] + 1 \pmod_1 n)$. Once we know how a right-to-left movement on string T projects into a movement on SA_T, we can reconstruct T from right to left, by starting from the position $x_1 = 1$ of suffix # in SA_T and by printing the sequence $\#L[x_1]L[x_2]\ldots L[x_{n-1}] = \underline{T}$, where $x_{i+1} = \mathsf{LF}(x_i)$ for $i \in [1..n-2]$. Similarly, we can reconstruct T from left to right by starting from the position $x_1 = \mathsf{LF}^{-1}(1)$ of T in SA_T, that is, from the position of # in BWT_T, and by printing the sequence $L[x_2]\ldots L[x_n]\# = T$, where $x_{i+1} = \mathsf{LF}^{-1}(x_i)$ for $i \in [1..n-1]$. More generally, we have the following lemma.

LEMMA 9.2 *Given a length m and a position i in SA_T, we can reconstruct the substring $T[\mathsf{SA}_T[i]-m \pmod_1 n)..\mathsf{SA}_T[i]-1 \pmod_1 n)]$ by iterating $m-1$ times the function LF, and we can reconstruct the substring $T[\mathsf{SA}_T[i]..\mathsf{SA}_T[i]+m-1 \pmod_1 n)]$ by iterating m times the function LF^{-1}.*

Note that we can compute all possible values of the function LF *without using* SA_T: indeed, we can apply Lemma 8.5 on page 134 to stably sort the list of tuples $(L[1], 1, 1)$, $(L[2], 2, 2), \ldots, (L[n], n, n)$ in $O(\sigma + n)$ time. Let L' be the resulting sorted list: it is easy to see that $\mathsf{LF}(L'[i].v) = i$ for all $i \in [1..n]$. In the next section we describe how to compute LF without even building and storing L'.

9.2   BWT index

9.2.1   Succinct LF-mapping

Let again L be the Burrows–Wheeler transform of a string $T = t_1 \cdots t_n$ such that $t_i \in [1..\sigma]$ for all $i \in [1..n-1]$, and $t_n = \#$, and assume that we have an array $C[0..\sigma]$ that stores in $C[c]$ the number of occurrences in T of all characters strictly smaller than c, that is, the sum of the frequency of all characters in the set $\{\#, 1, \ldots, c-1\}$. Note that $C[0] = 0$, and that $C[c] + 1$ is the position in SA_T of the first suffix that starts with character c. Recall that in Section 9.1 we defined $\mathsf{LF}(i) = j$ as the mapping from the position i of the kth occurrence of character c in L to the kth position inside the interval of character $c = L[i]$ in SA_T. It follows that $\mathsf{LF}(i) = C[L[i]] + k$, where the value $k = \mathtt{rank}_{L[i]}(L, i)$ can be computed in $O(\log \sigma)$ time if we represent L using a wavelet tree (see Section 3.3). Computing $\mathsf{LF}^{-1}(i)$ amounts to performing $\mathtt{select}_c(i - C[c])$, where $c \in [0..\sigma]$ is the character that satisfies $C[c] < i \leq C[c+1]$ and can be found in $O(\log \sigma)$ time by binary-searching array C, and `select` can be computed in $O(\log \sigma)$ time if we represent L using a wavelet tree (see Section 3.3). This setup of data structures is called a *BWT index*.

DEFINITION 9.3 *Given a string $T \in [1..\sigma]^n$, a* BWT index *on T is a data structure that consists of*

- $\mathsf{BWT}_{T\#}$, *encoded using a wavelet tree; and*
- *the integer array $C[0..\sigma]$, which stores in $C[c]$ the number of occurrences in $T\#$ of all characters strictly smaller than c.*

Combining Lemma 9.2 with the implementation of LF and of LF^{-1} provided by the BWT index, we get the following results, which will be useful in the following chapters.

LEMMA 9.4 *There is an algorithm that, given the BWT index of a string $T \in [1..\sigma]^n$, outputs the sequence $\mathsf{SA}_{T\#}^{-1}[1], \ldots, \mathsf{SA}_{T\#}^{-1}[n]$ or its reverse in $O(\log \sigma)$ time per value, in $O(n \log \sigma)$ total time, and in $O(\log n)$ bits of working space.*

Figure 9.2 illustrates Lemma 9.4.

LEMMA 9.5 *Suppose that the BWT index of a string $T \in [1..\sigma]^n$ has already been built. Given a position i in $\mathsf{SA}_{T\#}$ and a length m, we can extract in $O(m \log \sigma)$ total time and in $O(\log n)$ bits of working space the following strings:*

- $W_R = t_j \cdot t_{j+1 \, (\mathrm{mod}_1 \, n+1)} \cdots t_{j+m-1 \, (\mathrm{mod}_1 \, n+1)};$ *and*
- $W_L = t_{j-1 \, (\mathrm{mod}_1 \, n+1)} \cdot t_{j-2 \, (\mathrm{mod}_1 \, n+1)} \cdots t_{j-m \, (\mathrm{mod}_1 \, n+1)},$

Algorithm 9.1: Counting the number of occurrences of a pattern in a string using backward search

Input: Pattern $P = p_1 p_2 \cdots p_m$, count array $C[0..\sigma]$, function $\mathrm{rank}_c(L, i)$ on the $\mathsf{BWT}_T = L$ of string $T = t_1 \cdots t_n$, where $t_i \in [1..\sigma]$ for all $i \in [1..n-1]$ and $t_n = \#$.

Output: Number of occurrences of P in T.

1 $i \leftarrow m$;
2 $(sp, ep) \leftarrow (1, n)$;
3 **while** $sp \leq ep$ **and** $i \geq 1$ **do**
4     $c \leftarrow p_i$;
5     $sp \leftarrow C[c] + \mathrm{rank}_c(L, sp-1) + 1$;
6     $ep \leftarrow C[c] + \mathrm{rank}_c(L, ep)$;
7     $i \leftarrow i - 1$;
8 **if** $ep < sp$ **then**
9   **return** 0;
10 **else**
11   **return** $ep - sp + 1$;

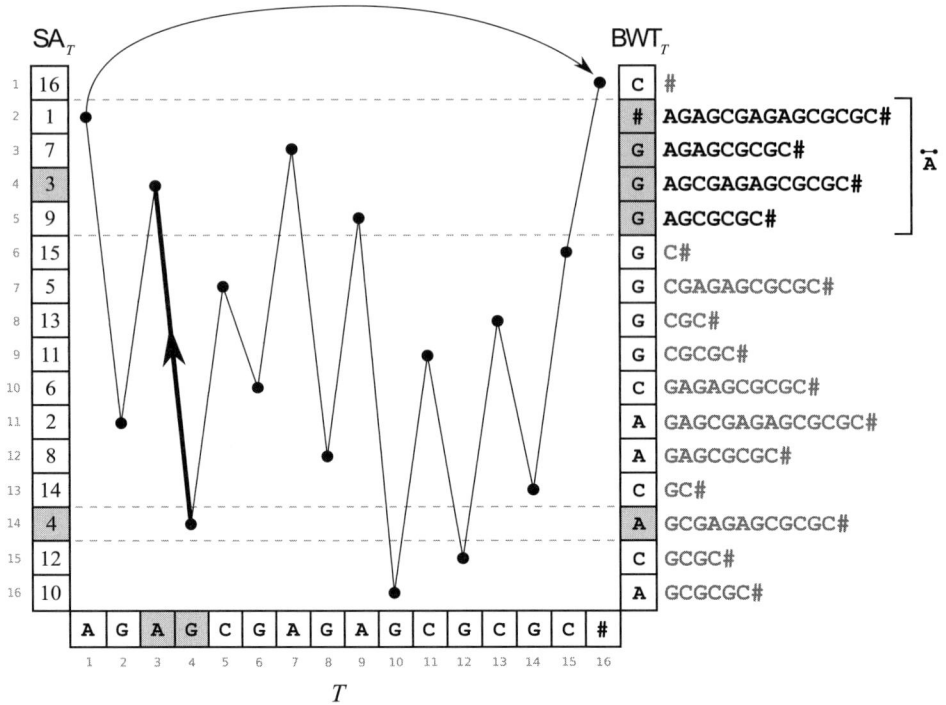

Figure 9.2 Function $\mathsf{LF}(i)$ for all $i \in [1..|T|]$, where $T = \mathtt{AGAGCGAGAGCGCGC\#}$. Dots represent suffixes of T, displayed at the same time in string order (horizontal axis) and in suffix array order (vertical axis). Lines represent the function LF, which connects suffixes that are consecutive in string order. Consider for example suffix $\mathtt{GCGAGAGCGCGC\#}$, which starts at position 4 in T, and corresponds to position 14 in SA_T: since $\mathsf{BWT}_T[14] = \mathtt{A}$, and since there are three occurrences of $\mathtt{A}$ in $\mathsf{BWT}_T[1..14]$, it follows that the previous suffix $\mathtt{AGCGAGAGCGCGC\#}$ that starts at position 3 in T is the third inside the interval of $\mathtt{A}$ in SA_T, therefore it corresponds to position 4 in SA_T. String T can be reconstructed from right to left by printing the characters that are met in BWT_T while following all arcs in the diagram, starting from position one in BWT_T.

where $j = \mathsf{SA}_{T\#}[i]$. The characters of W_R and of W_L are extracted one by one in left-to-right order, in $O(\log \sigma)$ time per character.

9.2.2   Backward search

Using the function LF, one can immediately implement Algorithm 9.1, which counts the number of occurrences in $T \in [1..\sigma]^{n-1}\#$ of a given pattern P, in $O(|P|)$ steps. Its correctness can be proved by induction, by showing that, at each step $i \in [1..|P|]$, the interval $[sp..ep]$ in the suffix array SA_T corresponds to all suffixes of T that are prefixed by $P[i..|P|]$ (see Figure 9.3). Algorithm 9.1 takes $O(|P|)$ time if the function $\mathtt{rank}$ can be computed in constant time on the Burrows–Wheeler transform of T: recall from Section 3.3 on page 24 that we can build a representation of BWT_T that supports $\mathtt{rank}$ queries in constant time using $\sigma|T|(1 + o(1))$ bits of space, or alternatively we can build a representation of BWT_T that supports $\mathtt{rank}$ queries in $O(\log \sigma)$ time using $|T|\log\sigma\,(1 + o(1))$ bits of space.

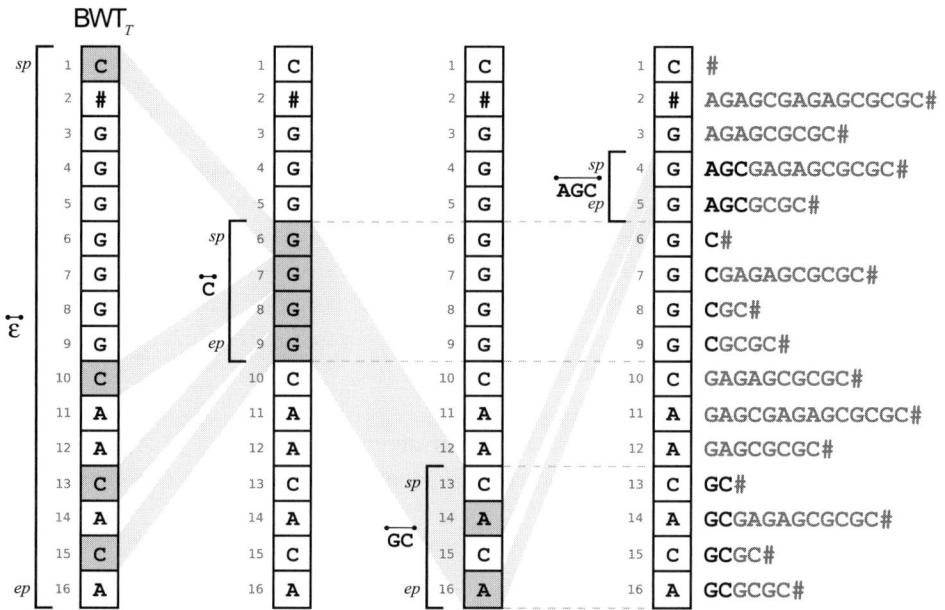

Figure 9.3 The backward search algorithm run on pattern $P = $ AGC in the Burrows–Wheeler transform of string $T = $ AGAGCGAGAGCGCGC# (see Algorithm 9.1). Assume that we are in the interval [13..16] of string GC: the first occurrence of A inside this interval is at position 14, and it is preceded by two other occurrences of A in BWT_T. It follows that $\mathsf{LF}(14) = 4$, that is, the third position in the interval of character A in BWT_T. The last occurrence of A in the interval of GC is at position 16, and it is preceded by three other occurrences of A in BWT_T. It follows that $\mathsf{LF}(16) = 5$, the fourth position in the interval of character A. See Figure 8.1 for a similar search performed on the suffix array of T.

9.2.3 Succinct suffix array

So far we have described only how to support *counting* queries, and we are still not able to *locate* the starting positions of a pattern P in string $T \in [1..\sigma]^{n-1}$#. One way to do this is to *sample* suffix array values, and to extract the missing values using the LF-mapping. Adjusting the sampling rate r gives different space/time tradeoffs.

Specifically, we sample all the values of $\mathsf{SA}[i]$ that satisfy $\mathsf{SA}[i] = rk$ for $0 \leq k \leq n/r$, and we store such samples consecutively, in the same order as in SA_T, in the array samples$[1..\lceil n/r \rceil]$. Note that this is equivalent to sampling every r positions *in string order*. We also mark in a bitvector $B[1..n]$ the positions of the suffix array that have been sampled, that is, we set $B[i] = 1$ if $\mathsf{SA}[i] = rk$, and we set $B[i] = 0$ otherwise. Then, if we are at a position i in SA_T such that $B[i] = 1$, we can extract the value of $\mathsf{SA}_T[i]$ by accessing samples$[\mathrm{rank}_1(B, i)]$. Otherwise, if $B[i] = 0$, we can set an auxiliary variable j to i, and then iteratively reset j to $\mathsf{LF}(j) = C[L[j]] + \mathrm{rank}_{L[j]}(L, j)$ until $B[j] = 1$, where L is again the Burrows–Wheeler transform of T: this corresponds to moving from right to left along string T until we find a sampled position. If we performed d such iterations of LF-mapping, we know that $\mathsf{SA}_T[i] = \mathsf{SA}_T[j] + d$.

Computing $\mathsf{SA}[i]$ with this strategy clearly takes $O(r \log \sigma)$ time, since each of the $d \leq r$ steps requires one $\texttt{rank}$ computation on L, which can be answered in $O(\log \sigma)$ time if L is represented as a wavelet tree. One can set $r = \log^{1+\epsilon} n / \log \sigma$ for any given $\epsilon > 0$ to have the samples fit in $(n/r) \log n = n \log \sigma / \log^{\epsilon} n = o(n \log \sigma)$ bits, which is asymptotically the same as the space required for supporting counting queries. This setting implies that the extraction of $\mathsf{SA}[i]$ takes $O(\log^{1+\epsilon} n)$ time. Bitvector B and its supporting data structure for $\texttt{rank}$ require $n + o(n)$ bits if implemented as described in Section 3.2. The resulting collection of data structures is called a *succinct suffix array*.

We leave the details of building the succinct suffix array of a string T from the BWT index of T to Exercise 9.2, and we just summarize the time and space complexities of the construction in the following theorem.

THEOREM 9.6   *The succinct suffix array of a string $T \in [1..\sigma]^n$ can be built from the BWT index of T in $O(n \log \sigma)$ time and $O(n \log \sigma + \texttt{out})$ bits of space, where $\texttt{out}$ is the size of the succinct suffix array.*

Succinct suffix arrays can be further extended into *self-indexes*. A self-index is a succinct representation of a string T that, in addition to supporting count and locate queries on arbitrary strings provided in input, allows one to access any substring of T by specifying its starting and ending position. In other words, a self-index for T completely replaces the original string T, which can be discarded.

Recall from Section 9.1 that we can reconstruct the whole string $T \in [1..\sigma]^{n-1}\#$ from BWT_T by applying the function LF iteratively. To reconstruct arbitrary substrings efficiently, it suffices to store, for every sampled position ri in string T, the position of suffix $T[ri..n]$ in SA_T: specifically, we use an additional array $\texttt{pos2rank}[1..\lceil n/r \rceil]$ such that $\texttt{pos2rank}[i] = j$ if $\mathsf{SA}_T[j] = ri$. Note that $\texttt{pos2rank}$ can itself be seen as a sampling of the *inverse suffix array* at positions ri. Clearly, $\texttt{pos2rank}$ takes the same amount of space as the array $\texttt{samples}$. Given an interval $[e..f]$ in string T, we can use $\texttt{pos2rank}[k+1]$ to go to the position i of suffix $T[rk..n]$ in SA_T, where $k = \lceil f/r \rceil$ and rk is the smallest sampling position bigger than f in T. We can then apply LF-mapping $rk - e - 1$ times starting from i: the result is the whole substring $T[e..rk]$ printed from right to left, thus we can return its proper prefix $T[e..f]$. The running time of this procedure is $O((f - e + r) \log \sigma)$.

Making a succinct suffix array a self-index does not increase its space complexity asymptotically. We can thus define the succinct suffix array as follows.

DEFINITION 9.7   *Given a string $T \in [1..\sigma]^n$, we define the* succinct suffix array *of T as a data structure that takes $n \log \sigma (1 + o(1)) + O(n/r \log n)$ bits of space, where r is the sampling rate, and that supports the following queries.*

- $\texttt{count}(P)$: *return the number of occurrences of string $P \in [1..\sigma]^m$ in T in $O(m \log \sigma)$ time.*
- $\texttt{locate}(i)$: *return $\mathsf{SA}_{T\#}[i]$ in $O(r \log \sigma)$ time.*
- $\texttt{substring}(e,f)$: *return $T[e..f]$ in $O\big(((f - e) + r) \log \sigma\big)$ time.*

9.2.4 Batched locate queries

Recall from Section 9.2.3 that the succinct suffix array of a string $T \in [1..\sigma]^{n-1}$ allow one to translate a given position i in $\mathsf{BWT}_{T\#}$ into the corresponding position $\mathsf{SA}_{T\#}[i]$ in $O(r \log \sigma)$ time, where r is the sampling rate of the succinct suffix array. Performing occ such conversions would take $O(\mathrm{occ} \cdot r \log \sigma)$ time, which might be too much for some applications. In this section we describe an algorithm that takes $O(n \log \sigma + \mathrm{occ})$ time and $O(\mathrm{occ} \cdot \log n)$ bits of space, and that answers a set of occ locate queries *in batch* using the BWT index of T, rather than the succinct suffix array of T.

Specifically, assume that an algorithm $\mathcal{A}$ selects some (possibly repeated) positions $i_1, i_2, \ldots, i_{\mathrm{occ}}$ in $\mathsf{SA}_{T\#}$, converts them into positions of T, and prints the set of pairs $\{(\mathsf{SA}_{T\#}[i_k], \mathrm{data}_k) : k \in [1..\mathrm{occ}]\}$, where data_k is an encoding of application-specific data that $\mathcal{A}$ associates with position i_k. We can rewrite algorithm $\mathcal{A}$ as follows. We use a bitvector $\mathrm{marked}[1..n+1]$ to mark the positions in $\mathsf{SA}_{T\#}$ which must be converted to positions in T, and we use a buffer $\mathrm{pairs}[1..N]$ with $N \geq \mathrm{occ}$ to store a representation of the output of $\mathcal{A}$. Whenever $\mathcal{A}$ computes $\mathsf{SA}_{T\#}[i_k]$, we modify the algorithm so that it sets instead $\mathrm{marked}[i_k] = 1$. Whenever $\mathcal{A}$ prints the pair $(\mathsf{SA}_{T\#}[i_k], \mathrm{data}_k)$ to the output, we modify the algorithm so that it appends instead pair (i_k, p_k) to pairs, where p_k is an integer that $\mathcal{A}$ uses to identify data_k uniquely.

Then, we invert $\mathsf{BWT}_{T\#}$ in $O(n \log \sigma)$ time, as described in Sections 9.1 and 9.2.1. During this process, whenever we are at a position i in $\mathsf{BWT}_{T\#}$, we also know the corresponding position $\mathsf{SA}_{T\#}[i]$ in $T\#$: if $\mathrm{marked}[i] = 1$, we append pair $(i, \mathsf{SA}_{T\#}[i])$ to an additional buffer $\mathrm{translate}[1..N]$ such that $N \geq \mathrm{occ}$. At the end of this process, the pairs in $\mathrm{translate}$ are in reverse string order, and the pairs in pairs are in arbitrary order: we thus sort both $\mathrm{translate}$ and pairs by suffix array position. Finally, we perform a linear, simultaneous scan of the two sorted arrays, matching a pair (i_k, p_k) in pairs to a corresponding pair $(i_k, \mathsf{SA}_{T\#}[i_k])$ in $\mathrm{translate}$, and printing $(\mathsf{SA}_{T\#}[i_k], \mathrm{data}_k)$ to the output by using p_k. See Algorithm 9.2 for the pseudocode of these steps.

Recall from Lemma 8.5 that we can sort a list of m triplets (p, s, v) by their primary key, breaking ties according to their secondary key, in $O(u + m)$ time and space, where $p \in [1..u]$ is a primary key, $s \in [1..u]$ is a secondary key, and v is an application-specific value. We can thus apply the radix sort algorithm of Lemma 8.5 to the array $\mathrm{pairs}[1..\mathrm{occ}]$, by interpreting each pair (i_k, p_k) as a triple $(\mathrm{msb}(i_k), \mathrm{lsb}(i_k), p_k)$, where $\mathrm{msb}(x)$ is a function that returns the most significant $\lceil (\log n)/2 \rceil$ bits of x and $\mathrm{lsb}(x)$ is a function that returns the least significant $\lfloor (\log n)/2 \rfloor$ bits of x. Since the resulting primary and secondary keys belong to the range $[1..2\sqrt{n}]$, sorting pairs takes $O(2\sqrt{n} + \mathrm{occ})$ time and space, and sorting $\mathrm{translate}$ takes $O(2\sqrt{n} + n) = O(n)$ time and space. We have thus proved the following lemma.

LEMMA 9.8 *Given the* **BWT** *index of a string* $T = t_1 \cdots t_n$, *where* $t_i \in [1..\sigma]$ *for all* $i \in [1..n]$, *a list* $\mathrm{pairs}[1..\mathrm{occ}]$ *of pairs* (i_k, p_k), *where* $i_k \in [1..n+1]$ *is a position in* $\mathsf{SA}_{T\#}$ *and* p_k *is an integer for all* $k \in [1..\mathrm{occ}]$, *and a bitvector* $\mathrm{marked}[1..n+1]$, *where* $\mathrm{marked}[i_k] = 1$ *for all* i_k *that appear in* pairs, *we can transform every pair*

$(i_k, p_k) \in$ pairs *into the corresponding pair* $(\mathsf{SA}_{T\#}[i_k], p_k)$, *possibly altering the order of pairs in list* pairs, *in* $O(n \log \sigma + \text{occ})$ *time and* $O(\text{occ} \cdot \log n)$ *space.*

Exercise 9.3 explores an alternative batch strategy that does away with radix sort.

Algorithm 9.2: Answering multiple locate queries in batch. We assume that radixSort(X, y) is a function that sorts a list of pairs X by component y. Symbols p and s in the code refer to the primary and secondary key of a pair, respectively.

Input: List pairs containing occ pairs (i_k, p_k), where $i_k \in [1..n]$ is a position in SA_T for all $k \in [1..\text{occ}]$. Bitvector marked$[1..n]$, where marked$[i_k] = 1$ for all i_k that appear at least once in pairs. Burrows–Wheeler transform of a string $T = t_1 \cdots t_n$ such that $t_i \in [1..\sigma]$ for all $i \in [1..n-1]$ and $t_n = \#$, with count array $C[0..\sigma]$ and support for rank$_c$ operations, $c \in [0..\sigma]$.

Output: List pairs, possibly permuted, in which i_k is replaced by $\mathsf{SA}_T[i_k]$ for all $k \in [1..\text{occ}]$.

```
1  i ← 1;
2  translate ← ∅;
3  for j ← n to 1 do
4      if marked[i] = 1 then
5          translate.append((i,j));
6      i ← LF(i);
7  radixSort(translate, 1);
8  radixSort(pairs, 1);
9  i ← 1;
10 j ← 1;
11 while i ≤ occ do
12     if pairs[i].p = translate[j].p then
13         pairs[i].p ← translate[j].s;
14         i ← i + 1;
15     else
16         j ← j + 1;
```

---

*9.3   Space-efficient construction of the BWT

Perhaps the easiest way to compute the Burrows–Wheeler transform of a string $T\#$ of length $n + 1$, where $T[i] \in [1..\sigma]$ for all $i \in [1..n]$ and $0 = \# \notin [1..\sigma]$, consists in building the suffix array of $T\#$ and in scanning the suffix array sequentially, setting $\mathsf{BWT}_{T\#}[i] = T[\mathsf{SA}_{T\#}[i] - 1]$ if $\mathsf{SA}_{T\#}[i] > 1$, and $\mathsf{BWT}_{T\#}[i] = \#$ otherwise. We could use the algorithm described in Theorem 8.12 on page 140 for building $\mathsf{SA}_{T\#}$, spending $O(\sigma + n)$ time and $O(n \log n)$ bits of space. Since $\mathsf{BWT}_{T\#}$ takes $(n + 1) \log \sigma$ bits of space, this algorithm would take asymptotically the same space as its output if

$\log \sigma \in \Theta(\log n)$. When this is not the case, it is natural to ask for a more space-efficient construction algorithm, and such an algorithm is developed next.

In this section we will repeatedly partition T into *blocks* of equal size B: this is a key strategy for designing efficient algorithms on strings. For convenience, we will work with a version of T whose length is a multiple of B, by appending to the end of T the smallest number of copies of character # such that the length of the resulting padded string is a multiple of B, and such that the padded string contains at least one occurrence of #. Recall that $B \cdot \lceil x/B \rceil$ is the smallest multiple of B that is at least x. Thus, we append $n' - n$ copies of character # to T, where $n' = B \cdot \lceil (n+1)/B \rceil$. To simplify the notation, we call the resulting string X, and we use n' throughout this section to denote its length.

We will interpret a partitioning of X into blocks as a new string X_B of length n'/B, defined on the alphabet $[1..(\sigma + 1)^B]$ of all strings of length B on the alphabet $[0..\sigma]$: the "characters" of X_B correspond to the blocks of X. In other words, $X_B[i] = X[(i-1)B + 1..iB]$. We assume B to be even, and we denote by left (respectively, right) the function from $[1..(\sigma + 1)^B]$ to $[1..(\sigma + 1)^{B/2}]$, such that left(W) returns the first (respectively, second) half of block W. In other words, if $W = w_1 \cdots w_B$, left$(W) = w_1 \cdots w_{B/2}$ and right$(W) = w_{B/2+1} \cdots w_B$.

We will also work with circular rotations of X: specifically, we will denote by $\overleftarrow{X}$ the string $X[B/2+1..n'] \cdot X[1..B/2]$, that is, the string X *circularly rotated to the left by $B/2$ positions*, and we will denote by $\overleftarrow{X}_B$ the string on the alphabet $[1..(\sigma + 1)^B]$ induced by partitioning $\overleftarrow{X}$ into blocks (see Figure 9.4).

Note that the suffix that starts at position i in $\overleftarrow{X}_B$ equals the half block $P_i = X[B/2 + (i-1)B + 1..iB]$, followed by the string $S_i = F_{i+1} \cdot X[1..B/2]$, where F_{i+1} is the suffix of X_B that starts at position $i + 1$ in X_B, if any (see Figure 9.4). Therefore the order of the suffixes of $\overleftarrow{X}_B$ can be obtained by sorting lexicographically all tuples (P_i, S_i) for $i \in [1..n'/B]$ by their first component and then by their second component. The lexicographic order of the second components coincides with the lexicographic order of the suffixes of X_B for all $i < n'/B$, while the second component of the tuple $(P_{n'/B}, S_{n'/B})$ is not used for determining the order of its tuple. It is thus not surprising that we can derive the BWT of string $\overleftarrow{X}_B$ from the BWT of string X_B.

LEMMA 9.9 *The BWT of string $\overleftarrow{X}_B$ can be derived from the BWT of string X_B in $O(n'/B)$ time and $O(\sigma^B \cdot \log(n'/B))$ bits of working space, where $n' = |X|$.*

Proof We use a method that is similar in spirit to the string sorting algorithm described in Lemma 8.7 on page 135. Recall that a character of X_B corresponds to a block of length B of X. First, we build the array C that contains the starting position of the interval of every character of X_B in the BWT of X_B. Specifically, we initialize an array $C'[1..(\sigma + 1)^B]$ to zeros, and then we scan the BWT of X_B: for every $i \in [1..n'/B]$ we increment $C'[\mathsf{BWT}_{X_B}[i]]$ by one. Then, we derive the array of starting positions $C[1..(\sigma + 1)^B]$ such that $C[c] = \sum_{i=1}^{c-1} C'[i] + 1$ by a linear scan of C'. We also build a similar array D, which contains the starting position of the interval of every *half block* in the BWT of string $\overleftarrow{X}_B$. Specifically, we build the count array $D'[1..(\sigma + 1)^{B/2}]$ by scanning the BWT of X_B and by incrementing $D'[\mathsf{right}(\mathsf{BWT}_{X_B}[i])]$, that is, we count

U W j

(a)

$X_4 = $ GATCCGGGCGGGGATCCCTCATGGATGGGATCCGGGCCTCC###

$\overleftarrow{X_4} = $ TCCGGGCGGGGATCCCTCATGGATGGGATCCGGGCCTCC###GA

V

i	C		BWT$_{X_4}$	
▷	ATGG ▶	1	CCTC	ATGG ATGG GATC CGGG CCTC C###
	ATGG	2	ATGG	ATGG GATC CGGG CCTC C###
	C### ▶	3	CCTC	C###
	CCTC ▶	4	GATC	CCTC ATGG ATGG GATC CGGG CCTC C###
		5	CGGG	CCTC C###
	CGGG ▶	6	GATC	CGGG CCTC C###
		7	GATC	CGGG CGGG GATC CCTC ATGG ATGG GATC CGGG CCTC C###
		8	CGGG	CGGG GATC CCTC ATGG ATGG GATC CGGG CCTC C###
	GATC ▶	9	CGGG	GATC CCTC ATGG ATGG GATC CGGG CCTC C###
		10	ATGG	GATC CGGG CCTC C###
		11	C###	GATC CGGG CGGG GATC CCTC ATGG ATGG GATC CGGG CCTC C###

D		BWT$_{\overleftarrow{X_4}}$	
## ▶	1		##GA
GG ▶	2		GGAT GGGA TCCG GGCC TCC# ##GA
	3		GGCC TCC# ##GA
	4		GGCG GGGA TCCC TCAT GGAT GGGA TCCG GGCC TCC# ##GA
	5		GGGA TCCC TCAT GGAT GGGA TCCG GGCC TCC# ##GA
	6		GGGA TCCG GGCC TCC# ##GA
TC ▶	7	TCCC	TCAT GGAT GGGA TCCG GGCC TCC# ##GA
	8		TCC# ##GA
	9		TCCC TCAT GGAT GGGA TCCG GGCC TCC# ##GA
	10		TCCG GGCC TCC# ##GA
	11		TCCG GGCG GGGA TCCC TCAT GGAT GGGA TCCG GGCC TCC# ##GA

U W j

(b)

$X_4 = $ GATCCGGGCGGGGATCCCTCATGGATGGGATCCGGGCCTCC###

$\overleftarrow{X_4} = $ TCCGGGCGGGGATCCCTCATGGATGGGATCCGGGCCTCC###GA

V

i	C		BWT$_{X_4}$	
	ATGG ▶	1	CCTC	ATGG ATGG GATC CGGG CCTC C###
▷	ATGG	2	ATGG	ATGG GATC CGGG CCTC C###
	C### ▶	3	CCTC	C###
		4	GATC	CCTC ATGG ATGG GATC CGGG CCTC C###
	CCTC ▶	5	CGGG	CCTC C###
	CGGG ▶	6	GATC	CGGG CCTC C###
		7	GATC	CGGG CGGG GATC CCTC ATGG ATGG GATC CGGG CCTC C###
		8	CGGG	CGGG GATC CCTC ATGG ATGG GATC CGGG CCTC C###
	GATC ▶	9	CGGG	GATC CCTC ATGG ATGG GATC CGGG CCTC C###
		10	ATGG	GATC CGGG CCTC C###
		11	C###	GATC CGGG CGGG GATC CCTC ATGG ATGG GATC CGGG CCTC C###

D		BWT$_{\overleftarrow{X_4}}$	
## ▶	1		##GA
GG ▶	2	TCAT	GGAT GGGA TCCG GGCC TCC# ##GA
	3		GGCC TCC# ##GA
	4		GGCG GGGA TCCC TCAT GGAT GGGA TCCG GGCC TCC# ##GA
	5		GGGA TCCC TCAT GGAT GGGA TCCG GGCC TCC# ##GA
	6		GGGA TCCG GGCC TCC# ##GA
	7	TCCC	TCAT GGAT GGGA TCCG GGCC TCC# ##GA
TC ▶	8		TCC# ##GA
	9		TCCC TCAT GGAT GGGA TCCG GGCC TCC# ##GA
	10		TCCG GGCC TCC# ##GA
	11		TCCG GGCG GGGA TCCC TCAT GGAT GGGA TCCG GGCC TCC# ##GA

Figure 9.4 The first step (a) and the second step (b) of Lemma 9.9 on a sample string X. The arrays of pointers C and D are represented as labeled gray triangles. The current iteration i is represented as a white triangle. The notation follows Lemma 9.9.

just the right half of every character of X_B, for every $i \in [1..n'/B]$. Then, we build the array $D[1..(\sigma + 1)^{B/2}]$ such that $D[c] = \sum_{i=1}^{c-1} D'[i] + 1$ by a linear scan of D'.

To build the BWT of $\overleftarrow{X}_B$, it suffices to scan the BWT of X_B from left to right, in a manner similar to Lemma 8.7. Specifically, consider the suffix of X_B at lexicographic rank one, and assume that it starts from position j in X_B (see Figure 9.4(a)). Since the suffix $X_B[j..n'/B]$ of X_B is preceded by the string $W = \mathsf{BWT}_{X_B}[1]$, it follows that the position of suffix $\overleftarrow{X}_B[j-1..n'/B]$ of $\overleftarrow{X}_B$ in the BWT of $\overleftarrow{X}_B$ belongs to the interval that starts at position $D[\mathtt{right}(W)]$: we can thus write string $V = \overleftarrow{X}_B[j-2]$ at this position in the BWT of $\overleftarrow{X}_B$, and increment the pointer $D[\mathtt{right}(W)]$ by one for the following iterations. Clearly $V = \mathtt{right}(U) \cdot \mathtt{left}(W)$, where $U = X_B[j-2]$, so we just need to compute string U.

Recall that $C[W]$ is the starting position of the interval of character W in the BWT of X_B. Since we are scanning the BWT of X_B from left to right, we know that the suffix S that corresponds to the current position i in BWT_{X_B} is the kth suffix that is preceded by W in BWT_{X_B}, where $k = \mathtt{rank}_W(\mathsf{BWT}_{X_B}, 1, i)$. It follows that the position of suffix WS in BWT_{X_B} equals the current value of $C[W]$, therefore we can set $U = \mathsf{BWT}_{X_B}[C[W]]$ and we can increment pointer $C[W]$ by one for the following iterations (see Figure 9.4).

All these operations take $O(n'/B)$ time and $O(\sigma^B \cdot \log(n'/B))$ bits of space. □

The second key observation that we will exploit for building the BWT of X is the fact that the suffixes of $X_{B/2}$ which start at odd positions coincide with the suffixes of X_B, and the suffixes of $X_{B/2}$ that start at even positions coincide with the suffixes of $\overleftarrow{X}_B$. We can thus reconstruct the BWT of $X_{B/2}$ from the BWT of X_B and of $\overleftarrow{X}_B$.

LEMMA 9.10 *Assume that we can read in constant time a block of size B. Then, the BWT of string $X_{B/2}$ can be derived from the BWT of string X_B and from the BWT of string $\overleftarrow{X}_B$, in $O(n' \log \sigma)$ time and $O(\sigma^B \log(n'/B))$ bits of working space, where $n' = |X|$.*

Proof We assume that both of the Burrows–Wheeler transforms provided in the input are represented using wavelet trees (see Section 3.3). The lexicographic rank of a suffix of $X_{B/2}$ equals its rank r with respect to the suffixes of X_B, plus its rank r' with respect to the suffixes of $\overleftarrow{X}_B$, minus one. Thus, we can backward-search string X_B in BWT_{X_B} and in $\mathsf{BWT}_{\overleftarrow{X}_B}$, reading each character of X_B in constant time, and we can determine the ranks r and r' of every suffix of X_B in both transforms. We can thus set $\mathsf{BWT}_{T_{B/2}}[r + r' - 1] = \mathtt{right}(\mathsf{BWT}_{X_B}[r])$. Similarly, we can backward-search $\overleftarrow{X}_B$ in BWT_{X_B} and in $\mathsf{BWT}_{\overleftarrow{X}_B}$, setting $\mathsf{BWT}_{T_{B/2}}[r + r' - 1] = \mathtt{right}(\mathsf{BWT}_{\overleftarrow{X}_B}[r'])$. These searches take in total $O((n'/B) \cdot \log(\sigma^B)) = O(n' \log \sigma)$ time, and they require arrays $C[1..(\sigma + 1)^B]$ and $C'[1..(\sigma + 1)^B]$ that point to the starting position of the interval of every character in the BWT of X_B and in that of $\overleftarrow{X}_B$, respectively. □

Lemmas 9.9 and 9.10 suggest that one should build the BWT of X in $O(\log B)$ steps: at every step i we compute the BWT of string $X_{B/2^i}$, stopping when $B/2^i = 1$ (see Figure 9.5). To start the algorithm we need the BWT of string X_B for some initial block size B: the most natural way to obtain it is by using a linear-time suffix array construction algorithm that works in $O(\sigma^B + n'/B)$ time and in $O((n'/B)\log(n'/B))$

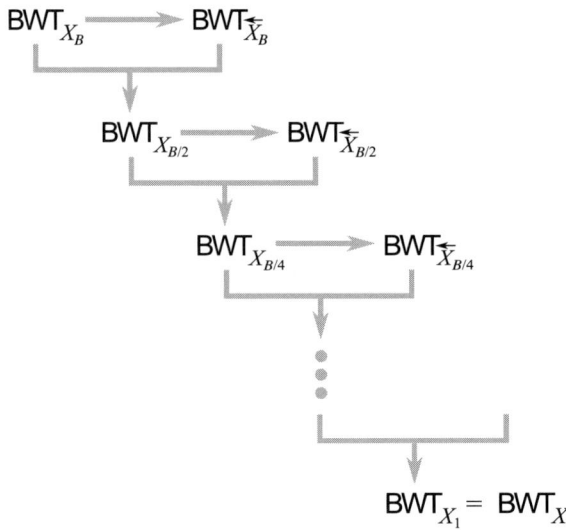

Figure 9.5 High-level structure of the algorithm for building the Burrows–Wheeler transform of a string. The notation follows Section 9.3. Horizontal arrows: Lemma 9.9. Vertical arrows: Lemma 9.10.

bits of space, for example the one described in Section 8.2. We want this first phase to take $O(n')$ time and $O(n' \log \sigma)$ bits of space, or in other words we want to satisfy the following constraints:

i. $\sigma^B \in O(n')$ and
ii. $(n'/B)\log(n'/B) \in O(n' \log \sigma)$, or more strictly $(n'/B)\log n' \in O(n' \log \sigma)$.

The time for every step is $O((n'/B)\log(\sigma^B)) = O(n' \log \sigma)$ for any choice of B, since it is dominated by the time taken by Lemma 9.10. The Burrows–Wheeler transforms of $X_{B/2^i}$ and of $\overleftarrow{X}_{B/2^i}$ at every step i take $O(n' \log \sigma)$ bits of space overall. We want the supporting arrays of pointers used by Lemmas 9.9 and 9.10 to take $O(n')$ bits of space overall, or in other words we want to satisfy the following constraint:

iii. $\sigma^B \log(n'/B) \in O(n')$, or even more strictly $\sigma^B \log n' \in O(n')$.

For simplicity, we also want B to be a power of two. Recall that $2^{\lceil \log x \rceil}$ is the smallest power of two that is at least x. Assume thus that we set $B = 2^{\lceil \log(\log n'/(c \log \sigma)) \rceil}$ for some constant c. Since $\lceil x \rceil \geq x$ for any x, we have that $B \geq \log n'/(c \log \sigma)$, thus constraint (ii) is satisfied by any choice of c. Since $\lceil x \rceil < x+1$, we have that $B < (2/c)\log n'/\log \sigma$, thus constraints (i) and (iii) are satisfied for any $c > 2$. For this choice of B the number of steps in the algorithm becomes $O(\log \log n')$, and we can read a block of size B in constant time as required by Lemma 9.10 since the machine word is assumed to be $\Omega(\log n')$ (see Section 2.2). We have thus proved that the BWT of X can be built in $O(n' \log \sigma \log \log n')$ time and in $O(n' \log \sigma)$ bits of space.

To complete the construction, recall that string X is the original input string T, padded at the end with $k = n' - n \geq 1$ copies of character #. We thus need to derive the

BWT of string $T\#$ from the BWT of X, or in other words we need to remove all the occurrences of character $\#$ that we added to T, except one. To do so, it suffices to note that $\mathsf{BWT}_X = T[n] \cdot \#^{k-1} \cdot W$ and $\mathsf{BWT}_{T\#} = T[n] \cdot W$, where string W is a permutation of $T[1..n-1]$: see Exercise 9.4. We are thus able to prove the key result of this section.

THEOREM 9.11 *The Burrows–Wheeler transform of a string $T\#$ of length $n+1$, where $T[i] \in [1..\sigma]$ for all $i \in [1..n]$ and $0 = \# \notin [1..\sigma]$, can be built in $O(n \log \sigma \log \log n)$ time and in $O(n \log \sigma)$ bits of space.*

9.4 Bidirectional BWT index

Given a string $T = t_1 t_2 \cdots t_n$ on the alphabet $[1..\sigma]$, consider two BWT indexes, one built on $T\#$ and one built on $\underline{T}\# = t_n t_{n-1} \cdots t_1 \#$. Let $\mathbb{I}(W, T)$ be the function that returns the interval in $\mathsf{BWT}_{T\#}$ of the suffixes of $T\#$ that are prefixed by string $W \in [1..\sigma]^+$. Note that the interval $\mathbb{I}(W, T)$ in the *suffix array* of $T\#$ contains all the starting positions of string W in T. Symmetrically, the interval $\mathbb{I}(\underline{W}, \underline{T})$ in the suffix array of $\underline{T}\#$ contains all those positions i such that $n - i + 1$ is an *ending position* of string W in T (see Figure 9.6 for an example). In this section we are interested in the following data structure.

DEFINITION 9.12 *Given a string $T \in [1..\sigma]^n$, a* bidirectional BWT index *on T is a data structure that supports the following operations:*

isLeftMaximal(i,j)	*Returns 1 if substring $\mathsf{BWT}_{T\#}[i..j]$ contains at least two distinct characters, and 0 otherwise.*
isRightMaximal(i,j)	*Returns 1 if substring $\mathsf{BWT}_{\underline{T}\#}[i..j]$ contains at least two distinct characters, and 0 otherwise.*
enumerateLeft(i,j)	*Returns all the distinct characters that appear in substring $\mathsf{BWT}_{T\#}[i..j]$, in lexicographic order.*
enumerateRight(i,j)	*Returns all the distinct characters that appear in $\mathsf{BWT}_{\underline{T}\#}[i..j]$, in lexicographic order.*
extendLeft $\big(c, \mathbb{I}(W, T), \mathbb{I}(\underline{W}, \underline{T})\big)$	*Returns the pair $\big(\mathbb{I}(cW, T), \mathbb{I}(\underline{Wc}, \underline{T})\big)$ for $c \in [0..\sigma]$.*
extendRight $\big(c, \mathbb{I}(W, T), \mathbb{I}(\underline{W}, \underline{T})\big)$	*Returns $\big(\mathbb{I}(Wc, T), \mathbb{I}(c\underline{W}, \underline{T})\big)$ for $c \in [0..\sigma]$.*

For completeness, we introduce here two specializations of the bidirectional BWT index, which will be used extensively in the following chapters.

DEFINITION 9.13 *Given a string $T \in [1..\sigma]^n$, a* forward BWT index *on T is a data structure that supports the following operations:* isLeftMaximal(i,j), enumerateLeft(i,j), *and* extendLeft (c, i, j), *where $[i..j]$ is an interval in $\mathsf{BWT}_{T\#}$ and $c \in [0..\sigma]$. Symmetrically, a* reverse BWT index *on T is a data structure that supports the following operations:* isRightMaximal(i,j), enumerateRight(i,j), *and* extendRight (c, i, j), *where $[i..j]$ is an interval in $\mathsf{BWT}_{\underline{T}\#}$ and $c \in [0..\sigma]$.*

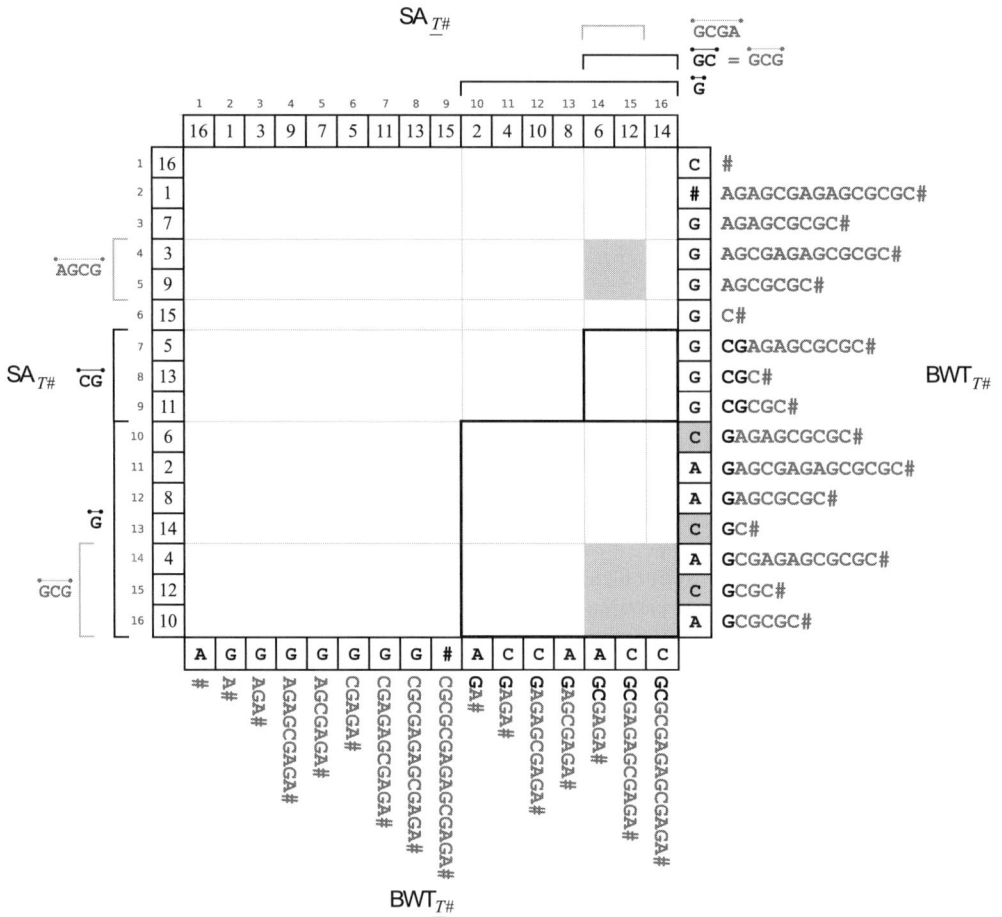

Figure 9.6 Visualizing `extendLeft` in the bidirectional BWT index of string $T = $ AGAGCGAGAGCGCGC. The BWT of string $T\#$ is shown on the rows, from top to bottom, and the BWT of string $\underline{T}\# = $ CGCGCGAGAGCGAGA# is shown on the columns, from left to right. Squares represent pairs of intervals $(\mathbb{I}(W, T), \mathbb{I}(\underline{W}, \underline{T}))$. Consider string G (the largest empty square), with interval [10..16] both in $\mathrm{BWT}_{T\#}$ and in $\mathrm{BWT}_{\underline{T}\#}$. A backward step in $\mathrm{BWT}_{T\#}$ with character C leads to interval [7..9] (the small empty square). Extending G to the left with character C in T corresponds to extending G to the right with character C in $\underline{T}$, therefore the interval of GC in $\mathrm{BWT}_{\underline{T}\#}$ is contained in the interval of G. Since there are four characters lexicographically smaller than C in the interval of G in $\mathrm{BWT}_{T\#}$, the interval of GC in $\mathrm{BWT}_{\underline{T}\#}$ must start four positions after the beginning of the interval of G. Gray squares show the intervals in $\mathrm{BWT}_{T\#}$ and in $\mathrm{BWT}_{\underline{T}\#}$ during the following steps of the backward search for pattern $P = $ AGCG. The numbers that correspond to the rows (respectively, columns) of a square are the starting (respectively, ending) positions in T of the substring associated with the square.

Clearly a forward BWT index can be implemented with a BWT index of T, and a reverse BWT index can be implemented with a BWT index of $\underline{T}$. In the bidirectional BWT index, `extendLeft` and `extendRight` are analogous to a standard backward step in $\mathrm{BWT}_{T\#}$ or $\mathrm{BWT}_{\underline{T}\#}$, but they keep the interval of a string W in one BWT *synchronized* with the interval of its reverse $\underline{W}$ in the other BWT. The following theorem describes how to achieve such synchronization, and Figure 9.6 gives a concrete example.

THEOREM 9.14 *Given a string $T \in [1..\sigma]^n$, there is a representation of the bidirectional BWT index of T that uses $2n \log \sigma (1 + o(1))$ bits of space, supports* isLeftMaximal, isRightMaximal, extendLeft, *and* extendRight *in $O(\log \sigma)$ time, and supports* enumerateLeft *and* enumerateRight *in $O(d \log(\sigma/d))$ time, where d is the size of the output of such operations.*

Proof We build a wavelet tree on $\mathsf{BWT}_{T\#}$ and $\mathsf{BWT}_{\underline{T}\#}$ (see Section 3.3), and we answer isLeftMaximal$(i,j) = 1$ if and only if isRangeUnary$(\mathsf{BWT}_{T\#}, i, j)$ returns false and isRightMaximal$(i,j) = 1$ if and only if isRangeUnary$(\mathsf{BWT}_{\underline{T}}, i, j)$ returns false. We implement enumerateLeft(i,j) by rangeList$(\mathsf{BWT}_{T\#}, i, j, 0, \sigma)$, and we implement enumerateRight(i,j) by rangeList$(\mathsf{BWT}_{\underline{T}\#}, i, j, 0, \sigma)$. The running times required by this theorem are matched by Corollary 3.5.

Consider now extendLeft$\big(c, \mathbb{I}(W, T), \mathbb{I}(\underline{W}, T)\big)$. For simplicity, let $\mathbb{I}(W, T) = [i..j]$ and let $\mathbb{I}(\underline{W}, \underline{T}) = [p..q]$. We use $[i'..j']$ and $[p'..q']$ to denote the corresponding intervals of cW and $\underline{W}c$, respectively. As in the standard succinct suffix array, we can get $[i'..j']$ in $O(\log \sigma)$ time using a backward step from interval $[i..j]$, implemented with rank$_c$ queries on the wavelet tree of $\mathsf{BWT}_{T\#}$.

Interval $[p..q]$ in $\mathsf{BWT}_{\underline{T}\#}$ contains all the suffixes of $\underline{T}\#$ that start with $\underline{W}$, or equivalently all the prefixes of $T\#$ that end with W: since $[p'..q']$ contains all the prefixes of $T\#$ that end with cW, it must be contained in $[p..q]$. The number k of prefixes of $T\#$ that end with a string aW with $a < c$ is given by $k = $ rangeCount$(\mathsf{BWT}_{\underline{T}\#}, i, j, 0, c - 1)$, therefore $p' = p + k$. The size of interval $[p'..q']$ equals the size of interval $[i'..j']$, since such intervals correspond to the starting and ending positions of string cW in $T\#$: it follows that $q' = p' + j' - i'$.

The function extendRight can be implemented symmetrically. $\square$

The function extendLeft allows one to know the interval of every suffix of a string W in $\mathsf{BWT}_{\underline{T}\#}$ while backward-searching W in $\mathsf{BWT}_{T\#}$: see again Figure 9.6 for an example of a backward search in the bidirectional BWT index. More importantly, the bidirectional BWT index allows one to traverse all nodes of the suffix tree of T, using Algorithm 9.3.

THEOREM 9.15 *Algorithm 9.3, executed on the bidirectional BWT index of a string $T \in [1..\sigma]^n$ represented as in Theorem 9.14, outputs all the intervals of the suffix array of $T\#$ that correspond to internal nodes of the suffix tree of T, as well as the length of the label of such nodes, in $O(n \log \sigma)$ time and in $O(\sigma \log^2 n)$ bits of working space.*

Proof Assume by induction that $[i..j]$ is the interval of some internal node v of ST_T at the beginning of the while loop in Algorithm 9.3, thus Σ' contains all characters a such that $a\ell(v)$ is a suffix of $T\#$. There is an internal node w in ST_T such that $\ell(w) = a\ell(v)$ if and only if there is a suffix of $T\#$ that starts with $a\ell(v)b$ and another suffix of $T\#$ that starts with $a\ell(v)c$, where $b \neq c$ are distinct characters. This condition can be tested by checking whether isRightMaximal(i', j') equals one, where $[i'..j'] = \mathbb{I}(\underline{\ell(v)a}, T)$.

Algorithm 9.3: Visiting all the internal nodes of the suffix tree of a string T of length n using the bidirectional BWT index of T

Input: Bidirectional BWT index `idx` of string T, and interval $[1..n+1]$ corresponding to the root of the suffix tree of T

Output: Pairs $(\vec{v}, |\ell(v)|)$ for all internal nodes v of the suffix tree of T, where $\vec{v}$ is the interval of v in the suffix array of $T\#$.

1 $S \leftarrow$ empty stack;

2 $S.\text{push}\big(([1..n+1], [1..n+1], 0)\big)$;

3 **while** *not* $S.\text{isEmpty}()$ **do**

4 $\quad$ $([i..j], [i'..j'], d) \leftarrow S.\text{pop}()$;

5 $\quad$ Output $([i..j], d)$;

6 $\quad$ $\Sigma' \leftarrow \text{idx.enumerateLeft}(i, j)$;

7 $\quad$ $I \leftarrow \emptyset$;

8 $\quad$ **for** $c \in \Sigma'$ **do**

9 $\quad\quad$ $I \leftarrow I \cup \big\{\text{idx.extendLeft}(c, [i..j], [i'..j'])\big\}$;

10 $\quad$ $x \leftarrow \text{argmax}\big\{j - i : ([i..j], [i'..j']) \in I\big\}$;

11 $\quad$ $I \leftarrow I \setminus \{x\}$;

12 $\quad$ $([i..j], [i'..j']) \leftarrow x$;

13 $\quad$ **if** $\text{idx.isRightMaximal}(i', j')$ **then**

14 $\quad\quad$ $S.\text{push}(x, d+1)$;

15 $\quad$ **for** $([i..j], [i'..j']) \in I$ **do**

16 $\quad\quad$ **if** $\text{idx.isRightMaximal}(i', j')$ **then**

17 $\quad\quad\quad$ $S.\text{push}\big(([i..j], [i'..j'], d+1)\big)$;

Thus Algorithm 9.3 pushes to stack S all and only the intervals that correspond to internal nodes in the suffix tree of T *that can be reached from node v via explicit Weiner links* (or, equivalently, via inverse suffix links). Recall from Section 8.3 that connecting all the internal nodes of the suffix tree of T with explicit Weiner links yields the *suffix-link tree* of T. Thus, Algorithm 9.3 implements a traversal of the suffix-link tree of T, which implies that it visits all the internal nodes of ST_T.

Cases in which $\text{isRightMaximal}(i', j') = 0$, where $[i'..j']$ is the interval in $\mathsf{BWT}_{T\#}$ of a string $a\ell(v)$, correspond to *implicit Weiner links* in ST_T: recall from Observation 8.14 that the total number of such links is $O(n)$. Note also that extending such a string $a\ell(v)$ to the left with additional characters cannot lead to an internal node of ST_T. It follows that Algorithm 9.3 runs in $O(n \log \sigma)$ time, assuming that the bidirectional BWT index of T is implemented as described in Theorem 9.14.

Note that the algorithm considers to push first to the stack the largest interval $[i..j]$. Every other interval is necessarily at most half of the original interval, thus stack S contains at any time intervals from $O(\log n)$ suffix-link tree levels. Every such level contains at most σ interval pairs, each taking $O(\log n)$ bits. $\qquad\square$

In practice it is more efficient to combine the implementations of `enumerateLeft` and of `extendLeft` (and symmetrically of `enumerateRight` and of `extendRight`) in Algorithm 9.3: see Exercise 9.7. Note that Algorithm 9.3 can be used to implement in $O(n \log \sigma)$ time and in $2n \log \sigma (1 + o(1)) + O(\sigma \log^2 n)$ bits of space *any algorithm* that

- traverses all nodes of ST_T;
- performs a number of additional operations which is linear in n or in the size of the output;
- does not depend on the *order* in which the nodes of the suffix tree are visited.

We will see in later chapters that a number of fundamental sequence analysis tasks, such as detecting maximal repeats and maximal unique matches, satisfy such properties. Some of the applications we will describe visit only nodes of the suffix tree whose label is of length at most k, where k is a constant: in such cases the size of the stack is $O(n \log \sigma)$ even without using the trick of Algorithm 9.3, and we can traverse the suffix-link tree of T depth-first.

*9.4.1   Visiting all nodes of the suffix tree with just one BWT

In order to implement Algorithm 9.3 we do not need all the expressive power of the bidirectional BWT index of T. Specifically, we do not need operations `enumerateRight` and `extendRight`, but we do need `isRightMaximal`: we can thus replace the BWT of $\underline{T}\#$ with a bitvector $\mathrm{runs}[1..n + 1]$ such that $\mathrm{runs}[i] = 1$ if and only if $i > 1$ and $\mathsf{BWT}_{\underline{T}\#}[i] \neq \mathsf{BWT}_{\underline{T}\#}[i - 1]$. We can build this vector in a linear scan of $\mathsf{BWT}_{\underline{T}\#}$, and we index it to support `rank` queries in constant time. We can then implement $\mathtt{isRightMaximal}(i, j)$ by checking whether there is a one in $\mathrm{runs}[i + 1..j]$, or equivalently by checking whether $\mathrm{rank}_1(\mathrm{runs}, j) - \mathrm{rank}_1(\mathrm{runs}, i) \geq 1$. In this section we show how to implement Algorithm 9.3 without even using runs, or equivalently *using just* $\mathsf{BWT}_{\underline{T}\#}$, without any penalty in running time.

Let v be a node of ST_T, and let $\gamma(v, c)$ be the number of children of v that have a Weiner link (implicit or explicit) labeled by character $c \in [1..\sigma]$. The following property, visualized in Figure 9.7, is key for using just one BWT in Algorithm 9.3. Its easy proof is left to Exercise 9.6.

PROPERTY 1   *There is an explicit Weiner link (v, w) labeled by character c in ST_T if and only if $\gamma(v, c) \geq 2$. In this case, node w has exactly $\gamma(v, c)$ children. If $\gamma(v, c) = 1$, then the Weiner link from v labeled by c is implicit. If $\gamma(v, c) = 0$, then no Weiner link from v is labeled by c.*

Thus, if we maintain at every step of Algorithm 9.3 the intervals of the children of a node v of ST_T in addition to the interval of v itself, we could generate all Weiner links

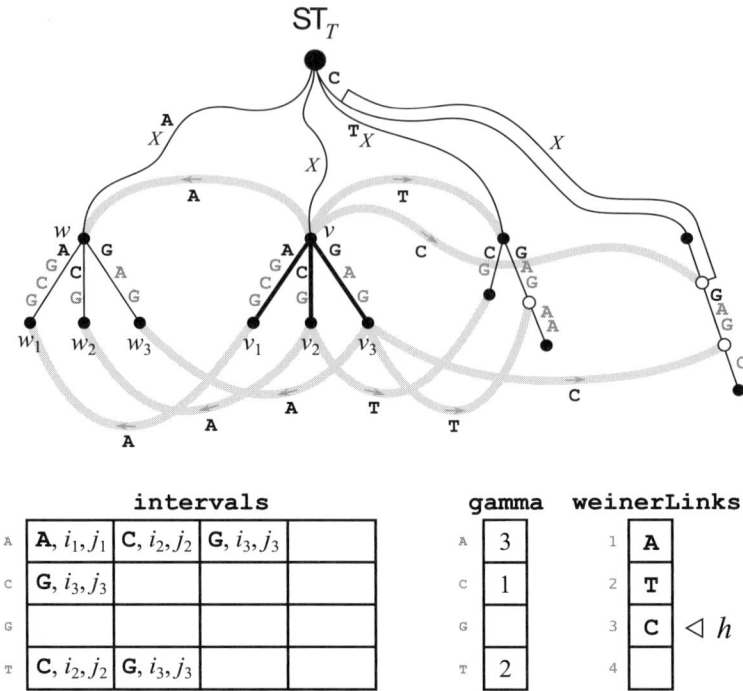

Figure 9.7 Theorem 9.16 applied to node v. Gray directed arcs represent implicit and explicit Weiner links. White dots represent the destinations of implicit Weiner links. Among all strings prefixed by string $\ell(v) = X$, only those prefixed by XGAG are preceded by C. It follows that CX is always followed by G and it is not a node in the suffix tree of T, thus the Weiner link from v labeled by C is implicit. Conversely, XAGCG, XCG, and XGAG are preceded by an A, so AX is a node in the suffix tree, and the Weiner link from v labeled by A is explicit.

from v and check whether they are explicit. The following theorem makes this intuition more precise.

THEOREM 9.16   *There is an algorithm that, given the BWT index of a string T, outputs all the intervals of the suffix array of $T\#$ that correspond to internal nodes of the suffix tree of T, as well as the length of the label of such nodes, in $O(n \log \sigma)$ time and $O(\sigma^2 \log^2 n)$ bits of working space.*

Proof   As anticipated, we adapt Algorithm 9.3 to use just $\mathsf{BWT}_{T\#}$ rather than the full bidirectional BWT index of T. Let v be an internal node of the suffix tree of T, let $v_1, v_2, \ldots, v_k$ be its children in the suffix tree in lexicographic order, and let $\phi(v_i)$ be the first character of the label of arc (v, v_i). Without loss of generality, we consider just characters in $[1..\sigma]$ in what follows. Assume that we know both the interval $\overleftrightarrow{v}$ of v in $\mathsf{BWT}_{T\#}$, and the interval $\overleftrightarrow{v_i}$ of every child v_i.

Let $\mathtt{weinerLinks}[1..\sigma]$ be a vector of characters, and let h be the number of non-zero characters in this vector. We will store in vector $\mathtt{weinerLinks}$ the labels of all (implicit and explicit) Weiner links that start from v. Let $\mathtt{intervals}[1..\sigma][1..\sigma]$ be a matrix, whose rows correspond to all possible labels of a Weiner link from v. We will store in row c of matrix `intervals` a set of triplets $(\phi(v_i), \overset{\cdot}{\overline{c\ell(v_i)}}, \overset{\cdot}{c\ell(v_i)})$, where intervals refer to $\mathsf{BWT}_{T\#}$. By encoding $\phi(v_i)$, every such triplet identifies a child v_i of v in the suffix tree, and it specifies the interval of the destination of a Weiner link from v_i labeled by c. We use array `gamma`$[1..\sigma]$ to maintain the number of children of v that have a Weiner link labeled by c, for every $c \in [1..\sigma]$. In other words, `gamma`$[c] = \gamma(v,c)$. See Figure 9.7 for an example of such arrays.

For every child v_i of v, we enumerate all the distinct characters that occur in $\mathsf{BWT}_{T\#}[\overline{v_i}]$: for every such character c, we compute $\overset{\cdot}{c\ell(v_i)}$, we increment counter `gamma`$[c]$ by one, and we store triplet $(\phi(v,i), \overset{\cdot}{\overline{c\ell(v_i)}}, \overset{\cdot}{c\ell(v_i)})$ in `intervals`$[c]$ $[$`gamma`$[c]]$. If `gamma`$[c]$ transitioned from zero to one, we increment h by one and we set `weinerLinks`$[h] = c$. At the end of this process, the labels of all Weiner links that start from v are stored in `weinerLinks` (not necessarily in lexicographic order), and by Property 1 we can detect whether the Weiner link from v labeled by character `weinerLinks`$[i]$ is explicit by checking whether `gamma`$[$`weinerLinks`$[i]] > 1$.

Let w be the destination of the explicit Weiner link from v labeled by character c. Note that the characters that appear in row c of matrix `intervals` are sorted lexicographically, and that the corresponding intervals are exactly the intervals of the children $w_1, w_2, \ldots, w_{\gamma(v,c)}$ of w in the suffix tree of T, in lexicographic order of $\ell(w_i)$. It follows that such intervals are adjacent in $\mathsf{SA}_{T\#}$, thus $\overline{w}$ equals the first position of the first interval in row c, and $\overset{\cdot}{w}$ equals the last position of the last interval in row c. We can then push on the stack used by Algorithm 9.3 intervals $\overline{w_1}, \overline{w_2}, \ldots, \overline{w_{\gamma(v,c)}}$ and number $\gamma(v,c)$. After having processed in the same way all destinations w of explicit Weiner links from v, we can reset h to zero, and `gamma`$[c]$ to zero for all labels c of (implicit or explicit) Weiner links of v.

The algorithm can be initialized with the interval of the root of ST_T and with the interval of the distinct characters in $\mathsf{SA}_{T\#}$, which can be derived using the C array of backward search. Using the same stack trick as in Algorithm 9.3, the stack contains $O(\log n)$ levels of the suffix-link tree of T at any given time. Every such level contains at most σ nodes, and each node is represented by the at most σ intervals of its children in the suffix tree. Every such interval takes $O(\log n)$ bits, therefore the total space required by the stack is $O(\log^2 n \cdot \sigma^2)$.   □

Note that the algorithm in Theorem 9.16 builds, for every internal node, its children in the suffix tree, thus it can be used to implement any algorithm that invokes `enumerateRight` and `extendRight` only on the intervals of $\mathsf{BWT}_{T\#}$ associated with internal nodes of the suffix tree. We will see in later chapters that this is the case for a number of fundamental sequence analysis tasks.

*9.5   BWT index for labeled trees

The idea of sorting lexicographically all the suffixes of a string, and of representing the resulting list in small space by storing just the character that precedes every suffix, could be applied to labeled trees: specifically, we could sort in lexicographic order all the *paths* that start from the root, and we could represent such a sorted list in small space by storing just the labels of the *children* of the nodes that correspond to every path. In this section we make this intuition more precise, by describing a representation technique for labeled trees that is flexible enough to generalize in Section 9.6 to labeled DAGs, and to apply almost verbatim in Section 9.7.1 to de Bruijn graphs.

Let $\mathcal{T} = (V, E, \Sigma)$ be a rooted tree with a set of nodes V, a root $r \in V$, and a set of *directed* arcs E, such that every node $v \in V$ is labeled by a character $\ell(v) \in \Sigma = [1..\sigma]$. Without loss of generality, we assume that $\ell(r) = \#$, where $\# = 0 \notin \Sigma$. Arcs are directed *from child to parent* (see Figure 9.8(a)): formally, $(u, v) \in E$ if and only if u is a child of v in $\mathcal{T}$. We also assume that V obeys a *partial order* $\prec$, in the sense that there is a total order among the children of every node. In particular, we extend the partial order $\prec$ to a *total order $\prec^*$ on the entire V*, which satisfies the following two properties:

- $u \prec v$ implies $u \prec^* v$;
- $u \prec^* v$ implies $u' \prec^* v'$ for every child u' of u and every child v' of v.

Such a total order can be induced, for example, by traversing $\mathcal{T}$ depth-first in the order imposed by $\prec$, and by printing the children of every traversed node. We denote as a *path* a sequence of nodes $P = v_1, v_2, \ldots, v_k$ such that $(v_i, v_{i+1}) \in E$ for $i \in [1..k-1]$. We lift the label operator ℓ to paths, by defining $\ell(P) = \ell(v_1) \cdot \ell(v_2) \cdots \cdot \ell(v_k)$, and we set $\bar{\ell}(v) = \ell(v_1, v_2, \ldots, v_k)$ for $v \in V$, where $v_1, v_2, \ldots, v_k$ is the path that connects $v = v_1$ to the root $v_k = r$. We call $\bar{\ell}(v)$ the *extended label* of node v. Note that, if $\mathcal{T}$ is a chain, $\bar{\ell}(v)$ corresponds to a *suffix* of $\mathcal{T}$. Given a string $W \in [0..\sigma]^*$, we use count(W) to denote the number of paths P in $\mathcal{T}$ such that $\ell(P) = W$. For brevity, we use the shorthands n and m to denote $|V|$ and the number of leaves of $\mathcal{T}$, respectively.

For a node $v \in V$, let last(v) be a binary flag that equals one if and only if v is the last child of its parent according to $\prec$, and let internal(v) be a binary flag that equals one if and only if v is an internal node of $\mathcal{T}$. Assume that for every node v in the order induced by $\prec^*$, and for every arc $(u, v) \in E$, we concatenate to a list L the quadruplet $(\ell(u), \bar{\ell}(v), \text{last}(u), \text{internal}(u))$ if v is an internal node, and the quadruplet $(\#, \bar{\ell}(v), 1, 1)$ if v is a leaf. The first type of quadruplet represents arc $(u, v) \in E$, and the second type of quadruplet represents an artificial arc $(r, v) \notin E$. Note that the same quadruplet can be generated by different arcs, and that every string W that labels a path of $\mathcal{T}$ is the prefix of the second component of at least count(W) quadruplets.

Assume now that we perform a *stable sort* of L in lexicographic order by the second component of each quadruplet. After such sorting, the incoming arcs of a node still form a consecutive interval of L in the order imposed by $\prec$, and the intervals of nodes v and w with $\bar{\ell}(v) = \bar{\ell}(w)$ still occur in L in the order imposed by $\prec^*$ (see Figure 9.8(b)). We denote by $L[i]$ the ith quadruplet in the sorted L, and by $L[i].\ell$, $L[i].\bar{\ell}$, $L[i].\text{last}$, and $L[i].\text{internal}$ the corresponding fields of quadruplet $L[i]$. Note that, if $\mathcal{T}$ is a chain, this process is analogous to printing all the suffixes of $\mathcal{T}$, along with their preceding characters. Thus, in the case of a chain, sorting the quadruplets in L by field $\bar{\ell}$ and printing the field ℓ of the sorted list yields precisely the Burrows–Wheeler transform of $\mathcal{T}$.

Recall that a suffix corresponds to a unique position in the Burrows–Wheeler transform of a string; analogously, in what follows we identify a node $v \in V$ with *the interval $\mathbb{I}(v)$ of L that contains all quadruplets* $(\ell(u), \bar{\ell}(v), \texttt{last}(u), \texttt{internal}(u))$, or equivalently all the incoming arcs of v in $\mathcal{T}$. List L can be represented implicitly, by just storing arrays $\texttt{labels}[1..n+m-1]$, $\texttt{internal}[1..n+m-1]$, $\texttt{last}[1..n+m-1]$, and $C[0..\sigma]$, where $\texttt{labels}[i] = L[i].\ell$, $\texttt{internal}[i] = L[i].\texttt{internal}$, $\texttt{last}[i] = L[i].\texttt{last}$, and $C[c]$ contains the number of nodes $v \in V$ whose extended label $\bar{\ell}(v)$ is lexicographically smaller than character $c \in [0..\sigma]$. We call this set of arrays the *Burrows–Wheeler index of tree $\mathcal{T}$* (see Figure 9.8(b)).

DEFINITION 9.17   *Given a labeled tree $\mathcal{T}$ with n nodes and m leaves, the* Burrows–Wheeler index of $\mathcal{T}$, *denoted by* $\mathsf{BWT}_{\mathcal{T}}$, *is a data structure that consists of*

- *the array of characters* labels, *encoded using a wavelet tree (see Section 3.3);*
- *bitvector* last, *indexed to support* rank *and* select *queries in constant time as described in Section 3.2;*
- *bitvector* internal;
- *the count array C.*

See the main text for the definition of such arrays. $\mathsf{BWT}_{\mathcal{T}}$ takes $(n+m)\log\sigma(1+o(1)) + 2(n+m) + o(n+m)$ bits of space.

*9.5.1   Moving top-down

$\mathsf{BWT}_{\mathcal{T}}$ shares most of the properties of its textual counterpart. In particular, nodes $v \in V$ such that $\bar{\ell}(v)$ starts with the same string $W \in [0..\sigma]^*$ form a contiguous interval $\overrightarrow{W}$ in $\mathsf{BWT}_{\mathcal{T}}$, and all arcs (u, v) with the same label $\ell(u)$ appear in lexicographic order of $\bar{\ell}(v)$ across all the intervals of nodes in $\mathsf{BWT}_{\mathcal{T}}$, with ties broken according to $\prec^*$. As already anticipated, all the $\texttt{inDegree}(v)$ incoming arcs of a node $v \in V$ are contained in its interval $\overrightarrow{v}$, they follow the total order imposed by $\prec$, and the subarray $\texttt{last}[\overrightarrow{v}]$ is a unary encoding of $\texttt{inDegree}(v)$. This immediately allows to compute $\texttt{inDegree}(v) = |\overrightarrow{v}|$ and to determine the number of children of v labeled by a given character $a \in [1..\sigma]$, by computing $\texttt{rank}_a(\texttt{labels}, \overrightarrow{v}) - \texttt{rank}_a(\texttt{labels}, \overrightarrow{v} - 1)$.

$\mathsf{BWT}_{\mathcal{T}}$ shares another key property with the Burrows–Wheeler transform of a string.

PROPERTY 2   *If two nodes u and v satisfy $\ell(u) = \ell(v)$, the order between intervals $\overrightarrow{u}$ and $\overrightarrow{v}$ is the same as the order between arcs $(u, \texttt{parent}(u))$ and $(v, \texttt{parent}(v))$.*

The proof of this property is straightforward and is left to Exercise 9.8. It follows that, if (u, v) is the ith arc in $\mathsf{BWT}_{\mathcal{T}}$ with $\ell(u) = a$, the interval of u is the ith sub-interval of $\overrightarrow{a}$, or equivalently the sub-interval of $\texttt{last}[\overrightarrow{a}]$ delimited by the $(i-1)$th and the ith ones.

Assume thus that we want to move from $\overrightarrow{v}$ to $\overrightarrow{u}$, where u is the kth child of v labeled by character $c \in [1..\sigma]$ according to order $\prec$. Arc (u, v) has lexicographic rank

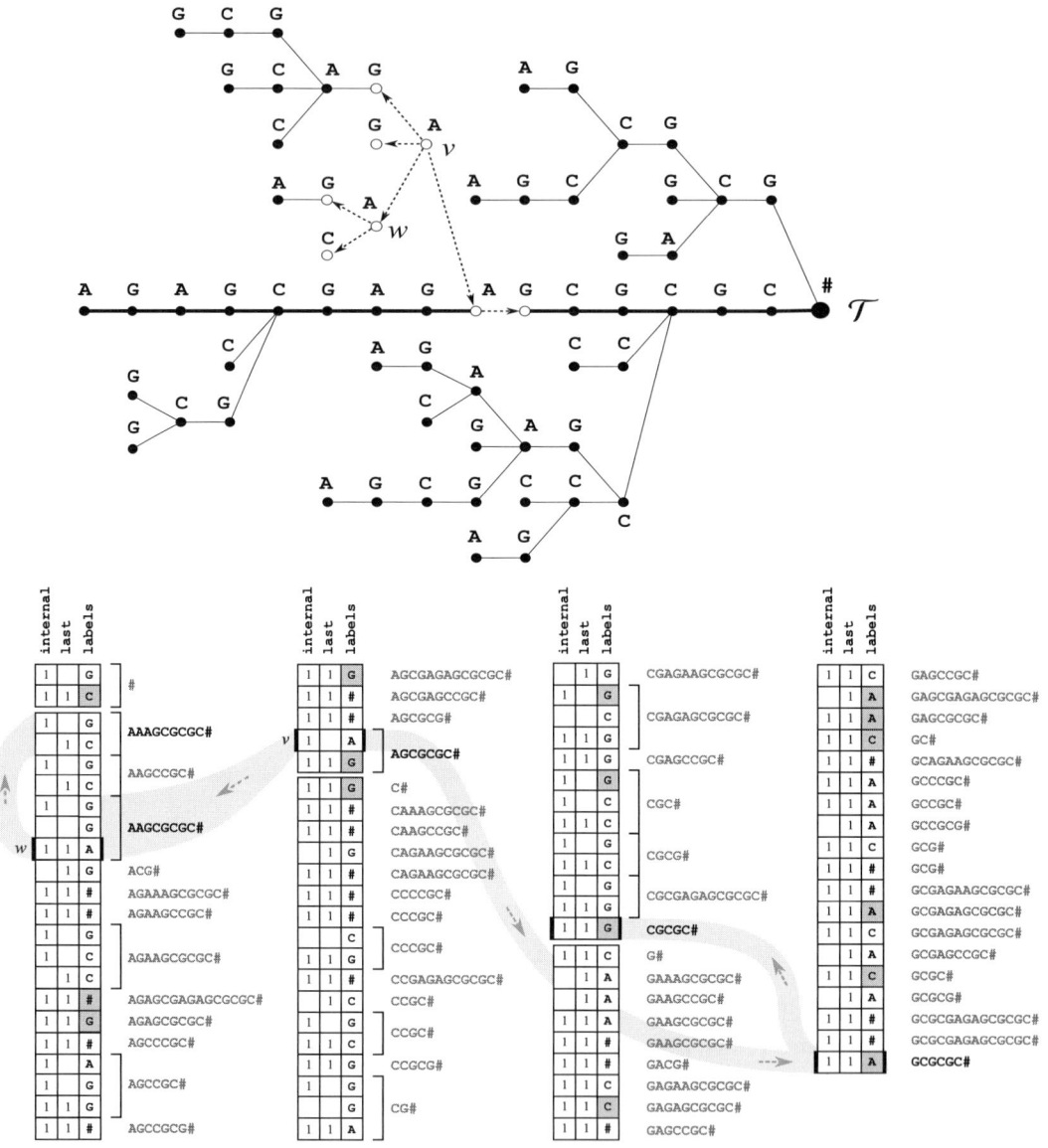

Figure 9.8 A labeled tree $\mathcal{T}$ (top panel) and its Burrows–Wheeler transform $\mathsf{BWT}_{\mathcal{T}}$ built using a pre-order traversal (bottom panel, laid out from top to bottom, from left to right). Arcs in the top panel are assumed to be directed from left to right. For clarity, zeros are omitted from bitvectors in the bottom panel. Spaces in $\mathsf{BWT}_{\mathcal{T}}$ mark the starting position of the interval of every character in $[1..\sigma]$. Black borders to the left and to the right of a row represent $\mathbb{I}'$ intervals. Gray cells correspond to the BWT of string $T = \text{AGAGCGAGAGCGCGC\#}$, which labels the thick path in the top panel. See Figure 9.1 for a representation of the BWT and of the suffix array of string T. Consider node v in the top panel, with $\bar{\ell}(v) = \text{AAGCGCGC\#}$: there are two occurrences of A before $\mathbb{I}'(v)$ in $\mathsf{BWT}_{\mathcal{T}}$, thus $\mathbb{I}(v)$ is the third block in the interval of character A. Similarly, AGCGCGC\# is the 16th sub-interval in lexicographic order inside the interval of character A, thus $\mathbb{I}'(\text{parent}(v))$ corresponds to the 16th A in $\mathsf{BWT}_{\mathcal{T}}$. The bottom panel continues the navigation of $\mathcal{T}$ to $\mathbb{I}(w)$ and to $\mathbb{I}'(\text{parent}(\text{parent}(v)))$ (dashed arrows).

$r = \text{rank}_c(\texttt{labels}, \overleftarrow{v} - 1) + k$ among all arcs with $\ell(u) = c$, thus $\overleftrightarrow{u}$ is the rth sub-interval of $\overrightarrow{c}$:

$$\texttt{child}(\overleftrightarrow{v}, c, k) = \Big[\texttt{select}_1\big(\texttt{last}, \textsf{LF}(\overleftarrow{v} - 1, c) + k - 1\big) + 1 \; .. $$
$$\texttt{select}_1\big(\texttt{last}, \textsf{LF}(\overleftarrow{v} - 1, c) + k\big) \Big], \qquad (9.2)$$
$$\textsf{LF}(i, c) = C[c] + \text{rank}_c(\texttt{labels}, i),$$

where $\textsf{LF}$ is a generalization of the function described in Section 9.2.1. See Figure 9.8(b) for a concrete example. Note that this operation takes $O(\log \sigma)$ time. Moving to the kth child of v according to order $\prec$ is equally easy, and it is left as an exercise.

Recall that, for a given string $W \in [0..\sigma]^+$, $\overrightarrow{W}$ contains the intervals of all nodes $v \in V$ whose extended label $\bar{\ell}(v)$ is prefixed by W, and for every such node it contains all its incoming arcs in the order imposed by $\prec$. Thus, $\text{rank}_1(\texttt{last}, \overrightarrow{W}) - \text{rank}_1(\texttt{last}, \overrightarrow{W}-1)$ is the number of paths (not necessarily ending at the root) labeled by W in $\mathcal{T}$, and $|\overrightarrow{W}|$ is the number of arcs $(u, v) \in E$ such that $\bar{\ell}(u) = cWQ$ for some $c \in [0..\sigma]$ and $Q \in [0..\sigma]^*$. We can thus lift the function $\texttt{child}$ to map $\overrightarrow{W}$ onto $\overrightarrow{cW}$ for any $c \in \texttt{labels}[\overrightarrow{W}]$, as follows:

$$\texttt{extendLeft}(c, \overrightarrow{W}, \overrightarrow{W}) = \Big[\texttt{select}_1\big(\texttt{last}, \textsf{LF}(\overrightarrow{W} - 1, c)\big) + 1 \; .. $$
$$\texttt{select}_1\big(\texttt{last}, \textsf{LF}(\overrightarrow{W}, c)\big) \Big]. \qquad (9.3)$$

This procedure clearly takes $O(\log \sigma)$ time, and it supports a generalization of the backward search algorithm described in Section 9.2.2 that returns the interval that contains the first nodes of all paths labeled by a string W of length k in $\mathcal{T}$: the search proceeds again from right to left along W, computing iteratively $\overrightarrow{W_{i..k}} = \texttt{extendLeft}\big(W[i], \overrightarrow{W_{i+1..k}}, \overrightarrow{W_{i+1..k}}\big)$. More generally, the following lemma holds.

LEMMA 9.18 $\textsf{BWT}_{\mathcal{T}}$ *is a* forward BWT index *(Definition 9.13). The operations* $\texttt{extendLeft}$ *and* $\texttt{isLeftMaximal}$ *can be implemented in* $O(\log \sigma)$ *time, and* $\texttt{enumerateLeft}$ *can be implemented in* $O(d \log(\sigma/d))$ *time, where d is the size of the output.*

*9.5.2   Moving bottom-up

Moving from a node u to $v = \texttt{parent}(u)$ requires one to introduce a different representation for nodes. Specifically, we assign to u *the unique position i of arc (u, v) in* L, or equivalently the position of the quadruplet $(\ell(u), \bar{\ell}(v), \texttt{last}(u), \texttt{internal}(u))$. We denote this unary interval by $\mathbb{I}'(u) = i$ to distinguish it from the interval $\mathbb{I}(v)$ of size $\texttt{inDegree}(v)$ used until now. To compute $\texttt{parent}(\mathbb{I}'(u)) = \mathbb{I}'(v)$, we proceed as follows. Let $\bar{\ell}(u) = abW$, where a and b are characters in $[1..\sigma]$ and $W \in [0..\sigma]^*$: we want to move to the position in $\textsf{BWT}_{\mathcal{T}}$ of the arc (v, w) such that $\bar{\ell}(v) = bW$. Character b satisfies $C[b] < \text{rank}_1(\texttt{last}, i - 1) + 1 \leq C[b + 1]$, which can be determined by a binary search over C, and the lexicographic rank of $\overrightarrow{bW}$ inside $\overrightarrow{b}$ is $\text{rank}_1(\texttt{last}, i - 1) - C[b] + 1$, therefore

$$\texttt{parent}(i,b) = \texttt{select}_b\big(\texttt{labels}, \texttt{rank}_1(\texttt{last}, i-1) - C[b] + 1\big). \quad (9.4)$$

See Figure 9.8(b) for a concrete example. This implementation of `parent` clearly takes $O(\log \sigma)$ time.

By Property 2, we can convert $\mathbb{I}'(v)$ to $\mathbb{I}(v)$ by simply computing $\mathbb{I}(v) = \texttt{child}(\mathbb{I}'(v), c)$, where $c = \texttt{labels}[\mathbb{I}'(v)]$. Indeed, if (v, w) is the ith arc in $\mathsf{BWT}_{\mathcal{T}}$ with $\ell(v) = c$, then all arcs (u, v) are clustered in the ith sub-interval of $\vec{c}$, which is the sub-interval of $\texttt{last}[\vec{c}]$ delimited by the $(i-1)$th and the ith ones. Converting $\mathbb{I}(v)$ into $\mathbb{I}'(v)$ is equally straightforward. We can thus navigate $\mathcal{T}$ in both directions and in any order using $\mathbb{I}(v)$ and $\mathbb{I}'(v)$. Note that using $\mathbb{I}(v)$ and $\mathbb{I}'(v)$ is equivalent to using as a unique identifier of a node v its rank in the list of all nodes, sorted lexicographically by $\bar{\ell}(v)$ and with ties broken according to the total order $\prec^*$.

*9.5.3 Construction and space complexity

It is easy to see that we can remove all occurrences of # from `labels`, since we are not interested in moving from a leaf to the root, or in moving from the root to its parent. The resulting index takes $2n + n \log \sigma (1 + o(1)) + o(n)$ bits of space overall, which matches the information-theoretic lower bound up to lower-order terms. Indeed, the number of distinct, unlabeled, full binary trees with n internal nodes is the *Catalan number* $C_n = \binom{2n}{n}/(n+1)$ (see also Section 6.6.1): thus, we need at least $\log(C_n) \approx 2n - \Theta(\log n)$ bits to encode the topology of $\mathcal{T}$, and we need at least $n \log \sigma$ bits to encode its labels. In contrast, a naive pointer-based implementation of $\mathcal{T}$ would take $O(n \log n + n \log \sigma)$ bits of space, and it would require $O(n|W|)$ time to find or count all paths labeled by $W \in [0..\sigma]^+$.

It is easy to build $\mathsf{BWT}_{\mathcal{T}}$ in $O(n \log n)$ time by extending the prefix-doubling technique used in Lemma 8.8 for building suffix arrays: see Exercise 9.10. The algorithm for filling the partially empty suffix array of a string initialized with the set of its small suffixes (Theorem 8.12) can be applied almost verbatim to $\mathsf{BWT}_{\mathcal{T}}$ as well, but it achieves linear time only when the topology and the labels of $\mathcal{T}$ are coupled: see Exercise 9.11.

*9.6 BWT index for labeled DAGs

As anticipated in Section 9.5, the technique used for representing labeled trees in small space applies with minor adaptations to labeled, directed, acyclic graphs. Genome-scale labeled DAGs are the natural data structure to represent millions of *variants* of a single chromosome, observed across hundreds of thousands of individuals in a population, as paths that share common segments in the same graph: see Figure 9.9(a) for a small example, and Section 10.7.2 for applications of labeled DAGs to read alignment.

Let $G = (V, E, \Sigma)$ be a directed acyclic graph with a set of vertices V, unique source $s \in V$, and unique sink $t \in V$, such that every vertex $v \in V \setminus \{s, t\}$ is labeled by a character $\ell(v)$ on alphabet $\Sigma = [1..\sigma]$. Without loss of generality, we assume that

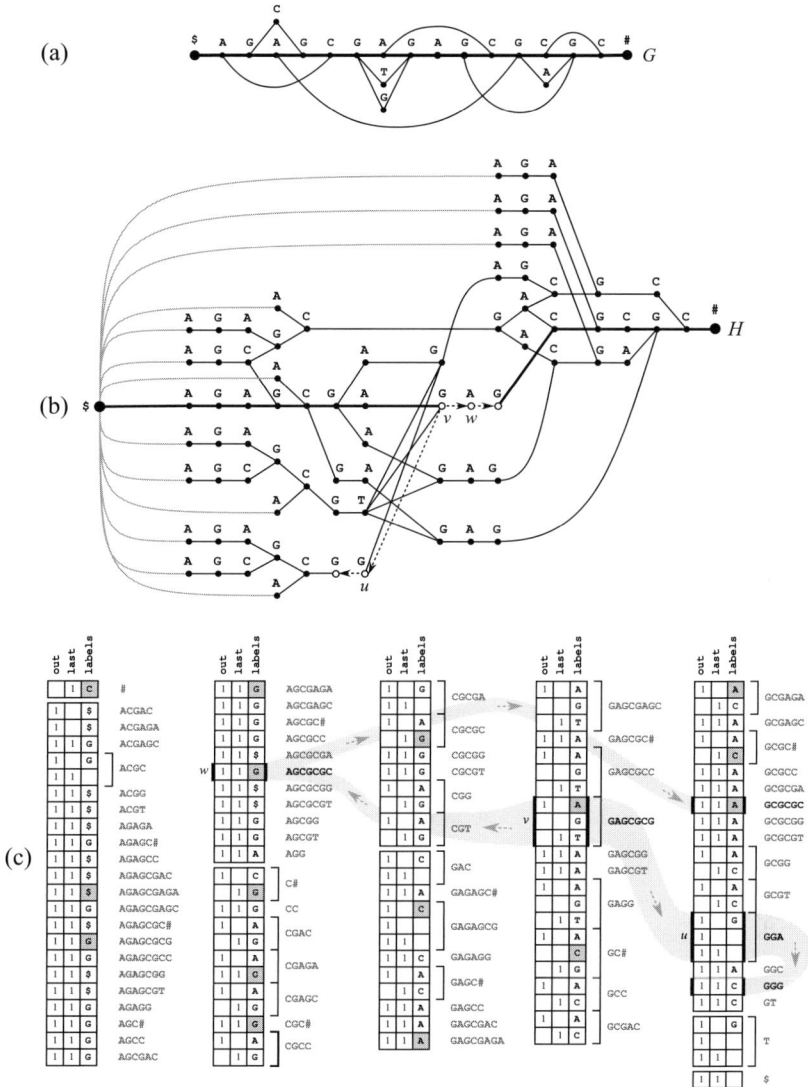

Figure 9.9 (a) A directed, acyclic, labeled, reverse-deterministic graph G, representing string $T = \$AGAGCGAGAGCGCGC\#$ with three single-character variations and five deletions. Arcs are directed from left to right. (b) The equivalent strongly distinguishable DAG H. (c) BWT_H, laid out from top to bottom, from left to right, with the corresponding set of distinguishing prefixes. For clarity, zeros are omitted from bitvectors. Spaces mark the starting position of the interval of every character in $[1..\sigma + 1]$. Black borders to the left and to the right of a row represent intervals that correspond to vertices. Gray cells correspond to the Burrows–Wheeler transform of string $T\#$, which labels the thick path in panels (a) and (b). See Figure 9.1 for a representation of the BWT and of the suffix array of string $T\#$. Consider vertex v in panel (b), corresponding to interval $[i..j]$ in BWT_H, and assume that we want to compute the interval of u by issuing $\text{inNeighbor}([i..j], G)$. There are 32 occurrences of G up to position i, and there are $C[G] = 48$ arcs (x, y) with P_x lexicographically smaller than G, therefore the interval of u contains the 80th one in out. Similarly, assume that we want to move from v to its only out-neighbor w by issuing $\text{outNeighbor}([i..j], 1)$. There are 62 arcs (x, y) where P_x is lexicographically smaller than P_v, 14 of which have P_x prefixed by $\ell(v) = $ G. Therefore, the interval of w contains the 15th G in labels. Panel (c) continues the navigation of H to the in-neighbor of vertex u and to the out-neighbor of vertex w (dashed arrows in panels (b) and (c)).

$\ell(s) = \$$ and $\ell(t) = \#$, where $\# = 0$ and $\$ = \sigma + 1$. As done in Section 9.5, we lift operator ℓ to paths, and we set $\bar{\ell}(v)$ for a vertex $v \in V$ to be the *set of path labels* $\{\ell(v_1, v_2, \ldots, v_k) \mid v_1 = v, v_k = t, (v_i, v_{i+1}) \in E \ \forall \ i \in [1..k-1]\}$. We say that G is *forward-deterministic* if every vertex has at most one out-neighbor with label a for every $a \in [1..\sigma]$. The notion of *reverse-deterministic* is symmetric. We use the shorthand n to denote $|V|$. We say that G is *distinguishable* if, for every pair of vertices v and w in V, $v \neq w$ implies either that P is lexicographically smaller than P' for all $P \in \bar{\ell}(v)$ and $P' \in \bar{\ell}(w)$, or that P is lexicographically larger than P' for all $P \in \bar{\ell}(v)$ and $P' \in \bar{\ell}(w)$. Note that distinguishability implies reverse-determinism. We say that vertex v is *strongly distinguishable* if there is a string $P_v \in [1..\sigma]^+$ such that all strings in $\bar{\ell}(v)$ are prefixed by P_v, and such that $v \neq w$ implies that no string in $\bar{\ell}(w)$ is prefixed by P_v. We call P_v the *distinguishing prefix* of vertex v. Although everything we describe in this section applies to distinguishable DAGs, we assume that every vertex in G is strongly distinguishable to simplify the notation.

Assume that, for every arc $e = (u, v) \in E$, we concatenate to a list L^- the triplet $(\ell(u), P_v, e)$, and assume that we sort L^- in lexicographic order by the second component of each triplet, breaking ties arbitrarily. We denote by $L^-.\ell$ the sequence built by taking the first component of each triplet in the sorted list L^-. Similarly, assume that for every arc $e = (u, v) \in E$, we concatenate to a list L^+ the pair (P_u, e), and that we sort L^+ in lexicographic order by the first component of each pair, breaking ties using the lexicographic order of P_v. In contrast to the case of labeled trees, let array $C[0..\sigma + 1]$ store at position c the number of *arcs* $(u, v) \in E$ with $\ell(u) < c$. Since G is strongly distinguishable, every vertex $v \in V$ can be assigned to the contiguous interval in L^- or L^+ that corresponds to P_v. The key property of L^- and L^+ is that we can easily move from the position of an arc in L^- to its position in L^+.

PROPERTY 3   *Let (u, v) be an arc with $\ell(u) = c$, and assume that it occurs at position i in L^-. The position of (u, v) in L^+ is $\mathsf{LF}(i, c) = C[c] + \mathrm{rank}_c(L^-.\ell, i)$.*

The proof of this property is straightforward: since (u, v) is the jth arc with $\ell(u) = c$ in L^-, where $j = \mathrm{rank}_c(L^-.\ell, i)$, the interval of string cP_v in L^+ starts at $C[c] + j$, and this is exactly the lexicographic rank of (u, v) according to P_u, breaking ties according to P_v. We represent L^- and L^+, aligned and in compact form, using three arrays, `labels`, `out`, and `last`, as follows. For a vertex $v \in V$, let $[i..j]$ and $[i'..j']$ be its intervals in L^- and in L^+, respectively. Such intervals are possibly different, but both of them are the kth interval in their list for some k. We assign to every $v \in V$ *exactly one interval* $\mathbb{I}(v)$ of size $\max\{j - i + 1, j' - i' + 1\}$, and we set

$$\mathtt{labels}[\bar{v} + k] = \begin{cases} L^-[i+k].\ell & \text{for } k \in [0..j-i] \\ 0 & \text{otherwise,} \end{cases}$$

$$\mathtt{out}[\bar{v} + k] = \begin{cases} 1 & \text{for } k \in [0..j'-i'] \\ 0 & \text{otherwise.} \end{cases}$$

We also set a one in `last` at $\vec{v}$ for every $v \in V$. Note that the arrays `labels`, `out`, and `last` have each m components, where $m = \sum_{v \in V} \max\{|N^+(v)|, |N^-(v)|\} \le 2|E| - (|V| - 2)$. The non-empty subarray of `labels` is an implicit representation of L^-, thus all arcs (u, v) with the same $\ell(u)$ appear in `labels` in lexicographic order of P_v. The non-empty subarray of `out` is the unary encoding of the *size* of the block of every P_v in L^+, which is the number of arcs $(v, w) \in E$. Analogously to L^- and L^+, we can easily move from the position i of an arc (u, v) in `labels` to its corresponding position in `out`, by computing

$$\mathsf{LF}(i, c) = \texttt{select}_1\big(\texttt{out}, C[c] + \texttt{rank}_c(\texttt{labels}, i)\big), \tag{9.5}$$

where $c = \texttt{labels}[i]$. We call this set of arrays the *Burrows–Wheeler index of the labeled, directed, acyclic graph G* (see Figure 9.9(c)).

DEFINITION 9.19   *Given a labeled, strongly distinguishable DAG G with n vertices and m arcs, the* Burrows–Wheeler index of G, *denoted by* BWT_G, *is a data structure that consists of*

- *the array of characters* `labels`, *encoded using a wavelet tree (see Section 3.3);*
- *bitvectors* `last` *and* `out`, *indexed to support* `rank` *and* `select` *queries in constant time as described in Section 3.2;*
- *the count array C.*

See the main text for the definition of such arrays. BWT_G takes $m \log \sigma (1 + o(1)) + 2m + o(m)$ bits of space.

Analogously to previous sections, we denote by $\vec{W}$ the range of BWT_G that contains the intervals of all vertices $v \in V$ whose distinguishing prefix P_v is prefixed by string $W \in [0..\sigma + 1]^+$, or the interval of the only vertex $v \in V$ whose P_v is a prefix of W. Note that, if G is a distinguishable tree, the arrays `labels` and `last` coincide with those of the Burrows–Wheeler index of a tree described in Section 9.5, up to the criterion used for breaking ties in the lexicographic order.

*9.6.1   Moving backward

Like its specialization for trees, BWT_G supports a number of navigational primitives. For example, assume that we know $\vec{v}$ for a specific vertex v, and that we want to determine its label $\ell(v)$. We know that the number of arcs (u, w) whose P_u is lexicographically smaller than P_v is $i = \texttt{rank}_1(\texttt{out}, \vec{v} - 1)$, thus $\ell(v)$ is the character $c \in [0..\sigma + 1]$ that satisfies $C[c] < i + 1 \le C[c + 1]$, which can be determined by a binary search over C. Thus, computing $\ell(v)$ takes $O(\log \sigma)$ time.

Similarly, let function $\texttt{inNeighbor}(\vec{v}, c) = \vec{u}$ return the interval of the only in-neighbor u of v with $\ell(u) = c$. Assume that i is the position of c in `labels`$[\vec{u}]$, or equivalently that $i \in \vec{u}$ and $\texttt{labels}[i] = c$. Then, position $\mathsf{LF}(i, c)$ belongs to $\vec{u}$ in `out`, thus we can implement the function `inNeighbor` as follows:

$$\text{inNeighbor}(\vec{v}, c) = \Big[\text{select}_1\Big(\text{last}, \text{rank}_1\big(\text{last}, \text{LF}(\vec{v}, c) - 1\big)\Big) + 1 ..$$

$$\text{select}_1\Big(\text{last}, \text{rank}_1\big(\text{last}, \text{LF}(\vec{v}, c) - 1\big) + 1\Big)\Big]. \quad (9.6)$$

See Figure 9.9(c) for a concrete example. Note that this operation takes $O(\log \sigma)$ time.

Recall that the interval of a string $W \in [0..\sigma + 1]^+$ in BWT_G contains the intervals of all vertices $v \in V$ such that W is a prefix of P_v, or the interval of the only vertex $v \in V$ such that P_v is a prefix of W. In the first case, the paths labeled by W in G are exactly the paths of length $|W|$ that start from a vertex v with $\vec{v} \subseteq \overrightarrow{W}$. In the second case, the paths labeled by W in G are a subset of all the paths of length $|W|$ that start from v. We can lift the function inNeighbor to map from $\overrightarrow{W}$ to $\overrightarrow{cW}$ for $c \in [1..\sigma + 1]$, as follows:

$$\text{extendLeft}(c, \overrightarrow{W}, \overleftarrow{W}) = \Big[\text{select}_1\Big(\text{last}, \text{rank}_1\big(\text{last}, \text{LF}(\overrightarrow{W} - 1, c)\big)\Big) + 1 ..$$

$$\text{select}_1\Big(\text{last}, \text{rank}_1\big(\text{last}, \text{LF}(\overrightarrow{W}, c) - 1\big) + 1\Big)\Big].$$

$$(9.7)$$

This procedure clearly takes $O(\log \sigma)$ time, and it supports a generalization of the backward search algorithm described in Section 9.2.2 that returns the interval of all vertices in V that are the starting point of a path labeled by a string W of length m: the search proceeds again from right to left along W, computing iteratively $\overrightarrow{W_{i..m}} = \text{extendLeft}(\overrightarrow{W_{i+1..m}}, W[i])$. More generally, the following lemma holds.

LEMMA 9.20 BWT_G *is a* forward BWT index *(Definition 9.13). The operations* extendLeft *and* isLeftMaximal *can be implemented in* $O(\log \sigma)$ *time, and* enumerateLeft *can be implemented in* $O(d \log(\sigma/d))$ *time, where d is the size of the output.*

*9.6.2   Moving forward

Let function $\text{outNeighbor}(\vec{v}, i) = \overrightarrow{w}$ be such that $w \in V$ is the out-neighbor of v whose P_w has lexicographic rank i among all the out-neighbors of v. Assume that $\ell(v) = c$. There are $j = \text{rank}_1(\text{out}, \vec{v} - 1)$ arcs (x, y) such that P_x is lexicographically smaller than P_v, and in particular there are $j - C[c]$ such arcs with P_x prefixed by c. It follows that position $k = \text{select}_c(\text{labels}, j - C[c] + i)$ in labels belongs to $\overrightarrow{w}$, thus

$$\text{outNeighbor}(\vec{v}, i) = \Big[\text{select}_1\Big(\text{last}, \text{rank}_1(\text{last}, k - 1)\Big) + 1 ..$$

$$\text{select}_1\Big(\text{last}, \text{rank}_1(\text{last}, k - 1) + 1\Big)\Big],$$

$$k = \text{select}_c\big(\text{labels}, \text{rank}_1(\text{out}, \vec{v} - 1) - C[c] + i\big).$$

$$c = \ell(v) \qquad (9.8)$$

See Figure 9.9(c) for a concrete example. Note that this operation takes $O(\log \sigma)$ time, and that the formula to compute k is identical to Equation (9.4), which implements the homologous function in the Burrows–Wheeler index of a tree. It is also easy to index numerical values assigned to vertices, such as identifiers or probabilities, by adapting the sampling approach described in Section 9.2.3: see Exercise 9.12.

*9.6.3 Construction

A labeled DAG might not be distinguishable, and it might not even be reverse-deterministic. It is possible to transform $G = (V, E, \Sigma)$ into a reverse-deterministic DAG $G' = (V', E', \Sigma)$ such that $\bar{\ell}(s') = \bar{\ell}(s)$, by applying the classical *powerset construction algorithm* for determinizing finite automata, described in Algorithm 9.4. Let A be an acyclic, nondeterministic finite automaton with n states that recognizes a finite language $L(A)$: in the worst case, the number of states of the minimal *deterministic* finite automaton that recognizes $L(A)$ is exponential in n, and this bound is tight. Therefore, Algorithm 9.4 can produce a set of vertices V' of size $O\left(2^{|V|}\right)$ in the worst case.

Algorithm 9.4: Powerset construction

Input: Directed labeled graph $G = (V, E, \Sigma)$, not necessarily acyclic.
Output: Directed, labeled, reverse-deterministic graph $G' = (V', E', \Sigma)$ such that $\bar{\ell}(s') = \bar{\ell}(s)$.

1 $t' \leftarrow \{t\}$;
2 $\ell(t') = \#$;
3 $V' \leftarrow \{t'\}$;
4 $E' \leftarrow \emptyset$;
5 toBeExpanded $\leftarrow \{t'\}$;
6 expanded $\leftarrow \emptyset$;
7 **while** toBeExpanded $\neq \emptyset$ **do**
8   Let $A \in$ toBeExpanded;
9   **for** $c \in \Sigma \cup \{\$\}$ **do**
10         $A' \leftarrow \{u \in V \mid (u, v) \in E, v \in A, \ell(u) = c\}$;
11         $\ell(A') \leftarrow c$;
12         $E' \leftarrow E' \cup \{(A', A)\}$;
13         $V' \leftarrow V' \cup \{A'\}$;
14   **if** $A' \notin$ expanded **then**
15   toBeExpanded $\leftarrow$ toBeExpanded $\cup \{A'\}$;
16   toBeExpanded $\leftarrow$ toBeExpanded $\setminus \{A\}$;
17   expanded $\leftarrow$ expanded $\cup \{A\}$;

To make a labeled, reverse-deterministic DAG G' strongly distinguishable (or, more concretely, to transform graph G into graph H in Figure 9.9), we can adapt the prefix-doubling approach used in Lemma 8.8 for building suffix arrays. Specifically,

G' is said to be *k-sorted* if, for every vertex $v \in V'$, one of the following conditions holds:

- v is strongly distinguishable with $|P_v| \leq k$;
- all paths that start from v share the same prefix of length exactly k.

Starting from G', which is 1-sorted, it is possible to build a *conceptual* sequence of 2^i-sorted graphs $G^i = (V^i, E^i)$ for $i = 1, 2, \ldots, \lceil \log(p) \rceil$, where p is the length of a longest path in G'. The final graph $G^{\lceil \log(p) \rceil} = G^*$ is p-sorted, thus it is also strongly distinguishable.

Specifically, we project each vertex $v \in V'$ into *a set of vertices* in V^*, corresponding to the distinct prefixes of paths that starts from v in G', as follows. Assume that a vertex $v \in V^i$ is annotated with the following elements:

- `path(v)`, the string of length 2^i that prefixes all paths of G^i that start from v;
- `source(v)`, the original vertex in V' that was projected to v;
- `targets(v)`, the set of all vertices in V' that are reachable from `source(v)` using a path of length 2^i labeled by `path(v)`.

Let L be the list that results from sorting set $\{\texttt{path}(v) : v \in V^i\}$ in lexicographic order, and let $R[1..|V^i|]$ be an array such that $R[v]$ is the position of the first occurrence of `path(v)` in L. If $R[u]$ is unique in R, then vertex u is strongly distinguishable, and it is projected onto a single vertex in V^{i+1} with the same values of `path`, `source`, and `targets`. Otherwise, we add to V^{i+1} a vertex uvw for every pair of vertices $v \in V'$ and $w \in V^i$ such that $v \in \texttt{targets}(u)$ and $\texttt{source}(w) = v$, where $\texttt{path}(uvw) = \texttt{path}(u)[1..2^i - 1] \cdot \texttt{path}(w)$ and $\texttt{targets}(uvw) = \texttt{targets}(w)$. We then merge all vertices in V^{i+1} with identical `source` and `path` into a single vertex, whose `targets` field is the union of the `targets` fields of the merged vertices. Finally, we build the set of arcs E^* just for the final graph G^*: this set consists of all pairs $v, w \in V^*$ such that $(\texttt{source}(v), \texttt{source}(w)) \in E$ and `path(v)` is a proper prefix of $\ell(\texttt{source}(v)) \cdot \texttt{path}(w)$. Note that this process expands the reverse-deterministic DAG G', and this DAG can in turn be exponential in the size of the input graph G. However, we still have that $|V^*| \in O(2^{|V|})$ in the worst case: see Exercise 9.15.

Note that G^* is strongly distinguishable only if graph G' is reverse-deterministic. Note also that a navigation on the original graph G can be simulated by a navigation on G^*. Building the vertices and arcs of G^* reduces to sorting, scanning, and joining lists of tuples of integers, thus it can be implemented efficiently in practice: see Exercise 9.16.

9.7   BWT indexes for de Bruijn graphs

A de Bruijn graph is a representation of the set of all distinct strings of a fixed length k, called *k-mers*, over a given alphabet $\Sigma = [1..\sigma]$.

DEFINITION 9.21   *Let k be an integer and let $\Sigma = [1..\sigma]$ be an alphabet. A de Bruijn graph with parameters Σ and k is a graph* $\mathsf{DBG}_{\Sigma, k} = (V, E)$ *with* $|V| = \sigma^{k-1}$ *vertices* *and* $|E| = \sigma^k$ *arcs, in which every vertex is associated with a distinct string of length* $k-1$ *on* $[1..\sigma]$, *and every arc is associated with a distinct string of length* k *on* $[1..\sigma]$. *An arc* (v, w) *associated with* k-*mer* $c_1 \cdots c_k$ *connects the vertex* v *associated with string* $c_2 \cdots c_{k-1}c_k \in [1..\sigma]^{k-1}$ *to the vertex* w *associated with string* $c_1 c_2 \cdots c_{k-1} \in [1..\sigma]^{k-1}$.

In this section we will systematically abuse notation, and identify a vertex of $\mathsf{DBG}_{\Sigma,k}$ with its associated $(k-1)$-mer, and an arc of $\mathsf{DBG}_{\Sigma,k}$ with its associated k-mer. Moreover, we will work with *subgraphs* of the full de Bruijn graph. Specifically, given a *cyclic string* T of length N on alphabet $[1..\sigma]$, which contains n distinct k-mers and m distinct $(k-1)$-mers, we are interested in the subgraph $\mathsf{DBG}_{T,k}$ of the de Bruijn graph $\mathsf{DBG}_{\Sigma,k}$ which contains m vertices and n arcs, such that every $(k-1)$-mer that appears in T has an associated vertex in $\mathsf{DBG}_{T,k}$, and every k-mer that appears in T has an associated arc in $\mathsf{DBG}_{T,k}$ (see the example in Figure 9.10(a)). Such graphs are the substrate of a number of methods to assemble fragments of DNA into chromosomes, a fundamental operation for reconstructing the genome of a new species from a set of reads (see Chapter 13). We will waive the subscripts from $\mathsf{DBG}_{T,k}$ whenever they are clear from the context, thus using DBG to refer to a subgraph of the full de Bruijn graph.

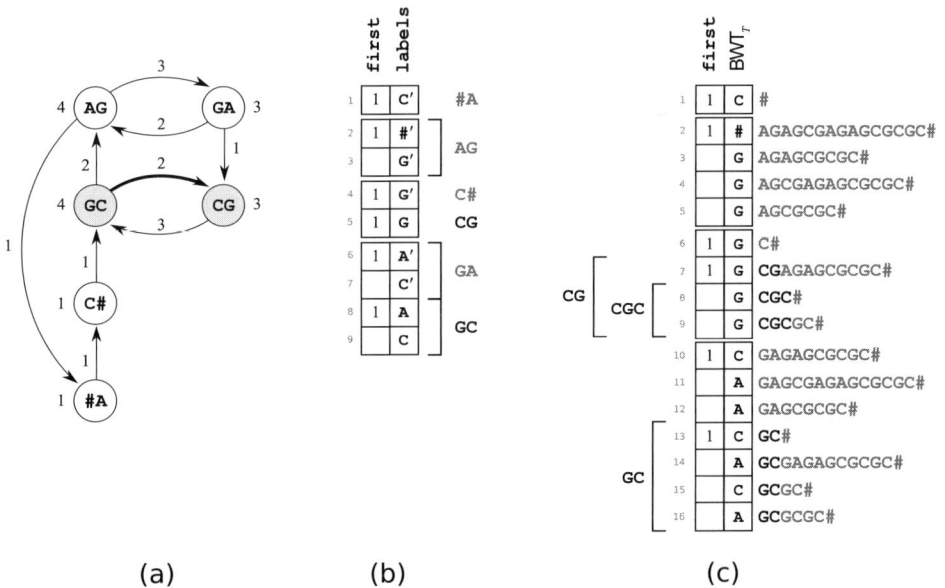

Figure 9.10 The de Bruijn graph $\mathsf{DBG}_{T,3}$ of the cyclic string $T = \mathtt{AGAGCGAGAGCGCGC\#}$ (a), with its frequency-oblivious (b) and its frequency-aware (c) representations. Vertices and arcs in (a) are labeled by the frequency of the corresponding dimers and trimers in T, respectively. In panels (b) and (c), zeros are omitted from bitvectors, and spaces mark the starting position of the interval of every character in $[1..\sigma]$. All panels highlight the transition from vertex GC to vertex CG.

Let v be a vertex of $\mathrm{DBG}_{T,k}$ associated with $(k-1)$-mer $c_2 \cdots c_k$, where $c_i \in [1..\sigma]$ for $i \in [2..k]$. In this section we are interested in the following operations on $\mathrm{DBG}_{T,k}$.

getArc(v, c_1)   Return the unique identifier of the arc from v associated with k-mer $c_1 c_2 \cdots c_k$, where $c_1 \in [1..\sigma]$, or -1 if no such arc exists.

followArc(v, c_1)   If getArc(v, c_1) $\neq -1$, return the unique identifier of the vertex associated with $(k-1)$-mer $c_1 c_2 \cdots c_{k-1}$.

outDegree(v)   Return the number of outgoing arcs from vertex v.

vertexFreq(v)   Return the number of occurrences in T of the $(k-1)$-mer associated with vertex v.

arcFreq(e)   Return the number of occurrences in T of the k-mer associated with arc e.

We assume that vertex and arc identifiers occupy a constant number of machine words, or equivalently $O(\log N)$ bits of space. A naive representation of $\mathrm{DBG}_{T,k}$ would clearly take $O(n \log n)$ bits: in this section we describe two alternative representations that take significantly less space, implement getArc and followArc in $O(\log \sigma)$ time, and can be built efficiently from the bidirectional Burrows–Wheeler index of T by adapting Algorithm 9.3.

Table 9.1 provides an overview of the two representations. Note that the frequency-aware representation takes more space than the frequency-oblivious representation whenever $n \ll N$, or equivalently whenever string T is highly repetitive. However, we will see that the frequency-aware representation is more flexible, since building it for different values of k requires just creating a different bitvector per value (see Section 9.7.3). In what follows we will also see that the two representations become essentially identical when every k-mer occurs at most once in T: see Exercise 9.21. It will also become apparent how to implement the function enumerateLeft, making both representations essentially equivalent to a forward Burrows–Wheeler index.

Without loss of generality, in what follows we will assume that the last character of T is the special character $0 = \# \notin [1..\sigma]$, and we will reset Σ to $\Sigma \cup \{\#\}$, σ to $\sigma + 1$, and N to $N + 1$. For brevity, we will also say that arc $e = c_1 c_2 \cdots c_{k-1}$ from vertex $v = c_2 \cdots c_{k-1}$ *is labeled by character* c_1. We will also focus on the case in which T is a single string: adapting the representations to the case in which T is the concatenation of a set of strings is left to Exercise 9.20.

9.7.1   Frequency-oblivious representation

In this representation the identifier of a vertex v is the lexicographic rank of the $(k-1)$-mer associated with v, among all distinct $(k-1)$-mers that occur in T. This is equivalent to what has been described for labeled trees in Section 9.5. Similarly, the identifier of an arc e is the lexicographic rank of the k-mer associated with e, among all distinct k-mers that occur in T. Note that the lexicographic order induces a total order on the vertices and on the arcs of $\mathrm{DBG}_{T,k}$: let $v_1, v_2, \ldots, v_m$ be the vertices of $\mathrm{DBG}_{T,k}$ in lexicographic order.

Table 9.1 Overview of the representations of $\text{DBG}_{T,k}$ described in Section 9.7. N is the length of T, n is the number of distinct k-mers that occur in T, σ is the cardinality of the alphabet, and d is the size of the output of operation `outDegree`.

	Frequency-oblivious	Frequency-aware
Space (bits)	$n(2 + \log \sigma (1 + o(1)))$	$N(1 + \log \sigma (1 + o(1)))$
Construction time	$O(N \log \sigma)$	$O(N \log \sigma)$
`getArc`	$O(\log \sigma)$	$O(\log \sigma)$
`followArc`	$O(\log \sigma)$	$O(\log \sigma)$
`outDegree`	$O(1)$	$O(d \log(\sigma/d))$
`vertexFreq`		$O(1)$
`arcFreq`		$O(1)$

To represent **DBG** we apply the same strategy as described in Section 9.5 for labeled trees: namely, we build the sorted list $v_1, v_2, \ldots, v_m$, we allocate an array `labels`$[1..n]$ of characters from alphabet $[1..\sigma]$, and we store the labels of the n_i arcs that start from vertex v_i, in arbitrary order, inside interval $\overleftarrow{v_i}$ of `labels`, where $\overleftarrow{v_i} = \sum_{p=1}^{i-1} n_p + 1$ and $\overrightarrow{v_i} = \sum_{p=1}^{i} n_p$. Thus, we assign to each vertex $v \in V$ an interval $\overleftrightarrow{v}$ in array `labels`. We represent `labels` as a wavelet tree, so that it supports `rank` and `select` queries, and we use again a bitvector `first`$[1..n]$ to mark the first position of every interval of a vertex: specifically, `first`$[i] = 1$ if and only if $i = 1$ or the arc with identifier i has a different source vertex from the arc with identifier $i-1$. We index `first` to support `rank` and `select` queries in constant time. Finally, we use an array $C_k[1..\sigma]$ that stores at position $C_k[c]$ the number of distinct k-mers prefixed by characters smaller than c, as well as an array $C_{k-1}[1..\sigma]$ that stores at position $C_{k-1}[c]$ the number of distinct $(k-1)$-mers prefixed by characters smaller than c.

Let i be the lexicographic rank of the $(k-1)$-mer $c_2 \cdots c_k$ associated with a vertex v of **DBG**. To compute $\overleftrightarrow{v}$, it suffices to perform

$$\overleftarrow{v} = \text{select}_1(\text{first}, i), \tag{9.9}$$

$$\overrightarrow{v} = \text{select}_1(\text{first}, i+1) - 1; \tag{9.10}$$

and to get `outDegree`$(\overleftrightarrow{v})$ it suffices to compute $|\overleftrightarrow{v}|$. To check whether there is an arc $e = (v, w)$ labeled by character c_1, we just need to check whether `labels`$[\overleftrightarrow{v}]$ contains character c_1, that is, whether $p = \text{rank}_{c_1}(\text{labels}, \overrightarrow{v})$ is greater than $\text{rank}_{c_1}(\text{labels}, \overleftarrow{v} - 1)$. If this is the case, we return

$$\text{getArc}(\overleftrightarrow{v}, c_1) = C_k[c_1] + \text{rank}_{c_1}(\text{labels}, \overleftarrow{v} - 1) + 1, \tag{9.11}$$

which is indeed the lexicographic rank of the k-mer $c_1 c_2 \cdots c_k$ associated with e among all k-mers that occur in T.

To implement `followArc`$(\overleftrightarrow{v}, c_1)$, we need to return the lexicographic rank of the $(k-1)$-mer $c_1 c_2 \cdots c_{k-1}$: this rank is $C_{k-1}[c_1] + q$, where q is the number of $(k-2)$-mers that are lexicographically smaller than or equal to $c_2 \cdots c_{k-1}$, *and that are the origin of an arc labeled by* c_1. Note that a string W of length $k-2$ corresponds to the contiguous interval $\overleftarrow{W}$ in labels that contains all vertices whose $(k-1)$-mer is prefixed by W. Consider thus $\overleftarrow{W}$, and let $i_1, i_2, \ldots, i_r$ be all the occurrences of character c_1 in labels$[\overleftarrow{W}]$, or equivalently all the arcs that start from a vertex prefixed by W and that end in a vertex prefixed by $c_1 W$. Assume that we preventively marked labels$[i_1]$ by setting it to an artificial character c_1' rather than to c_1, and that we performed this marking for every $(k-2)$-mer W and for every character in $[1..\sigma]$. Note that there are exactly m marked positions in total, where m is the number of distinct $(k-1)$-mers in T. Then, we return

$$\texttt{followArc}(\overleftarrow{v}, c_1) = C_{k-1}[c_1] + q,$$
$$q = \text{rank}_{c_1'}(\texttt{labels}, 1, \overleftarrow{v}). \tag{9.12}$$

Supporting followArc requires labels to store characters from alphabet $\Sigma \cup \Sigma'$, where Σ' is of size σ and contains a character c' for every character $c \in \Sigma$. Adapting the implementation of getArc to take characters in Σ' into account is left to Exercise 9.18.

The data structures described in this section represent just the *topology* of DBG, therefore all of the information about the frequency of $(k-1)$-mers and of k-mers is lost. A possible way to support vertexFreq and arcFreq consists in *indexing the original string T rather than* DBG, as described below.

9.7.2   Frequency-aware representation

In this representation the identifier of a vertex is the interval of the corresponding $(k-1)$-mer in BWT_T, and the identifier of an arc is the interval of the corresponding k-mer in BWT_T. Note that, for any k, the set of intervals of all k-mers is a *partition* of the range $[1..N]$, that is, each position in $[1..N]$ belongs to the interval of exactly one k-mer. Note also that the interval of a k-mer $c_1 c_2 \cdots c_k$ either coincides with the interval of $(k-1)$-mer $c_1 c_2 \cdots c_{k-1}$ (if $c_1 c_2 \cdots c_{k-1}$ is always followed by character c_k in T), or is included in the interval of $c_1 c_2 \cdots c_{k-1}$.

Suppose that we are given the BWT of string T represented as a wavelet tree. We build a bitvector first$[1..N]$ in which we mark with a one the first position of the interval of every $(k-1)$-mer W that occurs in T, and we keep the other positions to zero. That is, first$[\overleftarrow{W}] = 1$, first$[i] = 0$ for all $i \in [\overleftarrow{W} + 1..\overrightarrow{W}]$, and either $\overrightarrow{W} = N$ (if W is the last $(k-1)$-mer in lexicographic order) or first$[\overrightarrow{W} + 1] = 1$. We then index first to support rank and select queries in constant time. Our representation of DBG consists of BWT_T, first, and the usual array $C[1..\sigma]$ which stores in $C[c]$ the number of occurrences of characters smaller than c in T.

Associating vertices and arcs with intervals of BWT_T makes it easy to implement the following operations on $\text{DBG}_{T,k}$:

$$\texttt{vertexFreq}(\overleftarrow{v}) = |\overleftarrow{v}|, \tag{9.13}$$
$$\texttt{arcFreq}(\overleftarrow{e}) = |\overleftarrow{e}|, \tag{9.14}$$
$$\texttt{outDegree}(\overleftarrow{v}) = |\texttt{rangeList}(\text{BWT}_T, \overleftarrow{v}, \overrightarrow{v}, 1, \sigma)|, \tag{9.15}$$

where rangeList is the $O(d \log(\sigma/d))$-time function provided by the wavelet tree of BWT_T (see Section 3.3.2). Implementing getArc is equally straightforward, since it amounts to performing a $O(\log \sigma)$-time backward step on BWT_T with character c_1:

$$\mathtt{getArc}(\overleftrightarrow{v}, c_1) = \overleftrightarrow{e},$$

$$\overleftarrow{e} = C[c_1] + \mathtt{rank}_{c_1}(\mathsf{BWT}_T, \overleftarrow{v} - 1) + 1,$$

$$\overrightarrow{e} = C[c_1] + \mathtt{rank}_{c_1}(\mathsf{BWT}_T, \overrightarrow{v}). \tag{9.16}$$

Assume that vertex v corresponds to the $(k - 1)$-mer $c_2 \cdots c_k$. To implement $\mathtt{followArc}(\overleftrightarrow{v}, c_1)$ we first call $\mathtt{getArc}(\overleftrightarrow{v}, c_1)$ to obtain the interval of the arc $e = (v, w)$ labeled by c_1 that starts from v. Note that e corresponds to k-mer $c_1 \cdots c_k$, and we want to compute the interval of vertex w associated with the $(k - 1)$-mer $c_1 \cdots c_{k-1}$. Clearly $\overleftrightarrow{e} \subseteq \overleftrightarrow{w}$, therefore we know that $\mathtt{first}[i] = 0$ for all $i \in [\overrightarrow{e} + 1..\overrightarrow{e}]$. If $\mathtt{first}[\overrightarrow{e}] = 1$ and either $\overrightarrow{e} = N + 1$ or $\mathtt{first}[\overrightarrow{e} + 1] = 1$, then $\overleftrightarrow{e} = \overleftrightarrow{w}$: this happens whenever string $c_1 \cdots c_{k-1}$ is always followed by character c_k in T. Otherwise, $\overleftrightarrow{e} \subset \overleftrightarrow{w}$. If $\mathtt{first}[\overleftarrow{e}] = 1$ then clearly $\overleftarrow{w} = \overleftarrow{e}$. Otherwise $\overleftarrow{w} < \overleftarrow{e}$, the interval $\mathtt{first}[\overleftarrow{w} + 1..\overleftarrow{e} - 1]$ contains only zeros, and $\overleftarrow{w}$ is the position of the last one in $\mathtt{first}$ that precedes position $\overleftarrow{e}$:

$$\overleftarrow{w} = \mathtt{select}_1\big(\mathtt{first}, \mathtt{rank}_1(\mathtt{first}, \overleftarrow{e})\big). \tag{9.17}$$

Similarly, if $\overrightarrow{e} = N$ then $\overrightarrow{w} = \overrightarrow{e}$, otherwise $\overrightarrow{w} + 1$ is the first one which follows position $\overrightarrow{e}$ in $\mathtt{first}$:

$$\overrightarrow{w} = \mathtt{select}_1\big(\mathtt{first}, \mathtt{rank}_1(\mathtt{first}, \overrightarrow{e}) + 1\big) - 1. \tag{9.18}$$

9.7.3 Space-efficient construction

Both the frequency-aware and the frequency-oblivious representations of $\mathsf{DBG}_{T\#,k}$ for some string $T \in [1..\sigma]^{N-1}$ can be built from $\mathsf{BWT}_{T\#}$, augmented with the bitvector $\mathtt{first}_k[1..N]$ which marks the starting position of every interval in $\mathsf{BWT}_{T\#}$ that corresponds to a k-mer, for some value of k. It turns out that $\mathtt{first}_k$ itself can be derived from the bidirectional BWT index of T.

LEMMA 9.22 *Given an integer k, and the bidirectional BWT index of a string T of length $N - 1$ represented as in Theorem 9.14, we can compute array $\mathtt{first}_k$ in $O(N \log \sigma)$ time and in $O(\sigma \log^2 N)$ bits of working space.*

Proof To build $\mathtt{first}$ for a given value of k, we start by setting $\mathtt{first}[i] = 1$ for all $i \in [1..N]$. Then, we use Algorithm 9.3 to enumerate all intervals that correspond to an internal node v of the suffix tree of T, as well as the length of label $\ell(v)$, in overall $O(N \log \sigma)$ time. For every internal node v we check whether $|\ell(v)| \geq k$: if this is the case, we use operations $\mathtt{enumerateRight}$ and $\mathtt{extendRight}$ provided by the bidirectional BWT index of T to compute the intervals of all the children $w_1, w_2, \ldots, w_h$ of v in the suffix tree of T, in lexicographic order. Then, we flip bit $\mathtt{first}[\overleftarrow{w_i}]$ for all $i \in [2..h]$. Proving that this algorithm builds $\mathtt{first}_k$ is left to Exercise 9.19. $\square$

If the bidirectional BWT index of T is provided in the input, we already have $\mathsf{BWT}_{T\#}$ represented as a wavelet tree, as well as the vector $C[0..\sigma]$. Thus, to build the frequency-aware representation of $\mathsf{DBG}_{T\#,k}$, we just need to index $\mathtt{first}_{k-1}$ to answer $\mathtt{rank}$ and $\mathtt{select}$ queries in constant time.

THEOREM 9.23   *Given the bidirectional BWT index of a string T of length $N - 1$ represented as in Theorem 9.14, we can build the frequency-aware representation of the de Bruijn graph $\mathsf{DBG}_{T\#,k}$ in $O(N \log \sigma)$ time and in $O(\sigma \log^2 N)$ bits of working space.*

Building the frequency-oblivious representation of $\mathsf{DBG}_{T\#,k}$ requires just a little more effort.

THEOREM 9.24   *Given the bidirectional BWT index of a string T of length $N - 1$ represented as in Theorem 9.14, we can build the frequency-oblivious representation of the de Bruijn graph $\mathsf{DBG}_{T\#,k}$ in $O(N \log \sigma)$ time and in $O(N + \sigma \log^2 N)$ bits of working space.*

Proof   We build $\mathtt{first}_{k-1}$ as described in Lemma 9.22. To avoid ambiguity, we rename $\mathtt{first}_{k-1}$ to B_{k-1}. Then, we initialize the bitvector $\mathtt{first}$ of the frequency-oblivious representation of $\mathsf{DBG}_{T\#,k}$ to all zeros, except for $\mathtt{first}[1] = 1$. Let p_1, p_2, and p_3 be pointers to a position in B_{k-1}, in $\mathsf{BWT}_{T\#}$, and in $\mathtt{labels}$, respectively, and assume that they are all initialized to one. The first one in B_{k-1} is clearly at position one: we thus move p_1 from position two, until $B_{k-1}[p_1] = 1$ (or until $p_1 > N$). Let W be the first $(k-1)$-mer in lexicographic order. The interval of W in $\mathsf{BWT}_{T\#}$ is $[1..p_1 - 1]$. We can thus initialize a bitvector $\mathtt{found}[0..\sigma]$ to all zeros, move pointer p_2 between one and $p_1 - 1$ over $\mathsf{BWT}_{T\#}$, and check whether $\mathtt{found}[\mathsf{BWT}[p_2]] = 0$: if so, we set $\mathtt{labels}[p_3] = \mathsf{BWT}[p_2]$, we increment p_3 by one, and we set $\mathtt{found}[\mathsf{BWT}[p_2]]$ to one. At the end of this process, all distinct characters that occur to the left of W in $T\#$ have been collected in the contiguous interval $[1..p_3 - 1]$ of $\mathtt{labels}$, so we can set $\mathtt{first}[p_3] = 1$. To reinitialize $\mathtt{found}$ to all zeros, we scan again the interval $[1..p_3 - 1]$ by moving a pointer q, and we iteratively set $\mathtt{found}[\mathtt{labels}[q]] = 0$. We repeat this process with the following blocks marked in B_{k-1}.

The marking of the first occurrence of every distinct character that appears in the interval of a $(k-2)$-mer can be embedded in the algorithm we just described. Specifically, we first build the bitvector $\mathtt{first}_{k-2}$ using again the algorithm described in Lemma 9.22, and we rename it B_{k-2}. Then, we keep the first and last position of the current interval of a $(k - 2)$-mer in the variables x and y, respectively. Whenever we are at the end of the interval of a $(k-2)$-mer, or equivalently whenever $p_1 > N$ or $B_{k-2}[p_1] = 1$, we set $y = p_1 - 1$ and we scan again $\mathtt{labels}[x..y]$, using the array $\mathtt{found}$ as before to detect the first occurrence of every character. At the end of this scan, we reset $x = y + 1$ and we repeat the process.

Building arrays C_k and C_{k-1} is easy, and it is left to the reader.   □

9.8   Literature

The Burrows–Wheeler transform was first introduced by Burrows & Wheeler (1994) as a tool for text compression. It was immediately observed that the construction of the transform is related to suffix arrays, but it took some time for the tight connection that enables backward search in compressed space to be discovered in Ferragina & Manzini (2000) and Ferragina & Manzini (2005). In the literature the BWT index is often called the *FM-index*. Simultaneously, other ways of compressing suffix arrays were developed (Grossi & Vitter 2000, 2006; Sadakane 2000) building on the inverse of the function LF, called ψ therein. Our description of succinct suffix arrays follows Navarro & Mäkinen (2007).

The space-efficient construction of the Burrows–Wheeler transform given in Section 9.3 is a variant of an algorithm that appears in Hon *et al.* (2003). The original method allows one to build the Burrows–Wheeler transform in $O(n \log \log \sigma)$ time using $O(n \log \sigma)$ bits of space. Our version takes more time because it uses wavelet trees, while the original method applies more advanced structures to implement rank. The merging step of the algorithm can be further improved by exploiting the simulation of a bidirectional BWT index with just one BWT (Belazzougui 2014): this gives an optimal $O(n)$ time construction using $O(n \log \sigma)$ bits. The original result uses randomization, but this has been improved to deterministic in the full version of the article.

The bidirectional BWT index described in Section 9.4 is originally from Schnattinger *et al.* (2010). A similar structure was proposed also in Li *et al.* (2009). The variant that uses just one BWT is similar to the one in Beller *et al.* (2012) and Beller *et al.* (2013), and it is described in Belazzougui (2014). The latter version differs from the original bidirectional BWT index in that it uses a stack that occupies $o(n)$ bits, rather than a queue and an auxiliary bitvector that take $O(n)$ bits of space as proposed in the original work. The use of the so-called *stack trick* to enumerate intervals that correspond to suffix tree nodes with the bidirectional BWT index in small space is from Belazzougui *et al.* (2013): in this chapter we extended its use to the variant in Belazzougui (2014) which uses only one BWT index. The stack trick itself has been used as early as the quicksort algorithm (Hoare 1962). The idea of detecting right-maximality using a bitvector to encode runs of equal character in the Burrows–Wheeler transform is due to Kulekci *et al.* (2012).

Extensions of the Burrows–Wheeler index to trees and graphs were proposed by Ferragina *et al.* (2009) and by Sirén *et al.* (2011) and Sirén *et al.* (2014). Exercise 9.10 is solved in Ferragina *et al.* (2009), and Exercises 9.12, 9.13, 9.17, and partially 9.16, are solved in Sirén *et al.* (2011) and Sirén *et al.* (2014). The Rabin–Scott powerset construction related to the latter is from Rabin & Scott (1959). The frequency-oblivious and frequency-aware representations of the de Bruijn graph are from Bowe *et al.* (2012) and Välimäki & Rivals (2013), respectively. The space-efficient algorithm for constructing the frequency-oblivious representation is a new contribution of this book. In Välimäki & Rivals (2013), a different space-efficient construction for our frequency-aware representation is shown. There are at least two other small-space representations of de Bruijn graphs based on the Burrows–Wheeler transform in Rødland (2013) and Chikhi *et al.* (2014). Yet other small-space representations are based on hashing (Chikhi & Rizk 2012; Pell *et al.* 2012; Salikhov *et al.* 2013), and more precisely on Bloom filters: these are not as space-efficient as the ones based on the Burrows–Wheeler transform.

The size of the genome of *Picea glauca* was taken from the databases of the National Center for Biotechnology Information, US National Library of Medicine, and the estimate for the genome of *Paris japonica* was derived from Pellicer *et al.* (2010).

Exercises

9.1   Consider the original definition of the Burrows–Wheeler transform with the cyclic shifts, without a unique endmarker # added to the end of the string. Assume you have a fast algorithm to compute our default transform that assumes the endmarker is added. Show that you can feed that algorithm an input such that you can extract the original cyclic shift transform efficiently from the output.

9.2   Describe an algorithm that builds the succinct suffix array of a string T from the BWT index of T, within the bounds of Theorem 9.6.

9.3   Consider the strategy for answering a set of occ locate queries in batch described in Section 9.2.4. Suppose that, once all positions in $\mathsf{SA}_{T\#}$ have been marked, a bitvector marked$[1..n + 1]$ is indexed to answer rank queries. Describe an alternative way to implement the batch extraction within the same asymptotic time bound as the one described in Section 9.2.4, but *without resorting to radix sort*. Which strategy is more space-efficient? Which one is faster? The answer may vary depending on whether occ is small or large.

9.4   Recall that at the very end of the construction algorithm described in Section 9.3 we need to convert the BWT of the padded string X into the BWT of string $T\#$. Prove that $\mathsf{BWT}_X = T[n] \cdot \#^{k-1} \cdot W$ and $\mathsf{BWT}_{T\#} = T[n] \cdot W$, where $n = |T|$, $k = |X| - n$, and string W is a permutation of $T[1..n - 1]$.

9.5   Assume that you have a machine with P processors that share the same memory. Adapt Algorithm 9.3 to make use of all P processors, and estimate the resulting speedup with respect to Algorithm 9.3.

9.6   Recall that Property 1 is key for implementing Algorithm 9.3 with just one BWT. Prove this property.

9.7   In many applications, the operation enumerateLeft(i,j), where $[i..j] = \mathbb{I}(W, T)$, is immediately followed by operations extendLeft $\left(c, \mathbb{I}(W, T), \mathbb{I}(\underline{W}, \underline{T})\right)$ applied to every distinct character c returned by enumerateLeft(i,j).

    i.   Show how to use the operation rangeListExtended(T, i, j, l, r) defined in Exercise 3.12 to efficiently support a new operation called enumerateLeftExtended$(\mathbb{I}(W, T), \mathbb{I}(\underline{W}, \underline{T}))$ that returns the distinct characters that appear in substring $\mathsf{BWT}_{T\#}[i..j]$, and for each such character c returns the pair of intervals computed by extendLeft $\left(c, \mathbb{I}(W, T), \mathbb{I}(\underline{W}, \underline{T})\right)$. Implement also the symmetric operation enumerateRightExtended.

    ii.   Reimplement Algorithm 9.3 to use the operations enumerateLeftExtended and enumerateRightExtended instead of enumerateLeft, enumerateRight, extendLeft, and extendRight.

9.8 Recall that Property 2 is key for navigating the Burrows–Wheeler index of a labeled tree top-down (Section 9.5.1). Prove this property.

9.9 Consider the Burrows–Wheeler index of a tree $\mathcal{T}$ described in Section 9.5.

i.   Given a character $c \in [1..\sigma]$, let $\overleftarrow{c}$ be the starting position of its interval $\overleftarrow{c}$ in $\mathsf{BWT}_{\mathcal{T}}$, and let $C'[1..\sigma]$ be an array such that $C'[c] = \overleftarrow{c}$. Describe an $O(n)$-time algorithm to compute C' from C and $\mathtt{last}$.

ii. Recall that $\mathbb{I}(v) = [\overleftarrow{v}..\overrightarrow{v}]$ is the interval in $\mathsf{BWT}_{\mathcal{T}}$ that contains all the children of node v, and that $\mathbb{I}'(v)$ is the position of the arc $(v, \mathtt{parent}(v))$ in $\mathsf{BWT}_{\mathcal{T}}$. Let $D[1..n + m - 1]$ be an array such that $D[i] = \overleftarrow{v}$, where $\mathbb{I}'(v) = i$, or -1 if v is a leaf. Describe a $O(n + m)$-time algorithm to compute D from C', $\mathtt{last}$, and $\mathtt{labels}$. *Hint.* Use the same high-level strategy as in Lemmas 8.7 and 9.9.

iii. Give the pseudocode of an $O(n)$-time algorithm that reconstructs the original tree $\mathcal{T}$ from $\mathsf{BWT}_{\mathcal{T}}$ and D, in depth-first order.

9.10 In order to build the Burrows–Wheeler index of a tree $\mathcal{T}$ described in Section 9.5, we can adapt the prefix-doubling approach used in Lemma 8.8 for building suffix arrays. In particular, assume that we want to assign to every node v the order $R(v)$ of its path to the root, among all such paths in $\mathcal{T}$. We define the *ith contraction* of tree $\mathcal{T}$ as the graph $\widetilde{T}^i = (V, E^i)$ such that $(u, v) \in E^i$ iff $v_1, v_2, \ldots, v_{2i+1}$ is a path in $\mathcal{T}$ with $v_0 = u$ and $v_{2i+1} = v$. Clearly $\widetilde{T}^0 = \mathcal{T}$, and E^{i+1} is the set of all pairs (u, v) such that $P = u, w, v$ is a path in $\widetilde{T}^i$. Show how to iteratively refine the value $R(v)$ for every node, starting from the initial approximation $R^0(v) = \overrightarrow{\ell(v)} = C'[\ell(v)]$ computed as in Exercise 9.9. Describe the time and space complexity of your algorithm.

9.11 Assume that we want to generalize the recursive, $O(\sigma + n)$-time algorithm for building the suffix array $\mathsf{SA}_{S\#}$ of a string S to a similar algorithm for building the Burrows–Wheeler index of a labeled tree $\mathcal{T}$ described in Section 9.5. Do the results in Definition 8.9, Lemma 8.10, and Lemma 8.11 still hold? Does the overall generalization still achieve linear time for any $\mathcal{T}$? If not, for which topologies or labelings of $\mathcal{T}$ does the generalization achieve linear time?

9.12 Assume that a numerical value $\mathtt{id}(v)$ must be associated with every vertex v of a labeled, strongly distinguishable DAG G: for example, $\mathtt{id}(v)$ could be the probability of reaching vertex v from s. Assume that, whenever there is exactly one path P connecting vertex u to vertex v, we have that $\mathtt{id}(v) = f(\mathtt{id}(u), |P|)$, where f is a known function. Describe a method to augment the Burrows–Wheeler index of G described in Section 9.6 with the values of $\mathtt{id}$, and its space and time complexity. *Hint.* Adapt the strategy used for strings in Section 9.2.3.

9.13 Consider the Burrows–Wheeler index of a labeled, strongly distinguishable DAG described in Section 9.6, and assume that $\mathtt{labels}$ is represented with $\sigma + 1$ bitvectors $B_c[1..m]$ for all $c \in [1..\sigma + 1]$, where $B_c[i] = 1$ if and only if $\mathtt{labels}[i] = c$. Describe a more space-efficient encoding of $\mathtt{labels}$. Does it speed up the operations described in Section 9.6?

9.14 Given any directed labeled graph $G = (V, E, \Sigma)$, Algorithm 9.4 builds a reverse-deterministic graph $G' = (V', E', \Sigma)$ in which a vertex $v \in V'$ corresponds to *a subset of V*. Adapt the algorithm so that a vertex $v \in G'$ corresponds to *a subset of E*. Show an example in which the number of vertices in the reverse-deterministic graph produced by this version of the algorithm is smaller than the number of vertices in the reverse-deterministic graph produced by Algorithm 9.4.

9.15 Recall that in Section 9.6.3 a labeled DAG G is transformed into a reverse-deterministic DAG G', and then G' is itself transformed into a strongly distinguishable DAG G^*. Show that $|G^*| \in O(2^{|G|})$ in the worst case.

9.16 Consider the algorithm to enforce distinguishability described in Section 9.6.3.

i.   Show that this algorithm does not also enforce reverse-determinism.
ii.   Assume that the input to this algorithm contains two arcs (u, w) and (v, w) such that $\ell(u) = \ell(v)$: give an upper bound on the number of vertices that result from the projection of u and v to the output.
iii.   Modify the algorithm to build a distinguishable, rather than a strongly distinguishable, DAG.
iv.   Modify the algorithm to decide whether the input graph is distinguishable or not.
v.   Implement the algorithm using just sorts, scans, and joins of lists of tuples of integers, *without storing or manipulating strings* path *explicitly*. Give the pseudocode of your implementation.
vi.   Give the time and space complexity of the algorithm as a function of the size of its input and its output.

9.17 Describe a labeled, strongly distinguishable DAG that recognizes an infinite language. Describe an infinite language that cannot be recognized by any labeled, strongly distinguishable DAG.

9.18 Consider the frequency-oblivious representation of a de Bruijn graph described in Section 9.7.1. Show how to implement the function getArc when labels contains characters in $\Sigma \cup \Sigma'$.

9.19 Recall that marking the starting position of every interval in $\text{BWT}_{T\#}$ that corresponds to a k-mer is a key step for building both the frequency-aware and the frequency-oblivious representation of a de Bruijn graph. Prove that Lemma 9.22 performs this marking correctly.

9.20 Some applications in high-throughput sequencing require the de Bruijn graph of *a set of strings* $\mathcal{R} = \{R^1, R^2, \ldots, R^r\}$. Describe how to adapt the data structures in Sections 9.7.1 and 9.7.2 to represent the de Bruijn graph of string $R = R^1 \# R^2 \# \cdots \# R^r \#$, where $\# \notin [1..\sigma]$. Give an upper bound on the space taken by the frequency-oblivious and by the frequency-aware representation.

9.21 Describe the frequency-oblivious and the frequency-aware representations of the de Bruijn graph $\text{DBG}_{T,k}$ when every string of length k on a reference alphabet Σ occurs at most once in T.

Part IV

Genome-Scale Algorithms

10 Read alignment

Recall the high-throughput sequencing applications from Chapter 1. For example, in whole-genome resequencing we obtain a set of reads, that is, short extracts of DNA sequences from random positions in a donor genome. Assuming that the reference genome of the species has been assembled, one can align the reads to the reference in order to obtain an approximation of the content of the donor genome. We will consider this process in Chapter 14.

We remind the reader that reads come from both strands and, in diploid organisms, from both copies of each chromosome. This means that either the read or its reverse complement should align to the reference genome, but in what follows we do not explicitly repeat this fact. Let us now focus on how to efficiently align a set of (short) reads to a large genome.

The first observation is that, if a position in the reference genome is covered by several reads, one could try to exploit multiple alignment methods from Section 6.6 to obtain an accurate alignment. This gives an enormous semi-local multiple alignment problem with the reads and the reference sequence as input, and a multiple alignment as output (where reads can have an arbitrary amount of cost-free gaps before and after). Owing to the hardness of finding an optimal multiple alignment, shown for example in Section 6.6.3, we leave this approach just as a thought experiment. However, Chapter 14 deals with an application of this idea on a smaller scale.

Let us consider an incremental approach, which consists of aligning each read independently to the reference. This approach can be justified by the fact that, if the read is long enough, then it should not align to a random position in the genome, but exactly to that unique position from where it was sequenced from the donor. This assumes that this position is at least present in the reference genome, and that it is not inside a repeat. See Insight 10.1 for a more detailed discussion on this matter.

Insight 10.1 Read filtering, mapping quality

Since the incremental approach to read alignment is justified by the uniqueness of the expected occurrence positions, it makes sense to *filter* reads that are expected to occur inside a random sequence of length n. This filtering is aided by the fact that most sequencing machines give predictions on how accurately each base was measured. Let $\mathbb{P}(p_i)$ be such a probability for position i inside read P containing p_i. With probability $(1 - \mathbb{P}(p_i))$ this position contains an arbitrary symbol. We may view

---

P as a conditional *joker* sequence, with position i being a joker matching any symbol with probability $(1 - \mathbb{P}(p_i))$. Such a conditional joker sequence P matches a random sequence of length m with probability

$$\prod_{i=1}^{m} \left(\mathbb{P}(p_i)q_{p_i} + (1 - \mathbb{P}(p_i)) \right), \tag{10.1}$$

where q_c is the probability of symbol c in a random sequence. An approximation of the expected number of occurrences of a conditional joker sequence P in random sequence $T = t_1 t_2 \cdots t_n$ is

$$\mathbb{E}(P, T) = (n - m + 1) \prod_{i=1}^{m} \left(\mathbb{P}(p_i)q_{p_i} + (1 - \mathbb{P}(p_i)) \right). \tag{10.2}$$

One can use the criterion $\mathbb{E}(P, T) > f$, for some $f \le 1$, to filter out reads that are likely to match a random position, under our conditional joker sequence model.

Owing to repeats, some reads will match multiple positions in the genome, no matter how stringent the filtering which is used. In fact, it is impossible to know which of the positions is the one where the sequence originates. This phenomenon can be captured by the *mapping quality*; if a read has d equally good matches, one can assign a probability $1/d$ for each position being correct. This measure can be refined by taking into account also alignments that are almost as good as the best: see Exercise 10.1.

In Chapters 8 and 9 we learned that we can build an index for a large sequence such that exact pattern matching can be conducted in optimal time. In the read alignment problem, exact pattern matching is not enough, since we are in fact mainly interested in the *differences* between the donor and the reference. Furthermore, DNA sequencing may produce various measurement errors in the reads. Hence, a best-scoring semi-local alignment (in which an entire read matches a substring of the reference) is of interest. This was studied as the k-errors problem in Section 6.3.

In this chapter we will extend indexed exact pattern matching to *indexed approximate pattern matching*. For simplicity of exposition, we will focus on approximate string matching allowing mismatches only. Extending the algorithms to allow indels is left for the reader. In what follows, we describe several practical methods for this task, but none of them guarantees any good worst-case bounds; see the literature section on other theoretical approaches with proven complexity bounds.

10.1   Pattern partitioning

Assume we want to search for a read $P = p_1 p_2 \cdots p_m$ in a reference sequence $T = t_1 t_2 \cdots t_n$, allowing k errors.

Pattern partitioning exploits the pigeonhole principle, under the form of the following observation.

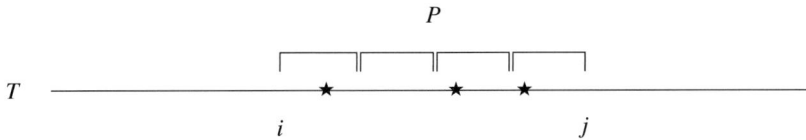

Figure 10.1 Pigeonhole principle in pattern partitioning. Here the pattern matches with three-errors whose positions in T are marked with $\star$ and partitioning of the pattern into four pieces is necessary and sufficient to ensure that one of the pieces does not contain an error.

The pigeonhole principle. If the pattern P is partitioned into $k+1$ pieces $P = P^1 P^2 \cdots P^{k+1}$, then one of the pieces must occur exactly (without errors) inside $T_{i\ldots j}$, if $T_{i\ldots j}$ is a k-error occurrence of P in T.

This observation follows from the fact that if all pieces $P^1 \cdots P^k$ have an occurrence with one or more errors in $T_{i\ldots j}$, then they can form only an occurrence with at least $k+1$ errors of P: see Figure 10.1.

Notice that the existence of an exact occurrence of a piece of P does not guarantee that there would be a k-error occurrence for P; the other pieces might not even occur one next to the other. Thus, pattern partitioning can be used only as a *lossless filter*. The final method needs to find the exact occurrences of each P^i, and then *verify* whether there is a k-error occurrence of P overlapping this exact occurrence position.

Since the sequence T is static, we can afford to build any of the indexes from the previous chapters on it. Then, we search for all the pieces P^i exactly. Using, for example, a succinct suffix array with a sampling factor $r = \log^{1+\epsilon} n / \log \sigma$ (for some constant $\epsilon > 0$), these searches and verifications take $O(m \log \sigma + (\log^{1+\epsilon} n + \lceil m/w \rceil m) \texttt{cand})$ time, where $\texttt{cand}$ is the number of *candidate* occurrences reported by the filter. The multiplicative factor comes from extracting each candidate position from the succinct suffix array, and verifying each of the occurrences in $O(\lceil m/w \rceil m)$ time, assuming that Myers' bitparallel algorithm from Section 6.1.3 is used for verification.

Pattern partitioning does not give good worst-case guarantees for approximate searches, and the expected case behavior is limited to small error levels; see Insight 10.2. For this reason, one could also partition the pattern into fewer pieces, but then one needs to search each piece allowing some errors: see Exercise 10.2. These approximate searches can be done as explained in the next section.

Insight 10.2 Average case bound for pattern partitioning

Obviously, pattern partitioning makes sense only if the number of candidates is not much bigger than the amount of actual occurrences. On a random sequence T it makes sense to partition a pattern $P[1..m]$ into as equally sized pieces as possible. Assuming a threshold k on the number of errors and a partitioning of the pattern into $k+1$ pieces of length at least d, the probability that a given piece of the pattern matches at a given position in the text is at most $1/\sigma^d$. Then the expected number of random matches between pieces of the pattern and positions of the text will be roughly $\texttt{cand} = (k + 1)n/\sigma^d \leq mn/\sigma^d$. Since the time needed to check every random match is $O(m)$ (assuming we allow only mismatches), we may wish to have $\texttt{cand} = O(1)$ in order for the expected checking time of those occurrences to be $O(m)$. Equal partitioning means (in our discrete world) a mixture of pieces of length $d = \lfloor m/(k + 1) \rfloor$ and pieces of length $d + 1$, from which it follows than that filter is expected to work well for

$$ k < \frac{m}{\log_{\sigma}(mn)} - 1. \qquad (10.3) $$

A corollary is that, for small enough k, a suffix tree built in $O(n)$ time can be used as a data structure for the k-errors problem such that searching for a pattern takes $O(m \log \sigma + (\texttt{cand} + \texttt{occ})m)$ time. Since the expected number of candidates is $O(1)$, the expected search time simplifies to $O(m(\log \sigma + \texttt{occ}))$. In order for the checking time per occurrence to be $O(m)$, one needs to avoid checking the same candidate position more than once (since the same candidate can match more than one piece). A simple solution for the problem is to use a bitvector of size n in which the positions of the candidate positions are marked. The stated time bound assumes mismatches, but a slightly worse bound can also be achieved when one allows both indels and mismatches. Allowing indels will increase the time needed to check occurrences, but in practice, for reasonable values of m, a significant speedup can be achieved using Myers' bitparallel algorithm.

10.2   Dynamic programming along suffix tree paths

We will now focus on the problem of an approximate search for a pattern P, using an index built on a text T.

Consider the suffix tree of the text T. Each path from the root to a leaf spells a suffix of T. Hence, a semi-local alignment of a read to each suffix of T can be done just by filling a dynamic programming matrix along each path up to depth $2m$. The bound $2m$ comes from the fact that the optimal alignment must be shorter than that obtained by deleting a substring of T of length m and inserting P. The work along prefixes of paths shared by many suffixes can be done only once: the already-computed columns of the dynamic programming matrix can be stored along the edges of the suffix tree while traversing it by a depth-first search. This simple scheme is illustrated in Figure 10.2, and Exercise 10.3 asks the reader to develop the details.

10.3   Backtracking on BWT indexes

The problem with our approach so far is that the suffix tree of a whole genome sequence requires a rather large amount of space. Luckily, just like with exact pattern matching,

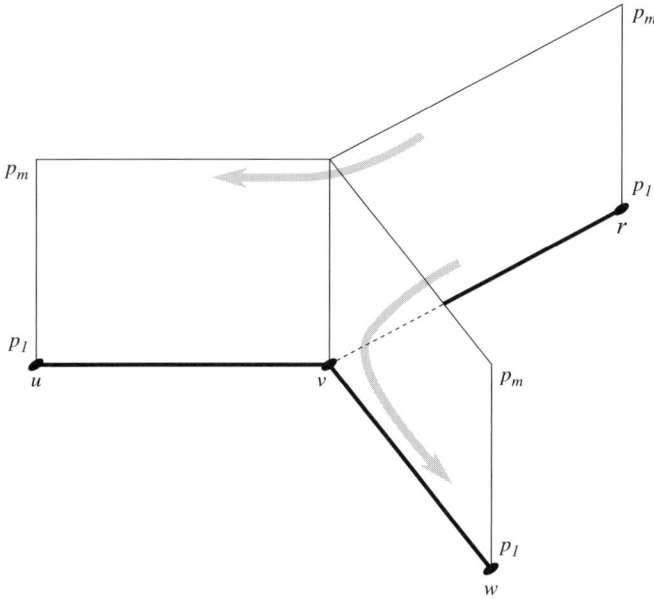

Figure 10.2 Dynamic programming along suffix tree paths.

there is a way to arrange the search with the more space-efficient Burrows–Wheeler indexes instead.

Consider the *backtracking* algorithm given in Algorithm 10.1 that uses the notions from Definition 9.12 at page 171.

Algorithm 10.1: Backtracking with the BWT index

Input: Forward BWT index `idx` of a sequence T.

Output: Search space states.

1 **def** *Branch(d,[i..j])***:**

2   **for** $c \in$ idx.enumerateLeft(i,j) **do**

3   Print (c,d);

4   Branch($d+1$,idx.extendLeft$(c,[i..j])$));

5 Branch(0,[1..n]);

Assume (by induction) that $[i..j]$ corresponds to a prefix α of length d of one or more suffixes of $\underline{T}$ in Algorithm 10.1. The operation idx.enumerateLeft(i,j) gives all those c such that αc is a prefix of length $d+1$ of one or more suffixes of $\underline{T}$.

> Observation. Backtracking on the succinct suffix array (the forward BWT index) of a sequence T explores the same search space as a depth-first traversal of the suffix tree of $\underline{T}$.

This means that we can almost verbatim implement dynamic programming along suffix tree paths by using backtracking on the BWT index, by reading the pattern backwards.

10.3.1   Prefix pruning

Dynamic programming along backtracking paths is nevertheless time consuming, and one needs to consider ways to *prune* the search space in order to obtain efficient algorithms in practice. Exercise 10.4 asks the reader to consider pruning under k-errors search.

For simplicity of exposition, we will now focus on the formulation of the k-mismatches problem as the problem of finding the suffixes of T that start with some string P', such that the Hamming distance between P and P' is at most k (formally, $D_H(P, P') \leq k$). The basic algorithm using backtracking on the BWT index is given in Algorithm 10.2. That algorithm prunes branches of the search tree as soon as the Hamming distance threshold is reached.

Algorithm 10.2: Backtracking with the BWT index for solving the k-mismatches problem

---

Input: Forward BWT index `idx` of a sequence T, pattern $P = p_1 p_2 \cdots p_m$, and error threshold k.

Output: Ranges of $\mathsf{BWT}(T)$ corresponding to the k-mismatch occurrences of P in T.

1 **def** *Branch(d,k,[i..j])*:
2   **for** $c \in$ `idx.enumerateLeft`(i,j) **do**
3   **if** $p_d \neq c$ **then**
4              $k \leftarrow k - 1$;
5   **if** $k \geq 0$ **then**
6   **if** $d = 1$ **then**
7   Print `idx.extendLeft`$(c, [i..j])$;
8   **else**
9   Branch($d - 1$,k,`idx.extendLeft`$(c, [i..j])$);

10 Branch(m,k,$[1..n]$);

It is possible to do a more effective *prefix pruning* by some preprocessing: precompute, for each prefix $P_{1..i}$ of the read, its partitioning into the maximum number of pieces such that no piece occurs in the text. Let us denote the maximum number of splits $\kappa(i)$. The value $\kappa(i)$ works as a lower bound for the number of mismatches that must be allowed in any approximate match for the prefix $P_{1..i}$: see Exercise 10.12. This estimate can be used to prune the search space as shown in Algorithm 10.3.

Algorithm 10.3: Prefix pruning backtracking with the BWT index for solving the k-mismatches problem

---

Input: Forward BWT index `idx` of a sequence T, pattern $P = p_1 p_2 \cdots p_m$, error threshold k, and prefix pruning function $\kappa()$.

Output: Ranges of $\mathsf{BWT}(T)$ corresponding to k-mismatch occurrences of P in T.

1 **def** *Branch(d,k,[i..j])*:
2   **for** $c \in$ `idx.enumerateLeft`(i,j) **do**
3   **if** $p_d \neq c$ **then**
4              $k \leftarrow k - 1$;
5   **if** $k - \kappa(d - 1) \geq 0$ **then**
6   **if** $d = 1$ **then**
7   Print `idx.extendLeft`$(c, [i..j])$;
8   **else**
9   *Branch(d* $- 1, k,$ `idx.extendLeft`$(c, [i..j]))$;

10 *Branch(m,k,[1..n])*;

The computation of $\kappa(i)$ for all prefixes is analogous to the backward search algorithm (Algorithm 9.1 on page 161) applied to the reverse of a pattern on a reverse BWT index: the backward search on the reverse BWT index is applied as long as the interval $[sp..ep]$ does not become empty. Then the process is repeated with the remaining suffix of the pattern until the whole pattern has been processed. In detail, if the backward step from $P_{1..i}$ to $P_{1..i+1}$ results in a non-empty interval $[sp..ep]$ with $sp \leq ep$, then $\kappa(i + 1) = \kappa(i)$. Otherwise, $\kappa(i + 1) = \kappa(i)$ and $\kappa(i + 2) = \kappa(i + 1) + 1$, and the backward search is started from the beginning with $P_{i+2..m}$ as pattern, and so on. Thus, all $\kappa(\cdot)$ values can be computed in $O(m \log \sigma)$ time.

Insight 10.3 discusses a different pruning strategy, based on a particular hash function.

Insight 10.3 Pruning by hashing and a BWT-based implementation

For a random text T, an alternative way of reducing the search space is the following one. Suppose we want to solve the k-mismatches search on strings of fixed length m, with $k = O(\log_\sigma n)$ and σ a power of 2. Suppose that we have indexed all the m-mers of the text. Then a naive search strategy for a pattern $P \in \Sigma^m$ is to generate the Hamming ball of distance k, $H_k(P) = \{X \mid D_\mathsf{H}(X, P) \leq k\}$, around the pattern P. For every $X \in H_k(P)$ we search for X in T, which generates `cand` $= \sum_{i=0}^{k}(\sigma - 1)^i \binom{m}{i}$ candidates. The search space can be reduced as follows. Let $B(a)$ be a function that returns a binary representation of $\log \sigma$ bits for any character $a \in \Sigma$. We assume that $B(X)$ applied to a string $X \in \Sigma^m$ is the concatenation $B(X[1]) \cdot B(X[2]) \cdot \cdots \cdot B(X[m])$ of length $m \log \sigma$. Consider the hash function $F : \Sigma^m \to \Sigma^h$ defined as

$$F(P) = B^{-1} \left(\left(\bigoplus_{i=1}^{\lceil m/h \rceil - 1} B(P[(i-1)h + 1..ih]) \right) \oplus B(P[m - h + 1..m]) \right),$$

where $x \oplus y$ denotes the standard bit-wise xor operator between binary strings.

It can be proved that the Hamming ball $H_{k'}(F(P))$ of distance $k' \leq 2k$ around the binary string $F(P)$ will be such that $F(H_k(P)) \subseteq H_{k'}(F(P))$, namely, for any $X \in \Sigma^m$ such that $D_H(P, X) \leq k$, it also holds that $D_H(F(P), F(X)) \leq k'$. This property guarantees that, instead of searching in T for all positions i such that $T[i..i+m-1] \in H_k(P)$, we can search for all positions i such that $F(T[i..i+m-1]) \in H_{k'}(F(P))$, and, for each such i, check whether $D_H(T[i..i+m-1], P) \leq k$. The number of candidates is now reduced to $\text{cand}' = \sum_{i=0}^{k'}(\sigma - 1)^i \binom{h}{i}$. Setting $h = \Theta(\log_\sigma n)$ ensures that the expected number of random matches of the candidates in T (false positives) is $O(1)$. If σ and k are constants, then $h = O(\log n)$ and $\text{cand}' = O(\log^c n)$, for some constant c that depends only on k and σ. A naive way to implement the index consists of storing in a hash table all the $n - m + 1$ pairs $(F(T[i..i+m]), i)$ for $i \in [1..n - m + 1]$. This would, however, use space $O(n \log n)$ bits. There exists an elegant way to reduce the space to $O(n \log \sigma)$ bits. The idea is to store a string T' of length $n - m + h$, where

$$T'[i] = B^{-1} \left(\left(\bigoplus_{j=0}^{\lceil m/h \rceil - 2} B(T[i + jh]) \right) \oplus B(T[i + m - h]) \right)$$

for all $i \in [1..n - m + h]$. It can then be easily seen that $T'[i..i + h - 1] = F[T[i..i+m-1]]$. Thus, searching in T for all positions i such that $F(T[i..i+m-1]) \in H_{k'}(F(P))$ amounts to searching for all exact occurrences of $X \in H_{k'}(F(P))$ as a substring in the text T'. See Figure 10.3. One can store T' as a succinct suffix array with sampling factor $r = (\log n)^{1+\epsilon}/\log \sigma$, for some constant $\epsilon > 0$, and obtain a data structure that uses $n \log \sigma (1 + o(1))$ bits of space and allows locate queries in $O(\log^{1+\epsilon} n)$ time. The original text is kept in plain form, occupying $n \log \sigma$ bits of space. Each of the cand' candidate hash values from $H_{k'}(F(P))$ can be searched in $O(m \log \sigma)$ time, their corresponding matching positions in T' (if any) extracted in $O(\log^{1+\epsilon} n)$ time per position, and finally their corresponding candidate occurrences in T verified in $O(m)$ time per candidate.

Overall, a k-mismatches search is executed in expected time $O(\text{cand}' \cdot \log n + \text{occ} \cdot (\log^{1+\epsilon} n + m)) = O(\log^{c+1} n + \text{occ} \cdot (\log^{1+\epsilon} n + m))$, where c is some constant that depends only on k and σ.

10.3.2   Case analysis pruning with the bidirectional BWT index

A search strategy closely related to pattern partitioning is to consider separately all possible distributions of the k mismatches to the $k + 1$ pieces, and to perform backtracking for the whole pattern for each case, either from the forward or from the reverse BWT

$a \in \Sigma$	$B(a)$
A	00
C	01
G	10
T	11

$$
\begin{array}{c}
10\ \boxed{11}\ 10 \\
10\ 01\ 00 \\
\hline
00\ \boxed{10}\ 10
\end{array} \oplus
$$

$F(\text{GTGCA}) = \ \text{A}\ \boxed{\text{G}}\ \text{G}$

$$
\begin{array}{c}
10\ \boxed{00}\ 10 \\
10\ 01\ 00 \\
\hline
00\ \boxed{01}\ 10
\end{array} \oplus
$$

$F(\text{GAGCA}) = \ \text{A}\ \boxed{\text{C}}\ \text{G}$

$T[5..9]$

$T = \text{ATCAG}\overline{\text{T}}\text{GCAGTA}$

$F[T[2..6]]$

$T' = \underset{F[T[1..5]]}{\underbrace{\text{C}\ \text{T}\ \text{T}}}\ \text{T}\ \underset{F[T[5..9]]}{\underbrace{\text{A}\ \text{G}\ \text{G}}}\ \text{T}\ \text{T}\ \text{G}$

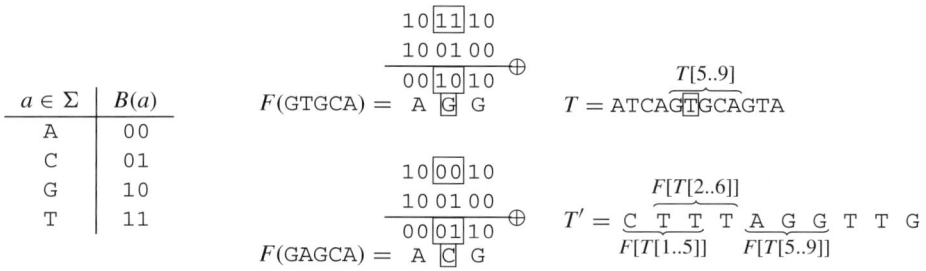

Figure 10.3 Illustrating the idea from Insight 10.3. We have chosen $h = 3$. The pattern $P = \text{GAGCA}$ matches the substring $T[5..9]$ with Hamming distance 1. We have $D_H(F(P), F(T[5..9])) = 1$.

index, depending on which one is likely to prune the search space better. Let us call this strategy *case analysis pruning*.

To see how case analysis pruning works, let us first consider the simplest case $k = 1$. The pattern P is split into two pieces, $P = \alpha\beta$. One error can be either (a) in α or (b) in β. In case (a), it is preferable to search for $P = \alpha\beta$ using backward backtracking on the forward BWT index, since β must appear exactly, and branching is needed only after reading the last $|\beta|$ symbols of P. In case (b), it is affordable to search for $\underline{P} = \beta\alpha$ using backward backtracking on the reverse BWT index, since α must appear exactly, and branching is needed only after reading the first $|\alpha|$ symbols of P. For obvious reasons, $|\alpha| \approx |\beta|$ is a good choice for an efficient pruning.

Let us then consider $k = 2$ to see the limitations of the approach. The different ways of distributing two errors into three pieces are (a) 002, (b) 020, (c) 200, (d) 011, (e) 101, and (f) 110. Obviously, in cases (a) and (d) it makes sense to use backtracking on the reverse BWT index and in cases (c) and (f) backtracking on the forward BWT index. For cases (b) and (e) neither choice is good or bad a priori. More generally, for any k there is always the bad case where both ends have at least $k/2$ errors. Hence, there is no strategy to start the backtracking with 0 errors, other than in the case $k = 1$.

This bottleneck can be alleviated using *bidirectional case analysis pruning*. For example, for the case 101 above, it is possible to search, say, with the forward BWT index, two first blocks backwards, an operation denoted $\overleftarrow{101}$, and then to continue the search from the reverse BWT index with the third block, an operation denoted $10\,\overrightarrow{1}$. For this, we need the synchronization provided by the bidirectional BWT: once an interval $[i..j]$ has been found to match a prefix of the pattern with the fixed distribution of errors using `idx.extendLeft(c, [i..j], [i'..j'])`, an operation analogous to the backtracking in Algorithm 10.2, one can continue the search from the original start position with the so-far-obtained interval $[i', j']$ using `idx.extendRight(c, [i..j], [i'..j'])` instead. Exercise 10.14 asks you to give the pseudocode of this algorithm.

10.4 Suffix filtering for approximate overlaps

In the absence of a reference sequence to align the reads to, one could try to align the reads to themselves in order to find overlaps, which could later be used for de novo fragment assembly. We studied the computation of exact overlaps in Section 8.4.4, but, due to measurement errors, computing approximate overlaps is also of interest.

Let use first describe an approach based on a *suffix filtering* designed for indexed approximate pattern matching. Let the pattern P be partitioned into pieces $P = P^1P^2\cdots P^{k+1}$. Then consider the set of suffixes $\mathcal{S} = \{P^1P^2\cdots P^{k+1}, P^2P^3\cdots P^{k+1},\ldots,$ $P^{k+1}\}$. Each $S \in \mathcal{S}$ is searched for in T, allowing zero errors before reaching the end of the first piece in S, one error before reaching the end of the second piece of S, and so on. Obviously this search can be done using backtracking on the reverse BWT index. This filter produces candidate occurrences that need to be verified. It produces fewer candidates than the pattern partition filter from Section 10.1, but it is nevertheless a lossless filter: see Exercise 10.15.

Let us now modify this approach for detecting approximate overlaps among a set of reads $\mathcal{R} = \{R^1, R^2, \ldots, R^d\}$. Consider the detection of all the suffix–prefix overlaps (R^i, R^j, o^{ij}) such that $D_{\mathrm{H}}(R^i_{|R^i|-o^{ij}+1..|R^i|}, R^j_{1..o^{ij}})/o^{ij} \leq \alpha$, where α is a relative error threshold, and $o^{ij} \geq \tau$ for some minimum overlap threshold τ.

Now, construct a reverse BWT index on $T = R^1\#^1R^2\#^2\cdots R^d\#^d\#$. Consider some $P = R^i$. First, fix some overlap length o, and partition the prefix $P_{1..o}$ into pieces $P^1P^2\cdots P^{k+1}$, for $k = \lfloor \alpha o \rfloor$. Suffix filtering could be executed as such, just leaving out candidates that are not followed by any $\#^j$ in T. Repeating this on all possible overlap lengths is, however, extremely ineffective, so we should find a way to partition the whole read so that the suffix filtering works correctly for any overlap length. However, any fixed partitioning will be bad for overlap lengths that cross a piece boundary only slightly. For this, we can use two partitionings that have maximal minimum distance between mutual piece boundaries; such equal partitioning is shown in Figure 10.4. The two suffix filters are executed independently, and overlaps are reported only after reaching the middle point of the current piece: the observation is that any proper overlap will be reported by one of the filters.

It remains to consider the maximum piece length in equal partitioning such that the suffix filters in Figure 10.4 work correctly for all overlap lengths. Such a length h is

$$h = \min_{1 \leq o \leq |P|} \lfloor o/(\lfloor \alpha o \rfloor + 1) \rfloor. \tag{10.4}$$

That is, pieces $P^1, \ldots, P^k$ are of length h and piece P^{k+1} is of length $|P| - kh$. The shifted filter in Figure 10.4 has then piece P^1 of length $h/2$, pieces $P^2, \ldots, P^k$ of length h, and piece P^{k+1} of length $|P| - kh + h/2$.

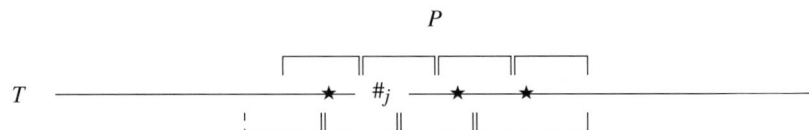

Figure 10.4 Two partitionings enabling suffix filtering to be used for approximate overlap computation. Searching with the second piece in the bottom-most partitioning yields an overlap with the read R^j.

10.5   Paired-end and mate pair reads

Many sequencing techniques provide *paired-end* and *mate pair* reads. A paired-end read
is a suffix–prefix pair of a fragment of DNA, with read length smaller than the fragment
length:

TGACGACTGCTAGCTA.......................AGCTACGTAGCATCGT

Mate pair reads are similar to paired-end reads but are produced with a different
sequencing technique. The most notable difference from the analysis perspective is
that the fragment corresponding to a mate pair can be longer. Also, depending on
the sequencing technique, the two ends can come from different strands, and can be
oriented in any of the four different relative orders. We may assume that the pairs are
normalized to the forward strand and forward reading direction.

Read alignment on these read pairs can be done independently, and often the purpose
is to detect larger-scale variants: see Exercise 10.18.

Another purpose is to detect the correct alignment in cases where one of the pairs
is inside a repetitive region; among the many occurrences of one of the reads, one can
select the one within fragment length distance of the (possibly) unique occurrence of its
pair. Often, this selection is done through a cross-product of occurrence lists. However,
it can be done more efficiently, as we discuss next.

Algorithm 10.4: Prefix pruning backtracking with the BWT index for solving the
k-mismatches problem inside a given range of T

Input: Forward BWT index idx of a sequence T, wavelet tree $\mathcal{T}$ of SA(T),
pattern $P = p_1 p_2 \cdots p_m$, error threshold k, prefix pruning function $\kappa(\cdot)$, and
query interval $[l..r]$ in T.

Output: Ranges of $\mathsf{BWT}(T)$ corresponding to k-mismatch occurrences of P in $T_{l..r}$.

1 **def** *Branch(d,k,[i..j])*:
2   **for** $c \in$ idx.enumerateLeft(i,j) **do**
3   **if** $p_d \neq c$ **then**
4          $k \leftarrow k - 1$;
5   **if** $k - \kappa(d-1) \geq 0$ **then**
6          $[i'..j'] =$ idx.extendLeft$(c, [i..j])$;
7   **if** $\mathcal{T}$.rangeCount$(i',j',l,r) > 0$ **then**
8   **if** $d = 1$ **then**
9   Print $[i'..j']$;
10   **else**
11   Branch($d-1$,k,idx.extendLeft($c, [i..j]$));

12 Branch(m,k,$[1..n]$);

Recall the query rangeCount(T, i, j, l, r) on wavelet trees (Corollary 3.5 on page 27). By building the wavelet tree on the suffix array SA of T instead, one can support the query rangeCount(SA, i, j, l, r). This can be used to check whether there are any suffix starting positions among the values SA[$i..j$] within the text fragment $T[l..r]$. With this information, one can modify any of the search algorithms considered in this chapter to target a genome range: Algorithm 10.4 gives an example.

The shortcoming of this improvement to paired-end alignment is that the wavelet tree of a suffix array requires $n \log n(1 + o(1))$ bits, where so far in this chapter we have used $O(n \log \sigma)$ bits, with a very small constant factor.

10.6   Split alignment of reads

In RNA-sequencing the reads represent fragments of messenger RNA. In higher organisms, such fragments may originate from different combinations of exons of the genome, as shown in Figure 10.5.

Assuming that we have the complete *transcriptome*, that is, all possible RNA transcripts that can exist in nature, the RNA-sequencing reads could be aligned directly to the transcriptome. This leads to a problem identical to the one of aligning DNA reads to a genome. Although databases approximating the transcriptome exist for widely studied organisms, it makes sense to ascertain whether an RNA-sequencing read could directly be aligned to the genome. This is even more relevant for studying various genetic diseases, in which the cell produces novel specific RNA transcripts. This leads us to the *split-read alignment* problem, which we also studied in Section 6.5 under the name *gene alignment with limited introns*. The alignment algorithm given there is too slow for high-throughput purposes, so we will now consider an indexed solution.

It turns out that there is no obvious clean algorithmic solution for split-read alignment. Insight 10.4 discusses an approach based on pattern partitioning and Insight 10.5 discusses an approach based on analysis of the whole read set simultaneously. We will come back to this problem in Section 15.4 with a third approach that is algorithmically clean, but it assumes a preprocessed input, and can thus lose information in that phase.

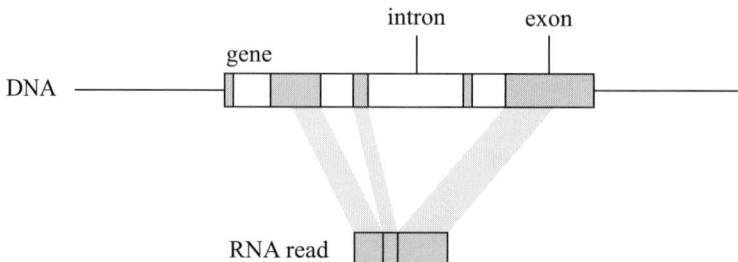

Figure 10.5 An RNA-sequencing read aligned to the genome.

Insight 10.4 Split-read alignment through pattern partitioning

Consider first the case when a read is aligned to the genome with one split. Then we know that at least half of the read should align to a contiguous region in the genome. With a large enough read length, we should not expect too many alignments for a half-read, so such alignments can be efficiently retrieved using the methods discussed earlier in this chapter. Then, we can extend the alignment of the half-read towards the actual split point, and search for the remaining head or tail. This can be done like in the paired-end alignment, by using Algorithm 10.4 with an estimate on the maximal possible intron length for forming the search region.

For longer reads containing more split points, we can keep splitting into more and more parts until we find a partitioning into s parts such that one of the pieces is successfully aligned to the genome. This piece, once extended to the left and to the right while maintaining a decent alignment, can be cut off, and the head and tail searched for recursively with the same approach.

The problem of this approach is that the pieces are becoming smaller and smaller, and at some point they are likely to occur in a random sequence. However, this position-restricted search helps somewhat, insofar as a random hit in a smaller region is less likely (see the exercises).

Insight 10.5 Split-read alignment through mutual information

Another approach to split-read alignment is to first consider the reads that map without any split, predict the exon boundaries from these initial alignments, and then align the rest of the reads such that they are allowed to jump from the end of one predicted exon to the start of a nearby predicted exon.

Consider first that an alignment needs to use only one split. Such 1-split alignments can be conducted efficiently, for example, by concatenating all pairs $\alpha\beta$ such that α is extracted from the end of some exon i and β is extracted from the start of some nearby exon j following i. The set of all these concatenated sequences can be indexed using BWT indexes, and now all unmapped reads can be aligned to this set. Choosing $|\alpha|$ and $|\beta|$ greater than the maximum read length and using some bookkeeping, one should be able to find most of the 1-split alignments, assuming that the read coverage is high enough that the exon prediction is accurate enough.

For longer reads containing more split points, we could extend the above approach by considering several exon combinations at once, but this would lead to a combinatorial explosion. There is a more principled approach that may avoid this explosion. Consider the genome as a labeled DAG with an arc from each position i to position $i + 1$, and from each i to all j, if i is an end position of an exon and j is the start position of a nearby exon following i. We can then build the BWT index of Section 9.6 on this labeled DAG and align unmapped reads on it; see Section 10.7 for how to conduct this alignment.

10.7   Alignment of reads to a pan-genome

So far we have assumed that we have only one reference genome to align all the reads to. For many species, the population-wide variation has been widely explored and there are databases containing variations common in (sub-)populations. One can also consider sets of individual genomes that have been assembled by read alignment or/and (partial) de novo assembly. Read alignment on such *pan-genomic* datasets is of interest, since one can hope to improve alignment accuracy by aligning directly to known variation; in this way, one can reduce the work required for variant calling (see Chapter 14) and improve its accuracy.

10.7.1   Indexing a set of individual genomes

Given a set of individual genomes from the same species, one can naturally build one of the indexes studied in this chapter for each genome, and then apply read alignment in each of them. A better approach for both compression and query efficiency is to concatenate all the genomes and build only one index. The concatenation is highly repetitive, so if the index uses internal compression, a significant amount of space can be saved. Such indexes exist, but we shall explore a more tailored approach that appears to be more efficient in practice.

Assume $T = t_1 t_2 \cdots t_n$ is the concatenation of individual genomes from a species. Apply Lempel–Ziv parsing on T (see Section 12.1) to obtain a partitioning $T = T^1 T^2 \cdots T^p$. Extract a set of *patches* around each phrase boundary to form the set

$$B = \left\{ T_{j-r-k..j+r+k} \mid j \in \{1\} \cup \left\{ 1 + \sum_{1 \leq i \leq p'} |T^i| \ : \ 1 \leq p' \leq p \right\} \right\}. \qquad (10.5)$$

Concatenating B into a sequence with some special symbols marking the boundaries of consecutive patches, one can build any index for B that allows one to perform read alignment with k errors for reads of length at most r. With some bookkeeping data structures, such alignment positions can be mapped back to the original sequence: see Exercise 10.19.

LEMMA 10.1   *A pattern $P = p_1 \cdots p_m$, $m \leq r$, has a k-errors occurrence in $T = t_1 \cdots t_n$, if and only if it has a k-errors occurrence in the set B defined in Equation (10.5).*

Proof   The direction from B to T is obvious from construction. Consider now an occurrence inside some T^i of the Lempel–Ziv partitioning $T = T^1 T^2 \cdots T^p$ such that the occurrence is not included in the corresponding boundary patches. This means that the occurrence substring is an exact copy of some substring occurring earlier in T. If this exact copy is not included inside the corresponding patches, it also must be an exact copy of some substring occurring earlier in T. Continuing this induction, one notices that there must be an exact copy of the occurrence overlapping a phrase boundary, which means that it is included in a patch in B.   □

The first occurrence found in the above proof is called the *primary occurrence* and the others are called *secondary occurrences*. Exercise 10.20 asks you to show how the wavelet tree can be used for reporting the secondary occurrences.

*10.7.2 Indexing a reference genome and a set of variations

Another approach to pan-genome indexing is to represent the reference and the known variants as a labeled DAG, as illustrated in Figure 10.6.

Then one can build the BWT index of Section 9.6 on this labeled DAG. The main difference of this approach from the one indexing a set of genomes is that now arbitrary (re)combinations of the underlying variants are allowed. That is, the labeled DAG represents all individuals that can be formed by combining the underlying variants.

In Section 9.6 we learned how to do exact pattern matching using the BWT index of a labeled DAG. Supporting basic backtracking is analogous to the corresponding algorithms on standard BWT indexes, but the various pruning techniques studied earlier are not easy to extend: there appears to be no bidirectional analog of the BWT index on a labeled DAG. Prefix pruning seems to be the only method that applies: Exercise 10.22 asks you how to compute values $\kappa(i)$ in this case. Assuming that these values have been computed, one can apply Algorithm 10.3 verbatim by just switching the index used. This algorithm returns intervals in BWT_G of the labeled DAG G, and one can then obtain the vertices of the DAG that have a path starting inside those intervals. Extracting the full alignments, that is, all the paths that match the pattern, is a nontrivial procedure, which we study next.

Let us assume that we have kept the *backtracking tree* $\mathcal{T}$ whose nodes are intervals in BWT_G, and whose edges are the backward steps taken from the parent intervals towards child intervals. That is, each edge e of $\mathcal{T}$ is labeled with a character $\ell(e)$. We shall explore how to extract the prefix of the lexicographically smallest path matching $P_{1..m}$. That is, we study the extraction of the first entry in the interval $\overrightarrow{u}$, where $\overrightarrow{u}$ is the left-most interval found during backtracking, and thus u is a leaf of $\mathcal{T}$. First, we need to find the vertex v of G whose interval $\overrightarrow{v}$ of BWT_G is the first belonging also to $\overrightarrow{u}$. This is given by

$$\overrightarrow{v} = [\mathtt{select}_1(\mathtt{out}, \mathtt{rank}(\mathtt{out}, \overrightarrow{u}))..\mathtt{select}_1(\mathtt{out}, \mathtt{rank}(\mathtt{out}, \overrightarrow{u}) + 1) - 1],$$

$$(10.6)$$

where $\mathtt{out}$ is the bitvector encoding the out-neighbors in the Burrows–Wheeler index of G of Definition 9.19. Knowing now $\overrightarrow{v}$, we want to move along arc (v, w) in G such that $\ell(w) = \ell((u, p))$, where p is the parent of u in $\mathcal{T}$, and w is the out-neighbor of v

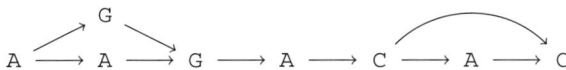

Figure 10.6 A labeled DAG representing a reference and a set of variants. Here the reference is AAGACAC and the set of variants consists of the second A substituted by G and the last A deleted.

with the first interval $\overleftarrow{w}$ included in $\overleftarrow{p}$. Recall that we considered how to find the interval of the ith out-neighbor of v in Section 9.6: see Equation (9.8). We can now carry out a binary search over all out-neighbor intervals to find the vertex w with the smallest interval included in $\overleftarrow{p}$. Continuing this way up in the backtracking tree $\mathcal{T}$, we can list one of the paths matching P. Exercise 10.23 asks you to extend this algorithm to list all such paths.

10.8 Literature

Pattern partitioning and more advanced filters for approximate pattern matching are surveyed in Navarro (2001); suffix filtering is a later development by Kärkkäinen & Na (2007).

Finding theoretical solutions to indexed approximate pattern matching was the Holy Grail of pattern matching for a long time, until in Cole *et al.* (2004) an index with optimal running time was found (assuming many parameters as constants). Their index requires superlinear space, making it less useful for genome-scale sequences. This bottleneck was soon overcome in Chan *et al.* (2006); they provide an $O(n \log n)$ bits index with running time $O(m + \text{occ} + \text{polylog}(n))$, where m is the length of the pattern and occ is the number of occurrences with up to k-errors, with k and the alphabet size σ assumed to be constants. The space can be further decreased to $O(n \log \sigma)$ bits by using a compressed suffix tree as part of the index. However, the result works only for very small values of k and σ, since the $O(\cdot)$ space terms hide 3^k, σ^k, and $\log^{k^2} n$ factors. In read alignment, the sequences are quite accurate and allowing only a few errors usually suffices in the search.

As far as we know, the above theoretical approaches have not been explored in practice; the tools used in read alignment are mostly based on backtracking on BWT indexes or k-mer indexes. Using dynamic programming on top of BWT indexes for read alignment was first proposed in Lam *et al.* (2008). Prefix pruning is from Li & Durbin (2009). The forward version of case analysis pruning is from Langmead *et al.* (2009) and its bidirectional improvement is from Li *et al.* (2009). The idea in Insight 10.3 is from Policriti & Prezza (2014a) and Policriti & Prezza (2014b), which adapt a hash function initially considered in Policriti *et al.* (2012).

The extension of suffix filtering to approximate overlaps was proposed in Välimäki *et al.* (2012). Better theoretical bounds were achieved in Välimäki (2012).

Split-read alignment using mutual information was proposed in Trapnell *et al.* (2009) and its extension to longer reads was experimented with in Sirén *et al.* (2014).

Pan-genome indexing for read alignment was first studied in Mäkinen *et al.* (2009, 2010) and Schneeberger *et al.* (2009); the first two use compressed variants of BWT indexes, and the last uses k-mer indexing. Theoretical advances to support linear-time exact pattern searches on arbitrary-length patterns with relative Lempel–Ziv compressed collections were achieved in Do *et al.* (2012) and Gagie *et al.* (2012); these results are significant since Lempel–Ziv compression is superior to BWT compression on highly repetitive collections, such as a set of individual genomes. Practical alternatives using a hybrid of Lempel–Ziv factorization and BWT indexes were given in Wandelt *et al.* (2013) and Ferrada *et al.* (2014). Our description follows Ferrada *et al.* (2014). An alternative to these is to extract directly the context around variations and build a BWT index on these patches as in Huang *et al.* (2013). These practical approaches work only on limited pattern length. Finally, the labeled DAG approach was proposed in Sirén *et al.* (2011) and Sirén *et al.* (2014). This approach works on any pattern length, but the index itself has exponential size in the worst case; the best-case and average-case bounds are similar to those of Mäkinen *et al.* (2010). Here one should observe that identical functionality with context extraction alone would result in the best-case exponential size index on the length of the read.

In Chapter 14 we study variation calling over pan-genome indexes. We shall learn that, while the labeled DAG approach can exploit recombinations of variants already during read alignment, this lack of functionality in the approach of indexing a set of individuals can be partly alleviated by a better control over the recombination model when predicting the novel variants.

Exercises

10.1 Recall mapping quality, Insight 10.1. Consider how it could be refined to take good alignments into account in addition to only the best. How would you in practice approximate such mapping quality under, for example, a *k*-errors search?

10.2 Consider splitting a pattern into fewer than $k+1$ pieces. How many errors should be allowed for the pieces in order to still obtain a lossless filter for the *k*-errors search?

10.3 Give the pseudocode for the algorithm depicted in Figure 10.2 to compute maximum-scoring semi-local alignment along suffix tree paths.

10.4 Modify the above approach to solve the *k*-errors problem. Do you always need to fill the full dynamic programming matrix on each path? That is, show how to prune branches that cannot contain an occurrence even if the rest of the pattern $P_{j..m}$ exactly matches a downward path.

10.5 Modify the above approach to solve the *k*-errors problem using Myers' bitparallel algorithm instead of standard dynamic programming.

10.6 Modify all the above assignments to work with backtracking on the BWT index.

10.7 Most sequencing machines give the probability that the measurement was correct for each position inside the read. Let $\mathbb{P}(p_i)$ be such a probability for position i containing p_i. Denote $M[c, i] = \mathbb{P}(p_i)$ if $p_i = c$ and $M[c, i] = (1 - \mathbb{P}(p_i))/(\sigma - 1)$ if $c \neq p_i$. Then we have a *positional weight matrix (PWM) M* representing the read (observe that this is a profile HMM without insertion and deletion states). We say that the matrix *M occurs* in position *j* in a genome sequence $T = t_1 t_2 \cdots t_n$ if $\mathbb{P}(M, T, j) = \prod_{i=1}^{m} M[t_{j+i-1}, i] > t$, where *t* is a predefined threshold. Give the pseudocode for finding all occurrences of *M* in *T* using backtracking on the BWT index. Show how to prune branches as soon as they cannot have an occurrence, even if all the remaining positions match $P_{1..j}$ exactly.

10.8   The goal in read alignment is typically to find the unique match if one exists. Consider how to optimize the backtracking algorithms to find faster a best alignment, rather than all alignments that satisfy a given threshold.

10.9   Some mate pair sequencing techniques work by having an adapter to which the two tails of a long DNA fragment bind, forming a circle. This circle is cut in one random place and then again X nucleotides apart from it, forming one long fragment and another shorter one (assuming that X is much smaller than the circle length). The fragments containing the adapter are fished out from the pool (together with some background noise). Then these adapters containing fragments are sequenced from both ends to form the mate pair. Because the cutting is a random process, some of the mate pair reads may overlap the adapter. Such overlaps should be cut before using the reads any further.

(a)   Give an algorithm to cut the adapter from the reads. Take into account that short overlaps may appear by chance and that the read positions have the associated quality values denoting the measurement error probability.

(b)   How can you use the information about how many reads overlap the adapter to estimate the quality of fishing?

10.10   Construct the Burrows–Wheeler transform of ACATGATCTGCATT and simulate the 1-mismatch backward backtracking search on it with the read CAT.

10.11   Give the pseudocode for computing the values $\kappa(i)$ for prefix pruning applied on the prefixes $P_{1..i}$ of the pattern P.

10.12   Show that the values $\kappa(i)$ in prefix pruning are correct lower bounds, that is, there cannot be any occurrence missed when using the rule $k' + \kappa(i) > k$ to prune the search space.

10.13   Show that the computation of the values $\kappa(i)$ in prefix pruning is also feasible using the forward BWT index alone by simulating the suffix array binary search.

10.14   Give the pseudocode for the k-mismatches search using case analysis pruning on the bidirectional BWT. You may assume that a partitioning of the pattern is given, together with the number of errors allowed in each piece. Start the search from a piece allowed to contain the fewest errors.

10.15   Show that suffix filtering is a lossless filter for a k-mismatches search.

10.16   Compute the minimal read length m such that the expected number of occurrences of the read in a random sequence of length n when allowing k mismatches is less than $1/2$.

10.17   Compute the minimal read length m such that the expected number of occurrences of the read in a random sequence of length n when allowing k errors is less than $1/2$ (use approximation, the exact formula is difficult).

10.18   Consider different large-scale variations in the genome, like gene duplication, copy-number variation, inversions, translocations, etc. How can they be identified using read alignment? Is there an advantage of using paired-end reads?

10.19   Consider the pan-genome read alignment scheme described in Section 10.7.1. Develop the bookkeeping data structures to map alignment position from the patched sequence B to the position in T.

10.20   Consider the pan-genome read alignment scheme described in Section 10.7.1. Show how the wavelet tree can be used to retrieve secondary occurrences, each in $O(\log n)$ time.

10.21   Consider the pan-genome read alignment scheme described in Section 10.7.1 and recall Exercise 3.6. Show how to use van Emde Boas trees to retrieve secondary occurrences, each in $O(\log \log n)$ time.

10.22   Show how to compute values $\kappa(i)$ for prefix pruning for prefixes $P_{1..i}$ of the pattern P in the case of the BWT index on a labeled DAG.

10.23   Give a pseudocode for listing all the subpaths matching a pattern using the BWT index of the labeled DAG. Extend the algorithm given in Section 10.7.2 for listing the matching prefix of a path starting at vertex v with smallest lexicographic order.

11 Genome analysis and comparison

Aligning whole genomes using optimal dynamic programming algorithms is a daunting task, being practically feasible only for very similar species, and conceptually incapable of capturing large-scale phenomena that alter the contiguity of homologous regions, like chromosome-level rearrangements, gene shuffling and duplication, translocations, and inversions of large areas.

We could circumvent these limits by first using good local alignments as anchors and then finding a set of large-scale edit operations that align such anchors as well as possible: Exercise 11.8 elaborates on how to find these optimal one-to-one correspondences.

Alternatively, we could try to detect a set of genes that are shared by two species, and we could compute the minimum set of rearrangements that transform one sequence of genes into the other. Despite having the advantage of providing a constructive explanation of the distance between two genomes, this approach is feasible only for closely related species, it discards information contained in non-coding regions, it assumes that a large-enough set of common genes can be reliably identified in each genome and mapped across genomes, and it is ineffective in cases in which gene order is preserved, like in mammalian mitochondrial DNA.

Yet another alternative could be aligning just a few conserved genes, building a phylogenetic tree for each such gene, and merging the trees into a common consensus: this is practically difficult in some viral and bacterial families with high rates of mutation or lateral gene transfer, and it is conceptually undesirable since different genes can tell different phylogenetic stories. Alignment costs, moreover, have an intrinsic ambiguity.

In *alignment-free genome comparison* the goal is to derive efficiently computable distance or similarity measures for whole genomes that capture the relatedness without resorting to alignments. Such distance measures are typically derived between sets of *local features* extracted from the genomes. This allows one to compare genomes on the basis of their local compositional biases, rather than on the basis of their large-scale sequential structure: such compositional biases are widespread in nature, and have been shown to correlate with accepted species classifications. These distance measures generalize to comparing sets of unassembled reads, or mixed datasets containing both assembled and unassembled reads.

The chapter starts with the analysis of simple (shared) local features such as maximal repeats, maximal exact matches (MEMs), and maximal unique matches (MUMs), which are common building blocks of practical whole-genome alignment tools: see

Insight 11.1 for a simple example. The second half extends the techniques to alignment-free genome comparison methods.

Insight 11.1 MUMs as anchors for multiple alignment

Let us consider a high-throughput heuristic method for multiple alignment exploiting the maximal unique matches (MUMs) studied in Section 8.4.2. We assume that the genomes to be aligned are quite similar; that is, we assume that there is no need for global genome rearrangement operations to explain the differences. Let $A^1, A^2, \ldots, A^d$ be the sequences of length n_i for $1 \leq i \leq d$. The approach uses a divide-and-conquer strategy. Find the maximum-length MUM, say α, shared by all sequences. This MUM has exactly one location in each sequence, say j_i for $1 \leq i \leq d$, so it can be used for splitting the set into two independent parts: $A^1_{1\ldots j_1}, A^2_{1\ldots j_2}, \ldots, A^d_{1\ldots j_d}$ and $A^1_{j_1+|\alpha|\ldots n_1}, A^2_{j_2+|\alpha|\ldots n_2}, \ldots, A^d_{j_d+|\alpha|\ldots n_d}$. Now apply the same recursively for each part until sufficiently short sequences are obtained in each part; then apply the optimal multiple alignment algorithm for each part. The MUMs calculated this way work as anchors for the multiple alignment.

Many of our algorithms are modifications of Algorithm 9.3 on page 174 for simulating traversal over all suffix tree nodes using the bidirectional BWT index.

11.1   Space-efficient genome analysis

Recall the suffix tree applications we covered in Section 8.4 (page 145), because we will now implement the same or similar algorithms using the bidirectional BWT index simulation approach rather than the explicit suffix tree. This leads to a significant saving in space usage.

11.1.1   Maximal repeats

Recall that a maximal repeat of a string T is any substring that occurs $k > 1$ times in T and that extending the substring to the left or right in any of its occurrences yields a substring that has fewer than k occurrences. Recall also from Section 8.4.1 the simple suffix tree algorithm for finding all maximal repeats of T in linear time. Algorithm 11.1 gives a more space-efficient variant of this algorithm using the bidirectional BWT index.

We already proved that Algorithm 9.3 on page 174 visits all internal suffix tree nodes, and here we just modified the algorithm into Algorithm 11.1 to output intervals corresponding to nodes that are left- and right-maximal. From an outputted interval $[i..j]$ corresponding to a node of depth d one can reconstruct the corresponding maximal repeat in $O(d \log \sigma)$ time by extracting $T_{\mathsf{SA}[i]..\mathsf{SA}[i]+d-1}$ using Lemma 9.2. If we want to just output starting and ending positions of the maximal repeats, then we could build a succinct suffix array and use the operation `locate` to determine each starting position $\mathsf{SA}[k]$ for $k \in [i..j]$. However, this will be potentially slow, since extracting a single

---

Algorithm 11.1: Computing maximal repeats using the bidirectional BWT index.

Input: Bidirectional BWT index idx and interval $[1..n+1]$ corresponding to the root of suffix tree of string $T = t_1 t_2 \cdots t_n$.

Output: Suffix array intervals associated with depth information representing the maximal repeats.

1 Let S be an empty stack;
2 S.push$(([1..n+1], [1..n+1], 0))$;
3 **while** *not* S.isEmpty() **do**
4     $([i..j], [i'..j'], d) \leftarrow S$.pop();
5   **if** idx.isLeftMaximal(i,j) **then**
6   Output $([i..j], d)$;
7     $\Sigma' \leftarrow$ idx.enumerateLeft(i,j);
8     $I \leftarrow \{$idx.extendLeft$(c, [i..j], [i'..j']) \mid c \in \Sigma'\}$;
9     $x \leftarrow \text{argmax} \{j - i : ([i..j], [i'..j']) \in I\}$;
10    $I \leftarrow I \setminus \{x\}$;
11    $([i..j], [i'..j']) \leftarrow x$;
12   **if** idx.isRightMaximal(i',j') **then**
13   S.push$(x, d+1)$;
14   **for** $([i..j], [i'..j']) \in I$ **do**
15   **if** idx.isRightMaximal(i',j') **then**
16   S.push$\big(([i..j], [i'..j'], d+1)\big)$;

---

position in the suffix array takes $O(r)$ time, where r is the sampling factor used for the succinct suffix array.

In order to achieve a better extraction time, we will use the batched locate routine (Algorithm 9.2) studied in Section 9.2.4.

Suppose that the algorithm associates a maximal repeat W with an integer identifier p_W. Let $\mathsf{SA}[\overrightarrow{W}]$ denote the $|\overrightarrow{W}|$ occurrences of maximal repeat W. Recall that Algorithm 9.2 receives as input the arrays marked and pairs, the former of which marks the suffix array entries to be extracted and the latter the marked entries associated with algorithm-dependent data. To construct the input, we set the bits marked$[\overrightarrow{W}]$ and append $|\overrightarrow{W}|$ pairs (k, p_W) to the buffer pairs, for all $k \in \overrightarrow{W}$. The batched locate routine will then convert each pair (k, i) into the pair $(\mathsf{SA}[k], i)$. Then the algorithm can group all the locations of the maximal repeats by sorting pairs according to the maximal repeat identifiers (the second component of each pair), which can be done in optimal time using radix sort.

THEOREM 11.1   *Assume a bidirectional BWT index is built on text $T = t_1 t_2 \cdots t_n$, $t_i \in [1..\sigma]$. Algorithm 11.1 solves Problem 8.2 by outputting a representation of maximal repeats in $O(n \log \sigma + \text{occ})$ time and $n + O(\sigma \log^2 n)$ bits of working space, where occ is the output size.*

Proof The proof of Theorem 9.15 on page 173 applies nearly verbatim to the first phase of the algorithm. The term n corresponds to the bitvector `marked` and the term $O(\sigma \log^2 n)$ corresponds to the space used by the stack. The batched locate incurs $O(n \log \sigma + \text{occ})$ additional time and uses $O(\text{occ} \log n)$ bits of additional space, which is counted as part of the output, since the output is also of size $O(\text{occ} \log n)$ bits.   □

We described in Section 9.3 a construction of the bidirectional BWT index in $O(n \log \sigma \log \log n)$ time using $O(n \log \sigma)$ bits of space. It was also mentioned that $O(n \log \sigma)$ time constructions exist in the literature.

11.1.2   Maximal unique matches

Recall that a maximal unique match (MUM) of two strings S and T is any substring that occurs exactly once in S and once in T and such that extending any of its two occurrences yields a substring that does not occur in the other string. Recall from Section 8.4.2 the simple suffix tree algorithm for finding all maximal unique matches in linear time. That algorithm works for multiple strings, but we will first consider the special case of two strings. Namely, Algorithm 11.2 gives a more space-efficient variant of this algorithm on two strings using the bidirectional BWT index, modifying the algorithm given above for maximal repeats.

The change to the maximal repeats algorithm consists in considering the concatenation $R = s_1 s_2 \cdots s_m \$ t_1 t_2 \cdots t_n \#$ of strings S and T, and in the addition of the indicator bitvector to check that the interval reached corresponding to a suffix tree node has a suffix of S as one child and a suffix from T as the other child. From an outputted interval $[i..i + 1]$, depth d, and an indicator bitvector $I[1..n + m + 2]$ one can reconstruct the corresponding maximal unique match: if, say, $I[i] = 0$, then $S_{\mathsf{SA}_R[i]..\mathsf{SA}_R[i]+d-1}$ is the maximal unique match substring, and it occurs at $\mathsf{SA}_R[i]$ in S and at $\mathsf{SA}_R[i] - m - 1$ in T.

THEOREM 11.2   *Assume a bidirectional BWT index is built on the concatenation R of strings $S = s_1 s_2 \cdots t_m$, $s_i \in [1..\sigma]$, and $T = t_1 t_2 \cdots t_n$, $t_i \in [1..\sigma]$, and a* `rank` *data structure is built on the indicator vector I such that $I[i] = 0$ iff the ith suffix of R in lexicographic order is from T, Algorithm 11.2 outputs a representation of all maximal unique matches of S and T in $O((m+n) \log \sigma)$ time and $(n+m)(2+o(1))+O(\sigma \log^2(m+ n))$ bits of working space.*

Proof Again, the proof of Theorem 9.15 on page 173 applies nearly verbatim. The `rank` queries take constant time by Theorem 3.2, with a data structure occupying $o(m+ n)$ bits on top of the indicator bitvector taking $m + n + 2$ bits; these are counted as part of the input.   □

The construction of the bidirectional BWT index in small space was discussed in the context of maximal repeats. Exercise 11.1 asks you to construct the indicator bitvector I in $O((m + n) \log \sigma)$ bits of space and in $O((m + n) \log \sigma)$ time.

Let us now consider how to extend the above approach to the general case of computing maximal unique matches on multiple strings.

Algorithm 11.2: Computing maximal unique matches using the bidirectional BWT index

---

Input: Bidirectional BWT index `idx` and interval $[1..m + n + 2]$ corresponding to the root of the suffix tree of string $R = s_1 s_2 \cdots s_m \$ t_1 t_2 \cdots t_n \#$, and `rank` data structure on the indicator bitvector I such that $I[i] = 0$ iff the ith suffix in the lexicographic order in R is from S.

Output: Suffix array intervals associated with depth information representing the maximal unique matches.

1 Let S be an empty stack;
2 $S.\text{push}(([1..m + n + 2], [1..m + n + 2], 0))$;
3 **while** *not* $S.\texttt{isEmpty}()$ **do**
4    $([i..j], [i'..j'], d) \leftarrow S.\text{pop}()$;
5    $\texttt{Scount} \leftarrow \text{rank}_0(I, j) - \text{rank}_0(I, i - 1)$;
6    $\texttt{Tcount} \leftarrow \text{rank}_1(I, j) - \text{rank}_1(I, i - 1)$;
7   **if** $\texttt{Scount} < 1$ *or* $\texttt{Tcount} < 1$ **then**
8 | **Continue**;
9   **if** $\texttt{Scount} = 1$ *and* $\texttt{Tcount} = 1$ *and* $\texttt{idx.isLeftMaximal}(i, j)$ **then**
10 | Output $([i..j], d)$;
11    $\Sigma' \leftarrow \texttt{idx.enumerateLeft}(i, j)$;
12    $I \leftarrow \{\texttt{idx.extendLeft}(c, [i..j], [i'..j']) \mid c \in \Sigma'\}$;
13    $x \leftarrow \text{argmax}\{j - i : ([i..j], [i'..j']) \in I\}$;
14    $I \leftarrow I \setminus \{x\}$;
15    $([i..j], [i'..j']) \leftarrow x$;
16   **if** $\texttt{idx.isRightMaximal}(i', j')$ **then**
17 | $S.\text{push}(x, d + 1)$;
18   **for** $([i..j], [i'..j']) \in I$ **do**
19   **if** $\texttt{idx.isRightMaximal}(i', j')$ **then**
20 | $S.\text{push}(([i..j], [i'..j'], d + 1))$;

We describe this algorithm on a more conceptual level, without detailing the adjustments to Algorithm 9.3 on a pseudocode level.

THEOREM 11.3 *Assume a bidirectional BWT index is built on the concatenation R of d strings $S^1, S^2, \ldots, S^d$ (separated by character #) of total length n, where $S_i \in [1..\sigma]^*$ for $i \in [1..d]$, then Problem 8.3 of finding all the maximal unique matches of the set of strings $S = \{S^1, S^2, \ldots, S^d\}$ can be solved in time $O(n \log \sigma)$ and $O(\sigma \log^2 n) + 2n + o(n)$ bits of working space.*

Proof The maximal unique matches between the d strings can be identified by looking at intervals of $\textsf{BWT}_R$ of size exactly d. The right-maximal substring that corresponds to an interval $[i..i+d-1]$ is a maximal unique match if and only if the following conditions hold.

i. $\mathsf{BWT}_R[i..i + d - 1]$ contains at least two distinct characters. This ensures left maximality.

ii. Every suffix pointed at by $\mathsf{SA}[i..i + d - 1]$ belongs to a different string. This condition ensures the uniqueness of the substring in every string.

We build a bitvector of intervals $\mathtt{intervals}[1..n + d]$ as follows. Initially all positions in $\mathtt{intervals}$ are set to zero. We enumerate all intervals corresponding to internal nodes of the suffix tree of R using Algorithm 9.3. For every such interval $[i..i + d - 1]$ such that $\mathsf{BWT}_R[i..i + d - 1]$ contains at least two distinct characters, we mark the two positions $\mathtt{intervals}[i] = 1$ and $\mathtt{intervals}[i + d - 1] = 1$. This first phase allows us to identify intervals that fulfill the first condition. For each such interval we will have two corresponding bits set to one in $\mathtt{intervals}$. We call the intervals obtained from the first phase *candidate intervals*, and we denote their number by t. In the second phase, we filter out all the candidate intervals that do not fulfill the second condition. For that purpose we allocate d bits for every candidate interval into a table $D[1..t][1..d]$ such that every position $D[i][j]$ will eventually be marked by a one if and only if one of the suffixes of string j appears in the candidate interval number i. Initially all the bits in D are set to zero. Then the marking is done as follows: we traverse the documents one by one from 1 to d, and, for each document S^j, we induce the position p in BWT_R of every suffix that appears in the document (by using *LF*-mapping) and we check whether p is inside a candidate interval. This is the case if either $\mathtt{intervals}[p] = 1$ or $\mathtt{rank}_1(\mathtt{intervals}, p)$ is an odd number. If that is the case, then the number associated with the interval will be $i = \lceil \mathtt{rank}_1(\mathtt{intervals}, p)/2 \rceil$. We then set $D[i][j] = 1$.

We can now scan the arrays D and $\mathtt{intervals}$ simultaneously and for every subarray $i \in [1..t]$ check whether $D[i][j] = 1$ for all $j \in [1..d]$. If that is the case, then we deduce that the candidate interval number i corresponds to a maximal unique match. If this is not the case, we unmark the interval number i by setting $\mathtt{intervals}[\mathtt{select}_1(\mathtt{intervals}, 2i - 1)] = 0$ and $\mathtt{intervals}[\mathtt{select}_1 (\mathtt{intervals}, 2i)] = 0$.

We finally use Algorithm 9.2 to extract the positions of the MUMs in a batch. For that, we scan the bitvector $\mathtt{intervals}$ and we write a bitvector $\mathtt{marked}[1..n]$ simultaneously. All the bits in $\mathtt{marked}$ are initially set to zero. During the scan of $\mathtt{intervals}$, we write d ones in $\mathtt{marked}[j..j+d-1]$ whenever we detect a MUM interval $[j..j+d-1]$ indicated by two ones at positions j and $j + d - 1$ in the bitvector $\mathtt{intervals}$, and we append the d pairs $(j, i), \ldots, (j + d - 1, i)$ to buffer $\mathtt{pairs}$, where i is the identifier of the MUM. With these inputs Algorithm 9.2 converts the first component of every pair into a text position, and then we finally group the text positions which belong to the same MUM by doing a radix sort according to the second components of the pairs. $\square$

11.1.3   Maximal exact matches

Maximal exact matches (MEMs) are a relaxation of maximal unique matches (MUMs) in which the uniqueness property is dropped: a triple (i, j, ℓ) is a maximal exact match (also called a *maximal pair*) of two strings S and T if $S_{i...i+\ell-1} = T_{j...j+\ell-1}$ and no extension of the matching substring is a match, that is, $s_{i-1} \neq t_{j-1}$ and $s_{i+\ell} \neq t_{j+\ell}$.

Problem 11.1   Maximal exact matches

Find all maximal exact matches of strings S and T.

The algorithm to solve Problem 11.1 using the bidirectional BWT index is also analogous to the one for MUMs on two strings, with some nontrivial details derived next.

Consider a left- and right-maximal interval pair $([i..j], [i'..j'])$ and depth d in the execution of Algorithm 11.2 on $R = S\$T\#$. Recall that, if $I[k] = 0$ for $i \leq k \leq j$, then $SA[k]$ is a starting position of a suffix of S, otherwise $SA[k] - |S| - 1$ is a starting position of a suffix of T. The cross-products of suffix starting positions from S and from T in the interval $[i..j]$ form all *candidates* for maximal pairs with $R[SA[i]..SA[i]+d-1]$ as the matching substring. However, not all such candidates are maximal pairs. Namely, the candidate pair (p, q) associated with $R[SA[i]..SA[i] + d - 1] = S[p..p + d - 1] = T[q..q + d - 1]$ is not a maximal pair if $s_{p-1} = t_{q-1}$ or $s_{p+d} = t_{q+d}$. To avoid listing candidates that are not maximal pairs, we consider next the *synchronized* traversal of two bidirectional BWT indexes, one built on S and one built on T.

Consider the interval pair $([i_0..j_0], [i'_0..j'_0])$ in the bidirectional BWT index of S corresponding to substring $R[SA_R[i]..SA_R[i] + d - 1]$. That is, $SA_R[k] = SA_S[\text{rank}_0(I, k)]$ for $i \leq k \leq j$, where I is the indicator bitvector in Algorithm 11.2. Analogously, consider the interval pair $([i_1..j_1], [i'_1..j'_1])$ in the bidirectional BWT index of T corresponding to substring $R[SA_R[i]..SA_R[i] + d - 1]$. That is, $SA_R[k] - |S| - 1 = SA_T[\text{rank}_1(I, k)]$ for $i \leq k \leq j$. Now, for each $a, b \in \Sigma$, we may check whether the interval pair $([i_0..j_0], [i'_0..j'_0])$ can be extended to the left with a and then to the right with b. If this is the case, $a \cdot R[SA_R[i]..SA_R[i] + d - 1] \cdot b$ is a substring of S. Let set A store such valid pairs a, b. Analogously, for each $c, d \in \Sigma$, we may check whether the interval pair $([i_1..j_1], [i'_1..j'_1])$ can be extended to the left with c and then to the right with d. If this is the case, $c \cdot R[SA_R[i]..SA_R[i]+d-1] \cdot d$ is a substring of T. Let set B store such valid pairs (c, d). We wish to compute $A \otimes B = \{(a, b, c, d) \mid (a, b) \in A, (c, d) \in B, a \neq c, b \neq d\}$, since this gives us the valid extension tuples to define the maximal pairs.

LEMMA 11.4   *Let Σ be a finite set, and let A and B be two subsets of $\Sigma \times \Sigma$. We can compute $A \otimes B = \{(a, b, c, d) \mid (a, b) \in A, (c, d) \in B, a \neq c, b \neq d\}$ in $O(|A| + |B| + |A \otimes B|)$ time.*

Proof   We assume without loss of generality that $|A| < |B|$. We say that two pairs (a, b) and (c, d) are *compatible* if $a \neq c$ and $b \neq d$. Clearly, if (a, b) and (c, d) are compatible, then the only elements of $\Sigma \times \Sigma$ that are incompatible with both (a, b) and (c, d) are (a, d) and (c, b). We iteratively select a pair $(a, b) \in A$ and scan A in $O(|A|) = O(|B|)$ time to find another compatible pair (c, d): if we find one, we scan B

and report every pair in B that is compatible with either (a, b) or (c, d). The output will be of size $|B| - 2$ or larger, thus the time taken to scan A and B can be charged to the output. Then, we remove (a, b) and (c, d) from A and repeat the process. If A becomes empty we stop. If all the remaining pairs in A are incompatible with our selected pair (a, b), that is, if $c = a$ or $d = b$ for every $(c, d) \in A$, we build subsets A^a and A^b, where $A^a = \{(a, x) : x \neq b\} \subseteq A$ and $A^b = \{(x, b) : x \neq a\} \subseteq A$. Then we scan B, and for every pair $(x, y) \in B$ different from (a, b) we do the following. If $x \neq a$ and $y \neq b$, then we report (a, b, x, y), $\{(a, z, x, y) : (a, z) \in A^a, z \neq y\}$ and $\{(z, b, x, y) : (z, b) \in A^b, z \neq x\}$. Pairs $(a, y) \in A^a$ and $(x, b) \in A^b$ are the only ones that do not produce output, thus the cost of scanning A^a and A^b can be charged to printing the result. If $x = a$ and $y \neq b$, then we report $\{(z, b, x, y) : (z, b) \in A^b\}$. If $x \neq a$ and $y = b$, then we report $\{(a, z, x, y) : (a, z) \in A^a\}$. □

Combining Lemma 11.4 and Theorem 9.15 yields the following result.

THEOREM 11.5   *Assume that two bidirectional BWT indexes are built, one on string $S = s_1 s_2 \cdots s_m$, $s_i \in [1..\sigma]$ and one on string $T = t_1 t_2 \cdots t_n$, $t_i \in [1..\sigma]$. Algorithm 11.3 solves Problem 11.1 by outputting a representation of maximal exact matches of S and T in $O((m + n)\log \sigma + \mathrm{occ})$ time and $m + n + O(\sigma \log^2(m + n))$ bits of working space, where occ is the output size.*

Proof   The proof of Theorem 9.15 on page 173 applies nearly verbatim to the main algorithm execution, without the conditions for left- and right-maximality, and without the calls to the cross-product subroutine.

For the first, we assume that the evaluation of conditions is executed left to right and terminated when the first condition fails; this guarantees that calls to `enumerateLeft()` and `enumerateRight()` inside conditions are returning sets of singleton elements, resulting in the claimed running time.

For the second, we need to show that the running time of the cross-product subroutine can be bounded by the input and output sizes. Indeed, each extension to the left is executed only twice, and corresponds to an explicit or implicit Weiner link in S or T. In Observation 8.14 we have shown that their amount is linear. Extensions to the right can be charged to the output. □

The construction of the bidirectional BWT index in small space was discussed in the context of maximal repeats and maximal unique matches. See Algorithms 11.3 and 11.4 for a pseudocode of the proof of Theorem 11.5. We have dropped the BWT index for the concatenation and the indicator bitvector, since the two added indexes contain all the information required in order to simulate the algorithm.

These algorithms can be used to derive all the *minimal absent words* of a single string S.

DEFINITION 11.6   *String W is a* minimal absent word *of a string $S \in \Sigma^+$ if W does not occur in S and if every proper substring of W occurs in S.*

For example, CGCGCG is a minimal absent word of string T in Figure 8.4. Therefore, to decide whether aWb is a minimal absent word of S, where $\{a, b\} \subseteq \Sigma$, it suffices to check that aWb does not occur in S, and that both aW and Wb occur in S. It is easy to see that only a maximal repeat of S can be the infix W of a minimal absent word aWb (Exercise 11.5). We can thus enumerate the children of every internal node v of the suffix tree of S, then enumerate the Weiner links of v and of each one of its children, and finally compare the corresponding sets of characters in a manner similar to Algorithm 11.3. We leave the proof of the following theorem to Exercise 11.4.

Algorithm 11.3: Computing maximal exact matches using two bidirectional BWT indexes

---

Input: Bidirectional BWT index $\mathtt{idx}_S$ and interval $[1..m + 1]$ corresponding to the root of the suffix tree of string $S = s_1 s_2 \cdots s_m$, and bidirectional BWT index $\mathtt{idx}_T$ and interval $[1..n + 1]$ corresponding to the root of the suffix tree of string $T = t_1 t_2 \cdots t_n$.

Output: Suffix array index pairs associated with depth information representing the maximal exact matches of S and T.

1 Let S be an empty stack;

2 $S.\text{push}(([1..m + 1], [1..m + 1], [1..n + 1], [1..n + 1], 0))$;

3 **while** *not* $S.\texttt{isEmpty()}$ **do**

4      $([i_0..j_0], [i'_0..j'_0], [i_1..j_1], [i'_1..j'_1], \text{depth}) \leftarrow S.\text{pop}()$;

5   **if** $j_0 - i_0 + 1 < 1$ *or* $j_1 - i_1 + 1 < 1$ **then**

6   **Continue**;

7   **if** $\mathtt{idx}_S.\texttt{isLeftMaximal}(i_0, j_0)$ *or* $\mathtt{idx}_T.\texttt{isLeftMaximal}(i_1, j_1)$ *or* $\mathtt{idx}_S.\texttt{enumerateLeft}(i_0, j_0) \neq \mathtt{idx}_T.\texttt{enumerateLeft}(i_1, j_1)$ **then**

8   Apply Algorithm 11.4 to output maximal pairs;

9      $\Sigma' \leftarrow \mathtt{idx}_S.\texttt{enumerateLeft}(i_0, j_0)$;

10      $I \leftarrow \{(\mathtt{idx}_S.\texttt{extendLeft}(c, [i_0..j_0], [i'_0..j'_0]), \mathtt{idx}_T.\texttt{extendLeft}(c, [i_1..j_1], [i'_1..j'_1])) \mid c \in \Sigma'\}$;

11      $x \leftarrow \text{argmax}\{\max(j_0 - i_0, j_1 - i_1) : ([i_0..j_0], [i'_0..j'_0], [i_1..j_1], [i'_1..j'_1]) \in I\}$;

12      $I \leftarrow I \setminus \{x\}$;

13      $([i_0..j_0], [i'_0..j'_0], [i_1..j_1], [i'_1..j'_1]) \leftarrow x$;

14   **if** $\mathtt{idx}_S.\texttt{isRightMaximal}(i'_0, j'_0)$ *or* $\mathtt{idx}_T.\texttt{isRightMaximal}(i'_1, j'_1)$ *or* $\mathtt{idx}_S.\texttt{enumerateRight}(i'_0, j'_0) \neq \mathtt{idx}_T.\texttt{enumerateRight}(i'_1, j'_1)$ **then**

15          $S.\text{push}(x, \text{depth} + 1)$;

16   **for** $([i_0..j_0], [i'_0..j'_0], [i_1..j_1], [i'_1..j'_1]) \in I$ **do**

17   **if** $\mathtt{idx}_S.\texttt{isRightMaximal}(i'_0, j'_0)$ *or* $\mathtt{idx}_T.\texttt{isRightMaximal}(i'_1, j'_1)$ *or* $\mathtt{idx}_S.\texttt{enumerateRight}(i'_0, j'_0) \neq \mathtt{idx}_T.\texttt{enumerateRight}(i'_1, j'_1)$ **then**

18              $S.\text{push}(([i_0..j_0], [i'_0..j'_0], [i_1..j_1], [i'_1..j'_1], \text{depth} + 1))$;

Algorithm 11.4: Cross-product computation for outputting maximal exact matches using two bidirectional BWT indexes

---

Input: Subroutine of Algorithm 11.3 getting an interval tuple
$([i_0..j_0], [i'_0..j'_0], [i_1..j_1], [i'_1..j'_1])$ and `depth` corresponding to a state of synchronized exploration of idx_S and idx_T.

Output: Suffix array index pairs associated with depth information representing the maximal exact matches.

1 $A \leftarrow \emptyset$;

2 for $a \in idx_S.\texttt{enumerateLeft}(i_0, j_0)$ **do**

3 $(i_a, j_a, i'_a, j'_a) \leftarrow idx_S.\texttt{extendLeft}(i_0, j_0, i'_0, j'_0, a)$;

4   **for** $b \in idx_S.\texttt{enumerateRight}(i'_a, j'_a)$ **do**

5 $A \leftarrow A \cup \{(a, b)\}$;

6 $B \leftarrow \emptyset$;

7 for $c \in idx_T.\texttt{enumerateLeft}(i_1, j_1)$ **do**

8 $(i_c, j_c, i'_c, j'_c) \leftarrow idx_T.\texttt{extendLeft}(i_1, j_1, i'_1, j'_1, c)$;

9   **for** $d \in idx_T.\texttt{enumerateRight}(i'_c, j'_c)$ **do**

10 $B \leftarrow B \cup \{(c, d)\}$;

11 for $(a, b, c, d) \in A \otimes B$ **do**

12 $(i_a, j_a, i'_a, j'_a) \leftarrow idx_S.\texttt{extendLeft}(i_0, j_0, i'_0, j'_0, a)$;

13 $(i_b, j_b, i'_b, j'_b) \leftarrow idx_S.\texttt{extendRight}(i_a, j_a, i'_a, j'_a, b)$;

14 $(i_c, j_c, i'_c, j'_c) \leftarrow idx_T.\texttt{extendLeft}(i_1, j_1, i'_1, j'_1, c)$;

15 $(i_d, j_d, i'_d, j'_d) \leftarrow idx_T.\texttt{extendRight}(i_c, j_c, i'_c, j'_c, d)$;

16   **for** $i \in [i_b..j_b]$ **do**

17   **for** $j \in [i_d..j_d]$ **do**

18   Output $(i, j, \texttt{depth})$;

THEOREM 11.7   *Assume that the bidirectional BWT index is built on a string* $S \in [1..\sigma]^n$. *All the minimal absent words of* S *can be computed in* $O(n \log \sigma + \texttt{occ})$ *time, and in* $O(\sigma \log^2 n)$ *bits of working space, where* $\texttt{occ}$ *is the output size.*

Note that $\texttt{occ}$ can be of size $\Theta(n\sigma)$ in this case: see Exercises 11.6 and 11.7.

11.2   Comparing genomes without alignment

Given two strings S and T, a *string kernel* is a function that simultaneously converts S and T into vectors $\{\mathbf{S}, \mathbf{T}\} \subset \mathbb{R}^n$ for some $n > 0$, and computes a similarity or a distance measure between $\mathbf{S}$ and $\mathbf{T}$, *without building and storing* $\mathbf{S}$ *and* $\mathbf{T}$ *explicitly*. The (possibly infinite) dimensions of the resulting vectors are typically all substructures of a specific type, for example all strings of a fixed length on the alphabet of S and T, and the value assigned to vectors $\mathbf{S}$ and $\mathbf{T}$ along dimension W corresponds to the frequency

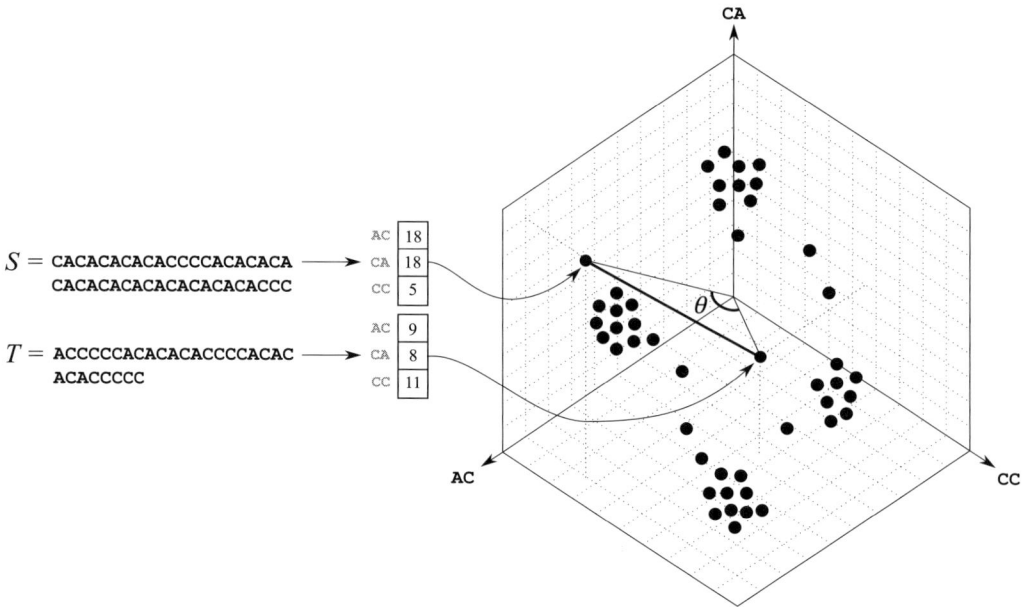

Figure 11.1 The k-mer kernel $\kappa(S, T)$ for $k = 2$, which coincides with the cosine of the angle θ shown. The bold segment is the norm $||\mathbf{S} - \mathbf{T}||_2$, or equivalently the Euclidean distance between $\mathbf{S}$ and $\mathbf{T}$ in $\mathbb{R}^3$.

of substructure W in strings S and T, respectively – often rescaled and corrected in domain-specific ways (see the example in Figure 11.1). For this reason, $\mathbf{S}$ and $\mathbf{T}$ are called *composition vectors*. A large number of components in $\mathbf{S}$ and $\mathbf{T}$ can be zero in practice: in this case, we say that $\mathbf{S}$ and $\mathbf{T}$ are *sparse*. The term *kernel* derives from learning theory: see Insight 11.2 for a general definition of kernels on vectors.

Conceptually, comparing genomes using the frequency of their substructures captures genome-wide *compositional biases* that are well known to correlate with accepted species classifications in biology. It thus allows one to estimate the similarity between two evolutionary distant species, for which a whole-genome alignment would not be meaningful, if a trace of their evolutionary history is still preserved by their compositional preferences. Composition-based kernels extend naturally to sets of reads and to metagenomic samples (see Section 16.3), estimating their similarity without the need for assembly, and they can be generalized to mixed settings in which sets of reads must be compared with sequenced genomes, and vice versa. Computationally, string kernels tend to be more time- and space-efficient than whole-genome alignments, and are thus the ideal candidates for applications that must estimate the similarity of two long strings very quickly, or that need to compute a similarity measure for all pairs of strings in a massive dataset.

This section walks the reader through a number of algorithms that compute string kernels, in roughly increasing order of expressive power and algorithmic sophistication. Most such algorithms can be implemented using suffix trees, and some of them can be mapped to the space-efficient genome analysis techniques described in Section 11.1: see

Table 11.1 Overview of the kernels between strings S and T presented in Section 11.2 and implemented using suffix trees. Each kernel is mapped to the data structures and algorithms used to implement it. Numbers indicate sections that describe a kernel explicitly. Bullets indicate solutions that we mention but do not describe explicitly. GST: generalized suffix tree of S and T. MS: suffix tree of T and matching statistics of S with respect to T. BWT: bidirectional BWT indexes of S and of T.

	GST	MS	BWT
k-mer kernel	•	11.2.3	11.2.1
Substring kernel	•	11.2.3	11.2.1
Generalized substring kernel	•	11.2.3	
Markovian substring kernel	11.2.2		•
Variable-memory Markov chain		11.2.3	

Table 11.1 for an overview. As will be apparent in what follows, the key to efficiency lies in the beautiful interplay between the probabilities of substrings and the combinatorial properties of suffix trees.

We will repeatedly call S and T the two strings on alphabet $\Sigma = [1..\sigma]$ which we want to compare, and we will call *k-mer* any string $W \in [1..\sigma]$ of fixed length $k > 0$. We will also denote by $f_S(W)$ the number of (possibly overlapping) occurrences of string W in S, using the shorthand $p_S(W) = f_S(W)/(|S| - k + 1)$ to denote the *empirical probability* of observing W in S. As is customary, we will denote the suffix tree of S by ST_S, dropping the subscript whenever it is implicit from the context.

Throughout this section we will be mostly interested in computing the *cosine of the angle between the composition vectors* $\mathbf{S}$ *and* $\mathbf{T}$, a natural measure of their similarity. In other words, the kernels we will consider have the form

$$\kappa(\mathbf{S}, \mathbf{T}) = \frac{\sum_W \mathbf{S}[W]\mathbf{T}[W]}{\sqrt{\left(\sum_W \mathbf{S}[W]^2\right)\left(\sum_W \mathbf{T}[W]^2\right)}}. \tag{11.1}$$

Exercise 11.16 gives some intuition on why $\kappa(\mathbf{S}, \mathbf{T})$ measures the cosine of the angle between $\mathbf{S}$ and $\mathbf{T}$. Note that $\kappa(\mathbf{S}, \mathbf{T})$, a measure of similarity in $[-1..1]$ that equals zero when $\mathbf{S}$ and $\mathbf{T}$ have no substructure in common, can be converted into a *distance* $d(\mathbf{S}, \mathbf{T}) = (1 - \kappa(\mathbf{S}, \mathbf{T}))/2$ that falls within the range $[0..1]$. Most of the algorithms described in this section apply to other commonly used norms of vector $\mathbf{S} - \mathbf{T}$, like the p-norm,

$$||\mathbf{S} - \mathbf{T}||_p = \left(\sum_W |\mathbf{S}[W] - \mathbf{T}[W]|^p\right)^{1/p},$$

and the infinity norm,

$$||\mathbf{S} - \mathbf{T}||_\infty = \max_W \{|\mathbf{S}[W] - \mathbf{T}[W]|\}.$$

Even more generally, most algorithms in this section apply to any function of the three arguments N, D_S, and D_T defined as follows:

$$N = \bigoplus_W g_1(\mathbf{S}[W], \mathbf{T}[W]),$$

$$D_S = \bigotimes_W g_2(\mathbf{S}[W]),$$

$$D_T = \bigodot_W g_3(\mathbf{T}[W]),$$

where g_1, g_2, and g_3 are arbitrary functions, and $\oplus$, $\otimes$, and $\odot$ are arbitrary *commutative* and *associative* binary operations, or equivalently operations (like sum and product) whose result depends neither on the order of the operands nor on the order in which the operations are performed on such operands.

11.2.1   Substring and k-mer kernels

Perhaps the simplest way to represent a string as a composition vector consists in counting the frequency of all its distinct substrings of a fixed length. This is known as the *k-mer spectrum* of the string.

DEFINITION 11.8   *Given a string $S \in [1..\sigma]^+$ and a length $k > 0$, let vector $\mathbf{S}_k = [1..\sigma^k]$ be such that $\mathbf{S}_k[W] = f_S(W)$ for every $W \in [1..\sigma]^k$. The k-mer complexity $C(S, k)$ of string S is the number of non-zero components of $\mathbf{S}_k$. The k-mer kernel between two strings S and T is Equation (11.1) evaluated on $\mathbf{S}_k$ and $\mathbf{T}_k$.*

Put differently, the k-mer complexity of S is the number of distinct k-mers that occur in S. Choosing the best value of k for a specific application might be challenging: Insight 11.3 describes a possible data-driven strategy. Alternatively, we could just remove the constraint of a fixed k, and use the more general notions of the *substring complexity* and the *substring kernel*.

Insight 11.2 Kernels on vectors

Given a set of vectors $\mathcal{S} \subset \mathbb{R}^n$ such that each vector belongs to one of two possible classes, assume that we have a *classification algorithm* $\mathcal{A}$ that divides the two classes with a plane in $\mathbb{R}^n$. If we are given a new vector, we can label it with the identifier of a class by computing the side of the plane it lies in. If the two classes cannot be separated by a plane, or, equivalently, if they are not *linearly separable* in $\mathbb{R}^n$, we could create a *nonlinear map* $\phi : \mathbb{R}^n \mapsto \mathbb{R}^m$ with $m > n$ such that they become linearly separable in $\mathbb{R}^m$, and then run algorithm $\mathcal{A}$ on the projected set of points $\mathcal{S}' = \{\phi(\mathbf{X}) \mid \mathbf{X} \in \mathcal{S}\}$. This is an ingenious way of building a *nonlinear* classification algorithm in $\mathbb{R}^n$ using a *linear* classification algorithm in $\mathbb{R}^m$.

Assume that algorithm $\mathcal{A}$ uses just inner products $\mathbf{X} \cdot \mathbf{Y}$ between pairs of vectors $\{\mathbf{X}, \mathbf{Y}\} \subseteq \mathcal{S}$. Building and storing the projected m-dimensional vectors, and computing inner products between pairs of such vectors, could be significantly more expensive than computing inner products between the original n-dimensional vectors. However, if ϕ is such that $\phi(\mathbf{X}) \cdot \phi(\mathbf{Y}) = \kappa(\mathbf{X}, \mathbf{Y})$ for some symmetric function κ that is fast to evaluate, we could run $\mathcal{A}$ on $\mathcal{S}'$ *without constructing $\mathcal{S}'$ explicitly*, by just replacing all calls to inner products by calls to function κ in the code of $\mathcal{A}$: this approach is called *kernelizing* algorithm $\mathcal{A}$, and κ is called a *kernel function*. The family of functions $\kappa : \mathbb{R}^n \times \mathbb{R}^n \mapsto \mathbb{R}$ for which there is a mapping ϕ such that $\kappa(\mathbf{X}, \mathbf{Y}) = \phi(\mathbf{X}) \cdot \phi(\mathbf{Y})$ is defined by the so-called *Mercer's conditions*. The symmetric, positive-definite matrix containing $\kappa(\mathbf{X}, \mathbf{Y})$ for all $\{\mathbf{X}, \mathbf{Y}\} \subset \mathcal{S}$ is called the *Gram matrix* of $\mathcal{S}$. Multiple kernel functions can be combined to obtain more complex kernels.

If the input set $\mathcal{S}$ contains objects other than vectors, kernelizing $\mathcal{A}$ means replacing all the inner products $\mathbf{X} \cdot \mathbf{Y}$ in $\mathcal{A}$ by calls to a function $\kappa(X, Y)$ that simultaneously converts objects $\{X, Y\} \in \mathcal{S}$ into vectors $\{\mathbf{X}, \mathbf{Y}\} \subset \mathbb{R}^m$ and computes the inner product $\mathbf{X} \cdot \mathbf{Y}$, without building and storing $\mathbf{X}$ and $\mathbf{Y}$ explicitly.

DEFINITION 11.9 *Given a string $S \in [1..\sigma]^+$, consider the infinite-dimensional vector* $\mathbf{S}_\infty$, *indexed by all distinct substrings* $W \in [1..\sigma]^+$, *such that* $\mathbf{S}_\infty[W] = f_S(W)$. *The substring complexity $C(S)$ of string S is the number of non-zero components of $\mathbf{S}_\infty$. The* substring kernel *between two strings S and T is Equation (11.1) evaluated on $\mathbf{S}_\infty$ and $\mathbf{T}_\infty$.*

Computing substring complexity and substring kernels, with or without a constraint on string length, is trivial using suffix trees: see Exercise 11.17. In this section we show how to perform the same computations space-efficiently, using the bidirectional BWT index described in Section 9.4. Recall that Algorithm 9.3 enumerates all internal nodes of the suffix tree of S using the bidirectional BWT index of S and a stack of bounded size. Recall also that, to limit the size of this stack, the nodes of ST_S are enumerated *in a specific order*, related to the topology of the suffix-link tree of S. In what follows, we describe algorithms to compute substring complexity and substring kernels *in a way that does not depend on the order in which the nodes of ST_S are enumerated*: we can thus implement such algorithms on top of Algorithm 9.3.

We first describe the simpler cases of k-mer complexity and substring complexity, and then we extend their solutions to kernels. Similar algorithms can compute the *kth-order empirical entropy* of S: see Exercise 11.13. The main idea behind all such algorithms consists in a *telescoping approach* that works by adding and subtracting terms in a sum, as described in the next lemma and visualized in Figure 11.2.

LEMMA 11.10 *Given the bidirectional BWT index of a string $S \in [1..\sigma]^n$ and an integer k, there is an algorithm that computes the k-mer complexity $C(S, k)$ of S in $O(n \log \sigma)$ time and $O(\sigma \log^2 n)$ bits of working space.*

Proof A k-mer of S can either be the label of a node of ST_S, or it could end in the middle of an edge (u, v) of the suffix tree. In the latter case, we assume that the k-mer is represented by its locus v, which might be a leaf. Let $C(S, k)$ be initialized to $|S| + 1 - k$, that is, to the number of leaves that correspond to suffixes of $S\#$ of length

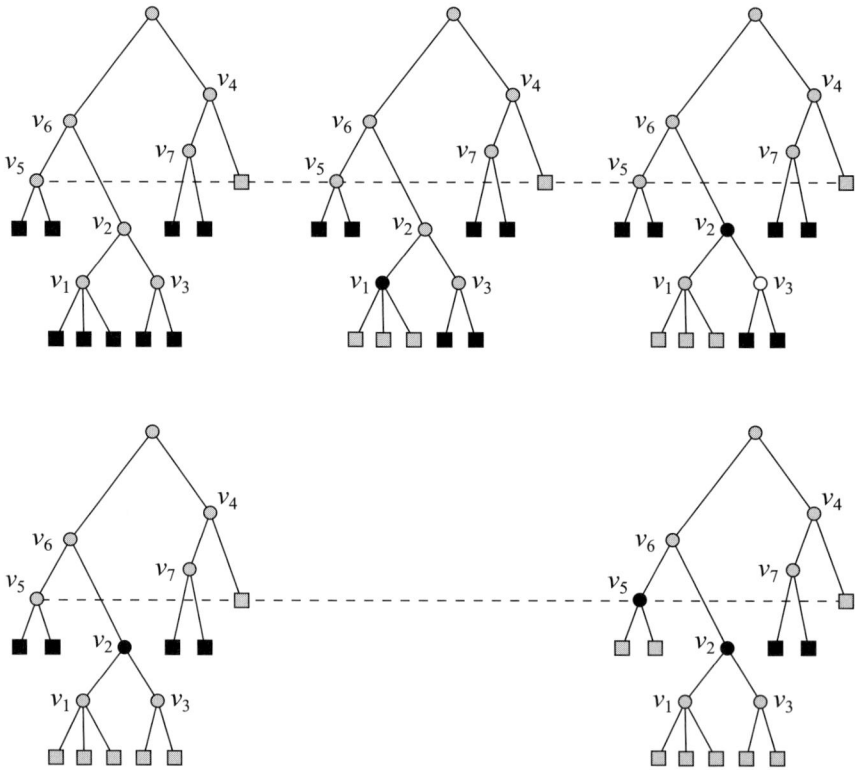

Figure 11.2 Lemma 11.10 illustrated from top to bottom, from left to right, on the suffix tree of string S, assuming that its internal nodes are enumerated in the order $v_1, v_2, \ldots, v_7$. Black indicates a $+1$ contribution to $C(S, k)$, gray indicates zero contribution, and white indicates a -1 contribution. Note that v_3 is white at step 3. The dashed line corresponds to k. The empty space without a tree at step 5 indicates that nothing happens when node v_4 is enumerated.

at least k, excluding suffix $S[|S| - k + 2..|S|]\#$. We use Algorithm 9.3 to enumerate the internal nodes of ST, and every time we enumerate a node v we proceed as follows. If $|\ell(v)| < k$ we leave $C(S, k)$ unaltered, otherwise we increment $C(S, k)$ by one and decrement $C(S, k)$ by the number of children of v in ST. This number is the size of the output of the operation enumerateRight provided by the bidirectional BWT index of S, and the value $|\ell(v)|$ is provided by Algorithm 9.3. It follows that every node v of ST that is located at depth at least k and that is not the locus of a k-mer is both added to $C(S, k)$ (when the algorithm visits v) and subtracted from $C(S, k)$ (when the algorithm visits parent(v)). Leaves at depth at least k are added by the initialization of $C(S, k)$, and subtracted during the enumeration. Conversely, every locus v of a k-mer of S (including leaves) is just added to $C(S, k)$, because $|\ell(\text{parent}(v))| < k$. $\square$

Computing the k-mer kernel between two strings S and T follows the same telescoping strategy.

LEMMA 11.11   *Given an integer k and the bidirectional BWT index of a string $S \in [1..\sigma]^n$ and of a string $T \in [1..\sigma]^m$, there is an algorithm that computes the k-mer kernel between S and T in $O((n + m)\log\sigma)$ time and $O(\sigma \log^2(m + n))$ bits of working space.*

Proof   Recall that Algorithm 11.3 in Section 11.1.3 enumerates all the internal nodes of the generalized suffix tree of S and T, using the bidirectional BWT index of S, the bidirectional BWT index of T, and a stack of bounded size. We compute the substring kernel by enumerating the internal nodes of the generalized suffix tree of S and T as done in Algorithm 11.3, and by simultaneously updating three variables, denoted by N, D_S, and D_T, which assume the following values after all the internal nodes of the generalized suffix tree have been enumerated:

$$N = \sum_{W \in [1..\sigma]^k} \mathbf{S}_k[W]\mathbf{T}_k[W],$$

$$D_S = \sum_{W \in [1..\sigma]^k} \mathbf{S}_k[W]^2,$$

$$D_T = \sum_{W \in [1..\sigma]^k} \mathbf{T}_k[W]^2.$$

Thus, after the enumeration, the value of the k-mer kernel is $N/\sqrt{D_S \cdot D_T}$. We initialize the variables as follows: $N = 0$, $D_S = |S| + 1 - k$, and $D_T = |T| + 1 - k$. These are the contributions of the $|S| + 1 - k + |T| + 1 - k$ leaves at depth at least k in the generalized suffix tree of S and T, excluding the leaves which correspond to suffixes $S[|S| - k + 2..|S|]\#$ and $T[|T| - k + 2..|T|]\#$. Whenever Algorithm 11.3 enumerates a node u of the generalized suffix tree of S and T, where $\ell(u) = W$, we keep all variables unchanged if $|W| < k$. Otherwise, we update them as follows:

$$N \leftarrow N + f_S(W)f_T(W) - \sum_{v \in \text{children}(u)} f_S(\ell(v)) \cdot f_T(\ell(v)),$$

$$D_S \leftarrow D_S + f_S(W)^2 - \sum_{v \in \text{children}(u)} f_S(\ell(v))^2,$$

$$D_T \leftarrow D_T + f_T(W)^2 - \sum_{v \in \text{children}(u)} f_T(\ell(v))^2,$$

where the values $f_S(W)$ and $f_T(W)$ are the sizes of the intervals which represent node u in the two BWT indexes, the values $f_S(\ell(v))$ and $f_T(\ell(v))$ can be computed for every child v of u using the operations enumerateRight and extendRight provided by the bidirectional BWT index of S and T, and the value $|\ell(v)|$ is provided by Algorithm 11.3. In analogy to Lemma 11.10, the contribution of the loci of the distinct k-mers of S, of T, or of both, is added to the three variables and never subtracted, while the contribution of every other node v at depth at least k in the generalized suffix tree of S and T (including leaves) is both added (when the algorithm visits v, or when N, D_S, and D_T are initialized) and subtracted (when the algorithm visits parent(v)).   □

This algorithm can easily be adapted to compute the *Jaccard distance* between the set of k-mers of S and the set of k-mers of T, and to compute a variant of the k-mer kernel in which $\mathbf{S}_k[W] = p_S(W)$ and $\mathbf{T}_k[W] = p_T(W)$ for every $W \in [1..\sigma]^k$: see Exercises 11.10 and 11.11. Computing the substring complexity and the substring kernel amounts to applying the same telescoping strategy, but with different contributions.

COROLLARY 11.12   *Given the bidirectional BWT index of a string $S \in [1..\sigma]^n$, there is an algorithm that computes the substring complexity $C(S)$ in $O(n \log \sigma)$ time and $O(\sigma \log^2 n)$ bits of working space.*

Proof   Note that the substring complexity of S coincides with the number of characters in $[1..\sigma]$ that occur on all edges of ST_S. We can thus proceed as in Lemma 11.10, initializing $C(S)$ to $n(n + 1)/2$, or equivalently to the sum of the lengths of all suffixes of S. Whenever we visit a node u of ST_S, we add to $C(S)$ the quantity $|\ell(u)|$, and we subtract from $C(S)$ the quantity $|\ell(u)| \cdot |\text{children}(v)|$. The net effect of all such operations coincides with summing the lengths of all edges of ST, discarding all occurrences of character # (see Figure 11.3). Note that $|\ell(u)|$ is provided by Algorithm 9.3, and $|\text{children}(v)|$ is the size of the output of operation enumerateRight provided by the bidirectional BWT index of S.   □

COROLLARY 11.13   *Given the bidirectional BWT index of a string $S \in [1..\sigma]^n$ and of a string $T \in [1..\sigma]^m$, there is an algorithm that computes the substring kernel between S and T in $O((n + m)\log\sigma)$ time and $O(\sigma \log^2(m + n))$ bits of working space.*

Proof   We proceed as in Lemma 11.11, maintaining the variables N, D_S, and D_T, which assume the following values after all the internal nodes of the generalized suffix tree of S and T have been enumerated:

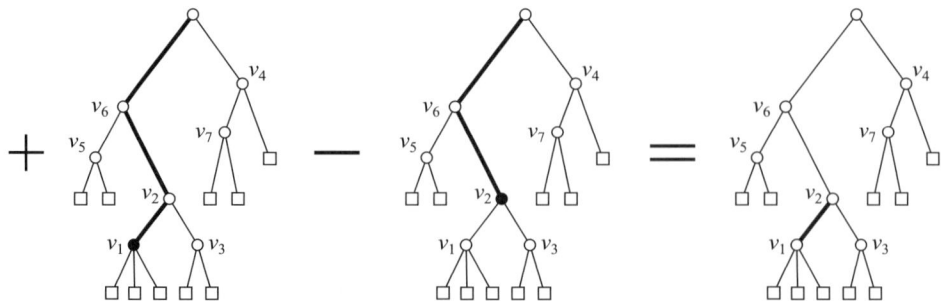

Figure 11.3 Illustrating Corollary 11.12 on the suffix tree of string S. When the algorithm enumerates v_1, it adds to $C(S)$ the number of characters on the path from the root to v_1. When the algorithm enumerates v_2, it subtracts the number of characters on the path from the root to v_2. The net contribution of these two operations to $C(S)$ is the number of characters on edge (v_2, v_1).

$$N = \sum_{W \in [1..\sigma]^+} \mathbf{S}_\infty[W] \mathbf{T}_\infty[W],$$

$$D_S = \sum_{W \in [1..\sigma]^+} \mathbf{S}_\infty[W]^2,$$

$$D_T = \sum_{W \in [1..\sigma]^+} \mathbf{T}_\infty[W]^2.$$

Thus, after the enumeration, the value of the k-mer kernel is $N / \sqrt{D_S \cdot D_T}$. We initialize such variables as follows: $N = 0$, $D_S = n(n + 1)/2$, and $D_T = m(m + 1)/2$. When we visit a node u of the generalized suffix tree of S and T, with $\ell(u) = W$, we update the variables as follows:

$$N \leftarrow N + |W| \cdot \left(f_S(W) f_T(W) - \sum_{v \in \text{children}(u)} f_S(\ell(v)) \cdot f_T(\ell(v)) \right),$$

$$D_S \leftarrow D_S + |W| \cdot \left(f_S(W)^2 - \sum_{v \in \text{children}(u)} f_S(\ell(v))^2 \right),$$

$$D_T \leftarrow D_T + |W| \cdot \left(f_T(W)^2 - \sum_{v \in \text{children}(u)} f_T(\ell(v))^2 \right).$$

□

The equations in the proof of Corollary 11.13 can be extended to compute a variant of the substring kernel in which $\mathbf{S}_\infty[W] = p_S(W)$ and $\mathbf{T}_\infty[W] = p_T(W)$ for every $W \in [1..\sigma]^k$: see Exercise 11.12. The telescoping technique described in this section can also be used to compute kernels and complexity measures that are restricted to substrings whose length belongs to a user-specified range.

Insight 11.3 Choosing k from the data

Given a string $S \in [1..\sigma]^n$, let $g : [1..n - 1] \mapsto [1..\lfloor n/2 \rfloor]$ be a map from the length ℓ of a substring of S to the number of distinct ℓ-mers that occur *at least twice* in S. In many genome-scale applications, k is set to be at least $k_1 = \text{argmax}\{g(\ell) \mid \ell \in [1..n - 1]\}$. This value, which is close to $\log_\sigma n$ for strings generated by independent, identically distributed random sources, can be derived exactly from S, by adapting the algorithm described in Lemma 11.10: see Exercise 11.15.

To determine an *upper bound* on k from the data, recall from Section 11.2.2 that, if S was generated by a Markov process of order $\ell - 2$ or smaller, the expected relative frequency of an ℓ-mer W in S can be approximated by Equation (11.3):

$$\tilde{p}_S(W) = \frac{p_S(W[1..\ell - 1]) \cdot p_S(W[2..\ell])}{p_S(W[2..\ell - 1])}.$$

It makes sense to disregard values of ℓ for which the probability distributions induced by p_S (the observed relative frequency) and by $\tilde{p}_S$ (the expected relative frequency) over all ℓ-mers are very similar. More formally, let $\mathbf{S}_\ell[1..\sigma^\ell]$ be the composition vector of S indexed by all possible ℓ-mers on alphabet $\Sigma = [1..\sigma]$, such that $\mathbf{S}_\ell[W] = p_S(W)$ for all $W \in [1..\sigma]^\ell$. Let $\tilde{\mathbf{S}}_\ell[1..\sigma^\ell]$ be the normalized composition vector of the estimates $\tilde{p}_S$, or equivalently $\tilde{\mathbf{S}}_\ell[W] = \tilde{p}_S(W)/\sum_{W \in [1..\sigma]^\ell} \tilde{p}_S(W)$ for all $W \in [1..\sigma]^\ell$. Recall from Section 6.4 that

$$\mathsf{KL}(\mathbf{S}_\ell, \tilde{\mathbf{S}}_\ell) = \sum_{W \in [1..\sigma]^\ell} \mathbf{S}_\ell[W] \cdot \Big(\log(\mathbf{S}_\ell[W]) - \log(\tilde{\mathbf{S}}_\ell[W])\Big)$$

is the Kullback–Leibler divergence of probability distribution $\tilde{\mathbf{S}}_\ell$ from probability distribution $\mathbf{S}_\ell$. Thus, let $Q(\ell) = \sum_{i=\ell}^{\infty} \mathsf{KL}(\mathbf{S}_\ell, \tilde{\mathbf{S}}_\ell)$. If $Q(\ell) \approx 0$ for some ℓ, then vector $\mathbf{S}_k$ can be derived almost exactly from vectors $\mathbf{S}_{k-1}$ and $\mathbf{S}_{k-2}$ for all $k \geq \ell$. Therefore, we can disregard values of k larger than $k_2 = \min\{\ell \mid Q(\ell) \approx 0\}$.

Given a collection of strings $\mathcal{S} = \{S^1, S^2, \ldots, S^m\}$, k should belong to the intersection of all intervals $[k_1^i..k_2^i]$ induced by strings $S^i \in \mathcal{S}$. However, note that both k_1^i and k_2^i depend on the length of S^i, and the intersection of all intervals might be empty. Without loss of generality, assume that S^1 is the shortest string in $\mathcal{S}$. A practical solution to this problem consists in partitioning every string in $\mathcal{S} \setminus \{S^1\}$ into blocks of size $|S^1|$, and in deciding the value of k by intersecting the intervals induced by all such blocks. Let $d : \mathcal{S} \times \mathcal{S} \mapsto \mathbb{R}$ be any k-mer distance between two strings in $\mathcal{S}$, and consider strings $S^i = S_1^i \cdot S_2^i \cdots \cdots S_p^i$ and $S^j = S_1^j \cdot S_2^j \cdots \cdots S_q^j$, where S_x^i and S_y^j are blocks of size $|S^1|$. We can replace $d(S^i, S^j)$ in practice by

$$d'(S^i, S^j) = \frac{1}{2} \left(\frac{1}{p} \sum_{a=1}^{p} \min\left\{ d(S_a^i, S_b^j) \mid b \in [1..q] \right\} \right.$$
$$\left. + \frac{1}{q} \sum_{b=1}^{q} \min\left\{ d(S_b^j, S_a^i) \mid a \in [1..p] \right\} \right).$$

*11.2.2 Substring kernels with Markovian correction

In some applications it is desirable to assign to component $W \in [1..\sigma]^+$ of composition vector $\mathbf{S}$ a score that measures the *statistical significance* of observing $f_S(W)$ occurrences of substring W in string $S \in [1..\sigma]^n$, rather than just the frequency $f_S(W)$ or the relative frequency $p_S(W)$. A string whose frequency departs from its expected value is more likely to carry biological signals, and kernels that take statistical significance into account are indeed more accurate when applied to genomes of evolutionarily distant species. In this section we focus on a score that is particularly effective in measuring the similarity of genomes.

We start by assuming that the input strings are generated by a Markov random process (see Chapter 7), which is a realistic assumption for genomes.

LEMMA 11.14 *Consider a Markov random process of order $k - 2$ or smaller that generates strings on alphabet $\Sigma = [1..\sigma]$ according to a probability distribution $\mathbb{P}$. Let $W \in [1..\sigma]^k$. The probability of observing W in a string generated by the random process is*

$$\mathbb{P}(W) = \frac{\mathbb{P}(W[1..k-1]) \cdot \mathbb{P}(W[2..k])}{\mathbb{P}(W[2..k-1])}. \tag{11.2}$$

Proof The probability of observing W in a string generated by the random process is

$$\mathbb{P}(W) = \mathbb{P}(W[1..k-1]) \cdot \mathbb{P}(W[k] \mid W[1..k-1])$$
$$= \mathbb{P}(W[1..k-1]) \cdot \mathbb{P}(W[k] \mid W[2..k-1]),$$

where the second equality comes from the fact that the process is Markovian with order $k - 2$ or smaller. The claim follows from combining this equation with

$$\mathbb{P}(W[2..k]) = \mathbb{P}(W[k] \mid W[2..k-1]) \cdot \mathbb{P}(W[2..k-1]). \qquad \square$$

We can estimate $\mathbb{P}(W)$ with the empirical probability $p_S(W)$, which might be zero for some W, obtaining the following approximation of Equation (11.2):

$$\tilde{p}_S(W) = \begin{cases} p_S(W[1..k-1]) \cdot p_S(W[2 \ldots k]) \big/ p_S(W[2 \ldots k-1]) & \text{if } p_S(W[2 \ldots k-1]) \neq 0, \\ 0 & \text{otherwise.} \end{cases} \tag{11.3}$$

Other estimators for the expected probability of a k-mer exist, with varying practical effectiveness for classifying genomes. We focus here on Equation (11.3), both because of its natural formulation in terms of Markov processes, and because it maximizes a notion of entropy detailed in Insight 11.4. We measure the significance of the event that substring W has empirical probability $p_S(W)$ in string S by the score

$$z_S(W) = \begin{cases} (p_S(W) - \tilde{p}_S(W))/\tilde{p}_S(W) & \text{if } \tilde{p}_S(W) \neq 0, \\ 0 & \text{if } \tilde{p}_S(W) = 0. \end{cases}$$

Note that, even though W does not occur in S, $z_S(W)$ can be -1 if both $f_S(W[1..k-1]) > 0$ and $f_S(W[2..k]) > 0$. Recall from Section 11.1.3 that such a W is a *minimal absent word* of S, and that there is an algorithm to compute all the minimal absent words of S in $O(n\sigma)$ time and $O(n(1 + o(1)) + \sigma \log^2 n)$ bits of working space, using the bidirectional BWT index of S (see Theorem 11.7).

After elementary manipulations, the formula for $z_S(W)$ becomes

$$z_S(W) = g(n, k) \cdot \frac{h_S(W[1..k])}{h_S(W[2..k])} - 1,$$

$$h_S(X) = \frac{f_S(X)}{f_S(X[1..|X| - 1])},$$

$$g(n, k) = \frac{(n - k + 2)^2}{(n - k + 1)(n - k + 3)}.$$

Let $\mathbf{S}_z$ be an infinite vector indexed by *all strings* in $[1..\sigma]^+$, and such that $\mathbf{S}_z[W] = z_S(W)$. As before, we want to compute Equation (11.1) on the composition vectors $\mathbf{S}_z$

and $\mathbf{T}_z$ of two input strings $S \in [1..\sigma]^n$ and $T \in [1..\sigma]^m$. To do so, we iterate over all substrings of S and T that contribute to Equation (11.1), including minimal absent words. For clarity, we describe our algorithms on the generalized suffix tree ST_{ST} of S and T, leaving to Exercise 11.18 the task of implementing them on the space-efficient enumeration of the internal nodes of ST_{ST} described in Algorithm 11.3.

Assume that ST_{ST} is augmented with explicit and implicit Weiner links, and that every destination of an implicit Weiner link is assigned a corresponding unary node in ST. We denote by ST^* this augmented generalized suffix tree. Recall from Observation 8.14 that the size of ST^* is still $O(n + m)$. Let $\mathtt{parent}(v)$ denote the (possibly unary) parent of a node v in ST^*, let $\mathtt{child}(v, c)$ denote the (possibly unary) child of node v reachable from an edge whose label starts by character $c \in [1..\sigma]$, and let $\mathtt{suffixLink}(v)$ denote the node that is reachable by taking the suffix link from v, or equivalently by following in the opposite direction the Weiner link that ends at node v. We organize the computation of Equation (11.1) in three phases, corresponding to the contributions of three different types of string.

In the first phase, we consider all strings that label the internal nodes and the leaves of ST^*.

Insight 11.4 Maximum-entropy estimators of k-mer probability

Let W be a $(k - 1)$-mer on alphabet $[1..\sigma]$. To simplify the notation, let $W = xVy$, where x and y are characters in $[1..\sigma]$ and $V \in [1..\sigma]^*$. Consider the estimator of the probability of W in Equation (11.3). Note that

$$\sum_{a \in [1..\sigma]} \tilde{p}_S(Wa) = \frac{p_S(W)}{p_S(Vy)} \cdot \sum_{a \in [1..\sigma]} p_S(Vya),$$

$$\sum_{a \in [1..\sigma]} \tilde{p}_S(aW) = \frac{p_S(W)}{p_S(xV)} \cdot \sum_{a \in [1..\sigma]} p_S(axV),$$

where the right-hand sides of these equations are constants that can be computed from counts of substrings in S. By applying these equations to every $(k - 1)$-mer W we get a system of $2\sigma^{k-1}$ constraints on function $\tilde{p}(S)$ evaluated at σ^k points, and the system is under-determined if $\sigma > 2$. Therefore, Equation (11.3) can be interpreted as *just one of the many possible functions that satisfy all such constraints*. Inside this set of feasible solutions, it is common practice to choose the one that maximizes the entropy $-\sum_{U \in [1..\sigma]^k} \tilde{p}_S(U) \log \tilde{p}_S(U)$, or equivalently the one that makes the weakest possible assumptions on the function beyond the constraints themselves. It is easy to see that this optimization problem, with the additional constraint that $\tilde{p}_S(U) \geq 0$ for all $U \in [1..\sigma]^k$, can be decoupled into σ^{k-2} subproblems that can be solved independently. Surprisingly, it turns out that Equation (11.3) already maximizes the entropy. However, this is not true for any function. For example, consider another popular estimator,

$$\tilde{p}_S(W) = \frac{p_S(x)p_S(Vy) + p_S(xV)p_S(y)}{2},$$

derived under the assumption that $k \geq 2$ and that x, V, and y occur independently in S. The maximum-entropy version of this function is

$$\tilde{p}_S(W) = \frac{(p_S(xV) + p_S(x)\alpha) \cdot (p_S(Vy) + p_S(y)\beta)}{2(\alpha + \beta)}$$

where $\alpha = \sum_{a \in [1..\sigma]} p_S(Va)$ and $\beta = \sum_{a \in [1..\sigma]} p_S(aV)$.

LEMMA 11.15   *Given* ST^*, *the values of* $z_S(W)$ *and of* $z_T(W)$ *for all strings* W *such that* $W = \ell(v)$ *for some node* v *of* ST^* *can be computed in overall* $O(n+m)$ *time.*

Proof   We store in every node v of ST^* two integers, $\mathtt{fs} = f_S(\ell(v))$ and $\mathtt{ft} = f_T(\ell(v))$, where $\ell(v)$ is the label of the path from the root to v in ST^*. This can be done in $O(n+m)$ time by traversing ST^* bottom-up. Note that $h_S(\ell(v)) = 1$ whenever the label of the arc $(\mathtt{parent}(v), v)$ has length greater than one, and $h_S(\ell(v)) = f_S(\ell(v))/f_S(\ell(\mathtt{parent}(v)))$ otherwise. We can thus store in every node v the two integers $\mathtt{hs} = h_S(\ell(v))$ and $\mathtt{ht} = h_T(\ell(v))$, using an $O(n+m)$-time top-down traversal of ST^*. At this point, we can compute the contribution of all internal nodes of ST^* in overall $O(n+m)$ time, by accessing $v.\mathtt{hs}$, $v.\mathtt{ht}$, $\mathtt{suffixLink}(v).\mathtt{hs}$ and $\mathtt{suffixLink}(v).\mathtt{ht}$ for every v, and by computing $g(|S|, |\ell(v)|)$ and $g(|T|, |\ell(v)|)$ in constant time.   □

In the second phase, we consider all substrings W that occur in S, in T, or in both, and whose label corresponds to a path in ST^* that *ends in the middle of an arc*. It is particularly easy to process strings that can be obtained by extending to the right by one character the label of an internal node of ST^*.

LEMMA 11.16   *Given* ST^*, *we can compute in* $O(n+m)$ *time the values of* $z_S(W)$ *and of* $z_T(W)$ *for all strings* W *such that*

- W *occurs in* S *or in* T;
- $W = \ell(u) \cdot c$ *for some* $c \in [1..\sigma]$ *and for some internal node* u *of* ST^*;
- $\ell(v) \neq W$ *for all nodes* v *of* ST^*.

Proof   Assume that $W = \ell(u) \cdot c$ ends in the middle of edge (u, v), and that the nodes of ST^* store the values described in Lemma 11.15. Then, $h_S(W) = v.\mathtt{fs}/u.\mathtt{fs}$ and $h_S(W[2..k]) = \mathtt{child}(\mathtt{suffixLink}(u), c).\mathtt{fs}/\mathtt{suffixLink}(u).\mathtt{fs}$. The same holds for computing $z_T(W)$.   □

Note that all the minimal absent words of S that do occur in T, and symmetrically all the minimal absent words of T that do occur in S, are handled by Lemma 11.16. All other strings W that end in the middle of an edge (u, v) of ST^* have the following key property.

Score of nonmaximal strings. Consider an edge (u, v) of ST^* with $|\ell(u, v)| > 1$, and let W be a string of length $k > |\ell(u)| + 1$ that ends in the middle of (u, v). Then, $z_S(W)$ is either zero or $g(n, k) - 1$.

Proof   If W does not occur in S, then neither does its prefix $W[1..k-1]$, thus W is not a minimal absent word of S and $z_S(W) = 0$. If W does occur in S, then $h_S(W) = 1$. Moreover, since $(\texttt{suffixLink}(u), \texttt{suffixLink}(v))$ is an edge of ST^* and $W[2..k]$ lies inside such an edge, we are guaranteed that $h_S(W[2..k]) = 1$ as well. $\square$

This proof works because, in ST^*, the destinations of the implicit Weiner links of ST have been assigned corresponding explicit nodes: in the original suffix tree ST, $f_S(W[2..k])$ is not guaranteed to equal $f_S(W[2..k-1])$, because $\texttt{suffixLink}(u)$ might be connected to $\texttt{suffixLink}(v)$ by a *path*, rather than by a single edge: see Figure 11.4. The following result derives immediately from the score of nonmaximal strings.

LEMMA 11.17   *Given ST^*, we can compute in $O(n+m)$ time and in $O(n+m)$ space the contribution to Equation (11.1) of all strings W such that*

- *W occurs in S or in T;*
- *$W = \ell(u) \cdot X$ for some internal node u of ST^* and for some string X of length at least two;*
- *$\ell(v) \neq W$ for all nodes v of ST^*.*

Proof   Consider again the variables N, D_S, and D_T, whose values at the end of the computation will be

$$N = \sum_{W \in [1..\sigma]^+} \mathbf{S}_z[W]\mathbf{T}_z[W],$$

$$D_S = \sum_{W \in [1..\sigma]^+} \mathbf{S}_z[W]^2,$$

$$D_T = \sum_{W \in [1..\sigma]^+} \mathbf{T}_z[W]^2.$$

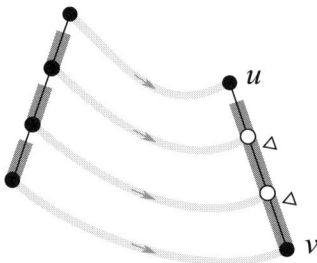

Figure 11.4 Illustrating the score of nonmaximal strings. Light gray arcs are Weiner links. Black dots are internal nodes in ST; white dots are destinations of implicit Weiner links. Both black and white dots are internal nodes in ST^*. Portions of edges of ST^* in which $h_S = 1$ are highlighted in dark gray. White triangles point to strings W for which $h_S(W)/h_S(W[2..|W|]) \neq 1$.

By the score of nonmaximal strings, edge (u, v) gives the following contributions to N, D_S, and D_T:

$$N(u, v) = \sum_{i=|\ell(u)|+2}^{|\ell(v)|} \big(g(|S|, i) - 1\big) \cdot \big(g(|T|, i) - 1\big),$$

$$D_S(u, v) = \sum_{i=|\ell(u)|+2}^{|\ell(v)|} \big(g(|S|, i) - 1\big)^2,$$

$$D_T(u, v) = \sum_{i=|\ell(u)|+2}^{|\ell(v)|} \big(g(|T|, i) - 1\big)^2.$$

Assume that we have constant-time access to arrays $\mathtt{cn}[1..n + m]$, $\mathtt{cs}[1..n]$, and $\mathtt{ct}[1..m]$ defined as follows:

$$\mathtt{cn}[i] = \sum_{j=1}^{i} \big(g(n, j) - 1\big) \cdot \big(g(m, j) - 1\big),$$

$$\mathtt{cs}[i] = \sum_{j=1}^{i} \big(g(n, j) - 1\big)^2,$$

$$\mathtt{ct}[i] = \sum_{j=1}^{i} \big(g(m, j) - 1\big)^2.$$

Such arrays take $O(n + m)$ space and can be built in $O(n + m)$ time overall. Then, we can compute the contribution of any edge (u, v) in constant time, as follows:

$$N(u, v) = \mathtt{cn}[|\ell(v)|] - \mathtt{cn}[|\ell(u)| + 1],$$
$$D_S(u, v) = \mathtt{cs}[|\ell(v)|] - \mathtt{cs}[|\ell(u)| + 1],$$
$$D_T(u, v) = \mathtt{ct}[|\ell(v)|] - \mathtt{ct}[|\ell(u)| + 1].$$

$\square$

In the third phase of our algorithm, we consider the contribution of all the minimal absent words of S and of T *which occur neither in S nor in T*, or in other words that do not have a locus in ST^*. These are all the strings aWb which occur neither in S nor in T, but such that aW and Wb occur in S, in T, or in both. Note that the contribution of every such string to N, D_1, and D_2 is either zero or one. Proving the following lemma amounts to applying the strategy used to solve Exercise 11.4.

LEMMA 11.18   *We can compute in $O((n+m)\sigma)$ time the values of $z_S(W)$ and of $z_T(W)$ for all the minimal absent words W of S that do not occur in T, and for all the minimal absent words of T that do not occur in S.*

By combining Lemmas 11.15–11.18, we obtain the key result of this section.

THEOREM 11.19   *Given two strings $S \in [1..\sigma]^n$ and $T \in [1..\sigma]^m$, there is an algorithm that computes Equation (11.1) on the composition vectors $\mathbf{S}_z$ and $\mathbf{T}_z$ in $O((n + m)\sigma)$ time and $O(n + m)$ space.*

11.2.3   Substring kernels and matching statistics

The *matching statistics* of string $S \in [1..\sigma]^n$ with respect to string $T \in [1..\sigma]^m$ is a vector $\mathsf{MS}_{S,T}[1..n]$ that stores at position i the length of the longest prefix of $S[i..n]$ that occurs at least once in T. Functions over $\mathsf{MS}_{S,T}$ can be used to measure the similarity between S and T without resorting to alignment and without specifying a fixed string length k, and $\mathsf{MS}_{S,T}$ allows one to estimate the cross-entropy of the random processes that generated S and T: see Insight 11.5. Computationally, knowing $\mathsf{MS}_{S,T}$ allows one to derive a number of more specific string kernels, by indexing just one of S and T and by carrying out a linear scan of the other string.

Computing matching statistics

We can compute $\mathsf{MS}_{S,T}$ by scanning S from left to right while simultaneously issuing child and suffix-link queries on ST_T (see Figure 11.5(a)). Recall that this was called *descending suffix walk* in Exercise 8.22.

THEOREM 11.20   *Let $S \in [1..\sigma]^n$ and let $T \in [1..\sigma]^m$. Given ST_T, there is an algorithm that computes $\mathsf{MS}_{S,T}$ in $O(n \log \sigma)$ time using just the operations* child *and* suffixLink *on ST_T.*

Insight 11.5 Matching statistics and cross-entropy

Consider a random process with finite memory that generates sequences of characters on alphabet $\Sigma = [1..\sigma]$ according to a probability distribution $\mathbb{P}$. Specifically, let $X = X_1 X_2 \cdots X_n$ be a sequence of random characters on the alphabet Σ, and for a given string $S = S_1 S_2 \cdots S_n \in [1..\sigma]^n$, let $\mathbb{P}(S)$ be the probability that $X_i = S_i$ for $i \in [1..n]$. The entropy of the probability distribution $\mathbb{P}$ is

$$H(\mathbb{P}) = \lim_{n \to \infty} \frac{-\sum_{S \in [1..\sigma]^n} \mathbb{P}(S) \log \mathbb{P}(S)}{n}.$$

Intuitively, it represents the average number of bits per character needed to identify a string generated by $\mathbb{P}$. Consider now a different random process that generates sequences of characters on the alphabet Σ with a probability distribution $\mathbb{Q}$. The *cross-entropy* of $\mathbb{P}$ with respect to $\mathbb{Q}$ is

$$H(\mathbb{P}, \mathbb{Q}) = \lim_{n \to \infty} \frac{-\sum_{S \in [1..\sigma]^n} \mathbb{P}(S) \log \mathbb{Q}(S)}{n}.$$

Intuitively, it measures the average number of bits per character that are needed to identify a string generated by $\mathbb{P}$, using a coding scheme based on $\mathbb{Q}$. Note that $H(\mathbb{P}, \mathbb{P}) = H(\mathbb{P})$. A related notion is the Kullback–Leibler divergence of two probability distributions (see Section 6.4):

$$KL(\mathbb{P}, \mathbb{Q}) = \lim_{n \to \infty} \frac{\sum_{S \in [1..\sigma]^n} \mathbb{P}(S) \cdot (\log \mathbb{P}(S) - \log \mathbb{Q}(S))}{n}.$$

Intuitively, it measures the expected number of *extra bits* per character required to identify a string generated by $\mathbb{P}$, using a coding scheme based on $\mathbb{Q}$. Note that $\mathsf{KL}(\mathbb{P}, \mathbb{Q}) = \mathsf{H}(\mathbb{P}, \mathbb{Q}) - \mathsf{H}(\mathbb{P})$. Since $\mathsf{KL}(\mathbb{P}, \mathbb{Q}) \neq \mathsf{KL}(\mathbb{Q}, \mathbb{P})$ in general, $d_{KL}(\mathbb{P}, \mathbb{Q}) = \mathsf{KL}(\mathbb{P}, \mathbb{Q}) + \mathsf{KL}(\mathbb{Q}, \mathbb{P})$ is typically preferred to quantify the dissimilarity between two probability distributions: indeed, $d_{KL}(\mathbb{P}, \mathbb{Q})$ is symmetric, non-negative, and it equals zero when $\mathbb{P} = \mathbb{Q}$.

Assume now that we have a finite sequence $Y = Y_1 \cdots Y_m$ of m observations of the random process with probability distribution $\mathbb{Q}$, and assume that we are receiving a sequence of observations $X = X_1 X_2 \cdots$ from the random process with probability distribution $\mathbb{P}$. Let ℓ be the length of the *longest prefix* of $X_1 X_2 \cdots$ which occurs at least once in Y, and let $\mathbb{E}_{\mathbb{P}, \mathbb{Q}}(\ell)$ be the expected value of ℓ under distributions $\mathbb{P}$ and $\mathbb{Q}$. It can be shown that

$$\lim_{m \to \infty} \left| \mathbb{E}_{\mathbb{P}, \mathbb{Q}}(\ell) - \frac{\log m}{\mathsf{H}(\mathbb{P}, \mathbb{Q})} \right| \in O(1).$$

Therefore, we can estimate $\mathsf{H}(\mathbb{P}, \mathbb{Q})$ by $\log m / \mathbb{E}_{\mathbb{P}, \mathbb{Q}}(\ell)$, and $\mathsf{H}(\mathbb{Q})$ by $\log m / \mathbb{E}_{\mathbb{P}}(\ell)$. If the sequence $X = X_1 X_2 \cdots X_m$ is finite but long enough, we can estimate ℓ by $\tilde{\ell} = \sum_{i=1}^{m} \mathsf{MS}_{X,Y}[i] / m$, where $\mathsf{MS}_{X,Y}$ is the vector of the matching statistics of X with respect to Y (clearly $\tilde{\ell} = (n+1)/2$ if X and Y coincide). These estimators for $\mathsf{H}(\mathbb{P}, \mathbb{Q})$, $\mathsf{KL}(\mathbb{P}, \mathbb{Q})$, and ultimately for $d_{KL}(\mathbb{P}, \mathbb{Q})$, are known as *average common substring* estimators. The estimator for $d_{KL}(\mathbb{P}, \mathbb{Q})$ has been used to build accurate phylogenetic trees from the genomes and proteomes of hundreds of evolutionarily distant species and thousands of viruses.

Proof Assume that we are at position i in S, and let $W = S[i..i + \mathsf{MS}_{S,T}[i] - 1]$. Note that W can end in the middle of an edge (u, v) of ST: let $W = aXY$, where $a \in [1..\sigma]$, $X \in [1..\sigma]^*$, $aX = \ell(u)$, and $Y \in [1..\sigma]^*$. Moreover, let $u' = \texttt{suffixLink}(u)$ and $v' = \texttt{suffixLink}(v)$. Note that suffix links can project arc (u, v) onto a *path* $u', v_1, v_2, \ldots, v_k, v'$. Since $\mathsf{MS}_{S,T}[i + 1] \geq \mathsf{MS}_{S,T}[i] - 1$, the first step to compute $\mathsf{MS}_{S,T}[i + 1]$ is to find the position of XY in ST: we call this phase of the algorithm the *repositioning phase*. To implement the repositioning phase, it suffices to take the suffix link from u, follow the outgoing arc from u' whose label starts with the first character of Y, and then iteratively jump to the next internal node of ST, choosing the next edge according to the corresponding character of Y. After the repositioning phase, we start matching the new characters of S on ST, or equivalently we read characters $S[i + \mathsf{MS}_{S,T}[i]], S[i + \mathsf{MS}_{S,T}[i] + 1], \ldots$ until the corresponding right-extension is impossible in ST. We call this phase of the algorithm the *matching phase*. Note that no character of S that has been read during the repositioning phase of $\mathsf{MS}_{S,T}[i + 1]$ will be read again during the repositioning phase of $\mathsf{MS}_{S,T}[i + k]$ with $k > 1$: it follows that every position j of S is read at most twice, once in the matching phase of some $\mathsf{MS}_{S,T}[i]$ with $i \leq j$, and once in the repositioning phase of some $\mathsf{MS}_{S,T}[k]$ with $i < k < j$. Since every mismatch can be charged to the position of which it concludes the matching statistics, the total number of mismatches encountered by the algorithm is bounded by the length of S. $\qquad\square$

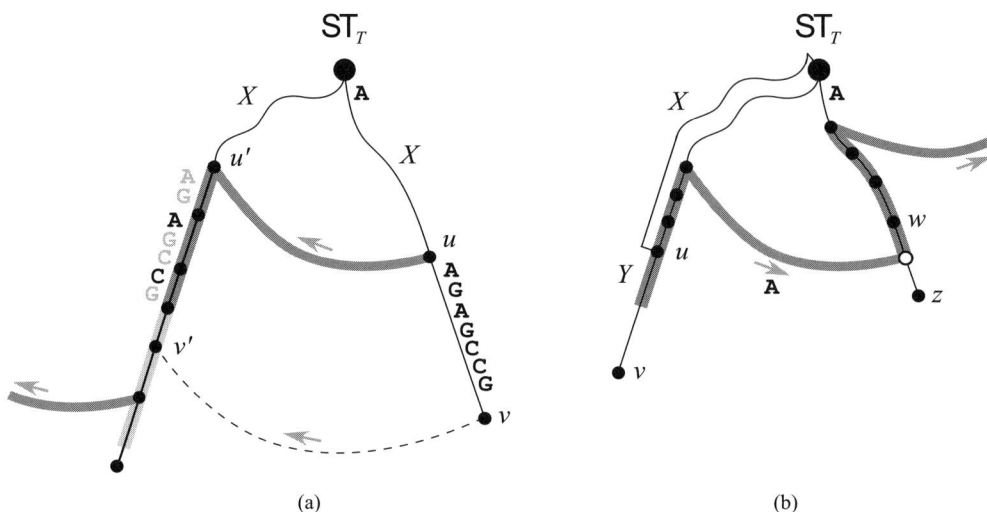

Figure 11.5 Two ways of computing $\mathsf{MS}_{S,T}$ on the suffix tree of T. (a) Using suffix links and child operations (Theorem 11.20). The picture shows substring $S[i..i + \mathsf{MS}_{S,T}[i] - 1] = \mathtt{A}XY$, where $Y = \mathtt{AGAGCCG}$. Dark gray lines mark the repositioning phase; light gray lines mark the matching phase. The characters of Y that are read during the repositioning phase are highlighted in black on edge (u', v'). Arrows indicate the direction of suffix links. (b) Using Weiner links and parent operations (Theorem 11.21). The picture shows substring $S[i..i + \mathsf{MS}_{S,T}[i] - 1] = XY$, where $S[i - 1] = \mathtt{A}$. Dark gray lines mark the traversal of Theorem 11.21. Arrows indicate the direction of Weiner links.

We can also compute $\mathsf{MS}_{S,T}$ symmetrically, by scanning S from right to left and by following parent links and Weiner links on ST_T (see Figure 11.5(b)).

THEOREM 11.21   *Let $S \in [1..\sigma]^n$ and let $T \in [1..\sigma]^m$. Given ST_T, there is an algorithm that computes $\mathsf{MS}_{S,T}$ in $O(n \log \sigma)$ time using just the operations* `parent` *and* `weinerLink` *on* ST_T.

Proof   We work with the extended version of ST_T described in Section 8.3, in which every destination of an implicit Weiner link is assigned to a corresponding artificial node. Every such node has exactly one child, and we further assume that it stores a pointer to its highest descendant with at least two children. We also assume that the operation $\mathtt{parent}(v)$ returns the lowest ancestor of node v with at least two children.

Assume that we are at position i in S, let u and v be defined as in Theorem 11.20, and let $S[i..i + \mathsf{MS}_{S,T}[i] - 1] = XY$, where $X = \ell(u)$ and $Y \in [1..\sigma]^*$. To compute $\mathsf{MS}_{S,T}[i - 1]$ we can proceed as follows. We read character $a = S[i - 1]$, and we check the existence of an explicit or implicit Weiner link $\mathtt{weinerLink}(v, a) = v'$ labeled by a from node v. If such a link exists, we set $\mathsf{MS}_{S,T}[i - 1] = \mathsf{MS}_{S,T}[i] + 1$. If the link is implicit, or equivalently if v' is a unary node that lies inside some edge (w, z) of the original suffix tree ST, then aXY lies inside edge (w, z) of ST as well, and we can find z in constant time by following the pointer stored in v'. If the link is explicit, or equivalently if v' has at least two children, then aXY lies inside the edge $(\mathtt{parent}(v'), v')$. If no Weiner link labeled by a exists from v, then aXY does not occur in T. Thus we iteratively reset v to $\mathtt{parent}(v)$ and we try a Weiner link $\mathtt{weinerLink}(v, a) = v'$ labeled by a from v: if such a link exists, we set $\mathsf{MS}_{S,T}[i-1] = |\ell(v)|+1$. Clearly every move from v to $\mathtt{parent}(v)$ can be charged to a distinct position in S, thus the algorithm performs at most n calls to $\mathtt{parent}$ in total. $\qquad\square$

As a byproduct, these approaches yield the nodes $v_i = \mathtt{locus}(W_i)$ and $u_i = \mathrm{argmax}\{|\ell(u')| : W_i = \ell(u') \cdot X, X \in [1..\sigma]^*\}$ for every $W_i = S[i..i + \mathsf{MS}_{S,T}[i] - 1]$ and $i \in [1..n]$. In what follows, we assume that pointers to such nodes of ST_T are stored in the arrays $\mathtt{floors}_{S,T}[1..n]$ and $\mathtt{ceils}_{S,T}[1..n]$, or equivalently that $\mathtt{floors}_{S,T}[i] = u_i$ and $\mathtt{ceils}_{S,T}[i] = v_i$ for every $i \in [1..n]$.

These algorithms on ST_T can also be adapted to compute the *shortest unique substring array* $\mathsf{SUS}_T[1..m]$, where $\mathsf{SUS}_T[i]$ contains the length of the shortest substring that starts at position i in T and that occurs exactly once in T, or equivalently the length of the shortest string that identifies position i uniquely among all positions in T: see Exercise 11.20. Shortest unique substrings have applications in a variety of sequence identification tasks, for example in the design of primer sequences for targeted resequencing (see Section 1.3), in the development of antibodies, and in comparative genomics.

Deriving substring kernels from matching statistics

Assume that $\alpha : [1..\sigma]^+ \mapsto \mathbb{R}$ is a function that assigns a weight to every string on alphabet $[1..\sigma]$. Given two strings $S \in [1..\sigma]^n$ and $T \in [1..\sigma]^m$, let $\beta : [1..n] \mapsto \mathbb{R}$ and $\gamma : [1..m] \mapsto \mathbb{R}$ be functions that assign a weight to every position of S and T, respectively. Assume that we want to compute the following generalization of the substring kernel between S and T described in Definition 11.9:

$$\kappa(S, T, \alpha, \beta, \gamma) = \sum_{W \in [1..\sigma]^+} \alpha(W) \cdot \left(\sum_{i \in \mathcal{L}(W,S)} \beta(i) \right) \cdot \left(\sum_{i \in \mathcal{L}(W,T)} \gamma(i) \right), \qquad (11.4)$$

where $\mathcal{L}(W, X)$ is the set of all starting positions of occurrences of string W in string X. Moreover, if W is a substring of T that ends inside edge (u, v) of ST_T, let

$$f(W, T) = \sum_{j=|\ell(u)|+1}^{|W|} \alpha(W[1..j]).$$

Computing Equation (11.4) is particularly easy using matching statistics.

THEOREM 11.22 *Assume that we are given* ST_T, *and that we can compute* $f(W, T)$ *in constant time for every substring* W *of* T. *Then, Equation (11.4) can be computed in* $O(n \log \sigma)$ *time.*

Proof Assume that a substring W of T occurs at position i in S. Observing W at position i in S contributes quantity $\beta(i) \cdot \texttt{contrib}(W)$ to $\kappa(S,T)$, where

$$
\texttt{contrib}(W) = \sum_{j=1}^{|W|} \alpha\big(W[1..j]\big) \cdot \left(\sum_{k \in \mathcal{L}(W[1..j],T)} \gamma(k) \right)
$$

$$
= \texttt{contrib}\big(\ell(u)\big) + \left(\sum_{k \in \mathcal{L}(\ell(v),T)} \gamma(k) \right) \cdot f(W,T). \tag{11.5}
$$

Therefore we can compute $\kappa(S,T)$ as follows:

$$
\kappa(S,T,\alpha,\beta,\gamma) = \sum_{i=1}^{n} \beta(i) \cdot \texttt{contrib}\big(S[i..i + \mathsf{MS}_{S,T}[i] - 1]\big). \tag{11.6}
$$

Recall that $\sum_{k \in \mathcal{L}(\ell(v),T)} \gamma(k)$ can be precomputed for every node v of ST_T in overall $O(m)$ time by a depth-first traversal of ST_T. Since we can compute $f(\ell(u),T)$ in constant time for every node u of ST, we can precompute $\texttt{contrib}\big(\ell(u)\big)$ and store it inside every node u of ST in overall $O(m)$ time, by a top-down traversal of ST that uses Equation (11.5). After this preprocessing, computing Equation (11.6) takes just $O(n \log \sigma)$ time using the algorithms described in Theorems 11.20 and 11.21. $\qquad \square$

If $\alpha(W)$ depends just on the length of W rather than on its characters, and if $\alpha(W) = 0$ for $|W| > \tau$ for some threshold τ, we can access $f(W,T)$ in constant time after a $O(\tau)$-time and $O(\tau \log \tau)$-space precomputation: see Exercise 11.21. Precomputing is not even necessary in the following length-based weighting schemes, since $f(W,T)$ has a closed form that can be evaluated in constant time (see Exercise 11.21):

- exponential decay, $\alpha(W) = \lambda^{-|W|}$ for some constant λ;
- constant weight, $\alpha(W) = \lambda|W|$;
- bounded range, $\alpha(W) = \lambda$ if $|W| \leq \tau$, and $\alpha(W) = 0$ otherwise;
- τ-spectrum, $\alpha(W) = \lambda$ if $|W| = \tau$, and $\alpha(W) = 0$ otherwise.

Theorem 11.22 can also be adapted to compute $\kappa(S,T,\alpha,\beta,\gamma)$ only on a user-specified subset of $[1..\sigma]^+$, with custom weights: see Exercise 11.22.

In most applications $\beta(i) = 1$ for every i, but varying β allows one to model some common kernels. For example, let $\{S^1, S^2, \dots, S^k\}$ be a fixed set of strings with weights $\{w_1, w_2, \dots, w_k\}$. Assume that we are given a query string T, and that we want to compute $\sum_{i=1}^{k} w_i \cdot \kappa(S^i, T, \alpha, \beta, \gamma)$, where $\beta(i) = 1$ for every i. This clearly coincides with $\kappa(S^1 \cdot S^2 \cdot \dots \cdot S^k, T, \alpha, \beta', \gamma)$, where $\beta'(i) = w_j$ for all positions i between $\sum_{h=1}^{j-1} |S^h| + 1$ and $\sum_{h=1}^{j} |S^h|$.

* Matching statistics and variable-memory Markov chains

Consider a Markov chain of order k (see Chapter 7). Given a string $W \in [1..\sigma]^h$ with $h < k$, it could happen in practice that $\mathbb{P}(a|UW) = \mathbb{P}(a|VW)$ for all strings $U \neq V$ in $[1..\sigma]^{k-h}$ and for all $a \in [1..\sigma]$. In other words, adding up to $k - h$ characters *before* W does not alter the probability of seeing any character *after* W. In such cases, the Markov chain is said to have *contexts of variable length*, and it can be represented in compact form by storing just a set of tuples $\mathcal{W} = \{(W, p_1, p_2, \ldots, p_\sigma)\}$, where W is a context and p_i is the probability of seeing character $i \in [1..\sigma]$ when W is the *longest suffix* of the generated string that equals the first component of a tuple in $\mathcal{W}$. We call such p_i *emission probabilities* of context W. In practice, set $\mathcal{W}$ is encoded as a trie $\mathsf{PST}_\mathcal{W}$, called a *probabilistic suffix trie*, in which contexts W are inserted *from right to left*, and in which every node of $\mathsf{PST}_\mathcal{W}$ that corresponds to a context W stores its emission probabilities.

Given a string $S \in [1..\sigma]^+$, the probability that S was generated by the variable-length Markov chain encoded by $\mathsf{PST}_\mathcal{W}$ is computed character by character, as follows: at every position j of S, the probability that $\mathsf{PST}_\mathcal{W}$ generated character $S[j]$ is determined by finding the *longest substring* $S[i..j-1]$ that labels a node v of $\mathsf{PST}_\mathcal{W}$ that corresponds to a context of $\mathcal{W}$, and by reading the emission probability of character $S[j]$ stored at v. Such a longest substring $S[i..j-1]$ is found by reading the characters $S[j-1], S[j-2], \ldots$ starting from the root of $\mathsf{PST}_\mathcal{W}$. The probability of S given $\mathsf{PST}_\mathcal{W}$ is the product of the probabilities of all of the characters of S. It can be shown that the probability distribution on $[1..\sigma]^+$ induced by a probabilistic suffix trie is equivalent to the probability distribution on $[1..\sigma]^+$ induced by the variable-length Markov chain of order k of which the trie is a compact representation. In what follows, we drop the subscript from $\mathsf{PST}_\mathcal{W}$ whenever $\mathcal{W}$ is implicit from the context.

Assume that we have a string or a concatenation of related strings T, for example genomes of similar species or proteins with the same function or structure. A PST for T can be interpreted as a *generative model* of T: given a new string S, measuring the probability that S was generated by PST is a way to measure the similarity between the random process that generated S and the random process that generated T, without resorting to alignment. This is conceptually analogous to Problem 7.4 in hidden Markov models.

In practice, it is natural to build PST just for the set of contexts defined as follows.

DEFINITION 11.23 *Let aW be a substring of a string $T \in [1..\sigma]^n$, where $a \in [1..\sigma]$ and $W \in [1..\sigma]^*$. We say that aW is a* context *if and only if all the following conditions hold.*

i.   *Given a user-specified threshold $\tau_1 > 0$,*

$$\frac{f_T(aW)}{|T| - |aW| + 1} \geq \tau_1.$$

ii.   *There is a character $b \in [1..\sigma]$ such that*
         (a)   *given a user-specified threshold $\tau_2 \leq 1$,*

$$\frac{f_T(aWb)}{f_T(aW)} \geq \tau_2;$$

(b)   given user-specified thresholds $\tau_3 > 1$ and $\tau_4 < 1$, either $\alpha \geq \tau_3$ or $\alpha \leq \tau_4$, where

$$\alpha = \frac{f_T(aWb)/f_T(aW)}{f_T(Wb)/f_T(W)}.$$

Surprisingly, the set of strings that satisfy condition ii is a subset of the nodes of the suffix tree of T.

LEMMA 11.24   *Given a string $T \in [1..\sigma]^n$ and the suffix tree ST_T, we can determine all context in Definition 11.23, and compute their emission probabilities, in overall $O(n)$ time.*

Proof   If string aW ends in the middle of an edge of ST_T, the left-hand side of Equation (11.7) equals one, therefore condition ii(a) is satisfied. However, if even W ends in the middle of an edge, that is, if aW is not the destination of an implicit Weiner link, then $\alpha = 1$ and condition ii(b) is not satisfied. It follows that, given the augmented version ST_T^* of ST_T in which all destinations of implicit Weiner links have been assigned a unary node, we can mark all nodes that correspond to contexts, and annotate them with emission probabilities, in overall $O(n)$ time.   □

The following equivalence between the PST of the contexts of T and the suffix-link tree of T is thus evident.

COROLLARY 11.25   *Let SLT_T^* be the suffix-link tree of T, extended with implicit Weiner links and their destinations. The PST of the set of contexts in Definition 11.23 is the subtree of SLT_T^* defined by the following procedure: starting from every node of ST_T^* that corresponds to a context, recursively follow suffix links up to the root of ST_T^*, marking all nodes met in the process.*

Corollary 11.25 enables a simple, linear-time algorithm for computing the probability of a query string S, based on matching statistics.

THEOREM 11.26   *Given the suffix tree $\mathsf{ST}_{\overline{T}}$ of the reverse of $T \in [1..\sigma]^n$, and the suffix-link tree SLT_T^* of T with nodes marked as in Lemma 11.24, we can compute the probability of a string $S \in [1..\sigma]^m$ with respect to the PST of T in $O(n + m \log \sigma)$ time.*

Proof   We add to $\mathsf{ST}_{\overline{T}}$ a node for every label W of a node in SLT_T^*, if string W does not already label a node in $\mathsf{ST}_{\overline{T}}$. Then, we add to every node in such an augmented suffix tree a pointer to its closest ancestor labeled by $\underline{W}$, where W is a context of T. Such an ancestor might be the node itself, or it might not exist. The size of the augmented suffix tree is still $O(n)$, and it can be built in $O(n)$ time: see Exercise 11.24. Then, we compute the matching statistics of $\underline{S}$ with respect to $\underline{T}$, obtaining for every position j in S a pointer to the locus $\mathtt{ceil}_{\underline{S},\underline{T}}[j]$ in $\mathsf{ST}_{\overline{T}}$ of the longest substring of T that ends at position j in S. From this node we can reach in constant time the node associated with the longest context of T that labels a suffix of $S[1..j]$, thus deriving the probability of character $S[j+1]$.   □

11.2.4   Mismatch kernels

Consider a string $S \in [1..\sigma]^+$, let $\Gamma(S,k)$ be the set of all distinct k-mers that occur exactly in S, and let $f_S(W,m)$ be the number of (possibly overlapping) occurrences of a k-mer W in S *with at most m mismatches*. Taking into account mismatches is of key importance for analyzing genomes, where sequential signals (like the transcription factor binding sites described in Section 1.1) are often inexact due to genetic variations. Given two strings S and T on alphabet $[1..\sigma]$, we want to compute the following generalization of the k-mer kernel described in Section 11.2.1:

$$\kappa(S,T,k,m) = \sum_{W \in [1..\sigma]^k} f_S(W,m) \cdot f_T(W,m). \tag{11.7}$$

Given two k-mers V and W on an alphabet $[1..\sigma]$, consider the Hamming distance $D_H(V,W)$ between V and W defined in Section 6.1, and let $B(W,m)$ be the ball of $[1..\sigma]^k$ with center W and radius m, or in other words the subset of $[1..\sigma]^k$ that lies at Hamming distance at most m from W. Note that the size of $B(V,m) \cap B(W,m)$ depends only on $D_H(V,W)$, so let $g(k,d,m)$ be the mapping from the Hamming distance d of two k-mers to the size of the intersection of their Hamming balls of radius m, where $g(k,d,m) = 0$ whenever $d > \min\{2m,k\}$. In this section we assume that we have a table $\texttt{kmersToDistance}_k[1..\sigma^{2k}]$ that contains $D_H(V,W)$ for every pair of k-mers V and W on alphabet $[1..\sigma]$, and a table $\texttt{distanceToSize}_{k,m}[1..\min\{2m,k\}]$ that contains $g(k,d,m)$ for every d that occurs in $\texttt{kmers2distance}_k$. Note that, for a specific choice of k and d, the value $g(k,d,m)$ has a closed-form representation that can be computed in $O(m)$ time: see Exercise 11.25.

To warm up, we consider an $O(|\Gamma(S,k)| \cdot |\Gamma(T,k)|)$-time algorithm to compute Equation (11.7).

LEMMA 11.27   *Given the sets $\Gamma(S,k)$ and $\Gamma(T,k)$ of two strings S and T on an alphabet $[1..\sigma]$, augmented with the frequency of their distinct k-mers, and given tables $\texttt{kmersToDistance}_k$ and $\texttt{distanceToSize}_{k,m}$, there is an algorithm that computes Equation (11.7) in $O(|\Gamma(S,k)| \cdot |\Gamma(T,k)|)$ time.*

Proof   After simple manipulations, Equation (11.7) becomes

$$\kappa(S,T,k,m) = \sum_{V \in \Gamma(S,k)} \sum_{W \in \Gamma(T,k)} f_S(V) \cdot f_T(W) \cdot |B(V,m) \cap B(W,m)|$$

$$= \sum_{V \in \Gamma(S,k)} \sum_{W \in \Gamma(T,k)} f_S(V) \cdot f_T(W) \cdot g(k,D_H(V,W),m), \tag{11.8}$$

where $f_S(W)$ is the number of *exact* occurrences of W in S (see the beginning of Section 11.2). We can thus compute $D_H(V,W)$ in constant time for every pair of k-mers V and W using the table $\texttt{kmersToDistance}_k$, and we can compute $g(k,D_H(V,W),m)$ in constant time using the table $\texttt{distanceToSize}_{k,m}$.   □

The time complexity of Lemma 11.27 can be improved by reorganizing the computation of Equation (11.8).

THEOREM 11.28 *Given the sets $\Gamma(S,k)$ and $\Gamma(T,k)$ of two strings S and T on alphabet $[1..\sigma]$, augmented with the frequency of their distinct k-mers, and given table* `distanceToSize`$_{k,m}$, *there is an algorithm that computes Equation (11.7) in $O(\sigma + |\Gamma(S,k)| + |\Gamma(T,k)|)$ time.*

Proof Recall that $g(k,d,m) = 0$ for $d > \min\{2m,k\}$. We can thus rewrite Equation (11.8) as

$$\kappa(S,T,k,m) = \sum_{d=0}^{\min\{2m,k\}} g(k,d,m) \cdot P_d,$$

$$P_d = \sum_{\substack{V \in \Gamma(S,k) \\ W \in \Gamma(T,k) \\ D_{\mathsf{H}}(V,W)=d}} f_S(V) \cdot f_T(W). \tag{11.9}$$

In other words, computing Equation (11.8) reduces to computing P_d for all $d \in [0..\min\{2m,k\}]$.

Assume that we remove a specific subset of d positions from all k-mers in $\Gamma(S,k)$ and $\Gamma(T,k)$. As a result, some distinct k-mers in $\Gamma(S,k)$ and in $\Gamma(T,k)$ can collapse into identical $(k-d)$-mers. We compute the inner product C_d between such lists of $|\Gamma(S,k)|$ and $|\Gamma(T,k)|$, potentially repeated $(k-d)$-mers, in overall $O(\sigma + (k-d) \cdot (|\Gamma(S,k)| + |\Gamma(T,k)|))$ time, by sorting the $(k-d)$-mers in each list into lexicographic order using Lemma 8.7, and by merging the sorted lists. Then, we repeat this process for all the possible selections of d positions out of k, adding all the $\binom{k}{d}$ possible inner products to C_d.

Note that, if $D_{\mathsf{H}}(V,W) = i < d$ for some k-mers V and W, $f_S(V)f_T(W)$ is added to C_d when processing $\binom{k-i}{d-i}$ subsets of d positions. It follows that $P_d = C_d - \sum_{i=0}^{d-1} \binom{k-i}{d-i}P_i$, therefore we can compute Equation (11.9) in overall $O(a_{k,m} \cdot \sigma + b_{k,m} \cdot (|\Gamma(S,k)| + |\Gamma(T,k)|) + c_{k,m})$ time, where

$$a_{k,m} = \sum_{d=0}^{\min\{2m,k\}} \binom{k}{d},$$

$$b_{k,m} = \sum_{d=0}^{\min\{2m,k\}} \binom{k}{d}(k-d),$$

$$c_{k,m} = \sum_{d=0}^{\min\{2m,k\}} \sum_{i=0}^{d-1} \binom{k-i}{d-i} + \min\{2m,k\}.$$

$\square$

Note that when $d = k$ we can avoid sorting and merging altogether, since $P_k = (|S|-k+1) \cdot (|T|-k+1) - \sum_{i=0}^{k-1} P_i$. The algorithm in Theorem 11.28 can be adapted to compute other popular inexact kernels for biosequence comparison, such as the *gapped kernel* and the *neighborhood kernel*: see Exercises 11.26 and 11.27.

11.2.5 Compression distance

Another way of quantifying the similarity of two strings S and T without resorting to alignment consists in using a *compressor*: intuitively, if S and T are very similar to each other, compressing them together should produce a smaller file than compressing each of them in isolation. This intuition can be formalized in the notion of *Kolmogorov complexity*, an algorithmic measure of information that defines the complexity of a string S as the length $K(S)$ of *the shortest program that outputs S* in a reference computer. The theory of Kolmogorov complexity is vast and theoretically deep: in what follows we avoid the abstraction of shortest-possible programs, and instead focus on real compressors. Measures of complexity and similarity based on real compressors have been used to detect low-complexity regions in genomes, and to build phylogenies of mitochondrial and viral genomes. Like string kernels, a compression-based measure of similarity allows one also to compare heterogeneous objects, like sequenced genomes and read sets.

Consider a real, *lossless* compressor C, that is, a compressor that allows one to reconstruct its input from its output *exactly*, like the Lempel–Ziv compressor described in Chapter 12. Let $C(S)$ be the number of bits in the output when the input of the compressor is string $S \in [1..\sigma]^n$, and assume that $C(S) \leq |S| + O(\log|S|)$. Given another string $T \in [1..\sigma]^m$, let $C(T|S) = C(ST) - C(S)$ be the number of bits that must be added to the compressed representation of S to encode the compressed representation of T. Similarly, let $C(T) - C(T|S)$ be the number of bits in the compressed representation of T that cannot be derived from the compressed representation of S.

In what follows we focus on *normal compressors*.

DEFINITION 11.29 *A compressor C is* normal *if it satisfies the following properties for all strings S, T, and U, up to an additive $O(\log n)$ term, where n is the size of the input in bits:*

- idempotency, $C(SS) = C(S)$;
- monotonicity, $C(ST) \geq C(S)$;
- symmetry, $C(ST) = C(TS)$;
- $C(S|T) \leq C(S|U) + C(U|T)$.

Symmetry might be violated by on-line compressors, like the Lempel–Ziv algorithm described in Section 12.1: while reading ST from left to right, for example, the Lempel–Ziv algorithm first adapts to the regularities of S; then, after switching to T, it adapts to the regularities of T only after a number of iterations. If S and T are sufficiently large, such violations of symmetry are negligible in practice. Compressors based on the Burrows–Wheeler transform tend to show even smaller departures from symmetry.

The last property of a normal compressor is intuitively desirable, and it immediately implies *distributivity* (see Exercise 11.28). In turn, distributivity implies *subadditivity*:

- *distributivity,* $C(ST) + C(U) \leq C(SU) + C(TU)$;
- *subadditivity,* $C(ST) \leq C(S) + C(T)$.

These properties are again valid up to an additive $O(\log n)$ term, where n is the size of the input in bits. Distributivity is satisfied by a number of real compressors, and it will be used repeatedly in what follows. Subadditivity is intuitively desirable, since a compressor should be able to use information present in S to better compress T. Again, minor imperfections to subadditivity might arise in practice, but they typically vanish when S and T become large enough.

We define the dissimilarity between two strings S and T using a normal compressor.

DEFINITION 11.30   *Let C be a normal compressor. The* normalized compression distance *between two strings S and T is*

$$\mathsf{NCD}(S, T) = \frac{C(ST) - \min\{C(S), C(T)\}}{\max\{C(S), C(T)\}}.$$

Note that $\mathsf{NCD}(S, T) \in [0..1 + \epsilon]$, where $\epsilon \in O(\log n)$ depends on specific features of the compressor. Note also that, if $C(S) < C(T)$, $\mathsf{NCD}(S, T) = \big(C(ST) - C(S)\big)/C(T)$ measures the improvement which accrues from compressing T together with S, with respect to compressing T in isolation. In the following theorem we show that NCD is a *metric*. Recall from Exercise 6.1 that a measure of dissimilarity D is a metric if it satisfies the following properties for all strings S, T, and U:

- *non-negativity*, $D(S, T) \geq 0$;
- *identity of indiscernibles*, $D(S, T) = 0$ if and only if $S = T$;
- *symmetry*, $D(S, T) = D(T, S)$;
- *triangle inequality*, $D(S, T) \leq D(S, U) + D(U, T)$.

THEOREM 11.31 NCD *is a metric up to an $O(\log n)$ additive term, where n is the number of bits in the input of C.*

Proof $\mathsf{NCD}(S, T) \geq 0$ by the monotonicity and symmetry of C. $\mathsf{NCD}(S, S) = 0$ by the idempotency of C. $\mathsf{NCD}(S, T) = \mathsf{NCD}(T, S)$ by the symmetry of C. We thus need to show only the triangle inequality $\mathsf{NCD}(S, T) \leq \mathsf{NCD}(S, U) + \mathsf{NCD}(U, T)$ for every set of strings $\{S, T, U\} \subset [1..\sigma]^+$.

Without loss of generality, assume that $C(S) \leq C(T) \leq C(U)$: since NCD is symmetric, we need to prove the triangle inequality just for $\mathsf{NCD}(S, T)$, $\mathsf{NCD}(S, U)$, and $\mathsf{NCD}(T, U)$. Here we prove just that $\mathsf{NCD}(S, T) \leq \mathsf{NCD}(S, U) + \mathsf{NCD}(U, T)$, leaving the other cases to Exercise 11.29. By distributivity, we have that

$$C(ST) + C(U) \leq C(SU) + C(TU)$$

and by symmetry

$$C(ST) + C(U) \leq C(SU) + C(UT).$$

Subtracting $C(S)$ from both sides of the inequality, we obtain

$$C(ST) + C(U) - C(S) \leq C(SU) + C(UT) - C(S),$$
$$C(ST) - C(S) \leq C(SU) - C(S) + C(UT) - C(U),$$

and on dividing both sides by $C(T)$ we have

$$\frac{C(ST) - C(S)}{C(T)} \leq \frac{C(SU) - C(S) + C(UT) - C(U)}{C(T)}.$$

By subadditivity, the left-hand side of this last inequality is at most one. If the right-hand side is at most one, then adding any positive number δ to both its numerator and its denominator can only increase the ratio. If the right-hand side is greater than one, then adding δ to the numerator and the denominator decreases the ratio, but the ratio still remains greater than one. It follows that adding a positive number δ to both the numerator and the denominator of the right-hand side keeps it greater than or equal to the left-hand side. We thus add the number δ that satisfies $C(U) = C(T) + \delta$, obtaining

$$\frac{C(ST) - C(S)}{C(T)} \leq \frac{C(SU) - C(S)}{C(U)} + \frac{C(UT) - C(T)}{C(U)}.$$

$\square$

11.3   Literature

The algorithms presented in this chapter for maximal repeats, maximal unique matches, and maximal exact matches are mostly from Belazzougui *et al.* (2013); the solution for MUMs on multiple strings was inspired by the non-compact solution in Section 8.4.2 and the cross-product computation for MEMs is analogous to the one in Baker *et al.* (1993). Alternative ways to solve some of these problems using a BWT index were given earlier in Beller *et al.* (2012). Another space-efficient solution to MUMs is sketched in Exercise 11.3, following ideas from Hon & Sadakane (2002) and Fischer *et al.* (2008). Maximal unique matches and the heuristic alignment method described in Insight 11.1 are from Delcher *et al.* (1999). Minimal absent words have been extensively studied: see for example Crochemore *et al.* (1998) for additional formal properties, and see Crochemore *et al.* (2000), Hampikian & Andersen (2007), Herold *et al.* (2008), Chairungsee & Crochemore (2012), Garcia & Pinho (2011), Garcia *et al.* (2011), and references therein for a small set of applications to data compression and to the analysis and comparison of genomes.

We just scratched the surface of the vast area of string kernels, without giving details about its large array of applications to molecular biology. The space-efficient computation of substring and k-mer kernels described in Section 11.2.1 was developed for this book: for more details, see Sobih *et al.* (2014). The suffix tree computation of the substring kernel with Markovian corrections is described in Apostolico & Denas (2008), and its estimate of k-mer probability was first introduced and tested in Qi *et al.* (2004). In addition to Equation (11.3), other formulas for computing $\tilde{p}_S(W)$ – the probability of observing k-mer W under a null model for string S – have been proposed: see Vinga & Almeida (2003), Song *et al.* (2014) and references therein for additional estimates and distance measures. The idea of converting an estimation formula into its maximum-entropy version described in Insight 11.4 is detailed in Chan *et al.* (2012).

An extensive description of the statistical properties of some alignment-free distance measures applied to strings and read sets can be found in Reinert *et al.* (2009), Wan *et al.* (2010), and Song *et al.* (2013), and a geometric treatment of a specific instance of alignment-free sequence comparison is described in Behnam *et al.* (2013). The data-driven approach for choosing k detailed in Insight 11.3 is taken from Sims *et al.* (2009). Statistical analyses related to the properties studied in Exercise 11.14 can be found in Reinert *et al.* (2009).

The connection between matching statistics and cross-entropy mentioned in Insight 11.5 is described in Farach *et al.* (1995) and references therein. The first genome classification algorithm based on this connection was introduced and tested in Ulitsky *et al.* (2006). The two symmetrical ways of computing matching statistics are described in Ohlebusch *et al.* (2010), and the method for deriving substring kernels from matching statistics that we presented in Section 11.2.3 is from Smola & Vishwanathan (2003). For clarity we presented the latter algorithm on suffix trees: an implementation on suffix arrays is described in Teo & Vishwanathan (2006). The connection between suffix-link trees, matching statistics, and variable-length Markov chains was established in Apostolico & Bejerano (2000). See Ziv (2008) and references therein for statistical properties of probabilistic suffix tries applied to compression. It is clear from Section 11.2 that substring and k-mer kernels have strong connections to maximal substrings: an overview of this connection appears in Apostolico (2010).

Kernels that take into account the frequency of occurrences with mismatches, gaps of bounded or unbounded length, rigid wildcards, and character substitutions according to a given probabilistic model, as well as other forms of gap or mismatch scores, are detailed in Leslie & Kuang (2003), Haussler (1999), Lodhi *et al.* (2002) and references therein. Corresponding statistical corrections have been defined, for example in Göke *et al.* (2012). The computation of such kernels follows a quadratic-time dynamic programming approach (Lodhi *et al.* 2002; Cancedda *et al.* 2003; Rousu *et al.* 2005), a traversal of ad-hoc tries (Leslie & Kuang 2003), or a combination of the two, with complexity that grows exponentially with the number of allowed gaps or mismatches. The algorithm to compute mismatch kernels described in Section 11.2.4 appears in Kuksa *et al.* (2008) and Kuksa & Pavlovic (2012), where the authors solve Exercise 11.25 and describe how to compute a closed-form expression for the size of the intersection of Hamming balls for every setting of m and $h(V, W)$. For an in-depth description of kernel methods in learning theory, see Shawe-Taylor & Cristianini (2004) and Cristianini & Shawe-Taylor (2000).

Finally, the notions of a normal compressor and of the normalized compression distance are detailed in Cilibrasi & Vitányi (2005). A more powerful *normalized information distance* based on Kolmogorov complexity is described in Li *et al.* (2004). For a detailed exposition of the foundations of Kolmogorov complexity and of its wide array of applications, see Li & Vitányi (2008).

Exercises

11.1   Give an algorithm to construct the indicator bitvector I of Algorithm 11.2 in $O((m + n) \log \sigma)$ time and $O((m + n) \log \sigma)$ bits of space.

11.2   Modify the algorithm for maximal unique matches on two strings to use two bidirectional indexes instead of the indicator bitvector as in our solution for maximal exact matches.

11.3   Recall the algorithm to compute maximal unique matches for multiple strings through document counting in Section 8.4.3. It is possible to implement this algorithm using $O(n \log \sigma)$ bits of space and $O(n \log \sigma \log n)$ time, but this requires some advanced data structures not covered in this book. Assume you have a data structure for solving the *dynamic partial sums* problem, that is, to insert, delete, and query elements of a list of non-negative integers, where the queries ask to sum values up to position i and to find the largest i such that the sum of values up to i is at most a given threshold x. There is a data structure for solving the updates and queries in $O(\log n)$ time using $O(n \log \sigma)$ bits of space, where n is the sum of the stored values. Assume also that you have a succinct representation of LCP values. There is a representation that uses $2n$ bits, such that the extraction of LCP[i] takes the same time as the extraction of SA[i]. Now, observe that the suffix array and the LCP array, with some auxiliary data structures, are sufficient to simulate the algorithm without any need for explicit suffix trees. Especially, differences between consecutive entries of string and node depths can be stored with the dynamic partial sums data structure. Fill in the details of this space-efficient algorithm.

11.4   Prove Theorem 11.7 assuming that in the output a minimal absent word aWb is encoded as a triplet $(i,j,|W|)$, where i (respectively j) is the starting position of an occurrence of aW in S (respectively Wb in S).

11.5   Prove that only a maximal repeat of a string $S \in [1..\sigma]^+$ can be the infix W of a minimal absent word aWb of S, where a and b are characters in $[1..\sigma]$.

11.6   Show that the number of minimal absent words in a string of length n over an alphabet $[1..\sigma]$ is $O(n\sigma)$. *Hint.* Use the result of the previous exercise.

11.7   A σ-ary de Bruijn sequence of order k is a circular sequence of length σ^k that contains all the possible k-mers over an alphabet $\Sigma = [1..\sigma]$. It can be constructed by spelling all the labels in an Eulerian cycle (a cycle that goes through all the edges) of a de Bruijn graph with parameters Σ and k. The sequence can be easily transformed into a string (non-circular sequence) of length $\sigma^k + k - 1$ that contains all the possible k-mers over alphabet Σ.

(a)   Describe the transformation from circular sequence to string.
(b)   Show that the number of minimal absent words in this string is σ^{k+1}.

11.8   Assume that you have a set of local alignments between two genomes A and B, with an alignment score associated with each alignment. Model the input as a bipartite graph where overlapping alignments in A and in B form a vertex, respectively, to two sides of the graph. Alignments form weighted edges. Which problem in Chapter 5 suits the purpose of finding anchors for rearrangement algorithms?

11.9   Recall that the formula $p_S(W) = f_F(W)/(|S| - |W| + 1)$ used in Section 11.2 to estimate the empirical probability of substring W of S assumes that W can occur at every position of T. Describe a more accurate expression for $p_S(W)$ that takes into account the *shortest period* of W.

11.10   The *Jaccard distance* between two sets $\mathcal{S}$ and $\mathcal{T}$ is defined as $J(\mathcal{S}, \mathcal{T}) = |\mathcal{S} \cap \mathcal{T}|/|\mathcal{S} \cup \mathcal{T}|$. Adapt the algorithms in Section 11.2.1 to compute $J(\mathcal{S}, \mathcal{T})$, both in the case where $\mathcal{S}$ and $\mathcal{T}$ are the sets of all distinct k-mers that occur in S and in T, respectively, and in the case in which $\mathcal{S}$ and $\mathcal{T}$ are the sets of all distinct *substrings*, of any length, that occur in S and in T, respectively.

11.11   Adapt Lemma 11.11 to compute a variant of the k-mer kernel in which $\mathbf{S}_k[W] = p_S(W)$ and $\mathbf{T}_k[W] = p_T(W)$ for every $W \in [1..\sigma]^k$.

11.12   Adapt Corollary 11.13 to compute a variant of the substring kernel in which $\mathbf{S}_\infty[W] = p_S(W)$ and $\mathbf{T}_\infty[W] = p_T(W)$ for every $W \in [1..\sigma]^k$.

11.13   Given a string S on alphabet $[1..\sigma]$ and a substring W of S, let $\mathtt{right}(W)$ be the set of characters that occur in S after W. More formally, $\mathtt{right}(W) = \{a \in [1..\sigma] \mid f_S(Wa) > 0\}$. The *$k$th-order empirical entropy* of S is defined as follows:

$$\mathsf{H}(S, k) = \frac{1}{|S|} \sum_{W \in [1..\sigma]^k} \sum_{a \in \mathtt{right}(W)} f_S(Wa) \log \left(\frac{f_S(W)}{f_S(Wa)} \right).$$

Intuitively, $\mathsf{H}(S, k)$ measures the amount of uncertainty in predicting the character that follows a context of length k, thus it is a lower bound for the size of a compressed version of S in which every character is assigned a codeword that depends on the preceding context of length k. Adapt the algorithms in Section 11.2.1 to compute $\mathsf{H}(S, k)$ for $k \in [k_1..k_2]$ using the bidirectional BWT index of S, and state the complexity of your solution.

11.14   Let S and T be two strings on alphabet $[1..\sigma]$, and let $\mathbf{S}$ and $\mathbf{T}$ be their composition vectors indexed by all possible k-mers. When the probability distribution of characters in $[1..\sigma]$ is highly nonuniform, the inner product between $\mathbf{S}$ and $\mathbf{T}$ (also known as the D_2 *statistic*) is known to be dominated by noise. To solve this problem, raw counts are typically corrected by *expected counts* that depend on $p(W) = \prod_{i=1}^{|W|} p(W[i])$, the probability of observing string W if characters at different positions in S are independent, identically distributed, and have empirical probability $p(a) = f_{ST}(a)/(|S| + |T|)$ for $a \in [1..\sigma]$. For example, letting $\tilde{\mathbf{S}}[W] = \mathbf{S}[W] - (|S| - k + 1)p(W)$, the following variants of D_2 have been proposed:

$$D_2^s = \sum_{W \in [1..\sigma]^k} \frac{\tilde{\mathbf{S}}[W]\tilde{\mathbf{T}}[W]}{\sqrt{\tilde{\mathbf{S}}[W]^2 + \tilde{\mathbf{T}}[W]^2}},$$

$$D_2^* = \sum_{W \in [1..\sigma]^k} \frac{\tilde{\mathbf{S}}[W]\tilde{\mathbf{T}}[W]}{\sqrt{(|S| - k + 1)(|T| - k + 1) \cdot p(W)}}.$$

Adapt the algorithms described in Section 11.2.1 to compute D_2^s, D_2^*, and similar variants based on $p(W)$, using the bidirectional BWT index of S and T, and state their complexity.

11.15   Show how to implement the computations described in Insight 11.3 using the bidirectional BWT index of $S \in [1..\sigma]^n$. More precisely, do the following.

(a)   Show how to compute the number of distinct k-mers that appear at least twice (repeating k-mers) in time $O(n \log \sigma)$ and space $O(n \log \sigma)$ bits.

(b)   Suppose that we want to compute the number of repeating k-mers for all values of k in a given range $[k_1..k_2]$. A naive extension of the algorithm in (a) would take time $O((k_1 - k_2)n \log \sigma)$ and it would use an additional $O((k_1 - k_2)\log n)$ bits of space to store the result. Show how to improve this running time to $O(n \log \sigma)$, using the same space.

(c)   Show how to compute the Kullback–Leibler divergence described in Insight 11.3 simultaneously for all values of k in $[k_1..k_2]$, in time $O(n \log \sigma)$ and an additional $O(k_1 - k_2)$ computer words of space to store the result.

11.16   Assume that vectors $\mathbf{S}$ and $\mathbf{T}$ have only two dimensions. Show that Equation (11.1) is indeed the cosine of the angle between $\mathbf{S}$ and $\mathbf{T}$.

11.17   Write a program that computes all kernels and complexity measures described in Section 11.2.1, given the suffix tree of the only input string, or the generalized suffix tree of the two input strings.

11.18   Adapt the algorithm in Section 11.2.2 to compute the substring kernel with Markovian corrections in $O(n \log \sigma)$ time and space, where n is the sum of the lengths of the input strings.

11.19   Given strings W and S, and characters a and b on an alphabet $\Sigma = [1..\sigma]$, consider the following expected probability of observing aWb in S, analogous to Equation (11.2):

$$\tilde{p}_{W,S}(a,b) = \frac{p_S(aW) \cdot p_S(Wb)}{p_S(W)}.$$

Moreover, consider the Kullback–Leibler divergence between the observed and expected distribution of $\tilde{p}_{W,S}$ over pairs of characters (a,b):

$$\mathsf{KL}_S(W) = \sum_{(a,b) \in \Sigma \times \Sigma} p_{W,S}(a,b) \cdot \ln\left(\frac{p_{W,S}(a,b)}{\tilde{p}_{W,S}(a,b)}\right),$$

where we set $\ln(p_{W,S}(a,b)/\tilde{p}_{W,S}(a,b)) = 0$ whenever $\tilde{p}_{W,S}(a,b) = 0$. Given strings S and T, let $\mathbf{S}$ and $\mathbf{T}$ be infinite vectors indexed by all strings on alphabet $[1..\sigma]$, such that $\mathbf{S}[W] = \mathsf{KL}_S(W)$ and $\mathbf{T}[W] = \mathsf{KL}_T(W)$. Describe an algorithm to compute the cosine of vectors $\mathbf{S}$ and $\mathbf{T}$ as defined in Equation (11.1).

11.20   Recall that the *shortest unique substring array* $\mathsf{SUS}_S[1..n]$ of a string $S \in [1..\sigma]^n$ is such that $\mathsf{SUS}_S[i]$ is the length of the shortest substring that starts at position i in S and that occurs exactly once in S.

i.   Adapt to SUS_S the left-to-right and right-to-left algorithms described in Section 11.2.3 for computing matching statistics.

ii.   Describe how to compute SUS_S by an $O(n)$ bottom-up navigation of the suffix tree of S.

11.21   Show how to compute $\sum_{j=|\ell(u)|+1}^{|W|} \alpha(W[1..j])$ in constant time for all the weighting schemes described in Section 11.2.3. More generally, show that, if $\alpha(W)$ depends just on the length of W rather than on its characters, and if $\alpha(W) = 0$ for $|W|$ bigger than some threshold τ, we can access $\sum_{j=|\ell(u)|+1}^{|W|} \alpha(W[1..j])$ in constant time after an $O(\tau)$-time and $O(\tau \log \tau)$-space precomputation.

11.22   Let S and T be strings, and let $D = \{d_1, d_2, \ldots, d_k\}$ be a fixed set of strings with weights $\{w_1, w_2, \ldots, w_k\}$. Show how to compute the kernel $\kappa(S, T)$ in Equation (11.4) restricted to the strings in D, using matching statistics and with the same time complexity as that of Theorem 11.22.

11.23   Given strings S and T and a window size $w \in [1..|S| - 1]$, describe an algorithm that computes the kernel $\kappa(S[i..i + w - 1], T)$ in Equation (11.4) for all $i \in [1..|S| - w + 1]$, performing fewer than m operations per position. *Hint.* Adapt the approach in Theorem 11.22.

11.24   Consider the suffix-link tree SLT_T of a string T, augmented with implicit Weiner links and their destinations. Describe an algorithm that adds to the suffix tree ST_T a node for every label W of a node in SLT_T (if string W is not already the label of a node in ST_T), and that adds to every node of this augmented suffix tree a pointer to its closest ancestor labeled by a string in SLT_T. The relationship between SLT_T and ST_T is further explored in Exercise 8.19.

11.25   Given two k-mers V and W on alphabet $[1..\sigma]$, recall that $D_\mathsf{H}(V, W)$ is the Hamming distance between V and W, and that $B(W, m)$ is the subset of $[1..\sigma]^k$ that lies at Hamming distance at most m from W. Give a closed-form expression for the size of $B(V, m) \cap B(W, m)$ for every possible value of $D_\mathsf{H}(V, W)$ when $m = 2$.

11.26   Given a string S on alphabet $[1..\sigma]$, and given a string W of length k, let $f_S(W, m)$ be the number of substrings of S of length $k + m$ that contain W as a *subsequence*. Consider the following *gapped kernel* between two strings S and T:

$$\kappa(S, T, k, m) = \sum_{W \in [1..\sigma]^k} f_S(W, m) \cdot f_T(W, m).$$

Adapt the approach described in Section 11.2.4 to compute the gapped kernel, and state its complexity.

11.27   In the *neighborhood kernel*, the similarity of two strings S and T is expressed in terms of the similarities of their *neighborhoods* $N(S)$ and $N(T)$, where $N(S)$ contains a given set of strings that the user believes to be similar or related to S. The neighborhood kernel is defined as

$$\kappa_N(S, T) = \sum_{U \in N(S)} \sum_{V \in N(T)} \kappa(U, V),$$

where $\kappa(U, V)$ is a given kernel. Describe how to adapt the approaches described in Section 11.2.4 to compute the neighborhood kernel.

11.28 Prove that $C(ST) + C(U) \leq C(SU) + C(TU)$, by using the symmetry of C.

11.29 Complete the proof of Theorem 11.31 with the two missing cases, using the distributivity of C.

12 Genome compression

A pragmatic problem arising in the analysis of biological sequences is that collections of genomes and especially read sets consisting of material from many species (see Chapter 16) occupy too much space. We shall now consider how to efficiently compress such collections. Observe that we have already considered in Section 10.7.1 how to support pattern searches on top of such compressed genome collections. Another motivation for genome compression comes from the compression distance measures in Section 11.2.5.

We shall explore the powerful *Lempel–Ziv* compression scheme, which is especially good for repetitive genome collections, but is also competitive as a general compressor: popular compressors use variants of this scheme.

In the *Lempel–Ziv parsing* problem we are given a text $T = t_1 t_2 \cdots t_n$ and our goal is to obtain a partitioning $T = T^1 T^2 \cdots T^p$ such that, for all $i \in [1..p]$, T^i is the longest prefix of $t_{|T^1 \ldots T^{i-1}|+1} \cdots t_n$ which has an occurrence starting somewhere in $T^1 \cdots T^{i-1}$. Such strings T^i are called **phrases** of T. Let us denote by L_i the starting position of one of those occurrences (an occurrence of T^i in T starting somewhere in $T^1 \cdots T^{i-1}$). In order to avoid the special case induced by the first occurrences of characters, we assume that the text is prefixed by a virtual string of length at most σ that contains all the characters that will appear in T. Then the string T can be encoded by storing the p pairs of integers $(|T^1|, L_1), (|T^2|, L_2), \ldots, (|T^p|, L_p)$. A naive encoding would use $2p \log n$ bits of space. Decoding the string T can easily be done in time $O(n)$, since, for every i, L_i points to a previously decoded position of the string, so recovering T^i is done just by copying the $|T^i|$ characters starting from that previous position.

In Section 12.2, we shall explore more sophisticated ways to encode the pairs of integers, allowing one to use possibly less space than the naive encoding. We will also explore variations of the Lempel–Ziv compression where the parse $T = T^1 T^2 \cdots T^p$ still has the property that every T^i has an occurrence in T starting in $T^1 \cdots T^{i-1}$, but T^i is not necessarily the longest prefix of $t_{|T^1 \ldots T^{i-1}|+1} \cdots t_n$ with this property.

The motivation for exploring these variations is that they have the potential to achieve better space bounds than achieved by the standard Lempel–Ziv parsing, if the parsing pairs are encoded in a non-naive encoding. Also the techniques to derive these variations include shortest-path computation on DAGs as well as usage of suffix trees, giving an illustrative example of advanced use of the fundamental primitives covered earlier in the book.

We start by showing algorithms for the standard Lempel–Ziv parsing, culminating in a fast and space-efficient algorithm. Those algorithms will make use of the

Knuth–Morris–Pratt (KMP) automaton, the suffix tree, and some advanced tools and primitives based on the BWT index.

To motivate the use of the standard Lempel–Ziv scheme in compressing genome collections, consider a simplistic setting with d length n genomes of individuals from the same species. Assume that these genomes contain s single-nucleotide substitutions with respect to the first one (the reference) in the collection. After concatenating the collection into one sequence of length dn and applying Lempel–Ziv compression, the resulting parse can be encoded in at most $2(n + 2s)\log(dn)$ bits, without exploiting any knowledge of the structure of the collection. With the real setting of the mutations containing insertions and deletions as well, the bound remains the same, when redefining s as the size of the new content with respect to the reference.

As mentioned above, this chapter applies heavily a large variety of techniques covered earlier in the book. Hence it offers a means to practice a fluent usage of all those techniques under one common theme.

12.1   Lempel–Ziv parsing

In this section, we focus on the following *Lempel–Ziv parsing* problem. We are given a text $T = t_1 t_2 \cdots t_n$. Our goal is to obtain a greedy partitioning $T = T^1 T^2 \cdots T^p$ such that for all $i \in [1..p]$, T^i is the longest prefix of $t_{|T^1...T^{i-1}|+1} \cdots t_n$ which has an occurrence in T that starts at a position in $[1..|T^1 \cdots T^{i-1}|]$. The following definition will be used throughout the presentation.

DEFINITION 12.1   *Given a string T and a position i, the* longest previous factor *at position a in T is the longest prefix of $t_a \cdots t_n$ that has an occurrence in T starting at a position in $[1..a-1]$ of T.*

Before studying algorithms to construct the parsing, we derive the following lemmas that characterize the power of Lempel–Ziv parsing.

LEMMA 12.2   *The number of phrases generated by the greedy Lempel–Ziv parsing cannot be more than $2n/\log_\sigma n + \sigma \sqrt{n}$.*

Proof   Let us fix a substring length $b = \lceil \log n / 2\log \sigma \rceil$. The lemma is clearly true whenever $\sigma \geq \sqrt{n}$, so let us assume that $\sigma < \sqrt{n}$. Obviously, there cannot be more than $\sigma^b < \sigma \sqrt{n}$ distinct substrings of length b over an alphabet of size σ. Each such substring could potentially occur once or more in T. Let us now consider the greedy parsing $T^1 \cdots T^p$. Consider Q^i, the prefix of $t_{|T^1...T^{i-1}|+1} \cdots t_n$ of length b. Then, either Q^i does have an occurrence that starts at some position in $[1..|T^1 \cdots T^{i-1}|]$ or it does not. If it does, then $|T^i| \geq b$. Otherwise $|T^i| < b$, and moreover we have that the first occurrence of Q^i in the whole string T is precisely at position $|T^1 \cdots T^{i-1}| + 1$.

Let us now count the number of phrases of length less than b. Since the number of distinct substrings of length b is upper bounded by $\sigma \sqrt{n}$, we deduce that the number of first occurrences of such substrings is also upper bounded by $\sigma \sqrt{n}$ and thus that the number of phrases of length less than b is also upper bounded by $\sigma \sqrt{n}$. It remains to bound the total number of phrases of length at least b. Since the total length of such phrases cannot exceed n and their individual length is at least b, we deduce that their number is at most $n/b \le 2n/\log_\sigma n$ which finishes the proof of the lemma. $\square$

LEMMA 12.3   *The number of phrases generated by the greedy Lempel–Ziv parsing cannot be more than $3n/\log_\sigma n$.*

Proof   We consider two cases, either $\sigma < \sqrt[3]{n}$ or $\sigma \ge \sqrt[3]{n}$. In the first case, by Lemma 12.2, we have that the number of phrases is at most $2n/\log_\sigma n + \sigma\sqrt{n} < 2n/\log_\sigma n + n^{5/6} < 3n/\log_\sigma n$. In the second case we have that $\log_\sigma n \le 3$ and, since the number of phrases cannot exceed the trivial bound n, we have that the number of phrases is at most $n = 3n/3 \le 3n/\log_\sigma n$. $\square$

With the fact that the two integers to represent each phrase take $O(\log n)$ bits, the lemmas above indicate that a naive encoding of the Lempel–Ziv parsing takes $O(n\log\sigma)$ bits in the worst case. However, the parsing can be much smaller on certain strings. Consider for example the string a^n, for which the size of the parsing is just 1.

We will next show our first algorithm to do Lempel–Ziv parsing. The algorithm is simple and uses only an augmented suffix tree. It achieves $O(n\log\sigma)$ time using $O(n\log n)$ bits of space. We then show a slower algorithm that uses just $O(n\log\sigma)$ bits of space. We finally present a more sophisticated algorithm that reuses the two previous ones and achieves $O(n\log\sigma)$ time and bits of space.

12.1.1   Basic algorithm for Lempel–Ziv parsing

Our basic algorithm for Lempel–Ziv parsing uses a suffix tree ST_T built on the text T. Every internal node v in the suffix tree maintains the minimum among the positions in the text of the suffixes under the node. We denote that number by N_v.

The algorithm proceeds as follows. Suppose that the partial parse $T^1 T^2 \cdots T^{i-1}$ has been constructed and that we want to determine the next phrase T^i which starts at position $a = |T^1 T^2 \cdots T^{i-1}| + 1$ of T.

We match the string $t_a t_{a+1} \cdots t_n$ against the path spelled by a top-down traversal of the suffix tree, finding the deepest internal node v in the suffix tree such that its path is a prefix of $t_a \cdots t_n$ and $N_v < a$. Then, the length of that path is the length of the longest previous factor of position a in T, which is precisely the length of T^i. Moreover, the pointer L_i can be set to N_v, since the latter points to an occurrence of T^i starting at a position in $T[1..a-1]$. We then restart the algorithm from position $a + |T^i|$ in T in order to find the length of T^{i+1} and so on.

Analyzing the time, we can see that traversing the suffix tree top-down takes time $O(|T^i|\log\sigma)$ for every phrase i, and then checking the value of N_v takes constant time. So the time spent for each position of the text is $O(\log\sigma)$ and the total time is $O(n\log\sigma)$. We leave to Exercise 12.3 the derivation of a linear-time algorithm that computes and stores the value N_v for every node v of the suffix tree.

The total space used by the algorithm is $O(n\log n)$ bits, which is needed in order to encode the suffix tree augmented with the values N_v for every node v. We thus have proved the following lemma.

LEMMA 12.4 *The Lempel–Ziv parse of a text* $T = t_1 t_2 \cdots t_n$, $t_i \in [1..\sigma]$, *can be computed in* $O(n \log \sigma)$ *time and* $O(n \log n)$ *bits of space. Moreover the pairs* $(|T^1|, L_1), (|T^2|, L_2), \ldots, (|T^p|, L_p)$ *are computed within the same time and space bound.*

12.1.2 Space-efficient Lempel–Ziv parsing

We now give a Lempel–Ziv parsing algorithm that uses space $O(n \log \sigma)$ bits, but requires $O(n \log n \log^2 \sigma)$ time; the time will be improved later. We first present the data structures used.

- A succinct suffix array (see Section 9.2.3) based on the Burrows–Wheeler transform $\mathsf{BWT}_{T\#}$ of T. We use a sampling rate $r = \log n$ for the sampled suffix array used in the succinct suffix array.
- An array $R[1..n/\log n]$ such that $R[i]$ stores the minimum of $\mathsf{SA}_{T\#}[(i-1)\log n + 1..i \log n]$.
- A range minimum query data structure (RMQ) on top of R (see Section 3.1), which returns $\min_{i' \in [l..r]} R[i']$ for a given range $[l..r]$.

Decoding an arbitrary position in the suffix array can be done in time $O(r \log \sigma) = O(\log n \log \sigma)$.

By analyzing the space, we can easily see that the succinct suffix array uses space $n \log \sigma (1 + o(1)) + O(n)$ bits and that both R and RMQ use $O(n)$ bits of space. Note that RMQ answers range minimum queries in $O(\log n)$ time. Note also that the succinct suffix array and the arrays R and RMQ can easily be constructed in $O(n \log \sigma)$ time and bits of space, from the BWT index, whose construction we studied in Theorem 9.11.

We are now ready to present the algorithm. Throughout the algorithm we will use Lemma 9.4 to induce the successive values $\mathsf{SA}_{T\#}^{-1}[1], \mathsf{SA}_{T\#}^{-1}[2], \ldots, \mathsf{SA}_{T\#}^{-1}[n]$, spending $O(\log \sigma)$ time per value. We thus assume that the value $\mathsf{SA}_{T\#}^{-1}[j]$ is available when we process the character t_j.

Suppose that we have already done the parsing of the first $i - 1$ phrases and that we want to find the length of the phrase T^i. Denote by a the first position of T^i in T, that is, $a = |T^1 T^2 \cdots T^{i-1}| + 1$. Using Lemma 9.4 we find the position j such that $\mathsf{SA}_{T\#}[j] = a$, or equivalently $\mathsf{SA}_{T\#}^{-1}[a] = j$.

We then find the right-most position ℓ such that $\ell < j$ and $\mathsf{SA}_{T\#}[\ell] < a$ and the left-most position r such that $r > j$ and $\mathsf{SA}_{T\#}[\ell] < a$. Assume, for now, that we know ℓ and r. Then we can use the following lemma, whose easy proof is left for Exercise 12.5, to conclude the algorithm.

LEMMA 12.5 *The length of the longest previous factor (recall Definition 12.1) of position* a *in* T *is the largest of* $\mathtt{lcp}(T[\ell..n], T[a..n])$ *and* $\mathtt{lcp}(T[r..n], T[a..n])$, *where* $\mathtt{lcp}(A, B)$ *denotes the length of the longest common prefix of two strings A and B, ℓ*

is the largest position such that $\ell < a$ *and* $\mathsf{SA}_{T\#}[\ell] < \mathsf{SA}_{T\#}[a]$, *and* r *is the smallest position such that* $r > a$ *and* $\mathsf{SA}_{T\#}[r] < \mathsf{SA}_{T\#}[a]$.

It remains to show how to find ℓ and r in time $O(\log^2 n \log \sigma)$. We let $b_a = \lceil a/\log n \rceil$. We use a binary search over the interval $R[1..b_a - 1]$ to find the right-most position b_ℓ in R such that $b_\ell < b_a$ and $R[b_\ell] < a$. This binary search takes time $O(\log^2 n)$. It is done using $O(\log n)$ queries on the RMQ. We then similarly do a binary search over $R[b_a + 1..n/\log n]$ to find the left-most position b_r in R such that $b_r > b_a$ and $R[b_r] < a$ in time $O(\log^2 n)$.

In the next step, we extract the regions $\mathsf{SA}_{T\#}[(b_a - 1)\log n + 1..b_a \log n]$, $\mathsf{SA}_{T\#}[(b_\ell - 1)\log n + 1..b_\ell \log n]$, and $\mathsf{SA}_{T\#}[(b_r - 1)\log n + 1..b_r \log n]$ using the succinct suffix array. The extraction of each suffix array position takes time $O(\log n \log \sigma)$. The extraction of all the $3\log n$ positions takes time $O(\log^2 n \log \sigma)$. By scanning $\mathsf{SA}_{T\#}[a + 1..b_a \log n]$ left-to-right we can determine the smallest position r such that $a < r \le b_a \log n$ and $\mathsf{SA}_{T\#}[r] < a$ if it exists. Similarly, by scanning $\mathsf{SA}_{T\#}[b_a \log n..a - 1]$ in right-to-left order, we can determine the largest position ℓ such that $(b_a - 1)\log n + 1 \le \ell < a$ and $\mathsf{SA}[\ell] < a$ if it exists. If ℓ is not in the interval $[(b_a - 1)\log n + 1..a - 1]$, then we scan $\mathsf{SA}_{T\#}[(b_\ell - 1)\log n + 1..b_\ell \log n]$ right-to-left and ℓ will be the largest position in $[(b_\ell - 1)\log n + 1..b_\ell \log n]$ such that $\mathsf{SA}_{T\#}[\ell] < a$. Similarly, if r is not in the interval $[a + 1..b_a \log n]$, then the value of r will be the smallest position in $\mathsf{SA}_{T\#}[(b_r - 1)\log n + 1..b_r \log n]$ such that $\mathsf{SA}_{T\#}[r] < a$. The final step is to determine $l_1 = \mathrm{lcp}(T[\mathsf{SA}_{T\#}[\ell]..n], T[a..n])$ and $l_2 = \mathrm{lcp}(T[\mathsf{SA}_{T\#}[r]..n], T[a..n])$. This can be done in time $O((l_1 + l_2)\log \sigma)$ by using Lemma 9.5, which allows one to extract the sequence of characters in T that start from positions $\mathsf{SA}_{T\#}[\ell]$ and $\mathsf{SA}_{T\#}[r]$, one by one in $O(\log \sigma)$ time per character. By Lemma 12.5, the length of T^i is precisely the maximum of l_1 and l_2, and moreover $L_i = \ell$ if $l_1 > l_2$, and $L_i = r$, otherwise.

The determination of L_i is done in time $O(|T^i|\log \sigma)$. Overall, we spend $O(|T^i|\log \sigma + \log^2 n \log \sigma)$ time to find the factor T^i. The terms $|T^i|\log \sigma$ sum up to $O(n \log \sigma)$. The terms $\log^2 n \log \sigma$ sum up to $O(n \log n \log^2 \sigma)$, since we have $O(n/\log_\sigma n)$ such terms.

Together with Theorem 9.11, we thus have proved the following lemma.

LEMMA 12.6 *The Lempel–Ziv parse* $T = T^1 T^2 \cdots T^p$ *of a text* $T = t_1 t_2 \cdots t_n$, $t_i \in [1..\sigma]$ *can be computed in* $O(n \log n \log^2 \sigma)$ *time and* $O(n \log \sigma)$ *bits of space. Moreover, the pairs* $(|T^1|, L_1), (|T^2|, L_2), \ldots, (|T^p|, L_p)$ *are computed within the same time and space bounds and the determination of every phrase* T^i *is done in time* $O(|T^i|\log \sigma + \log^2 n \log \sigma)$.

*12.1.3 Space- and time-efficient Lempel–Ziv parsing

Now we are ready to prove the main theorem of Section 12.1.

THEOREM 12.7 *The Lempel–Ziv parse of a text* $T = t_1 t_2 \cdots t_n$, $t_i \in [1..\sigma]$ *can be computed in* $O(n \log \sigma)$ *time and* $O(n \log \sigma)$ *bits of space, given the Burrows–Wheeler transform of* T *and of the reverse of* T. *Moreover the pairs* $(|T^1|, L_1), (|T^2|, L_2), \ldots, (|T^p|, L_p)$ *can be computed within the same time and space bound.*

The algorithm to match the claim of the theorem consists of several cases, where we exploit the earlier lemmas as well as some new constructions. In addition to all the structures shown in Section 12.1.2 (the succinct suffix array, the vector R, and the RMQ), we require the BWT index on the reverse of the text T (that is, $\mathsf{BWT}_{\overline{T\#}}$).

We also make use of a bitvector V of length n bits, indexed to answer rank and select queries in constant time. This takes $n + o(n)$ bits of space: see Section 3.2. The overall space for all the required structures is $O(n\log\sigma)$ bits.

In what follows, we give a high-level overview of the algorithm, dividing it into several cases, which will then be detailed in the forthcoming subsections. In addition to the global data structures listed above, the algorithm uses some local data structures related to the block division defined next and to the cases into which the algorithm is divided.

We start by partitioning the text into $\log n$ equally sized blocks of size $B = \lceil n/\log n\rceil$ characters, except possibly for the last one, which might be of smaller size.

We process the blocks left-to-right. When we are processing the Ith block, we say that we are in *phase I* of the algorithm. In phase I, we first build the augmented suffix tree shown in Section 12.1.1 on block I. We then scan block I in left-to-right order and determine the lengths of the phrases that start inside the block. Suppose that the partial parse $T^1 T^2 \cdots T^{i-1}$ has already been obtained, and that the starting position of the next phrase in the parse is $a = |T^1 T^2 \cdots T^{i-1}| + 1 \in [(I-1)B + 1..I \cdot B]$. The length of T^i will be the length of the longest previous factor at position a in T.

A phrase can be allocated to one of the following categories according to the position of the phrase and the position of its previous occurrence.

(1) The phrase does not cross a block boundary, and its previous occurrence is within the same block.

(2) The phrase does not cross a block boundary, and its previous occurrence starts in a previous block. This case can be decomposed into two cases:

    (a) the previous occurrence ends in a previous block;

    (b) the previous occurrence ends in the current block.

(3) The phrase crosses a block boundary.

The cases are illustrated in Figure 12.1. In order to find the phrases from different categories, the algorithm considers two separate cases.

In the first case, the algorithm searches for the longest prefix of $t_a \cdots t_{I \cdot B - 1}$ which has an occurrence starting in $T[(I-1)B + 1..a - 1]$. In the second case, the algorithm searches for the longest prefix of $t_a \cdots t_{I \cdot B - 1}$ that has an occurrence starting inside $T[1..(I-1)B]$ and returns a starting position of such a prefix. Then the length of the phrase T^i will be the larger of those two lengths. We have a special case with the last phrase of the block. We now show how to find the length for the first case, and then for the second case, except in the case of the last phrase in the block, which will be described at the end, since it is common to both cases. Finally, we conclude by putting the pieces together. The space and time analyses are summarized at the end of each subsection.

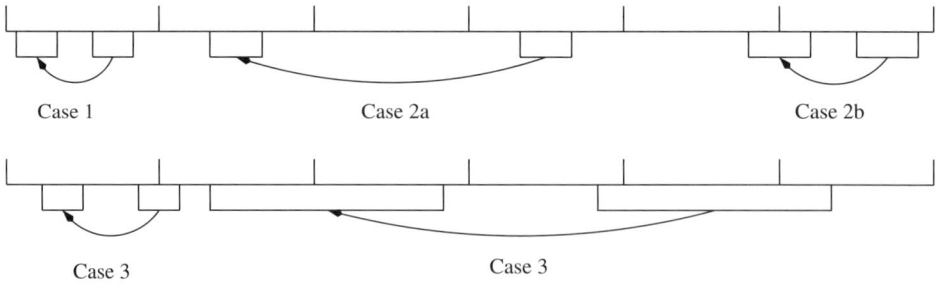

Figure 12.1 Illustration of the different cases of the algorithm in Section 12.1.3. Case 1 is a phrase whose previous occurrence is fully contained in the same block and the phrase does not cross a block boundary. Case 2a is a phrase that does not cross a block boundary and whose previous occurrence occurs before the current block. Case 2b is the same as Case 2a, except that the previous occurrence ends inside the current block. Finally, Case 3 is for a phrase that crosses a block boundary.

Finding occurrences inside the block

Case 1 in Figure 12.1 of finding the longest prefix of $t_a \cdots t_n$ which has an occurrence starting in $T[(I-1)B+1..a-1]$ is easily done using the structure shown in Section 12.1.1 and then using the algorithm described in Section 12.1.2 to determine the length of the longest prefix $t_a \cdots t_n$ which has an occurrence starting in $T[(I-1)B+1..a-1]$ and the starting position of one of those occurrences. The space used by the structure is $O((n/\log n)\log n) = O(n)$ bits and the time is $O((n/\log n)\log \sigma)$.

Finding occurrences before the block

We now consider Case 2 in Figure 12.1, that is, we show how to search for the longest prefix of $t_a \cdots t_n$ which has an occurrence starting in $T[1..(I-1)B]$. We first show how to find the longest prefix of $t_a \cdots t_n$ that has an occurrence fully contained in $T[1..(I-1)B]$ (Case 2a). Let such a longest prefix be of length m. We then show how to find the length of the longest prefix of $t_a \cdots t_n$ which has an occurrence starting in $T[1..(I-1)B]$ and ending at some position in $T[(I-1)B+1..n]$, *but only if that length exceeds m* (Case 2b).

As mentioned before, the algorithm uses a bitvector $V[1..n]$. Initially, all the bits are set to zero. Recall that the bitvector is augmented so as to support rank and select operations in constant time. We assume throughout the algorithm that we can get successively for every position $j \in [1..n]$ the number $\mathrm{SA}_{T\#}^{-1}[j]$ which is the lexicographic order of the prefix $T[1..j]$. This number can be computed using Lemma 9.4: this takes $O(\log \sigma)$ time. Before processing block I, we assume that the bits in V that represent the lexicographic order of all prefixes that end at positions in $[1..I \cdot B]$ in T have all been set to one. We later show how these bits are updated.

We start by doing successive backward steps on $T[a], T[a+1], T[a+2], \ldots, T[I \cdot B]$ on $\mathrm{BWT}_{T\#}$. After each backward step on a character $T[a+k]$, we check whether the interval $V[s_k..e_k]$ contains at least one bit set to one, where $[s_k..e_k]$ is the interval returned by the backward step on character $a+k$. Note that all backward steps are successful, since we are testing only substrings that appear in the text. If there is a bit set to one, we continue taking backward steps and doing the test again. We stop exactly at the moment where the test fails, because of the following observation.

OBSERVATION 12.8   *Range $[s_k..e_k]$ in bitvector V contains a bit set to one if and only if the substring $T[a..a + k]$ has already appeared in $T[1..(I - 1)B]$.*

Proof   If $V[p] = 1$ for some $p \in [s_k..e_k]$, then $V[p] = 1$ implies that the prefix of lexicographic rank p ends somewhere in $T[1..(I - 1)B]$ and $p \in [s_k..e_k]$ implies that the prefix has $T[a..a + k]$ as a suffix and thus the factor $T[a..a + k]$ occurs inside $T[1..(I - 1)B]$.

If $V[p] = 0$ for all $p \in [s_k..e_k]$, then we conclude that none of the prefixes that end in $T[1..(I - 1)B]$ can have $T[a..a + k]$ as a suffix and thus that $T[a..a + k]$ does not occur in $T[1..(I - 1)B]$.   □

Whenever we stop at some position $a + k \leq I \cdot B - 1$, we deduce that the longest prefix of $T[a..I \cdot B]$ that occurs in $T[1..(I - 1)B]$ is $T[a..a + k - 1]$. Moreover one possible ending position of such a factor is given by $\mathsf{SA}_{T\#}[p]$, where p is any position in $[s_{k-1}, e_{k-1}]$ such that $V[p] = 1$. Such a position can be found by rank and select on bitvector V.

We can now set $m = k$ for the k at which the test fails. Now that we have determined the length m of the longest prefix of $T[a..n]$ which has an occurrence fully contained in previous blocks, that is, Case 2a of Figure 12.1, it remains to handle Case 2b of Figure 12.1, where we have an occurrence that ends in the current block. We are interested in such an occurrence only if its length is at least m.

We make use of the Knuth–Morris–Pratt (KMP) algorithm described in Example 2.1 in Section 2.1. Suppose that we have found a factor $T[a..a + m - 1]$ of length m such that the length of its previous occurrence in the previous blocks is m. We then build the KMP automaton on $T[a..a + 2m - 1]$: this takes time $O(m)$. We then consider the window $T[(I - 1)B + 1 - m..(I - 1)B + 2m - 1]$. We use the automaton to find the longest prefix of $T[a..a + 2m - 1]$ that occurs inside $T[(I - 1)B + 1 - m..(I - 1)B + 2m - 1]$ and that starts inside $T[(I - 1)B + 1 - m..(I - 1)B]$. Both the construction of the KMP automaton and the matching of the text fragment $T[(I - 1)B + 1 - m..(I - 1)B + 2m - 1]$ take time $O(m)$. If the length of the found prefix of $T[a..a + 2m - 1]$ is less than $2m$, we are done, since we could not find an occurrence of $T[a..a + 2m - 1]$ that starts inside $T[(I - 1)B + 1 - m..(I - 1)B]$. Otherwise, if the length of the found prefix is exactly $2m$, we build the KMP automaton on $T[a..a + 4m - 1]$, scan the text fragment $T[(I - 1)B + 1 - m..(I - 1)B + 4m - 1]$, and use the automaton to determine the longest prefix of $T[a..a + 4m - 1]$ that occurs inside $T[(I - 1)B + 1 - m..(I - 1)B + 4m - 1]$ and starts inside $T[(I - 1)B + 1 - m..(I - 1)B]$. We continue in this way as long as $m' = 2^k m$ for $k = 2, 3, \ldots$ and $m' \leq \log^3 n$. Note that the total time spent to compute m' is $O(2^k m) = O(m')$, since we are spending time $O(2^{k'} m)$ at every value of k' increasing from 1 to k.

Now, if $m' < \log^3 n$, $T[a..a + m']$ will be the longest factor whose current occurrence starts at position a and whose previous occurrence starts inside $[1..(I - 1)B]$. Otherwise, we use the algorithm described in Section 12.1.2 to determine the length of the phrase starting at position a in T. This takes time $O(m' \log \sigma + \log^2 n \log \sigma)$.

Handling the last phrase in the block

It remains to handle the case of the last phrase of the block (Case 3 in Figure 12.1). During the course of the algorithm, whenever we detect that $t_a \cdots t_{I \cdot B}$ occurs before or inside the block, we deduce that the phrase T^i starting at position a is actually the last one in the block. In such a case we use directly the algorithm described in Section 12.1.2 to determine the length of T^i in time $O(|T^i|\log\sigma + \log^2 n\log\sigma)$. Since we have only one last phrase per block, the total additional time incurred by the application of the algorithm is $O(\log n\log^2 n\log\sigma) = O(\log^3 n\log\sigma)$.

Postprocessing the block

At the end of the processing of block I, we set the bits that represent the lexicographic order of all prefixes that end at positions in $[(I-1) \cdot B + 1..I \cdot B]$ in T. That is, for every $j \in [(I-1) \cdot B + 1..I \cdot B]$, we set position $V[\mathsf{SA}_{T\#}^{-1}[j]]$ to one.

At the end, we update the bitvector V such that the positions that correspond to the ranks in $\mathsf{BWT}_{T\#}$ of the prefixes of $T[1..B]$ are set to 1. We can now scan the bitvector V and update the auxiliary rank support structures. The time taken to update the auxiliary structures is $O(n/\log n)$, by taking advantage of bit-parallelism (see Exercise 3.8).

Concluding the proof

The length of each phrase will be the maximum of the lengths found from the two first cases and the special case. It can easily be seen that the total time spent for all phrases is $O(n\log\sigma)$. In all cases, the algorithms will also return a pointer to a previous occurrence, except in Case 2a, in which only a position in a suffix array is returned. We can then use Algorithm 9.2 to transform the suffix array positions into text positions in $O(n\log\sigma + p) = O(n\log\sigma)$ time. This concludes the proof of Theorem 12.7.

*12.2   Bit-optimal Lempel–Ziv compression

In practice it is desirable to engineer a given compression scheme C to encode its input string $S \in [1..\sigma]$ *with the minimum possible number of bits*: we call such a version of C *bit-optimal*. Minimizing the size of the output makes $C(S)$ a better approximation of the Kolmogorov complexity of S, and it might improve the accuracy of the compression-based measure of dissimilarity introduced in Section 11.2.5. Here we focus on making the Lempel–Ziv compression scheme introduced in Section 12.1 bit-optimal. Conceptually, the construction presented here relies on the careful interplay of a number of data structures and algorithms borrowed from different parts of the book, including the shortest paths in a DAG, the construction and traversal of suffix trees, the construction of suffix arrays and LCP arrays, and range-minimum queries.

We denote by LZ the Lempel–Ziv compression scheme and by LZ(S) its output, that is, a sequence of pairs (d, ℓ), where $d > 0$ is the distance, from the current position i in S, of a substring to be copied from a previous position $i - d$, and ℓ is the length of such a copied substring. Note that we use distances here, rather than absolute positions in S, because we focus on making the output as small as possible. We call both the copied substring represented by a pair (d, ℓ), and the pair (d, ℓ) itself a *parse*. We call any representation of S as a sequence of phrases a *parse*. Recall that, at every step, LZ copies the *longest substring* that occurs before the current position i: this greedy choice, called the *longest previous factor* (see Definition 12.1), is guaranteed to minimize the number of *phrases* over all possible parses of S. If pairs are encoded with integers of fixed length, then the greedy choice of LZ minimizes also the number of *bits* in the output, over all possible parses. However, if pairs are encoded with a *variable-length encoder*, LZ is no longer guaranteed to minimize the number of bits in the output.

DEFINITION 12.9   A variable-length integer encoder *on a range* $[1..n]$ *is a function* $f : [1..n] \mapsto \{0, 1\}^+$ *such that* $x \leq y \iff |f(x)| \leq |f(y)|$ *for all* x *and* y *in* $[1..n]$.

Equal-length codes, as well as the gamma and delta codes described in Section 8.1, are variable-length encoders. An encoder f partitions its domain $[1..n]$ into contiguous intervals $\overrightarrow{b_1}, \overrightarrow{b_2}, \ldots, \overrightarrow{b_m}$, such that all integers that belong to interval $\overrightarrow{b_i}$ are mapped to binary strings of length exactly b_i, and such that two integers that belong to intervals $\overrightarrow{b_i}$ and $\overrightarrow{b_j}$ with $i \neq j$ are mapped to binary strings of length $b_i \neq b_j$. Clearly $m \in O(\log n)$ in practice, since n can be encoded in $O(\log n)$ bits. In some cases of practical interest, m can even be constant.

The greedy scheme of LZ can easily be optimized to take variable-length encoders into account: see Insight 12.1. Moreover, specific non-greedy variants of LZ can work well in practice: see Insights 12.2 and 12.3. In this section, we consider the set of *all possible* LZ *parses of* S, which can be represented as a directed acyclic graph.

Insight 12.1 Bit-optimal greedy Lempel–Ziv scheme

A natural way of reducing the number of bits in the output of LZ, without altering its greedy nature, consists in choosing the smallest possible value of d for every pair (d, ℓ) in the parse. This amounts to selecting *the right-most occurrence of each phrase*, a problem that can be solved using the suffix tree of S and its LZ parse, as follows.

Let $\mathsf{ST}_S = (V, W, E)$, where V is the set of internal nodes of ST_S, W is the set of its leaves, and E is the set of its edges. Recall from Section 12.1 that every phrase in the parse corresponds to a node of ST: we say that such a node is *marked*, and we denote the set of all marked nodes of ST by $V' \subseteq V$. A marked node can be associated with more than one phrase, and with at most σ phrases. At most m nodes of ST are marked, where m is the number of phrases in the parse. Let $\mathsf{ST}' = (V', W, E')$ be the *contraction* of ST induced by its leaves and by its marked nodes. In other words, ST' is a tree whose leaves are exactly the leaves of ST, whose internal nodes are the marked nodes of ST, and in which the parent of a marked node (or of a leaf) in ST' is its lowest marked ancestor in ST. Given the LZ parse of S, marking ST takes $O(|S|\log \sigma)$ time, and ST' can be built with an $O(|S|)$-time top-down traversal of ST.

With ST' at hand, we can scan S from left to right, maintaining for every node $v \in V'$ the largest identifier of a leaf that has been met during the scan, and that is *directly attached to* v in ST'. Specifically, for every position i in S, we go to the leaf w with identifier i in ST', and we set the value of $\texttt{parent}(v)$ to i. If i corresponds to the starting position of a phrase, we first jump in constant time to the node $v \in V'$ associated with that phrase, and we compute the maximum value over all the internal nodes in the subtree of ST' rooted at v. The overall process takes $O(|S| + \sigma \cdot \sum_{v \in V'} f(v))$ time, where $f(v)$ is the number of internal nodes in the subtree of ST' rooted at v. Discarding multiplicities (which are already taken into account by the σ multiplicative factor), a node $v \in V'$ is touched exactly once by each one of its ancestors in ST'. Since v has at most $|\ell(v)|$ ancestors in ST', it follows that $\sum_{v \in V'} f(v) \leq \sum_{v \in V'} |\ell(v)| \leq |S|$, thus the algorithm takes $O(\sigma|S|)$ time overall.

Insight 12.2 Practical non-greedy variants of the Lempel–Ziv scheme

Let W_i be the *longest previous factor* that starts at a position i in string S. In other words, W_i is the longest substring that starts at position i in S, and that occurs also at a position $i' < i$ in S. Let $\mathsf{LPF}[i] = |W_i|$, and let $D[i] = i - i'$, where i' is the right-most occurrence of W_i before i. Recall that the bit-optimal version of LZ described in Insight 12.1 outputs the following sequence of pairs: $(D[i], \mathsf{LPF}[i]), (D[i + \mathsf{LPF}[i]], \mathsf{LPF}[i + \mathsf{LPF}[i]]), \ldots$. Rather than encoding at every position i a representation of its longest previous factor, we could set a domain-specific *lookahead threshold* $\tau \geq 1$, find the position $i^* = \mathrm{argmax}\{\mathsf{LPF}[j] : j \in [i..i + \tau]\}$, and output substring $S[i..i^* - 1]$ explicitly, followed by pair $(D[i^*], \mathsf{LPF}[i^*])$. We could then repeat the process from position $i^* + \mathsf{LPF}[i^*]$.

Using a *dynamic threshold* might be beneficial in some applications. Given a position i in S, we say that the *maximizer of suffix* $S[i..|S|]$ is the smallest position $i^* \geq i$ such that $\mathsf{LPF}[j] < \mathsf{LPF}[i^*]$ for all $j \in [i..i^* - 1]$ and for all $j \in [i^* + 1..i^* + \mathsf{LPF}[i^*] - 1]$. Assume that i^* is initially set to i. We can check whether there is a position $j \in [i^* + 1..i^* + \mathsf{LPF}[i^*] - 1]$ such that $\mathsf{LPF}[j] > \mathsf{LPF}[i^*]$: we call such j a *local maximizer*. If a local maximizer exists, we reset i^* to j and repeat the process, until we find no local maximizer. The resulting value of i^* is the maximizer of the suffix $S[i..|S|]$: we thus output the pair $(D[i^*], \mathsf{LPF}[i^*])$, and we encode the substring $S[i..i^* - 1]$ using a sequence of nonoverlapping local maximizers and of plain substrings of S.

Insight 12.3 Practical relative Lempel–Ziv scheme

In practice we often want to compress a string S with respect to a *reference string* T, for example to implement the conditional compression measure $C(S|T)$ described in Section 11.2.5, or to archive a large set of genomes from similar species (see also Section 10.7.1). In the latter case, we typically store a concatenation T of reference genomes in plain text, and we encode the LZ factorization of every other genome S with respect to T. Note that the LPF vector described in Insight 12.2 becomes in this case the matching statistics vector $\mathsf{MS}_{S,T}$ of Section 11.2.3, and the array D encodes *positions in T* rather than distances. The space taken by the D component of any LZ parse of S becomes thus a large portion of the output.

However, if genomes S and T are evolutionarily related, the sequence of D values output by any LZ parse should contain a *long increasing subsequence*, which corresponds to long substrings that preserve their order in S and T. We could thus reduce the size of the compressed file by computing the longest increasing subsequence of D values, using for example the algorithm in Example 2.2, and by encoding *differences* between consecutive D values in such increasing subsequence.

The fixed or dynamic lookahead strategies described in Insight 12.2 are also useful for compressing pairs of similar genomes, since they allow one to encode SNPs, point mutations, and small insertions in plain text, and to use positions and lengths to encode long shared fragments.

DEFINITION 12.10 *Given a string S, its* parse graph *is a DAG $G = (V, E)$ with $|S| + 1$ vertices, such that the first $|S|$ such vertices, $v_1, v_2, \ldots, v_{|S|}$, are in one-to-one correspondence with the positions of S. Set E contains arc (v_i, v_{i+1}) for all $i \in [1..|S|]$, and every such arc is labeled by the pair $(0, 1)$. For every vertex $v_i \in V$ and for every string $S[i..j]$ with $j > i + 1$ that starts also at a position $i' < i$ in S, there is an arc (v_i, v_{j+1}) with label (d_e, ℓ_e), where $\ell_e = j - i + 1$ is the length of the arc, $d_e = i - i^*$ is its copy distance from i, and i^* is the starting position of the* right-most *occurrence of $S[i..j]$ before i. G has a unique source $s = v_1$ and a unique sink $t = v_{|S|+1}$.*

Note that G encodes all possible LZ parses of S as *paths* from s to t. Note also that the set of arcs E is *closed*, in the sense that the existence of an arc $e = (v_i, v_{j+1})$ implies the existence of all possible *nested arcs* whose start and end vertices belong to the range $[i..j + 1]$: see Figure 12.2(b). Formally, we can express this as follows.

Arc closure. $(v_i, v_{j+1}) \in E \Rightarrow (v_x, v_y) \in E$ for all $x < y$ such that $x \in [i..j - 1]$ and $y \in [i + 2..j + 1]$.

In other words, given an arc $(v_i, v_{j+1}) \in E$, the subgraph of G induced by vertices $v_i, v_{i+1}, \ldots, v_j, v_{j+1}$ is an orientation of a complete undirected graph. Recall from Exercise 4.9 that such a graph is called a *tournament*. Since there can be at most λ arcs starting from the same vertex of G, where $\lambda \in O(|S|)$ is the length of the longest repeated substring of S, we have that $|E| \in O(|S|^2)$.

Consider now two variable-length encoders f and g used to represent copy distances and lengths, respectively. We denote by ϕ the number of intervals induced by f on the range $[1..|S|]$, and we denote by γ the number of intervals induced by g on the range $[1..\tau + \lambda]$, where $\tau = \overline{|g(\sigma)|}$. Assume that we assign a cost $c(e) = |f(d_e)| + |g(\tau + \ell_e)|$ to every arc e of length greater than one, and a cost $c(e) = |g(S[i])|$ to every arc e of length one. Note that we are reserving the integers $[1..\sigma]$ to encode the *label* of arcs of

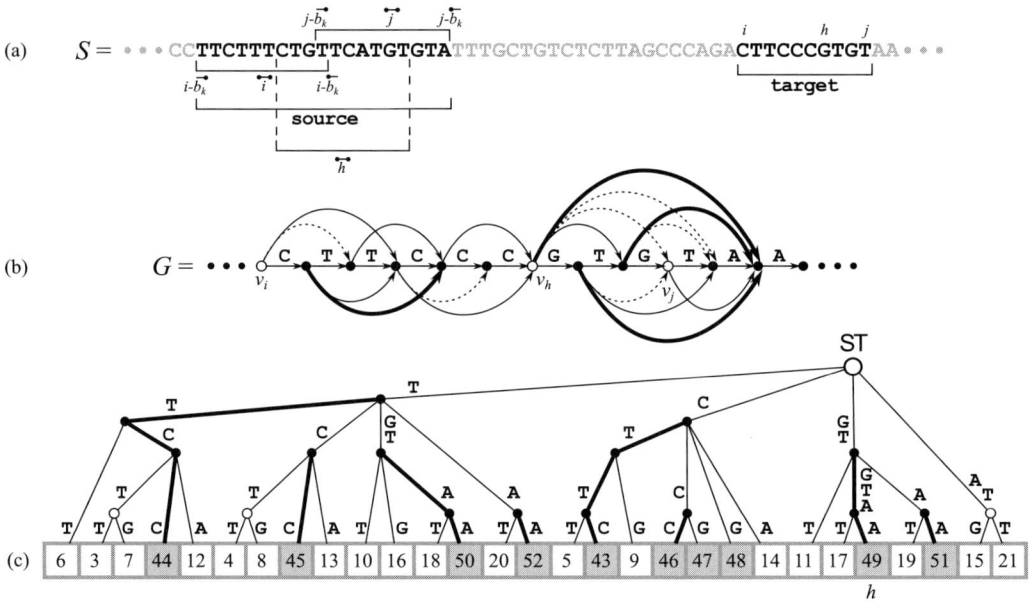

Figure 12.2 Building distance-maximal arcs: a sub-iteration of iteration b_k. In this example, the distances that are encoded in exactly b_k bits are in the range [31..40]. (a) The relationship between the intervals `target` and `source` in S. (b) The portion of the graph G that corresponds to `target`. To reduce clutter, arc labels are not shown, and characters are attached to arcs of length one. Dashed arcs are not distance-maximal, solid arcs are distance-maximal, and bold arcs are distance-maximal with a distance encoded in exactly b_k bits. (c) The compact trie ST of the suffixes that start inside `source` ∪ `target`. For every edge that connects an internal node to a leaf, only the first character of its label is shown. The children of every node are sorted according to the order T < C < G < A, which is the order in which the characters in $[1..\sigma]$ first appear when `source` is scanned from left to right. Numbers at leaves indicate their corresponding positions in S. Gray leaves are positions in `target`. The bold path that starts from every gray leaf v (if any) is the set of internal nodes u of ST such that the smallest gray leaf in the subtree rooted at u is v itself.

length one, and we are mapping lengths greater than one to the interval $[\tau + 1..\tau + \lambda]$. We use the shorthand $c(v_i, v_j)$ to mean $c(e)$ where $e = (v_i, v_j)$.

The cost function obeys the following monotonicity property, whose proof is left to Exercise 12.7.

Monotonicity of cost. $c(v_i, v_j) \leq c(v_i, v_{j'})$ for all $j < j'$. Similarly, $c(v_{i'}, v_j) \leq c(v_i, v_j)$ for all $i < i'$.

We are interested in arcs for which $c(v_i, v_j)$ is *strictly smaller* than $c(v_i, v_{j+1})$.

DEFINITION 12.11 *Consider arcs* $e = (v_i, v_j)$ *and* $e' = (v_i, v_{j+1})$. *We say that e is* maximal *if* $c(e) < c(e')$. *In particular, e is* distance-maximal *if* $|f(d_e)| < |f(d_{e'})|$, *and e is* length-maximal *if* $|g(\ell_e)| < |g(\ell_{e'})|$.

See Figure 12.2(b) for an example of distance-maximal arcs. Note that the cost we assigned to arcs of length one makes them both distance-maximal and length-maximal.

It turns out that maximal arcs suffice to compute the smallest number of bits in the output over all LZ parses of S that use encoders f and g.

LEMMA 12.12 *There is a shortest path from v_1 to $v_{|S|+1}$ that consists only of maximal arcs.*

The easy proof of this lemma is left to Exercise 12.8. We denote by $G' = (V, E')$ the subgraph of G induced by maximal arcs. The following fact is immediately evident from Lemma 12.12, and its proof is developed in Exercise 12.9.

THEOREM 12.13 *Given G', we can compute an LZ parse of S that minimizes the number of bits over all possible LZ parses of S, in $O(|S| \cdot (\phi + \gamma))$ time.*

Note that $\phi + \gamma$ can be significantly smaller than $O(|V| + |E|)$ in practice. We are thus left with the task of building G' from S, *without representing G explicitly*.

Before proceeding, let us make the problem a little bit easier. Note that all the arcs that start from the same vertex can be partitioned into equivalence classes $b_{i_1}, b_{i_2}, \ldots, b_{i_k}$ according to the number of bits required to encode their distances using f, where $k \leq \phi$. Once the distance-maximal arc of every equivalence class is known, it is easy to compute all the length-maximal arcs inside each class b_{i_p}, *in constant time per length-maximal arc*: see Exercise 12.10. Thus, in what follows we focus exclusively on building the set of distance-maximal arcs.

*12.2.1 Building distance-maximal arcs

Given a position i in S and a number of bits b, we denote by $\overrightarrow{i_b} = [i - \overrightarrow{b}..i - \overleftarrow{b}]$ the range of all positions in S smaller than i that are addressable from i using exactly b bits (we omit the subscript whenever b is clear from the context). We organize the construction of distance-maximal arcs into ϕ iterations, one iteration for each interval $\overrightarrow{b_k}$ induced by f on the range of distances $[1..|S|]$, in the increasing order $\overrightarrow{b_1}, \overrightarrow{b_2}, \ldots, \overrightarrow{b_\phi}$. Every iteration k is itself organized into $\lceil |S|/|\overrightarrow{b_k}| \rceil$ sub-iterations: in each sub-iteration we examine a distinct interval `target` $= [i..j]$ of S spanning $|\overrightarrow{b_k}|$ consecutive positions, and we compute all the distance-maximal arcs that start from vertices $\{v_h \in V : h \in \texttt{target}\}$ and whose copy distance is encoded in exactly b_k bits. We will engineer each sub-iteration of iteration k to take $O(|\overrightarrow{b_k}| \log |S|)$ time and $O(|\overrightarrow{b_k}|)$ space, and we will use global auxiliary data structures that take $O(|S|)$ space. Combined with Theorem 12.13, this approach will provide the key result of this section.

THEOREM 12.14 *The bit-optimal LZ parse of a string S on alphabet $[1..\sigma]$ can be computed in $O(\phi|S| \log |S| + \gamma|S|)$ time and in $O(|S|)$ space.*

Performing all the ϕ iterations in parallel takes just $O\left(\sum_{k=1}^{\phi} |\overrightarrow{b_k}|\right) = O(|S|)$ temporary space in addition to the input and to the auxiliary data structures, and it allows one to build G' and to run the single-source shortest-path algorithm *on the fly, without storing the entire G' in memory*.

Consider the interval `source` $= [i - \overrightarrow{b_k}..j - \overrightarrow{b_k}] = \overrightarrow{i} \cup \overrightarrow{j}$ of size $2|\overrightarrow{b_k}| - 1$ (see Figure 12.2(a)), and note that `source` and `target` might intersect. All arcs that start from a vertex v_h with $h \in$ target and whose distance is encoded in exactly b_k bits correspond to substrings of S that start both inside target and inside source. Moreover, such substrings cannot start *between* source and target, that is between position $j - \overleftarrow{b_k} + 1$ and position $i - 1$, otherwise the distance of the corresponding arc that starts from v_h could be represented in strictly less than b_k bits. Distance-maximal arcs enjoy one more property: their length depends on the range of positions of S that are addressable using their distance.

Length of distance-maximal arcs. If arc $e = (v_h, v_{h+\ell_e+1})$ is distance-maximal with copy distance d_e, then its length ℓ_e equals $\max\{\mathsf{LCP}(h, h') : h' \in \overleftarrow{h_b}, b = |f(d_e)|\}$, where $\mathsf{LCP}(h, h')$ is the length of the longest common prefix between suffix $S[h..|S|]\#$ and suffix $S[h'..|S|]\#$.

The easy proof of this property is left to Exercise 12.11. Consider thus the trie ST that contains all suffixes $S[h..|S|]\#$ for $h \in$ source $\cup$ target, and in which unary paths have been compacted into single edges whose labels contain more than one character (see Figure 12.2(c)). In other words, ST is the subgraph of the suffix tree of S induced by the suffixes that start inside source $\cup$ target. Let ST_u denote the subtree of ST rooted at node u. Given a position $h \in$ target, there is an easy algorithm to detect whether there is a distance-maximal arc that starts from vertex v_h and whose distance is encoded in b_k bits.

LEMMA 12.15   *Let h be a position in* target, *let v be the leaf of* ST *associated with position h, and let x be the length of the longest distance-maximal arc that starts from v_h and that has been detected in a previous iteration b_q with $q < k$. Assume that we can answer the following queries for any internal node u of* ST:

- allSmaller$(u, \overrightarrow{h})$, *return true if all leaves of* ST_u *are smaller than* $\overrightarrow{h}$;
- oneLarger$(u, \overrightarrow{h})$, *return true if* ST_u *contains a leaf whose position is larger than* $\overrightarrow{h}$.

Then, we can compute the distance-maximal arc from v_h whose distance is encoded in b_k bits, or decide that no such distance-maximal arc exists, in $O(|\overleftarrow{b_k}|)$ time.

Proof Consider node $u = $ parent(v): if ST_u contains a leaf that corresponds to a position of S greater than $\overrightarrow{h}$, then no arc of G that starts from v_h can possibly encode its distance with b_k bits, because the longest substring that starts both at h and inside $\overrightarrow{h}$ has a distance that can be encoded with strictly fewer than b_k bits.

Conversely, if all leaves in ST_u correspond to positions in S that are smaller than $\overrightarrow{h}$, then the distance d_e of arc $e = (v_h, v_{h+|\ell(u)|+1})$ requires either strictly more or strictly fewer than b_k bits to be encoded. Note that we cannot yet determine which of these two cases applies, because substring $\ell(u)$ might start inside the interval $[j - \overleftarrow{b_k} + 1..i - 1]$, and these positions are not in ST. The same observation holds for all substrings of S that end in the middle of the edge (parent$(u), u$) in ST. However, it might be that the distance of arc $e' = (v_h, v_{h+|\ell(\mathrm{parent}(u))|+1})$ is encoded in b_k bits and that e' is distance-maximal: we thus repeat the algorithm from parent(u).

Finally, if ST_u does contain a leaf $p \in \overleftrightarrow{h}$ (where p might itself be in `target`), then $\ell(u)$ is the longest substring that starts both at h and at a position inside $\overleftrightarrow{h}$: it follows that arc $e = (v_h, v_{h+|\ell(u)|+1})$ does exist in G, and by the property of the length of distance-maximal arcs, it *might* be distance-maximal. If its copy distance d_e is $h-p$, and thus if $|f(d_e)|$ equals b_k (see for example arc $(v_h, v_{h+5}) = (v_{49}, v_{54})$ in Figures 12.2(b) and (c)), then e is indeed distance-maximal. Clearly d_e might be smaller than $h - p$ if substring $\ell(u)$ occurs *after* the last position of `source` (for example, for arc $(v_i, v_{i+3}) = (v_{43}, v_{46})$ in Figures 12.2(b) and (c)), but in this case the maximality of e has already been decided in a previous iteration b_q with $q < k$. We thus need to compare the length of $\ell(u)$ with the length x of the longest distance-maximal arc in E' that starts from v_h. □

Supporting the queries required by Lemma 12.15 turns out to be particularly easy.

LEMMA 12.16 *After an $O(|\overrightarrow{b_k}|)$-time preprocessing of* ST, *we can answer in constant time queries* allSmaller$(u,\overrightarrow{h})$ *and* oneLarger$(u,\overrightarrow{h})$ *for any internal node u of* ST *and for any $h \in$* target.

Proof Since $|\overrightarrow{h}| = |\overrightarrow{b_k}|$, we know that $\overrightarrow{h}$ overlaps either $\overleftrightarrow{i}$ or $\overleftrightarrow{j}$, or both. We can thus store in every internal node u of ST the following values:

- largestI, the largest position in $\overleftrightarrow{i}$ associated with a leaf in ST_u;
- largestJ, the largest position in $\overleftrightarrow{j}$ associated with a leaf in ST_u;
- minTarget, the smallest position in `target` associated with a leaf in ST_u.

Note that some of these values might be undefined: in this case, we assume that they are set to zero. Given $\overrightarrow{h}$, we can implement the two queries as follows:

$$\text{allSmaller}(u,\overrightarrow{h}) = \begin{cases} \text{false} & \text{if } h = i, \\ (u.\text{largestI} < \overrightarrow{h}) \wedge (u.\text{largestJ} = 0) & \text{if } h \neq i, \end{cases}$$

$$\text{oneLarger}(u,\overrightarrow{h}) = \begin{cases} \text{false} & \text{if } h = j, \\ (u.\text{largestJ} > \overrightarrow{h}) \vee (u.\text{minTarget} \in [\overrightarrow{h}+1..h-1]) & \text{if } h \neq i. \end{cases}$$

Computing $u.\text{largestI}$, $u.\text{largestJ}$ and $u.\text{minTarget}$ for every internal node u of ST takes $O(|ST|) = O(|\overrightarrow{b_k}|)$ time. □

We would like to run the algorithm in Lemma 12.15 for all $h \in$ target. Surprisingly, it turns out that this does not increase the asymptotic running time of Lemma 12.15.

LEMMA 12.17 *Running the algorithm in Lemma 12.15 on all leaves that correspond to positions in* target *takes $O(|\overrightarrow{b_k}|)$ time overall.*

Proof It is easy to see that the upward path traced by Lemma 12.15 from a leaf v that corresponds to a position $h \in$ target is a (not necessarily proper) prefix of the upward path $P_v = u_1, u_2, \ldots, u_{p-1}, u_p$ defined as follows:

- $u_1 = v$;
- $u_x = \text{parent}(u_{x-1})$ for all $x \in [2..p]$;

- h is smaller than the position of all leaves in `target` that descend from u_x, for all $x \in [2..p-1]$;
- there is a leaf in `target` that descends from u_p whose position is smaller than h.

The subgraph of ST induced by all paths P_v such that v corresponds to a position $h \in$ `target` is $O(|ST|) = O(|source \cup target|) = O(|\vec{b_k}|)$, since whenever paths meet at a node, only one of them continues upward (see Figure 12.2(c)). This subgraph contains the subgraph of ST which is effectively traced by Lemma 12.15, thus proving the claim.   □

*12.2.2   Building the compact trie

The last ingredient in our construction consists in showing that ST can be built efficiently. For clarity, we assume that `source` and `target` intersect: Exercise 12.13 explores the case in which they are disjoint.

LEMMA 12.18   *Let $S \in [1..\sigma]^n$ be a string. Given an interval $[p..q] \subseteq [1..n]$, let ST be the compact trie of the suffixes of S# that start inside $[p..q]$, in which the children of every internal node are sorted lexicographically. Assume that we are given the suffix array $SA_{S\#}$ of S, the longest common prefix array $LCP_{S\#}$ of S, the range-minimum data structure described in Section 3.1 on array $LCP_{S\#}$, and an empty array of σ integers. Then, we can build in $O((q-p)\log|S|)$ time and $O(q-p)$ space a compact trie ST' that is equivalent to ST, except possibly for a permutation of the children of every node.*

Proof   First, we scan $S[p..q]$ exactly once, mapping all the distinct characters in this substring onto a contiguous interval of natural numbers that starts from one. The mapping is chronological, thus it does not preserve the lexicographic order of $[1..\sigma]$. We denote by S' a *conceptual* copy of S in which all the characters that occur in $S[p..q]$ have been globally replaced by their new code.

Then, we build the artificial string $W = W_1 W_2 \cdots W_{q-p+1}\#$, where each character W_h is a *pair* $W_h = (S'[p+h-1], r_h)$, and r_h is a flag defined as follows:

$$r_h = \begin{cases} -1 & \text{if } S'[p+h..|S|]\# < S'[q+1..|S|]\#, \\ 0 & \text{if } h = q, \\ 1 & \text{if } S'[p+h..|S|]\# > S'[q+1..|S|]\#. \end{cases}$$

The inequality signs in the definition refer to lexicographic order. In other words, we are storing at every position $p+h-1$ inside interval $[p..q]$ its character in S', along with the lexicographic relationship between the suffix of S' that starts at the *following* position $p+h$ and the suffix that starts at position $q+1$. Pairs are ordered by first component and by second component. It is easy to see that sorting the set of suffixes $S'[h..|S|]\#$ with $h \in [p..q]$ reduces to computing the suffix array *of the artificial string W*, and thus takes $O(q-p)$ time and space: see Exercise 12.12.

String W can be built by issuing $|W|$ range-minimum queries on $LCP_{S\#}$ centered at position $q+1$, taking $O((q-p)\log|S|)$ time overall (see Section 3.1). Recall from

Exercise 8.14 that we can build the longest common prefix array of W in linear time from the suffix array of W. Given the sorted suffixes of S' in the range $[p..q]$, as well as the corresponding LCP array, we can build their compact trie in $O(q - p)$ time: see Section 8.3.2.

The resulting tree is topologically equivalent to the compact trie of the suffixes that start inside range $[p..q]$ in the original string S: the only difference is the *order* among the children of each node, caused by the transformation of the alphabet (see the example in Figure 12.2(c)). $\qquad\qquad\square$

Recall that the order among the children of an internal node of ST is not used by the algorithm in Lemma 12.15, thus the output of Lemma 12.18 is sufficient to implement this algorithm. To complete the proof of Theorem 12.14, recall that the RMQ data structure on array $\text{LCP}_{S\#}$ takes $O(|S|)$ space, and it can be constructed in $O(|S|)$ time (see Section 3.1).

12.3   Literature

The Lempel–Ziv parsing problem we described in this chapter is also called Lempel–Ziv 77 (Ziv & Lempel 1977). Given a string of length n and its Lempel–Ziv parse of size z, one can construct a context-free grammar of size $O(z\log(n/z))$ that generates S and only S, as shown by Rytter (2003) and Charikar *et al.* (2002). This is the best approximation algorithm for the *shortest grammar problem*, which is known to be NP-hard to approximate within a small constant factor (Lehman & Shelat 2002). Short context-free grammars have been used to detect compositional regularities in DNA sequences (Nevill-Manning 1996; Gallé 2011).

Space-efficient Lempel–Ziv parsing is considered in Belazzougui & Puglisi (2015). The version covered in the book is tailored for our minimal setup of data structures. The computation of the smallest number of bits in any Lempel–Ziv parse of a string described in Section 12.2 is taken from Ferragina *et al.* (2009) and Ferragina *et al.* (2013), where the complexity of Theorem 12.14 is improved to $O(|S| \cdot (\phi + \gamma))$ by using an RMQ data structure that can be built in $O(|S|)$ space and time, and that can answer queries in constant time. The same paper improves the $O(\sigma|S|)$ complexity of the greedy algorithm mentioned in Insight 12.1 to $O(|S| \cdot (1 + \log \sigma / \log \log |S|))$. The non-greedy variants of LZ described in Insights 12.2 and 12.3 are from Kuruppu *et al.* (2011a) and references therein, where they are tested on large collections of similar genomes. Other solutions for large collections of genomes appear in Deorowicz & Grabowski (2011) and Kuruppu *et al.* (2011b), and applications to webpages are described in Hoobin *et al.* (2011).

Except for the few insights mentioned above, we did not extensively cover algorithms tailored to DNA compression: see Giancarlo *et al.* (2014) for an in-depth survey of such approaches.

Exercises

12.1   Recall from Exercise 11.7 that a σ-ary de Bruijn sequence of order k is a string of length $\sigma^k + k - 1$ over an alphabet of size σ that contains as substrings all the σ^k possible strings of length k. What is the number of phrases in the Lempel–Ziv parse of such a sequence? What does this tell you about the asymptotic worst-case complexity of a Lempel–Ziv parse? *Hint.* Combine your result with Lemma 12.2.

12.2   Consider the solution of Exercise 12.1. Is the size of the Lempel–Ziv parse of a string a good approximation of its Kolmogorov complexity, roughly defined in Section 11.2.5?

12.3   Show how to build in linear time the augmented suffix tree used in Section 12.1.1.

12.4   Show how to build the vector R described in Section 12.1.2 in $O(n\log\sigma)$ time and in $O(n\log\sigma)$ bits of space.

12.5   Prove Lemma 12.5 shown in Section 12.1.2.

12.6   Consider a generalization of Lempel–Ziv parsing where a phrase can point to its previous occurrence as a *reverse complement*. Show that the space-efficient parsing algorithms of Section 12.1 can be generalized to this setting. Can you modify also the bit-optimal variant to cope with reverse complements?

12.7   Prove the monotonicity of cost property for parse graphs described in Section 12.2.

12.8   Prove Lemma 12.12 by contradiction, working on the shortest path $P = v_{x_1} v_{x_2} \ldots v_{x_m}$ of the parse graph G such that its first nonmaximal arc $(v_{x_i}, v_{x_{i+1}})$ occurs as late as possible.

12.9   Prove Theorem 12.13, using the algorithm for computing the single-source shortest path in a DAG described in Section 4.1.2.

12.10   Consider all arcs in the parse graph G of Definition 12.10 that start from a given vertex, and let $b_{i_1}, b_{i_2}, \ldots, b_{i_k}$ be a partition of such arcs into equivalence classes according to the number of bits required to encode their copy distances using encoder f. Describe an algorithm that, given the length of the distance-maximal arc of class b_{i_p} and the length of the distance-maximal arc of the previous class $b_{i_{p-1}}$, where $p \in [2..k]$, returns all the length-maximal arcs inside class b_{i_p}, in constant time per length-maximal arc.

12.11   Recall from Section 12.2 the relationship between the length ℓ_e of a distance-maximal arc e and the LCP of the suffixes of S that start at those positions which are addressable with the number of bits used to encode the copy distance d_e of arc e. Prove this property, using the definition of a distance-maximal arc.

12.12   With reference to Lemma 12.18, show that sorting the set of suffixes $S'[h..|S|]\#$ with $h \in [p..q]$ coincides with computing the suffix array of string W.

12.13   Adapt Lemma 12.18 to the case in which `source` and `target` do not intersect. *Hint.* Sort the suffixes that start inside each of the two intervals separately, and merge the results using LCP queries.

13 Fragment assembly

In the preceding chapters we assumed the genome sequence under study to be known. Now it is time to look at strategies for how to *assemble* fragments of DNA into longer contiguous blocks, and eventually into chromosomes. This chapter is partitioned into sections roughly following a plausible workflow of a de novo assembly project, namely, *error correction*, *contig assembly*, *scaffolding*, and *gap filling*. To understand the reason for splitting the problem into these realistic subproblems, we first consider the hypothetical scenario of having error-free data from a DNA fragment.

13.1 Sequencing by hybridization

Assume we have separated a single DNA strand spelling a sequence T, and managed to measure its *k-mer spectrum*; that is, for each k-mer W of T we have the frequency $\mathrm{freq}(W)$ telling us how many times it appears in T. *Microarrays* are a technology that provides such information. They contain a slot for each k-mer W such that fragments containing W hybridize to the several copies of the complement fragment contained in that slot. The amount of hybridization can be converted to an estimate on the frequency count $\mathrm{freq}(W)$ for each k-mer W. The *sequencing by hybridization* problem asks us to reconstruct T from this estimated k-mer spectrum.

Another way to estimate the k-mer spectrum is to use high-throughput sequencing on T; the k-mer spectrum of the reads normalized by average coverage gives such an estimate.

Now, assume that we have a perfect k-mer spectrum containing no errors. It turns out that one can find in linear time a sequence T' having exactly that k-mer spectrum. If this problem has a unique solution, then of course $T' = T$.

The algorithm works by solving the Eulerian path problem (see Section 4.2.1) on a de Bruijn graph (see Section 9.7) representing the k-mers.

Here we need the following *expanded* de Bruijn graph $G = (V, E)$. For each k-mer $W = w_1 w_2 \cdots w_k$ we add $\mathrm{freq}(W)$ copies of the arcs (u, v) to E such that vertex $u \in V$ is labeled with the $(k - 1)$-mer $\ell(u) = w_1 w_2 \cdots w_{k-1}$ and vertex $v \in V$ is labeled with the $(k-1)$-mer $\ell(v) = w_2 w_3 \cdots w_k$. Naturally, the arc (u, v) is then labeled $\ell(u, v) = W$. The set V of vertices then consists of these $(k - 1)$-mer prefixes and suffixes of the k-mers, such that no two vertices have the same label. That is, the arcs form a multiset,

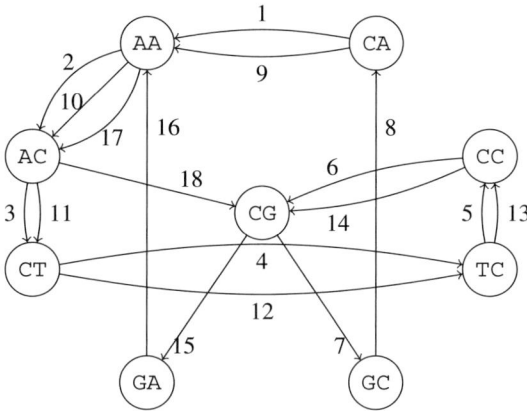

Figure 13.1 The expanded de Bruijn graph of the k-mer spectrum freq(AAC) = 3, freq(ACG) = 1, freq(ACT) = 2, freq(CAA) = 2, freq(CCG) = 2, freq(CGA) = 1, freq(CGC) = 1, freq(CTC) = 2, freq(GAA) = 1, freq(GCA) = 1, and freq(TCC) = 3. The numbers on the arcs denote the order in which they appear in an Eulerian path spelling $T' = $ CAACTCCGCAACTCCGAACG.

and the vertices form a set. Figure 13.1 illustrates the construction of this expanded de Bruijn graph.

An Eulerian path $v_1, v_2, \ldots, v_{n-k+1}$ in the expanded de Bruijn graph G visits all arcs, and it can be interpreted as a sequence $T' = \ell(v_1) \cdot \ell(v_2)[k-1] \cdot \ell(v_3)[k-1] \cdots \ell(v_{n-k+1})[k-1]$. Each arc of G has its label associated with a unique position in T'. Hence, T' has the same k-mer spectrum as the one from which the expanded de Bruijn graph was built. As shown in Section 4.2.1, this path can be found in linear time. One such path is shown in Figure 13.1.

Obviously, one issue with the above approach is that the k-mer spectrum is extremely hard to estimate perfectly. Having an inaccurate spectrum, one could formulate a problem asking how to correct the graph minimally so that it contains an Eulerian path. However, this formulation makes the problem hard (see the literature section).

Another possibility is to formulate a problem where one is given for every k-mer (that is, an arc of the expanded de Bruijn graph) a lower bound on the number of times it must appear in the reconstructed string T. Exercise 13.1 asks the reader to show that this problem, called the Chinese postman problem, can be reduced to a network flow problem (recall Chapter 5).

Another issue is that there can be several sequences having the same k-mer spectrum. Moreover, if one estimates the spectrum with high-throughput sequencing, one should also take into account the facts that reads come from both DNA strands and from both haplotypes of a diploid individual.

We shall next explore practical assembly strategies that take into account the above realistic constraints. We will come back to a combinatorial modeling, and give a proof of the computational hardness of the assembly problem when studying *scaffolding*, since this subproblem best captures the nature of the entire fragment assembly problem.

13.2   Contig assembly

Instead of trying to estimate an entire sequence using a k-mer spectrum, a de Bruijn graph, or read overlaps, the *contig assembly* problem has a more modest goal: find a set of paths in a given *assembly graph* such that the contiguous sequence content (called *contig*) induced by each path is likely to be present in the genome under study. Naturally, the goal is to find as few and as long contigs as possible, without sacrificing the accuracy of the assembly.

Here an assembly graph refers to any directed graph whose vertices represent plausible substrings of the genome and whose arcs indicate overlapping substrings. The de Bruijn graph constructed from the k-mers of the sequencing reads is one example of an assembly graph. Another example is an *overlap graph* based on maximal exact/approximate overlaps of reads, whose construction we covered in Section 10.4. A characteristic feature of an assembly graph is that a path from a vertex s to a vertex t can be interpreted as a plausible substring of the genome, obtained by merging the substrings represented by the vertices on the path. In the previous section, we saw how this procedure was carried out for de Bruijn graphs, and we will later detail this procedure on other kinds of assembly graphs.

The simplest set of contigs to deduce from an assembly graph is the one induced by *maximal unary paths*, also known as *unitigs* for *unambiguous contigs*.

DEFINITION 13.1   *A path* $P = (s, v_1, \ldots, v_n, t)$ *from s to t in a directed graph G is called* maximal unary *if all of the v_i have exactly one in-neighbor and one out-neighbor (on P), and s and t do not have this property.*

Unfortunately, the contigs extracted from maximal unary paths from an assembly graph built on a set of *raw* reads will be too short and redundant due to substructures caused by sequencing errors and alleles. *Cleaning* an assembly graph is discussed in Insight 13.1. Exercise 13.22 asks the reader under which assumptions a unary path in an assembly graph correctly spells a contig. Exercise 13.23 asks for an algorithm to extract all maximal unary paths in linear time by a simple graph traversal. Exercise 13.24 asks for an alternative algorithm based on a compaction process of the directed graph. Finally, Exercise 13.25 further formalizes the notion of a contig.

Insight 13.1 Tips, bubbles, and spurious arcs

Consider a de Bruijn graph built on the k-mers of a set of reads that may contain sequencing errors. A sequencing error may cause an otherwise unary path to branch at one point with some short unary path that ends in a sink (or, symmetrically, a short unary path starting from a source ending inside an otherwise unary path). Such substructures are called *tips*; these are easy to remove, given a threshold on the maximum length of such paths. Sequencing error in the middle of a read may also cause a *bubble*, in which two vertices are connected by two parallel unary paths. Contig extraction through maximal unary paths would report both parallel paths of a bubble, although they might represent the same area in the genome. These substructures are visualized below:

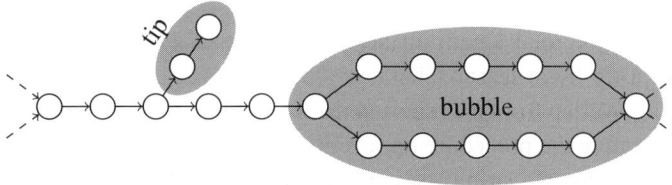

In addition to *erroneous bubbles* caused by sequencing errors, there can be *redundant bubbles*. These are caused by alleles where reads (or k-mers) containing the same allele type form one path of a bubble. In diploid organisms, one should then expect to see a bubble in the de Bruijn graph for each heterozygous mutation. This gives in fact a way to do *de novo variation calling* without a reference genome, as opposed to the alignment-based variation calling that we will cover in Section 14.1.

Finally, bubbles can also appear accidentally when source and sink k-mers appear at many places in the genome. Such bubbles are likely to have different sequence content in the two parallel paths, and hence one can try to separate them from erroneous and redundant bubbles.

For fragment assembly, it is useful to clean the graph of erroneous and redundant bubbles, by removing one of their parallel paths. Since we can label each arc by the amount of reads containing that k-mer, it is easy to select the lower-coverage path for removal. This avoids some redundancy in the resulting contigs. After the assembly has been finished, the variants can be reconstructed, for example by read alignment (see Section 14.1).

Sequencing errors can also create more complex substructures that are difficult to detect. A possible strategy to clean the graph more aggressively is to remove *spurious arcs*, that is, arcs with low support compared with their neighbors.

13.2.1 Read error correction

A practical problem in cleaning an assembly graph, as discussed in Insight 13.1, is that one should modify the assembly graph dynamically. The succinct structures described in Section 9.7 are, however, static. A workaround is to do *read error correction/removal* based on the identified substructures of the assembly graph and build the assembly graph again using the corrected reads. The advantage is that one can do contig assembly using a different assembly graph than the one used for correcting the reads.

To fix ideas, consider that we have identified tips, bubbles, and spurious arcs in a de Bruijn graph, as defined in Insight 13.1. We wish to correct the reads so that the identified tips and spurious arcs do not appear in the de Bruijn graph built on the corrected reads. Moreover, each erroneous or redundant bubble should be replaced by one of its two unary paths, selected by some criterion, as discussed in Insight 13.1.

A tip spells a string that must overlap either prefixes of some reads or suffixes of some reads by at least k characters. In Section 8.4.4, we studied how to find such exact overlaps, and in Section 13.2.3 we will study a space-efficient algorithm for this problem. That is, we can build a BWT index on the set of reads, occupying $O(N \log \sigma)$ bits, where N is the total length of the reads, so that each read whose prefix matches the suffix of a tip by at least k characters will be reported. It is then easy to cut the corresponding overlap from the reported reads. Building the BWT index on the reversed reads and repeating the procedure on reversed tips solves the symmetric case.

A bubble b spells two strings, say A^b and B^b, where A^b has been identified as erroneous or redundant and B^b as the consensus. That is, we have a set of replacement rules $S = \{(A^b \to B^b) : b \text{ is a bubble}\}$. A BWT index on the set of reads can again be used to implement these replacement rules: search each A^b, locate all its occurrences, and replace substring A^b with B^b inside each read found.

A spurious arc is a locally under-represented k-mer, meaning that the source of the corresponding arc has at least one other outgoing arc with significantly more support. Hence, one can consider a replacement rule like in the case of bubbles.

Exercise 13.3 asks you to consider how to implement the replacement rules in detail. Exercise 13.4 asks you to develop a linear-time algorithm to detect all bubbles in an assembly graph.

13.2.2   Reverse complements

One aspect we have omitted so far is that reads can come from both strands of a DNA fragment. This is addressed in Insight 13.2 for the case of overlap graphs. We will later learn how to structurally compress overlap graphs, and such compaction is also defined on the special structure designed in Insight 13.2.

With de Bruijn graphs, one could use the same reduction as in the case of overlap graphs. We will not consider this variant in detail, since the succinct representations of Section 9.7 do not easily extend to this special de Bruijn graph structure. A trivial solution is, however, to build the de Bruijn graph on the set of reads and their reverse complements. This will result in redundant contigs representing opposite strands that need to be detected afterwards.

In the next subsections, we assume for simplicity that reads come from one strand, but one can modify those approaches similarly to what is done in Insight 13.2 to cope with the real setting. Later, in the scaffolding step we need to explicitly consider reverse complementation again.

Insight 13.2 Reverse complements in assembly graphs

Let us consider an overlap graph, whose vertices represent reads, and whose arcs represent their overlaps. Each read can originate from either one of the two DNA strands, and recall that the strands are sequenced in opposite directions. Let us consider all eight ways in which two reads R and S can overlap (denote by $\underline{R}$ and $\underline{S}$ the reverse complements of R and S, respectively):

(i)   a suffix of R overlaps a prefix of S;

(i′)   a suffix of $\underline{S}$ overlaps a prefix of $\underline{R}$;

(ii)   a suffix of R overlaps a prefix of $\underline{S}$;

(ii′)   a suffix of S overlaps a prefix of $\underline{R}$;

(iii)   a suffix of $\underline{R}$ overlaps a prefix of S;

(iii′)   a suffix of $\underline{S}$ overlaps a prefix of R;

(iv)   a suffix of $\underline{R}$ overlaps a prefix of $\underline{S}$;

(iv′)   a suffix of S overlaps a prefix of R.

Observe that, if two reads overlap as in case (i) on one strand, then on the other strand we have an overlap as in (i′). Symmetrically, if two reads overlap as in (i′) on one strand, then on the other strand we have the overlap (i). A similar analogy holds for the overlaps (ii)–(ii′), (iii)–(iii′), and (iv)–(iv′).

A trivial solution for dealing with these eight cases is to add, for each read, two separate vertices: one for the read as given and one for its reverse complement. However, this has the problem that both of these vertices can appear in some assembled fragment, meaning that the read is interpreted in two ways.

Another solution is to use a modified graph structure that has some added semantics. Split the vertex corresponding to each read R into two vertices R_{start} and R_{end} and connect these vertices with the undirected edge $(R_{\text{start}}, R_{\text{end}})$, which we call the *read edge*. If a path reaches R_{start} and then continues with R_{end}, then this path spells the read R as given; if it first reaches R_{end} and then continues with R_{start}, then this path spells the reverse complement of R.

The eight overlap cases now correspond to the following *undirected* edges:

i. $(R_{\text{end}}, S_{\text{start}})$ for an overlap as in the above cases (i) or (i′);

ii. $(R_{\text{end}}, S_{\text{end}})$ for an overlap as in the above cases (ii) or (ii′);

iii. $(R_{\text{start}}, S_{\text{start}})$ for an overlap as in the above cases (iii) or (iii′);

iv. $(R_{\text{start}}, S_{\text{end}})$ for an overlap as in the above cases (iv) or (iv′).

A path in this mixture graph must start and end with a read arc, and read arcs and overlap arcs must alternate. The direction of the read arcs visited by a path determines whether a read is interpreted as given or as reverse complemented.

13.2.3   Irreducible overlap graphs

Succinct representations for de Bruijn graph were studied in Section 9.7. They enable space-efficient contig assembly. However, overlap graphs contain more contiguous information about reads, possibly resulting in more accurate contigs. So it makes sense to ascertain whether they also could be represented in small space in order to obtain a scalable and accurate contig assembly method. Some general graph compression data structures could be applied on overlap graphs, but it turns out that this does not significantly save space. We opt for *structural compression*; some overlaps are redundant and can be removed. This approach will be developed in the following.

Recall that we studied in Section 8.4.4 how to efficiently find maximal exact overlaps between reads using the suffix tree. A more space-efficient approach using the BWT index was given to find approximate overlaps in Section 10.4. We can hence build an overlap graph with reads as vertices and their (approximate) overlaps as arcs. For large genomes, such an approach is infeasible due to the size of the resulting graph. Now that we have studied how to correct sequencing errors, we could resort back to an overlap graph defined on exact overlaps. We will next study how to build such a graph on maximal exact overlaps with the suffix tree replaced (mostly) by the BWT index, to achieve a practical space-efficient algorithm. Then we proceed with a practical algorithm to build directly an *irreducible overlap graph*, omitting arcs that can be derived through an intermediate vertex.

Space-efficient computation of maximal overlaps

Consider a set of reads $\mathcal{R} = \{R^1, R^2, \ldots, R^d\}$. We form a concatenation $C = \#R^1\#R^2\#\cdots\#R^d\#$ and build a BWT index on it. Note that every position i of an occurrence of $\#$ can easily be mapped to the lexicographic order of the corresponding read through the operation $\mathtt{rank}_\#(\mathsf{BWT}_C, i) - 1$. We also store the array $\mathsf{RA}[1..d] = \mathsf{SA}[2..d+1]$, which gives the starting positions of the reads in the concatenation, given their lexicographic order. In some cases, it might be more convenient that the reads are stored in lexicographic order, so that the lexicographic number returned by the BWT index represents the sequential order in which the read is stored. We can put the reads into lexicographic order, by scanning the array RA, to retrieve the starting position of the reads in lexicographic order and concatenate them together. We then rebuild the BWT index and the array RA. Exercise 13.10 asks you for an alternative way to sort the reads in $O(N \log \sigma)$ time and bits of space, where N is the length of the reads concatenation.

Consider a backward search with read R^i on the BWT index of C. We obtain an interval $[p_k..q_k]$ in the BWT of C associated with each suffix of $R^i_{k..|R^i|}$. If a backward step with $\#$ from $[p_k..q_k]$ results in a non-empty interval $[p_k^\#..q_k^\#]$, there is at least one suffix of C in the interval $[p_k..q_k]$ of the suffix array of C that starts with an entire read R^j.

Given a threshold τ on the minimum overlap, we should consider only intervals $[p_k^\#..q_k^\#]$ with $k \leq |R^i| - \tau$. One could list all pairs (i, j) as arcs of an overlap graph with $j \in [p_k^\# - 1..q_k^\# - 1]$. However, there are cases where the same (i, j) can be reported multiple times; this problem was solved using a suffix tree in Section 8.4.4, but now we aim for a more space-efficient solution. We exploit the following observation on the relationship between the problematic intervals.

LEMMA 13.2   *If reads R^i and R^j have two or more suffix–prefix overlaps, the shorter overlapping strings are prefixes of the longer overlapping strings.*

Proof   It is sufficient to consider two overlaps of length o_1 and o_2 with $o_1 < o_2$. We have that $R^i[|R^i| - o_1 + 1..|R^i|] = R^j[1..o_1]$ and $R^i[|R^i| - o_2 + 1..|R^i|] = R^j[1..o_2]$. Hence, $R^i[|R^i| - o_2 + 1..|R^i| - o_2 + o_1] = R^i[|R^i| - o_1 + 1..|R^i|]$. $\qquad\square$

An even stronger result holds, namely that a shorter suffix–prefix overlap is a border (see Section 2.1) of the longer ones. A consequence of the above lemma is that the intervals of the BWT containing occurrences of R^j that overlap R^i must be nested.

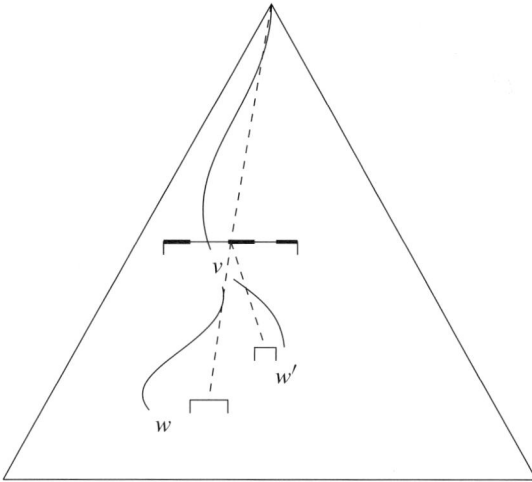

Figure 13.2 Conceptual suffix tree of read R^i associated with nested intervals in the BWT of the concatenation of reads. Curved lines denote suffix tree paths connecting the root to v, v to w, and v to w'. The corresponding associated intervals are shown and connected by dashed arrows to form the tree of nested intervals. The thick-line segments inside the interval of v show its sub-intervals to be reported.

To report only the maximal overlap, it suffices to associate each SA[r] with the deepest interval in the hierarchy of nested intervals that contains position r. This hierarchy can be constructed with the aid of the suffix tree of R^i as follows. For each suffix $R^i[k..|R^i|]$, we follow the path from the root to the first node v such that a prefix of the path to v spells $R^i[k..|R^i|]$ (that is, v is a proper locus of $R^i[k..|R^i|]$ using the notation introduced in Section 8.3). We store interval $[p_k^\#..q_k^\#]$ at node v. The tree with nested intervals is then constructed by linking each stored interval to its closest ancestor in the suffix tree having also an interval stored, or to the root if no such ancestor is found. This construction is illustrated in Figure 13.2.

Consider now a node v in the tree of nested intervals with a set of children $N^+(v)$. Let us denote the associated interval by $\vec{v}$ and the overlap length by depth(v). We report the overlap of length depth(v) between reads R^i and R^j iff $j \in [p_{\text{depth}(v)}^\# - 1..q_{\text{depth}(v)}^\# - 1]$ and $j \notin [p_{\text{depth}(w)}^\# - 1..q_{\text{depth}(w)}^\# - 1]$ for all $w \in N^+(v)$. By examining the intervals of the children, the interval $\vec{v}$ is easy to splice into at most $|N^+(v)| + 1$ parts containing only the entries to be reported.

THEOREM 13.3   *Let $\mathcal{R} = \{R^1, R^2, \ldots, R^d\}$ be a set of reads with overall length $\sum_{i=1}^{d} |R^i| = N$, maximum read length $\max_{i=1}^{d} |R^i| = M$, and each $R^i \in [1..\sigma]^*$. There is an algorithm to solve Problem 8.5, that is, to report all triplets (i, j, ℓ) such that a suffix of R^i overlaps a prefix of R^j by $\ell \geq \tau$ characters, where ℓ is the maximal such overlap length for the pair R^i and R^j. This algorithm uses $O(d \log N + M \log M + N \log \sigma)$ bits of space and reports all c valid triplets in $O(N \log \sigma + c)$ time.*

Proof   The $O(N \log \sigma)$ time and space come from supporting backtracking on the BWT index. The array RA[1..d] (suffix array values only for suffixes starting with #)

contributes $d \log N$ bits of space. The suffix tree built for every read and the nested intervals stored in it take $O(M \log M)$ bits. The construction takes time linear in the read length; Exercise 13.11 asks you to detail the linear-time procedure to deduce the tree of nested intervals from the suffix tree. Finally, only the maximal overlaps are reported, and the value ℓ for each reported triplet (i, j, ℓ) is stored at the corresponding node in the tree of nested intervals.   □

The graph constructed by maximal exact overlaps can still be considerably large, so one could think of removing arcs that can be deduced through some other vertices. We call an arc of an overlap graph *redundant* if it represents an overlap triplet $(i, j', \ell_{i,j'})$ and there also exist arcs of the overlap graph representing triplets $(i, j, \ell_{i,j})$ and $(j, j', \ell_{j,j'})$, where $\ell_{j,j'} > \ell_{i,j'}$. That is, the overlap length $\ell_{i,j'}$ between R^i and R^j can be deduced from overlaps of length $\ell_{i,j}$ between R^i and R^j and of length $\ell_{j,j'}$ between R^j and $R^{j'}$. After all redundant arcs have been removed, we have an *irreducible overlap graph*.

Constructing the graph and removing redundant arcs is not a good solution because the peak memory consumption is not reduced. It turns out that one can directly construct the irreducible overlap graph by extending the previous algorithm for maximal overlaps. We sketch the solution in Insight 13.3 and study the algorithm further in the exercises.

Insight 13.3 Direct construction of irreducible overlap graphs

Consider the algorithm leading to Theorem 13.3, until it reports the arcs. That is, we have at each vertex v of the tree of nested intervals the main interval $\bar{v}$ spliced into at most $|N^+(v)| + 1$ parts containing the maximal overlaps, with each such subinterval associated with the overlap length $\texttt{depth}(v)$. In addition, we replace the BWT index with the bidirectional BWT index, so that for each interval in $\mathsf{BWT}(C)$ we also know the corresponding interval in $\mathsf{BWT}(\underline{C})$; notice that the splicing on $\mathsf{BWT}(\underline{C})$ is defined uniquely by the splicing of $\mathsf{BWT}(C)$.

Let $\mathcal{I}$ collect all these (interval in $\mathsf{BWT}(C)$, interval in $\mathsf{BWT}(\underline{C})$, overlap length) triplets throughout the tree. One can then do backtracking on $\mathsf{BWT}(\underline{C})$ starting from intervals in $\mathcal{I}$ using the bidirectional BWT index, updating along all backtracking paths all intervals in $\mathcal{I}$ until they become empty on the path considered, until one finds a right extension of some interval with symbol #, or until there is only one interval left of size one. In the two latter cases, one has found a non-redundant arc, and, in the first of these two cases, the other entries in the intervals that have survived so far are redundant.

Exercise 13.12 asks you to detail the above algorithm as a pseudocode. We complement the above informal description with a conceptual simulation of the algorithm on an example. Consider the following partially overlapping reads:

        ACGATGCGATGGCTA
               GATGGCTAGCTAG
                  GGCTAGCTAGAGGTG
                  GGCTAGacgagtagt

Consider the overlap search for the top-most read with the bidirectional BWT index of $C = $ #ACGATGCGATGGCTA#GATGGCTAGCTAG#GGCTAGCTAGAGGTG# GGCTAGACGAGTAGT#. After left extensions, we know the two intervals of #GATGGCTA and #GGCTA in $\mathsf{BWT}(C)$. Backtracking with right extensions starts only with G, with both intervals updated into non-empty intervals corresponding to #GATGGCTAG and #GATGGCTAG; other options lead to non-empty intervals. After this, backtracking branches, updating two intervals into ones corresponding to #GATGGCTAGC and #GATGGCTAGC and into one interval corresponding to #GATGGCTAGA. The former branch continues, with the two intervals updated until the interval of #GATGGCTAGCTAG is found to have right extension with #: the non-redundant arc between the first and the second read is reported. The latter branch has one interval of size one: the non-redundant arc between the first and the last read is reported. The redundant arc between the first and the third read is correctly left unreported.

13.3 Scaffolding

In this section we are concerned with finding the relative order of, orientation of, and distance between the contigs assembled in the previous section (a *scaffolding* of the contigs). We will do this with the help of paired-end reads aligning in different contigs. In the next section we will see a method for filling the gaps between contigs.

Let us summarize below some assumptions that this scaffolding step is based on.

- **Double-stranded DNA.** One strand is the complement of the other, and the two strands are sequenced in opposite directions.
- **Paired-end reads.** Recall from Section 10.5 that a paired-end read is a pair consisting of two reads, such that the second read occurs in the genome at a known distance after the first one (in the direction of the DNA strand). We refer to the missing sequence between the occurrences of the two reads as *gap*.
- **Same-strand assumption.** Both reads of a paired-end read originate from the same strand, but which strand is unknown. The input paired-end reads can always be normalized to satisfy this assumption, by reverse complementing one of the two reads of a paired-end read.
- **Contigs.** Each assembled contig entirely originates from one unknown strand. This assumption should be guaranteed by the previous contig assembly step. However, note that this might not always be the case in practice, but such errors will be tolerated to some extent in the scaffolding formulation to be studied next.

To start with, assume that we have a paired-end read (R^1, R^2) with a known gap length, and assume that R^1 aligns to a position inside a contig A, and that R^2 aligns to a position inside a contig B. This alignment of (R^1, R^2) implies that

- on one strand, contig A is followed by contig B, with a gap length given by the alignment of (R^1, R^2);
- on the other strand, the reverse complement of B, denoted $\underaccent{\sim}{B}$, is followed by the reverse complement $\underaccent{\sim}{A}$ of A, with the same gap length estimate given by the alignment of (R^1, R^2).

In Figure 13.3 we show all possible alignment cases of (R^1, R^2) into contigs A and B, and then indicate the corresponding alternatives in the double-stranded DNA we are trying to reconstruct.

A scaffold will be a list of contigs appearing in the *same strand* (but which strand is unknown) of the assembled DNA. For this reason, with each contig C in a scaffold, we need to associate a Boolean variable o, which we call the *orientation* of C, such that, if $o = 0$, then we read C as given, and if $o = 1$, then we read C as reverse complemented.

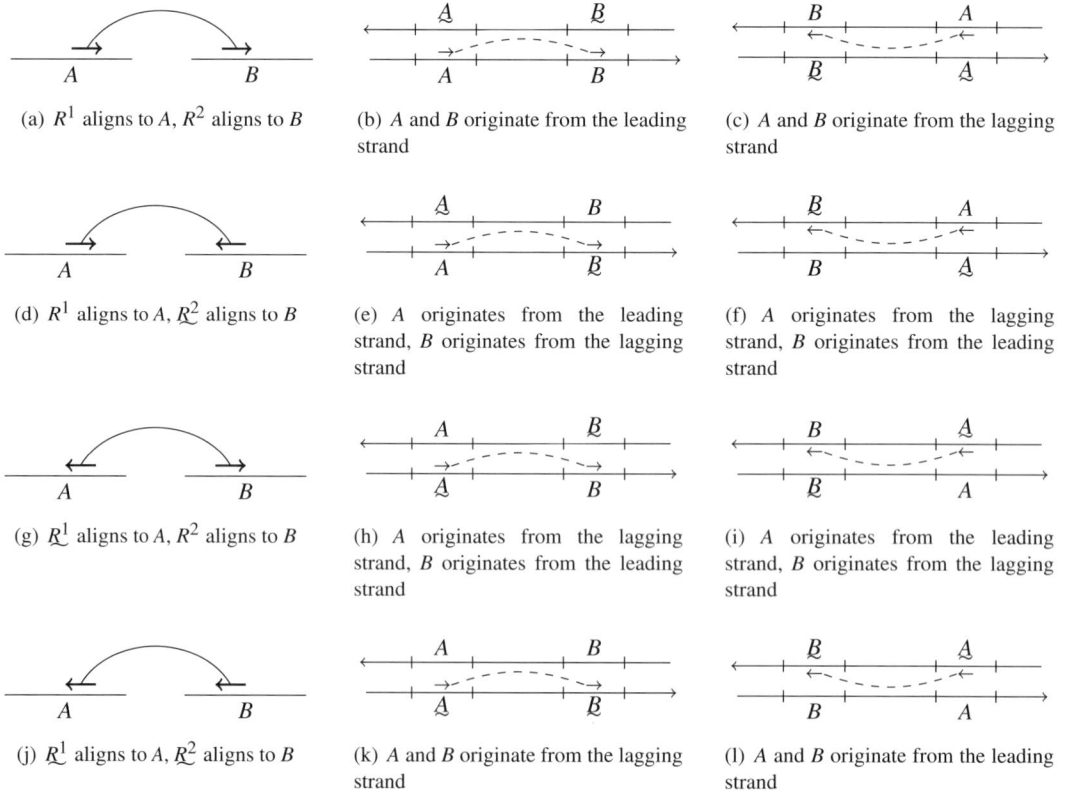

(a) R^1 aligns to A, R^2 aligns to B

(b) A and B originate from the leading strand

(c) A and B originate from the lagging strand

(d) R^1 aligns to A, $\underaccent{\sim}{R^2}$ aligns to B

(e) A originates from the leading strand, B originates from the lagging strand

(f) A originates from the lagging strand, B originates from the leading strand

(g) $\underaccent{\sim}{R^1}$ aligns to A, R^2 aligns to B

(h) A originates from the lagging strand, B originates from the leading strand

(i) A originates from the leading strand, B originates from the lagging strand

(j) $\underaccent{\sim}{R^1}$ aligns to A, $\underaccent{\sim}{R^2}$ aligns to B

(k) A and B originate from the lagging strand

(l) A and B originate from the leading strand

Figure 13.3 All fours cases of aligning a paired-end read (R^1, R^2) to two contigs A and B. For each case, the two corresponding placements of the contigs in a double-stranded DNA are shown.

DEFINITION 13.4 *A scaffold S is a sequence of tuples* $S = ((C^1, o_1, d_1),$
$(C^2, o_2, d_2), \ldots, (C^{n-1}, o_{n-1}, d_{n-1}), (C^{n-1}, o_{n-1}, d_n = 0))$, *where each* C^i *is a contig,*
o_i *is its orientation, and* d_i *is the length of the gap between contigs* C^i *and* C^{i+1} *(by convention, we set* $d_n = 0$*).*

For simplicity, we will implicitly assume that each d_i is greater than or equal to 0, implying that in a scaffold there can be no overlap between two contigs.

In the scaffolding problem, which we will introduce in Problem 13.1, we would like to report (1) as few scaffolds as possible, while at the same time (2) making sure that a large proportion of the paired-end read alignments will be consistent with at least one scaffold. For addressing (1), we need to introduce the following notion of a *connected* scaffold with respect to a collection $\mathcal{R}$ of paired-end reads.

DEFINITION 13.5 *Given a set $\mathcal{R}$ of paired-end reads, we say that a scaffold S is* connected *with respect to $\mathcal{R}$ if the contigs of S form a connected graph G_S, in the sense that if we interpret each C^i as a vertex of a graph, and each paired-end read in $\mathcal{R}$ mapping to some contigs C^i and C^j as an edge between the vertices C^i and C^j, then we obtain a connected graph.*

For addressing (2), let us formally define this notion of consistency. For every two contigs C^i and C^j in S, we let $d(C^i, C^j)$ denote the distance between them in S, namely

$$d(C^i, C^j) = d_{\min(i,j)} + \sum_{k=\min(i,j)+1}^{\max(i,j)-1} \left(\ell(C^k) + d_k \right),$$

where $\ell(C^k)$ denotes the length of a contig C^k.

DEFINITION 13.6 *Given a scaffold S, a paired-end read (R^1, R^2) aligning to contigs C^i and C^j, and an integer threshold t, we say that (R^1, R^2) is* t-consistent *with S if its alignment is consistent with the cases shown in Figure 13.3, that is,*

- *if R^1 aligns to C^i and R^2 aligns to C^j, then either*
  $i < j$ and $o_i = o_j = 0$ (the case in Figure 13.3(b)), or
  $j < i$ and $o_i = o_j = 1$ (the case in Figure 13.3(c));
- *if R^1 aligns to C^i and $\underline{R}^2$ aligns to C^j, then either*
  $i < j$ and $o_i = 0$, $o_j = 1$ (the case in Figure 13.3(e)), or
  $j < i$ and $o_i = 1$, $o_j = 0$ (the case in Figure 13.3(f));
- *if $\underline{R}^1$ aligns to C^i and R^2 aligns to C^j, then either*
  $i < j$ and $o_i = 1$, $o_j = 0$ (the case in Figure 13.3(h)), or
  $j < i$ and $o_i = 0$, $o_j = 1$ (the case in Figure 13.3(i));
- *if $\underline{R}^1$ aligns to C^i and $\underline{R}^2$ aligns to C^j, then either*
  $i < j$ and $o_i = o_j = 1$ (the case in Figure 13.3(k)), or
  $j < i$ and $o_i = o_j = 0$ (the case in Figure 13.3(l)).

Moreover, we require that the distance between C^i and C^j indicated by the alignment of (R^1, R^2) differ by at most t from $d(c_i, c_j)$.

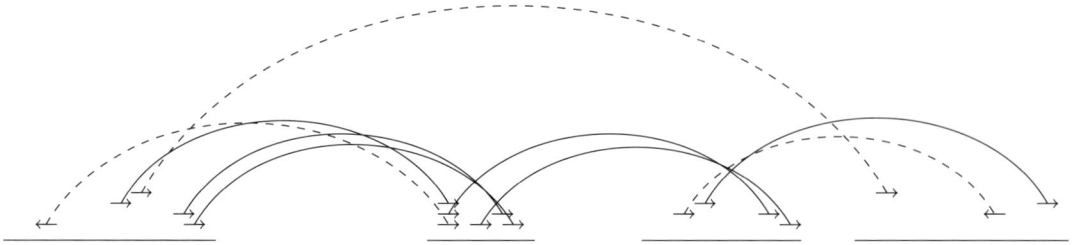

Figure 13.4 A scaffold. Solid read pairs are consistent and dashed read pairs are inconsistent. Here the distance constraint is not explicitly visualized, but the gap of each read pair is assumed to be the same. The alignments of all solid read pairs are t-consistent, with some small t, since their distance is approximately the same and their orientation is consistent with the choice made for the scaffold. The right-most dashed read pair would be consistent in terms of the distance, but the orientation differs from the choice made for the scaffold, making it inconsistent. The middle dashed read pair is inconsistent due to there being too large a distance. The left-most dashed read pair violates the same orientation constraint as the right-most gray read pair, while also the distance constraint could be violated with some choice of t.

See Figure 13.4 for an example of a connected scaffold and some consistent and inconsistent read alignments.

In practice, some paired-end reads are erroneously aligned to the contigs. Therefore, we formulate the scaffolding problem as Problem 13.1 below by asking to remove at most a given number of paired-end reads, such that the remaining alignments give rise to the least number of connected scaffolds. An alternative formulation is considered in Exercise 13.16.

Problem 13.1   Scaffolding

Given a set $\mathcal{C}$ of contigs, the alignments of a set $\mathcal{R}$ of paired-end reads in these contigs, and integers t and k, find a subset $\mathcal{R}' \subseteq \mathcal{R}$ of paired-end reads such that

- $|\mathcal{R}| - |\mathcal{R}'| \leq k$,

and, among all such subsets, $\mathcal{R}'$ has the property that the contigs in $\mathcal{C}$ can be partitioned into the minimum number of scaffolds such that

- every read in $\mathcal{R}'$ is t-consistent with some scaffold, and
- each scaffold is connected with respect to $\mathcal{R}'$.

We prove next that the scaffolding problem is NP-hard.

THEOREM 13.7   *Problem 13.1 is NP-hard.*

Proof We reduce from the Hamiltonian path problem. Given an undirected graph $G = (V = \{1, \ldots, n\}, E = \{e_1, \ldots, e_m\})$, we construct the following instance of the scaffolding problem. For every $i \in V$, we add a unique contig C^i to $\mathcal{C}$. For every $e_s \in E$, with $e_s = (i, j)$, we add the two paired-end reads (C^i, C^j) and (C^j, C^i) to $\mathcal{R}$. Moreover, we set the length of the gap between all $2m$ paired-end reads to some fixed non-negative integer, say 1. We prove that G has a Hamiltonian path if and only if this instance for the scaffolding problem admits a solution with $t = 0$, $k = 2m - (n-1)$, and only one scaffold.

For the forward direction, let $P = (v_1, \ldots, v_n)$ be a Hamiltonian path in G. Then we keep in $\mathcal{R}'$ only the paired-end reads $(C^{v_i}, C^{v_{i+1}})$, for all $i \in \{1, \ldots, n-1\}$. This implies that $|\mathcal{R}| - |\mathcal{R}'| = 2m - (n-1) = k$. Moreover, the set $\mathcal{C}$ of contigs admits one scaffold, connected with respect to $\mathcal{R}'$, formed by following the Hamiltonian path P, namely $S = ((C^{v_1}, 0, 1), (C^{v_2}, 0, 1), \ldots, (C^{v_{n-1}}, 0, 1), (C^{v_n}, 0, 0))$.

For the reverse implication, let $S = ((C^{v_1}, o_{v_1}, d_{v_1}), (C^{v_2}, o_{v_2}, d_{v_2}), \ldots, (C^{v_{n-1}}, o_{v_{n-1}}, d_{n-1}), (C^{v_n}, o_{v_n}, 0))$ be the resulting scaffold. We argue that $P = (v_1, \ldots, v_n)$ is a Hamiltonian path in G, or, more precisely, that there is an edge in G between every v_i and v_{i+1}, for all $1 \le i \le n-1$. Because the length of the gap in any paired-end read is 1 and $t = 0$, we get that each paired-end read (C^i, C^j) is such that C^i and C^j are consecutive contigs in S. Since we added paired-end reads only between adjacent vertices in G, and the scaffold graph G_S of S is connected, P is a Hamiltonian path. $\square$

Observe that the above proof actually shows that the problem remains hard even if it is assumed that all contigs and paired-end reads originate from the same DNA strand. Moreover, the bound k on the size $|\mathcal{R}| - |\mathcal{R}'|$ is not crucial in the proof (particularly in the reverse implication). Exercise 13.14 asks the reader to show that the scaffolding problem remains NP-hard even if no constraint k is received in the input. Finally, notice also that we allow each contig to belong to exactly one scaffold. Thus we implicitly assumed that no contig is entirely contained in a repeated region of the genome. Exercise 13.15 asks the reader to generalize the statement of Problem 13.1 to include such multiplicity assumptions.

Owing to this hardness result, we will next relax the statement of the scaffolding problem, and derive an algorithm to find at least some reasonable scaffolding, yet without any guarantee on the optimality with respect to Problem 13.1. This algorithm is similar to the minimum-cost disjoint cycle cover from Section 5.4.1. We show this mainly as an exercise on how problems from the previous chapters can elegantly deal with the complications arising from reverse complements.

We proceed in the same manner as in Insight 13.2 for reverse complements in overlap graphs. We construct an undirected graph $G = (V, E)$ by modeling each contig similarly to the approach from Insight 13.2. For every $C^i \in \mathcal{C}$, we introduce two vertices, C^i_{start} and C^i_{end}. For every paired-end read (R^1, R^2) aligning to contigs C^i and C^j, we analogously introduce an *undirected* edge in G as follows (see Figure 13.5 and notice the analogy with the edges added in Insight 13.2):

i.   if R^1 aligns to C^i and R^2 aligns to C^j, we add the edge $(C^i_{\text{end}}, C^j_{\text{start}})$;
ii. if R^1 aligns to C^i and $\underline{R}^2$ aligns to C^j, we add the edge $(C^j_{\text{end}}, C^i_{\text{end}})$;

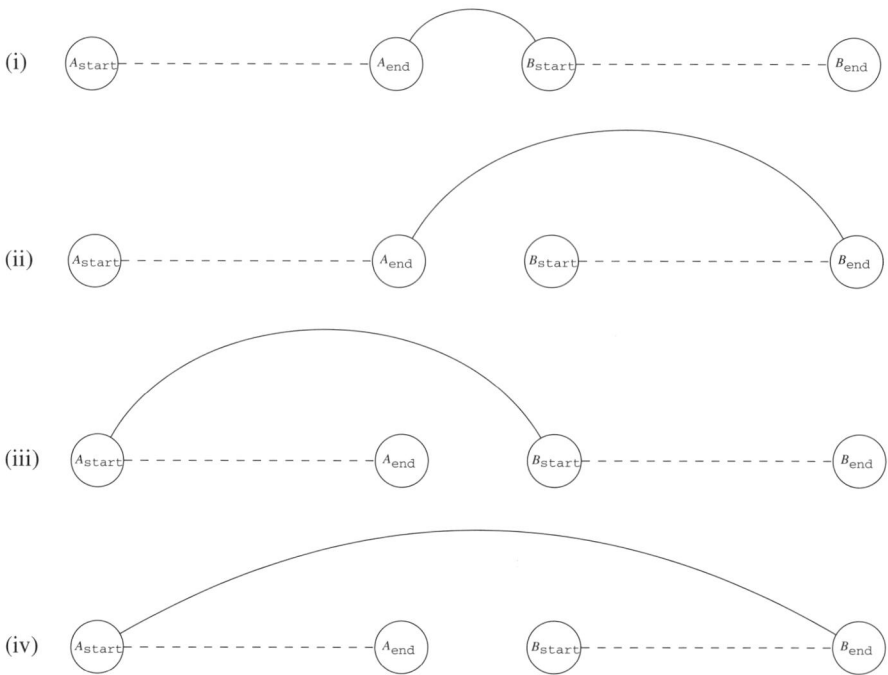

Figure 13.5 Modeling the scaffolding problem as a maximum matching problem in an undirected graph.

iii.   if R^1 aligns to C^i and R^2 aligns to C^j, we add the edge $(C^i_{\text{start}}, C^j_{\text{start}})$;

iv.   if R^1 aligns to C^i and R^2 aligns to C^j, we add the edge $(C^i_{\text{start}}, C^j_{\text{end}})$.

We also add a cost to each edge e of G, modeling the *uncertainty* of the edge indicated by the alignments of all paired-end reads corresponding to e. This cost can be defined, for example, as the absolute deviation of the contig distances predicted by the alignments. We then find a minimum-cost maximum matching M in G (recall Exercise 5.15).

The matching M then induces a collection of scaffolds as follows. Consider the undirected graph G^* with the same vertex set as G, but whose edge set consists of M and all contig edges $(C^i_{\text{start}}, C^i_{\text{end}})$, for every $C^i \in \mathcal{C}$. Say that a path or a cycle in G^* is *alternating* if its edges alternate between contig edges and edges of M. The edges of G^* can be decomposed into a collection of alternating paths or cycles. An alternating path P in G^* naturally induces a scaffold S_P. The contigs of S_P are taken in the order in which they appear on P, oriented as given if the path first uses the `start` vertex followed by the `end` vertex, or oriented as reverse complemented if the path first uses the `end` vertex followed by the `start` vertex. The distance between consecutive contigs is taken as the average distance indicated by the alignments between the contigs. Finally, every alternating cycle can be transformed into a path by removing its edge from M with highest cost; the resulting alternating path induces a scaffold as above. See Figure 13.6 for an example.

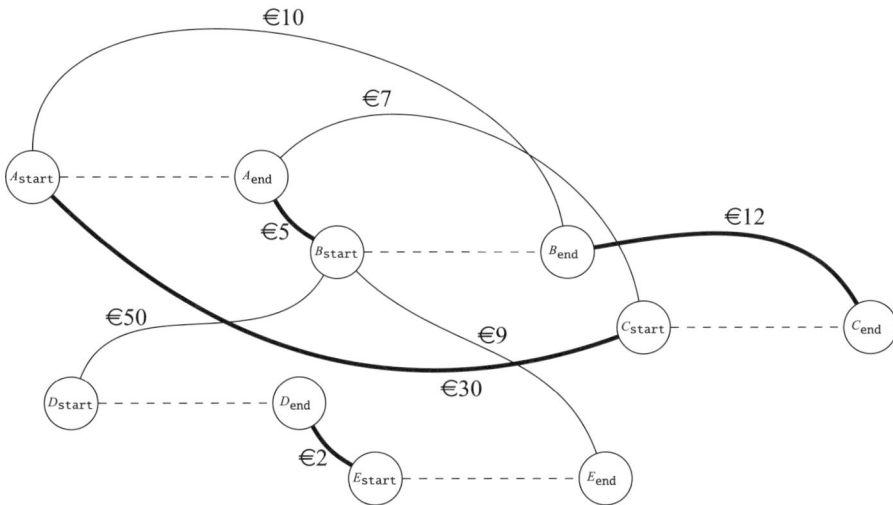

Figure 13.6 Transforming a matching (the thick edges) into scaffolds. The alternating dashed and thick edges give rise to two paths, one of which arises from a cycle cut at the highest-cost edge (€30). Two scaffolds are formed: $A \to B \to C$ and $D \to E$ (here we omit the distances between contigs).

13.4 Gap filling

In this section we are concerned with reconstructing the missing sequence between every two consecutive contigs in a scaffold. Consider the set $\mathcal{R} = \{R^1, \ldots, R^n\}$ of all reads that do not entirely align to any of the contigs. From these reads, and a pair of consecutive contigs S and T, one can build an overlap graph, and then try to find a path between the two contigs. This reconstruction phase can be guided by the constraint that the path length should match the gap length estimated in the scaffolding step. This problem is called *gap filling*.

To state this problem more formally, we let $R^0 = S$ and $R^{n+1} = T$. We build an overlap graph G having a vertex i for every R^i. For every overlap between some R^i and R^j, we add an arc (i, j). This arc is associated with the cost $c(i, j) = |R^i| - \ell_{i,j}$, where $\ell_{i,j}$ is the length of the longest suffix–prefix overlap between R^i and R^j. In other words, $c(i, j)$ is the length of the prefix of R^i obtained by removing the longest overlap with R^j. Observe that we can assume that there are no 0-cost arcs in prefix graph G, since this would indicate a read included in another read, which can be removed without changing the solution. Because of this particular definition of arc costs, this graph is also called a *prefix graph*.

A path $v_1, v_2, \ldots, v_k$ in G *spells* a string of length

$$\sum_{i=1}^{k-1} c(v_i, v_{i+1}) + |R^{v_k}|,$$

obtained by concatenating, for i from 1 to $k - 1$, the prefixes of length $c(v_i, v_{i+1}) = |R^{v_i}| - \ell_{v_i, v_{i+1}}$ of R^i, followed by the last read R^{v_k}.

Given a path P from the start contig S to the end contig T, made up of $S = v_1, v_2, \ldots, v_k = T$, we say that the *cost* of P is

$$\text{cost}(P) = \sum_{i=1}^{k-1} c(v_i, v_{i+1}),$$

namely, the cost of P is equal to the length of the string spelled by P starting with the string S, until the position immediately preceding T.

We formulate the gap-filling problem below, by requiring that the cost of a solution path belong to a given interval. In practice, d' and d should be chosen such that the midpoint $(d' + d)/2$ reflects the same distance as the length of the gap between S and T, estimated from the previous scaffolding step.

Problem 13.2   Gap filling with limited path cost

Given a prefix graph $G = (V, E)$, with a cost function $c : E \to \mathbb{Z}_+$, two of its vertices s and t, and an interval of possible path costs $[d'..d]$, decide if there is a path $P = v_1, v_2, \ldots, v_k$ such that $v_1 = s$, $v_k = t$, and

$$\text{cost}(P) = \sum_{i=1}^{k-1} c(v_i, v_{i+1}) \in [d'..d],$$

and return such a path if the answer is positive.

This gap-filling problem can be solved by a dynamic programming algorithm similar to the Bellman–Ford algorithm, which we studied in Section 4.2.2, and to the Viterbi algorithm studied in Section 7.2. By using as an additional parameter the cost of a path ending in each vertex, we can decompose the problem and break possible cycles of G. More precisely, we define, for all $v \in V(G)$ and $\ell \in [0..d]$,

$$b(v, \ell) = \begin{cases} 1, & \text{if there is a path from } s \text{ to } v \text{ of cost } \textit{exactly } \ell, \\ 0, & \text{otherwise.} \end{cases}$$

We initialize $b(s, 0) = 1$, $b(v, 0) = 0$ for all $v \in V(G) \setminus \{s\}$, and $b(v, \ell) = 0$ for all $v \in V(G)$ and $\ell < 0$. The values $b(\cdot, \cdot)$ can be computed by dynamic programming using the recurrence

$$b(v, \ell) = \begin{cases} 1, & \text{if there is some in-neighbor } u \in N^-(v) \text{ such that } b(u, \ell - c(u, v)) = 1, \\ 0, & \text{otherwise,} \end{cases}$$

$$(13.1)$$

which we could even write more compactly as

$$b(v, \ell) = \max\{b(u, \ell - c(u, v)) \mid u \in N^-(v)\}.$$

The values $b(\cdot, \cdot)$ can be computed by filling a table $B[1..|V|, 0..d]$ column-by-column. This can be done in total time $O(dm)$, where m is the number of arcs of G, since, for each $\ell \in [0..d]$, each arc is inspected only once. The gap-filling problem admits a solution if there exists some $\ell \in [d'..d]$ such that $b(t, \ell) = 1$. One solution path can be traced back by repeatedly selecting the in-neighbor of the current vertex which allowed setting a value to 1 in (13.1). Therefore, we have the following result.

THEOREM 13.8   *Problem 13.2 can be solved in time $O(dm)$, where m is the number of arcs in the input prefix graph and d is the maximum path cost.*

Observe that in practice we may construct a prefix graph in which we use longest overlaps with Hamming distance at most h, where h is a threshold given in the input (this task was solved in Section 10.4). We can then associate the Hamming distance of the overlap with every arc of the input graph. In this case, we are interested in reporting, among all possible paths of cost belonging to the interval $[d'..d]$, the one with minimum total Hamming distance. Exercise 13.18 asks the reader to modify the algorithm presented here for solving this generalization, while maintaining the same time complexity.

Moreover, in this section we constructed a graph made up of *all* the reads that do not entirely align to any of the contigs. However, it makes sense to restrict the problem only to those vertices to which there is a path from s of cost at most d. Exercise 13.19 asks the reader to modify this algorithm to exploit this practical assumption.

Finally, recall that in Section 4.2.2 we interpreted the Bellman–Ford dynamic programming algorithm as a shortest-path problem on a particular layered DAG. Exercise 13.21 asks the reader to interpret the dynamic programming algorithm presented in this section as a problem in a similarly constructed DAG.

13.5   Literature

The first algorithmic approaches to genome assembly came about through the *shortest common supersequence* problem, where one tries to find the shortest sequence that has each read as its substring. This is an NP-hard problem (Räihä & Ukkonen 1981), but good approximation algorithms exist for this problem that are nicely described in a textbook (Vazirani 2001, Chapter 7). In fact, the main part of the approximation algorithms described in Vazirani (2001, Chapter 7) is very similar to the scaffolding reduction given here for turning a matching into a set of paths and cycles. Here we omitted the derivation of the approximation algorithms for the shortest common supersequence problem due to the difficulty in making such algorithms work on realistic datasets with reverse complements, measurement errors, and different alleles present.

The Eulerian path approach to sequencing by hybridization was first studied by Pevzner (1989). The NP-hardness of the problem under measurement errors is considered in Blazewicz & Kasprzak (2003). See Medvedev *et al.* (2007) and Nagarajan & Pop (2009) for applications of the Chinese postman problem to assembly.

Contig assembly with cleaning of substructures such as tips and bubbles is common to many assemblers based on de Bruijn graphs. One of the most popular such assemblers is described in Zerbino & Birney (2008). Read error correction is not typically described through cleaning of tips and bubbles, but quite similar methods based on k-mer indexes exist (Pevzner *et al.* 2001; Chaisson *et al.* 2004; Salmela & Schröder 2011; Yang *et al.* 2012). We used read error correction for indirectly implementing the cleaning so that static data structures could be used for contig assembly. For other applications, such as variation calling, error correction should be done more sensitively, as is partly considered in Exercise 13.2.

We used here a definition for bubbles that involves two unary parallel paths, and these are sometimes called *simple bubbles* in the literature. One can extend the definition to include a pair of non-unary paths. The algorithmics around this general definition is studied in Sacomoto (2014). A further generalization is that of *superbubbles*, introduced in Onodera *et al.* (2013). Exercise 13.8 defines a superbubble and asks the reader to write an algorithm for listing all superbubbles of a directed graph.

The overlap graph and its extension to manage with reverse complements is analogous to the ones in Kececioglu (1991) and Kececioglu & Myers (1993); the representations are not exactly the same, but they are functionally equivalent. An analogous modification for de Bruijn graphs is studied in Medvedev *et al.* (2007). The NP-hardness of many variants of the assembly problem is studied in Kececioglu (1991), Kececioglu & Myers (1993), Medvedev *et al.* (2007), and Nagarajan & Pop (2009).

For the scaffolding problem the NP-hardness of a very similar formulation was first given in Huson *et al.* (2001). As with the contig assembly and graph cleaning, many practical approaches aim to remove spurious edges in the scaffolding graph such that the scaffolding can be solved independently and even exactly on the remaining small components using solvers for integer programming formulations of the problem (Dayarian *et al.* 2010; Salmela *et al.* 2011; Lindsay *et al.* 2012). It is also possible to derive and apply fixed-parameter tractable algorithms (Gao *et al.* 2011; Donmez & Brudno 2013). Accurate estimation of the distances between consecutive contigs inside scaffolds is studied in Sahlin *et al.* (2012).

The direct and space-efficient construction of irreducible overlap graphs is from Simpson & Durbin (2010); there the resulting graph is called an *irreducible string graph*. String graphs and their reduction by removal of transitive edges were first proposed in Myers (2005).

The problem of finding paths in graphs with an exact sum of edge weights was studied in Nykänen & Ukkonen (2002). They showed that this problem is NP-hard: Exercise 13.17 asks you to rediscover this reduction. They also gave a pseudo-polynomial-time algorithm by bounding the edge weights and the requested sum. Our gap-filling problem is easier because the weights in the prefix graph are strictly positive, and therefore we could solve this restricted problem with the simple dynamic programming algorithm. This algorithm is further optimized in Salmela *et al.* (2015).

Exercises

13.1   Consider the following problem, called the Chinese postman problem. Given a directed graph $G = (V, E)$, a demand $d : E \to \mathbb{Z}_+$ for every arc, and a cost $c : E \to \mathbb{Q}_+$ for every arc, find a cycle in G (possibly with repeated vertices) visiting every arc (x, y) at least $d(x, y)$ times, and minimizing the sum of the costs of the arcs it traverses (note that, if the cycle uses an arc (x, y) t times, then the cost associated with this arc is $tc(x, y)$). *Hint.* Reduce the problem to a minimum-cost flow problem.

13.2   An alternative formulation of correcting reads is to directly study the k-mer spectrum. Each k-mer having unexpectedly low coverage could be marked as *erroneous*. Consider a bipartite graph with erroneous k-mers on one side and the remaining k-mers on the other side. Add all edges between the two sides and define the edge cost between two k-mers as their edit distance. Find the minimum-cost maximum matching on this graph: recall Exercise 5.14 on page 66. How can you interpret this matching in order to correct the reads? Can you find a better formulation to exploit matching or network flows for error correction?

13.3   Fix some concrete implementation capable of implementing the replacement rules related to the read error correction after detection of bubbles in an assembly graph. What time and space complexities do you achieve?

13.4   Give a linear-time algorithm to find all bubbles in a directed graph.

13.5   Modify the linear-time bubble-finding algorithm from the previous assignment to work in small space using the succinct de Bruijn graph representations of Section 9.7. *Hint.* Use a bitvector to mark the already-visited vertices. Upon ending at a vertex with no unvisited out-neighbors, move the finger onto the bitvector for the next unset entry.

13.6   Consider replacing each bubble with one of its parallel unary paths. This can introduce new bubbles. One can iteratively continue the replacements. What is the worst-case running time of such an iterative process?

13.7   Consider the above iterative scheme combined with the read error correction of Exercise 13.3. What is the worst-case running time of such an iterative read error correction scheme?

13.8   Given a directed graph $G = (V, E)$, let s and t be two vertices of G with the property that t is reachable from s, and denote by $U(s, t)$ the set of vertices reachable from s without passing through t. We say that $U(s, t)$ is a *superbubble* with *entrance s* and *exit t* if the following conditions hold:

- $U(s, t)$ equals the set of vertices from which t is reachable without passing through s;
- the subgraph of G induced by $U(s, t)$ is acyclic;
- no other vertex t' in $U(s, t)$ is such that $U(s, t')$ satisfies the above two conditions.

Write an $O(|V||E|)$-time algorithm for listing the entrance and exit vertices of all super-bubbles in G.

13.9   Insight 13.2 introduced a special graph structure to cope with reverse comple-ment reads in assembly graphs. Consider how maximal unary paths, tips, bubbles, and superbubbles should be defined on this special graph structure. Modify the algorithms for identifying these substructures accordingly.

13.10   Suppose you are given a set of d variable-length strings (reads) of total length N over the integer alphabet $[1..\sigma]$. Can you induce the sorted order of the strings in $O(N \log \sigma)$ time and bits of space (notice that the algorithm presented in Section 8.2.1 can sort in $O(N)$ time but uses $O(N \log N)$ bits of space)? *Hint.* Use some standard sorting algorithm, but without ever copying or moving the strings.

13.11   Recall the algorithm for computing maximal overlaps in Section 13.2.3. Fill in the details for constructing in linear time the tree of nested intervals from the suffix tree associated with intervals. *Hint.* First prove that the locus of $R^i[k..|R^i|]$ for any k is either a leaf or an internal node of depth $|R^i| - k + 1$.

13.12   Give the pseudocode for the algorithm sketched in Insight 13.3.

13.13   Analyze the running time and space of the algorithm in Insight 13.3. For this, you might find useful the concept of a *string graph*, which is like an overlap graph, but its arcs are labeled by strings as follows. An arc (i, j, ℓ) of an overlap graph becomes a bidirectional arc having two labels, $R^i[1..|R^i| - \ell]$ for the forward direction and $R^j[\ell + 1..|R^j|]$ for the backward direction. Observe that one can bound the running time of the algorithm in Insight 13.3 by the total length of the labels in the string graph before redundant arcs are removed.

13.14   Show that the following relaxation of Problem 13.1 remains NP-hard. Given a set $\mathcal{C}$ of contigs, the alignments of a set $\mathcal{R}$ of paired-end reads in these contigs, and integer t, find a subset $\mathcal{R}' \subseteq \mathcal{R}$ of paired-end reads such that $\mathcal{R}'$ has the property that the contigs in $\mathcal{C}$ can be partitioned into the minimum number of scaffolds such that

- every read in $\mathcal{R}'$ is t-consistent with some scaffold, and
- each scaffold is connected with respect to $\mathcal{R}'$.

13.15   Formulate different problem statements, similar to Problem 13.1, for taking into account the fact that a contig may belong to multiple scaffolds. What can you say about their computational complexity?

13.16   Problem 13.1 could be modified to ask for the minimum k read pairs to remove such that the remaining read pairs form a consistent scaffolding. Study the computa-tional complexity of this problem variant.

13.17   We gave a pseudo-polynomial algorithm for gap filling (Problem 13.2), in the sense that, if the gap length parameter d is bounded (by say a polynomial function in n and m), then the algorithm runs in polynomial time (in n and m). We left open the hardness of the problem when d is unbounded. Consider a generalization of the problem when the input graph is any weighted graph (with any non-negative arc costs), not just a prefix graph constructed from a collection of strings. Show that this generalized problem is NP-complete, by giving a reduction from the *subset sum* problem (recall its definition from the proof of Theorem 5.3) to this generalization (you can use some arcs with 0 cost). Why is it hard to extend this reduction to the specific case of prefix graphs?

13.18 Consider the following generalization of Problem 13.2 in which we allow approximate overlaps. Given a prefix graph $G = (V, E)$, with a cost function $c : E \rightarrow \mathbb{Z}_+$ and $h : E \rightarrow \mathbb{Z}_+$, two of its vertices s and t, and an interval of possible path costs $[d'..d]$, decide whether there is a path P made up of $s = v_1, v_2, \ldots, v_k = t$ such that

$$\text{cost}(P) = \sum_{i=1}^{k-1} c(v_i, v_{i+1}) \in [d'..d],$$

and among all such paths return one P minimizing

$$\sum_{i=1}^{k-1} h(v_i, v_{i+1}).$$

Show how to solve this problem in time $O(dm)$. What is the space complexity of your algorithm?

13.19 Modify the solution to Problem 13.2 so that it runs in time $O(m_r(d + \log r))$, where r is the number of vertices reachable from s by a path of cost at most d, and m_r is the number of arcs between these vertices.

13.20 Modify the solution to Problem 13.2 so that we obtain also the number of paths from s to t whose cost belongs to the interval $[d'..d]$.

13.21 Interpret the dynamic programming algorithm given for Problem 13.2 as an algorithm working on a DAG, constructed in a similar manner to that in Section 4.2.2 for the Bellman–Ford method.

13.22 Consider a de Bruijn graph G built from a set $\mathcal{R}$ of reads sequenced from an unknown genome, and let P be a unary path in G. Under which assumptions on $\mathcal{R}$ does the unary path P spell a string occurring in the original genome?

13.23 Let $G = (V, E)$ be a directed graph. Give an $O(|V| + |E|)$-time algorithm for outputting all maximal unary paths of G.

13.24 Let $G = (V, E)$ be a directed graph, and let G' be the graph obtained from G by compacting its arcs through the following process. First, label every vertex v of G with the list $l(v) = (v)$, that is, $l(v)$ is made up only of v. Then, as long as there is an arc (u, v) such that $N^+(u) = \{v\}$ and $N^-(v) = \{u\}$ remove the vertex v, append v at the end of $l(u)$, and add arcs from u to every out-neighbor of v. Explain how, given only G', it is possible to report all maximal unary paths of G in time $O(|V| + |E|)$. What is the time complexity of computing G'?

13.25   Let $G = (V, E)$ be a directed graph. A more formal way of defining a contig is to say that a path P is a *contig* if P appears as a subpath of any path in G which covers all vertices of G.

- Let P be a unary path of length at least 2. Show that P is a conting.
- Let $P = (s, v_1, \ldots, v_k, \ldots, v_n, t)$ be a path such that v_k is a unary vertex, for all $1 \leq i < k$, we have $|N^-(v_i)| = 1$, and for all $k < j \leq n$, we have $|N^+(v_i)| = 1$. Show that P is a contig.

Part V

Applications

14 Genomics

We shall now explore how the techniques from the previous chapters can be used in studying the genome of a species. We assume here that we have available the so-called finalized genome assembly of a species. Ideally, this means the complete sequences of each chromosome in a species. However, in practice, this consists of chromosome sequences containing large unknown substrings, and some unmapped contigs/scaffolds.

To fix our mindset, we assume that the species under consideration is a diploid organism (like a human). We also assume that the assembled genome is concatenated into one long sequence T with some necessary markers added to separate its contents, and with some auxiliary bookkeeping data structures helpful for mapping back to the chromosome representation.

We start with a *peak detection* problem in Insight 14.1 that gives a direct connection to the segmentation problem motivating our study of hidden Markov models in Chapter 7. Then we proceed into variation calling, where we mostly formulate new problems assuming read alignment as input. The results of variant calling and read alignments are then given as the input to haplotype assembly.

Insight 14.1 Peak detection and HMMs

The *coverage profile* of the genome is an array storing for each position the amount of reads aligned to cover that position. In ChIP-sequencing and bisulfite sequencing only some parts of the genome should be covered by reads, so that clear peak areas should be noticeable in the coverage profile. In targeted resequencing, the areas of interest are known beforehand, so automatic detection of peak areas is not that relevant (although usually the targeting is not quite that accurate). Peak detection from a signal is a classical signal processing task, so there are many existing general-purpose peak detectors available. In order to use the machinery we developed earlier, let us consider how HMMs could be used for our peak detection task. The input is a sequence $S_{1..n}$ of numbers, where $s_j \geq 0$ gives the number of reads *starting* at position j in the reference. Our HMM has two states: Background and Peak area. Assume that r_b reads come from the background and r_p from peak areas, where $r = r_b + r_p$ is the number of reads. Assume that the background covers n_b positions and the peak areas cover n_p positions, where $n = n_b + n_p$ is the length of the genome. Then the probability that there are s_j reads starting at position j inside the peak area is given by the probability of an event in the binomial distribution

---

$$\mathbb{P}(s_j \mid j \in \texttt{Peak area}) = \binom{r_p}{s_j}\left(\frac{1}{n_p}\right)^{s_j}\left(\frac{n_p - 1}{n_p}\right)^{r_p - s_j}. \tag{14.1}$$

An analogous equation holds for positions j in the background area. To fix the HMM, one needs to set realistic a-priori values for transition probabilities so that switching from peak area to background, and vice versa, is expensive. One can use Baum–Welch or Viterbi training to find values for the parameters n_p, n_b, r_p, and r_b and for the transition probabilities. The peak areas found using ChIP-sequencing data should contain a common motif that explains the binding of the protein targeted by the experiment. *Motif discovery* algorithms are beyond the scope of this book; see references in the literature section.

14.1   Variation calling

We now recall some concepts from Chapter 1. A *reference* genome represents an average genome in a population, and it is of general interest to study what kind of variants are frequent in the population. However, one is also interested in the genome of a certain individual, called the *donor*. For example, in cancer genetics it is routine practice to sequence a family carrying an inherited disease, and to look for those variants unique to the family, when compared with the rest of the population. Such variants are also called *germline mutations*. *Somatic mutations* are those variations which do not affect germ cells. They sometimes affect the cell metabolism, and spread their effects to nearby cells, causing new somatic mutations that may, possibly together with existing germline mutations, cause cancer.

The process of identifying variants from sequencing data is called *variation calling*; the terms variant/variation detection/analysis are also frequently used. The approaches employed to discover these variants differ significantly depending on the scale of the sought-for variants. Most of them are based on the read alignment techniques we covered in Chapter 10, and on the subsequent analysis of the *read pileup*, that is, of the read alignments over a certain position in T. We now discuss some practical variation calling schemes.

14.1.1   Calling small variants

Let us first consider *single-nucleotide polymorphisms, SNPs*. Read pileup provides direct information about SNPs in the donor genome: if position j in T is covered by r_j reads, out of which p percent say that T contains a nucleotide a on position j, and the rest say that T contains a nucleotide b, one can reason whether this is because of heterozygous polymorphism or because of a measurement error. Measurement errors are typically easy to rule out; they are independent events, so the probability of observing many in the same position decreases exponentially: see Exercise 14.1.

Hence, by sequencing with a high enough coverage, one can easily obtain the SNP profile of the donor genome with reasonable accuracy. This assumes that most reads are aligned to the correct position. The analysis can also take into account the mapping accuracy, to put more weight on trusted alignments.

The same straightforward idea can basically be applied to short indels, but more care needs to be taken in this case. First, if the read alignments are conducted in terms of the k-errors search problem, one can observe only indels of size at most k. Second, since the reads are aligned independently of each other, their alignments can disagree on the exact location of the indel. To tackle the latter problem, one can take the r_j reads covering position j and compute a multiple alignment; this phase is called *re-alignment* and it is likely to set the indels to a consensus location. This multiple alignment instance is of manageable size, so even optimal algorithms to each subproblem may be applied.

14.1.2   Calling large variants

Larger variations need a more global analysis of the read pileup. If there is a deletion in the donor genome longer than the error threshold in read alignment, then there should be an un-covered region in the read pileup with the same length as the deletion. If there is an insertion in the donor genome longer than the error threshold in read alignment, then there should be a pair of consecutive positions $(j, j+1)$ in the read pileup such that no read covers both j and $j+1$ as shown in Figure 14.1.

Moreover, there should be reads whose prefix matches some $T_{j'..j}$, and reads whose suffix matches some $T_{j+1..j''}$; in the case of deletions, both conditions should hold for some reads. To "fill in" the inserted regions, one can apply fragment assembly methods from Chapter 13: consider all the reads that are not mapped anywhere in the genome, together with substrings $T_{j-m+1..j}$ and $T_{j+1..j+m+1}$ of length m (m being the read length); try to create a contig that starts with $T_{j-m+1..j}$ and ends with $T_{j+1..j+m+1}$. This is just a simplistic view of indel detection, since in practice one must prepare for background noise.

Another approach for larger-scale variation calling is to use paired-end, or mate pair, reads. Consider paired-end read pairs $(L^1, R^1), (L^2, R^2), \ldots, (L^r, R^r)$ such that each read

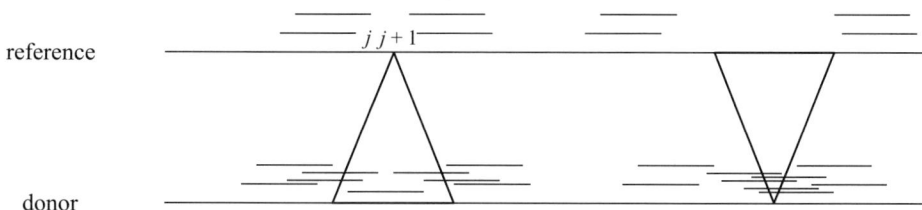

Figure 14.1 Analysis of read pileup for indels. On the left, an insertion in the donor; on the right, a deletion in the donor.

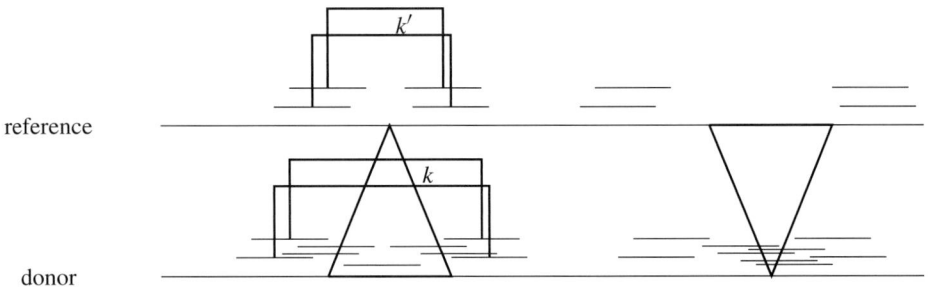

Figure 14.2 Analysis of the expected indel length. On the left, an insertion in the donor; on the right, a deletion in the donor.

L^i covers the same position j of the reference. Suppose that k is the expected distance between the input paired-end reads, with some variance, both known in advance from the library preparation protocol. One can compute the average, k', and the variance of the observed distances between the mapping positions of all L^i and R^i. The distance deviation can be tested to ascertain whether it is statistically significant; in such a case, the expected indel length is $|k - k'|$. This estimation is illustrated in Figure 14.2.

The length of the insertion is useful also for the fragment assembly approach mentioned above, since it imposes a constraint on the contig to be assembled. This problem is identical to the *gap-filling* problem, see Section 13.4.

In the case of insertions longer than the paired-end read length, one should be able to find the boundaries of the inserted area by analyzing read pairs having one end that is not mapped anywhere, but in this case the insert size remains unpredicted.

The third approach is to use split-read alignments: a read spanning an area deleted in the donor sequence could be aligned to the reference so that its prefix maps before the deleted area and its suffix maps after the deleted area. When many such split-read alignments vote for the same boundaries, one should be able to detect the deleted area accurately.

Let us now discuss how the paired-end approach gives rise to a graph optimization problem. Observe that in diploid organisms also indels can be homozygous or heterozygous, and, moreover, heterozygous indels can overlap. This means that two paired-end reads whose alignments overlap may in fact originate from different haplotypes. We wish to assign each paired-end read to support exactly one variant hypothesis, and to select a minimum set of hypotheses (the most parsimonious explanation) to form our prediction.

Assume we have a set H of hypotheses for the plausible variants, found by split-read mapping, by drops in read-pileup, or by paired-end alignments. Each hypothesis is a pair (j, g), where j is the starting position of the indel and g is its expected length. We say that an alignment of a paired-end read r *supports* a hypothesis $(j, g) \in H$, if the left part of r aligns left from j, the right part of r aligns right from j, and the difference between the alignment length and the expected indel size is g' ($g' = k - k'$ in Figure 14.2) such that $|g - g'| \leq t$, where t is a given threshold. We obtain a weighted bipartite graph

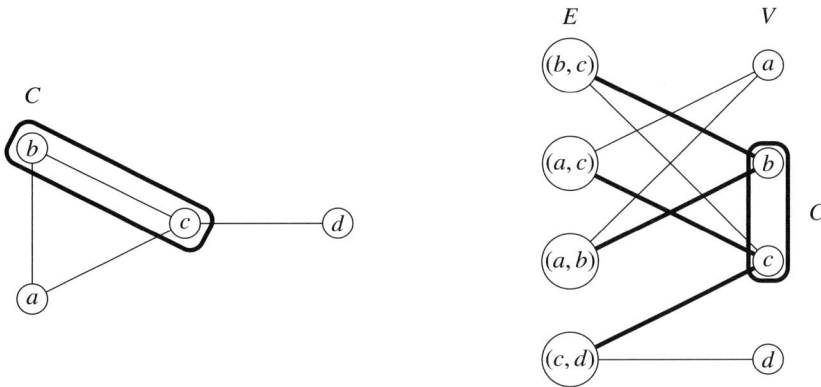

Figure 14.3 On the left, an undirected graph $G = (V, E)$ and a vertex cover C in G. On the right, the graph $G^* = (E \cup V, A)$ constructed in the proof of Theorem 14.1; a solution M for Problem 14.1 on G^* is highlighted.

$G = (R \cup H, E)$, where $r \in R$ is a paired-end read, $h = (j, g) \in H$ is a hypothesis, $(r, h) \in E$ if there is an alignment of r supporting hypothesis h, and the cost of the edge (r, h) is $c(r, h) = |g - g'|$, where g' is defined as above.

We start with a problem formulation (Problem 14.1) which captures the features of this problem. We show that Problem 14.1 is NP-hard, and then briefly discuss how this biological problem can be transformed into another NP-hard problem, called minimum-cost set cover. This alternative formulation is a standard problem, and can thus benefit from other results in the literature; see the further references in the last section of this chapter.

Problem 14.1 Minimum-cost minimum covering assignment

Given a bipartite graph $G = (R \cup H, E)$ and a cost function $c : E \to \mathbb{Q}_+$, find a subset $M \subseteq E$ of the edges of G such that

- each $r \in R$ is incident to exactly one $e \in M$, and
- the number of vertices of H incident to an edge of M is minimum.

Moreover, among all sets M satisfying the above two conditions, find one minimizing the total cost $\sum_{e \in M} c(e)$.

In practice, to make sure that our bipartite graph of reads and hypotheses always admits at least one solution for Problem 14.1, one can add a dummy hypothesis for each read, having a cost amounting to the sum of all existing edge costs; such costs allow the selection of these edges only when there is no solution for Problem 14.1 to the original input graph.

THEOREM 14.1   *Problem 14.1 is NP-hard.*

Proof   We reduce from the NP-hard vertex cover problem (recall Example 2.3 on page 16). Given an undirected graph $G = (V, E)$, create a bipartite incidence graph $G^* = (E \cup V, A)$ such that $A = \{(e, v) \mid e \in E$ is incident to $v \in V$ in $G\}$, and set the cost to all its edges to zero (see Figure 14.3). Given a set $M \subseteq A$, denote by $d_M(V)$ the number of vertices in V incident in G^* to an edge of M, namely

$$d_M(V) = |\{v \in V \mid \text{ there exists } (e, v) \in M\}|.$$

We claim that $G = (V, E)$ has a vertex cover of size at most k if and only if Problem 14.1 on G^* has a solution M satisfying $d_M(V) \le k$.

For the forward implication, let $C \subseteq V$ be a vertex cover of G. Since C touches all edges of G in at least one endpoint, we can construct a solution M for Problem 14.1 on G^* by selecting, for every $e \in E$, one arbitrary edge $(e, v) \in A$ such that $v \in C$. Thus, $d_M(V) \le |C|$.

For the reverse implication, let $M \subseteq A$ be a solution to Problem 14.1 on instance G^*, and let $C \subseteq V$ be the set of vertices in V incident to one element of M (observe that $|C| = d_M(V)$). Since each $e \in E$ is incident in G^* to one $a \in M$, we get that C is a vertex cover of G.   □

Exercise 14.3 asks the reader to show that a greedy algorithm for this problem can produce a solution with a number of hypotheses greater by a factor $O(|H|)$ than the optimal one.

Observe that, if we relax Problem 14.1 by dropping the requirement that each read should support exactly only one hypothesis, then we obtain the following classical problem.

Problem 14.2   Minimum-cost set cover

Given a collection of subsets $H = \{H_1, \ldots, H_n\}$ of a set $\{r_1, \ldots, r_m\}$ and a cost function $c : \{H_1, \ldots, H_n\} \to \mathbb{Q}_+$, find a subset $C \subseteq H$ such that

- each r_i belongs to at least one H_j in C,

and it minimizes

$$\sum_{H_j \in C} c(H_j).$$

For our variant calling problem, we can assign as the cost for every hypothesis the sum of the costs of the reads supporting it. This problem admits a polynomial-time algorithm producing a solution whose total cost is at most $O(\log|H|)$ times the optimal one; see Exercise 14.4 and the literature section for references to the minimum-cost set cover problem.

14.2   Variation calling over pan-genomes

In Section 10.7 we discussed two approaches for aligning reads, by taking into account variants already discovered in a population. Let us now consider two approaches to how a donor sequence can be predicted, on the basis of these alignments. For both of them, the idea is to first predict which known variants from the population are included in the donor sequence, and then predict the donor-specific variants. This donor-specific prediction is done using the scheme from the previous section, by taking as reference the predicted donor sequence.

14.2.1   Alignments on a set of individual genomes

Assume that we have the read alignments on top of a collection of d reference sequences, whose multiple alignment we already know. Namely, we have a $d \times n$ matrix as defined in Section 6.6. We can then create another $d \times n$ matrix C, which contains now the read coverage counts; that is, each $c_{i,j}$ tells us how many reads cover the ith genome at position j of the multiple alignment. Figure 14.4 gives an example of the coverage counts.

Consider the simple strategy of predicting a donor sequence by choosing, at each column j, the row i with the maximum value $c_{i,j}$. Extracting the corresponding nucleotides from multiple alignment and removing the gap symbols gives us a predicted donor genome.

Observe that such a predicted genome recombines the underlying sequences arbitrarily, by choosing the recombination having most support by read alignments. However, in order to model more closely the biological process, we can limit the number of recombinations by a given parameter. This leads to Problem 14.3 below, whose setting is illustrated in Figure 14.4.

Problem 14.3   Heaviest path with recombination constraint

Given a $d \times n$ matrix C, with $c_{i,j} \geq 0$ for all i,j, and a threshold $K < n$, find a path in C that traverses at most K rows of C, namely a sequence P of cells of C, $P = (i_1, 1), (i_1, 2), \ldots, (i_1, j_1), (i_2, j_1 + 1), (i_2, j_1 + 2), \ldots, (i_2, j_2), \ldots, (i_K, j_{K-1} + 1), (i_K, j_{K-1} + 2), \ldots, (i_K, n)$, maximizing

$$\sum_{(i,j) \in P} c_{i,j}.$$

Problem 14.3 can be solved by dynamic programming, as follows. Let $v(i, j, k)$ denote the value of the optimum solution for the submatrix $C_{1..i,1..j}$, among all paths recombining rows $k - 1$ times. The matrix v can be computed with the following recurrence:

1244444443333225778888888887744232112
6677788998887664232222222333333211111
3334444444333110000011113333225588899

Figure 14.4 Illustration of a heaviest path in a matrix C with recombination constraint $K = 3$. Each row of the matrix stands for a different reference sequence, and each column corresponds to a position in the multiple alignment of the three reference sequences. The pieces corresponding to the optimal path through the matrix are underlined.

$$v(i,j,k) = \max \left(v(i,j-1,k) + c_{i,j}, \max\{v(i',j-1,k-1) + c_{i,j} \mid 1 \leq i' \leq d \text{ and } i' \neq i\} \right),$$
(14.2)

where we initialize $v(i,1,1) = c_{i,1}$, for all $1 \leq i \leq d$. The value of the optimal solution for the entire matrix C is obtained as

$$\max\{v(i,n,K) \mid 1 \leq i \leq d\}.$$

Under a suitable evaluation order, the matrix v can be computed in time $O(dnK)$. The corresponding optimal path can be extracted by a traceback in the matrix v. A straightforward implementation of this traceback can be implemented in time $O(dnK)$ and space $O(dnK)$. However, using the same idea as the reader was asked for in Exercise 7.2 (about a space-efficient implementation of the Viterbi algorithm), this traceback can be implemented with the same time complexity, but using only $O(d\sqrt{n}K)$ space. We have thus proved the following.

THEOREM 14.2 *Problem 14.3 can be solved in $O(dnK)$ time and $O(d\sqrt{n}K)$ space.*

14.2.2   Alignments on the labeled DAG of a population

Recall the approach to pan-genome indexing in which the reference, and the known variants, are represented as a labeled DAG G (illustrated in Figure 10.6 on page 215). When extracting the path to which each read is aligned, as shown in Section 10.7, we can increase a counter $c(v)$, for each vertex v covered by the read.

Observe that in the previous section, since we had available each individual genome of the population, we could impose a limit on the number of recombinations. In the case of G, this information is no longer available, so we also assume that the donor sequence is a heaviest path in G (from a source to a sink), namely one maximizing the sum of the weights of its vertices.

The dynamic programming recurrences that we derived for the shortest-path problem on a DAG with arc weights (Problem 4.1) carry over to this problem, resulting in an algorithm running in time linear in the number of arcs of G. Equivalently, one can apply any algorithm for solving Problem 4.1, by creating a DAG G^* obtained from G by subdividing each vertex v into two vertices v_{in} and v_{out}, with all in-neighbors of v now in-neighbors of v_{in}, and all out-neighbors of v now out-neighbors of v_{out}, and adding the arc (v_{in}, v_{out}) with weight $-c(v)$. All other arcs of G^* have weight equal to zero. A shortest path in the DAG G^* from a source to a sink is a heaviest path in G.

14.2.3   Evaluation of variation calling results

Variation calls can be validated by using targeted sequencing, but this is an expensive process, and partly manual wet-lab work. Hence, simulation studies mimicking a real setting are needed for guiding the variation calling tool development.

For such a study, one can create a diploid ground-truth by taking a reference sequence, create homozygous random mutations to it, and then obtain two haplotypes by carrying out this process twice independently, in order to create the heterozygous mutations. The result can be represented as a pair-wise alignment of an artificial diploid individual. One can then generate reads from these two sequences from random positions, in order to mimic sequencing.

In a pan-genomic setting, one can add another layer at the beginning by taking a random sample of already observed variants in the population. In this way the generated ground-truth models the likely germline mutations and some rarer germline mutations, as well as somatic mutations. We discuss in Insight 14.2 how one could evaluate the predictions of several variation calling tools with such data.

Insight 14.2 Variation calling evaluation and DAG alignment

Consider a tool producing a set of variant predictions and the evaluation problem of checking how close the predictions are to the ground-truth artificially generated diploid genome. A one-to-one mapping of these predictions to the true variants is not always possible, since predictions may be inaccurate (especially for longer indels). One way to overcome this difficulty is to apply the predicted variants to the reference to form a predicted haploid donor genome. This predicted donor sequence does not necessarily need to be close to either of the two sequences forming the diploid ground-truth, but it should be close to some recombination of them; otherwise it is a bad prediction.

That is, we should align a haploid prediction to a diploid ground-truth allowing recombinations. We solved this problem already in Section 6.6.5: we can convert the diploid pair-wise alignment into a DAG, and represent the haploid as a trivial DAG (one path from source to sink labeled by the haploid). Although the running time of this approach is quadratic, in practice, with some assumptions on the underlying DAG, one can apply the shortest-detour method of Section 6.1.2 to obtain an $O(dn)$ time algorithm, where d is the edit distance from the predicted haploid to the best recombination of the diploid ground-truth: see Exercise 14.6.

14.3   Haplotype assembly and phasing

With diploid organisms, one is interested not only in discovering variants, but also in discovering to which of the two haplotypes each variant belongs. One would thus like to identify the variants which are co-located on the same haplotype, a process called *haplotype phasing*. The variants present in an individual which are clearly homozygous can be trivially assigned to both haplotypes. Thus, in this section we limit ourselves to heterozygous variants, and, furthermore, just to SNP variants.

Thanks to high-throughput sequencing technologies, we can guide this process by read alignments and read overlaps. Consider the problem formulation below, and see also Example 14.5.

Problem 14.4   Haplotype assembly under minimum error correction

Given an $r \times s$ matrix M, with values $\{0, 1, -\}$, where reads are rows, columns are SNPs, and

$$
m_{i,j} = \begin{cases} 0, & \text{if the } i\text{th read does not support the } j\text{th SNP,} \\ 1, & \text{if the } i\text{th read supports the } j\text{th SNP,} \\ -, & \text{if the } i\text{th read does not overlap the } j\text{th SNP,} \end{cases}
$$

find

- a partition of the rows of M (the reads) into two subsets R^0 and R^1, and
- two binary sequences A and B of length s (the two haplotypes),

which minimize the number of bit flips in M that are required to make the reads in R^0 compatible with A, and the reads in R^1 compatible with B.

We say that a read i is compatible with a haplotype $X = x_1 \cdots x_s$ if, for all $j \in \{1, \ldots, s\}$, $m_{i,j} = -$ or $m_{i,j} = x_j$.

Observe that, for clarity, in Problem 14.4 we ask both for a partition of the reads and for the two haplotypes. However, having an optimal partition (R^0, R^1) of the rows of M, we can easily find the two optimal haplotypes in a column-wise manner. For every $j \in \{1, \ldots, s\}$, we either set $A[j] = 0, B[j] = 1$, by flipping all bits in R^0 on column j equal to 1 and all bits in R^1 on column j equal to 0, thus with a total of

$$
|\{i \in R^0 \mid m_{i,j} = 1\}| + |\{i \in R^1 \mid m_{i,j} = 0\}| \tag{14.3}
$$

bit flips, or we set $A[j] = 1, B[j] = 0$ with a total of

$$
|\{i \in R^0 \mid m_{i,j} = 0\}| + |\{i \in R^1 \mid m_{i,j} = 1\}| \tag{14.4}
$$

bit flips. We set $A[j]$ and $B[j]$ by checking which of the two costs (14.3) or (14.4) is minimum. Moreover, it will be convenient to denote the minimum of the two by $c(R^0, R^1, j)$, namely

$$
c(R^0, R^1, j) = \min \Big(|\{i \in R^0 \mid m_{i,j} = 1\}| + |\{i \in R^1 \mid m_{i,j} = 0\}|,
$$
$$
|\{i \in R^0 \mid m_{i,j} = 0\}| + |\{i \in R^1 \mid m_{i,j} = 1\}| \Big).
$$

Conversely, having the two haplotypes A and B, it is easy to find the optimal partition (R^0, R^1) row-wise by setting, for each $i \in \{1, \ldots, r\}$,

$$i \in \begin{cases} R^0, & \text{if } |\{j \mid m_{i,j} \neq - \text{ and } m_{i,j} \neq A[j]\}| \leq |\{j \mid m_{i,j} \neq - \text{ and } m_{i,j} \neq B[j]\}|, \\ R^1, & \text{otherwise.} \end{cases}$$

Example 14.5 Consider the following setting in which six reads have been aligned to a reference, resulting in the SNPs C at position s_1, T at position s_2, and G at position s_3. We assume the two unknown haplotypes to be ACGTATAGATATACAC (from which the three reads on the top originate) and ACGTACAGAGATGCAC (from which the three reads on the bottom originate).

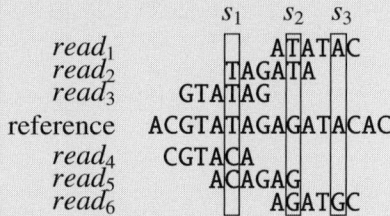

$$\begin{array}{rl}
 & \quad s_1 \quad s_2 \quad s_3 \\
read_1 & \qquad \text{ATATAC} \\
read_2 & \quad \text{TAGATA} \\
read_3 & \text{GTATAG} \\
\text{reference} & \text{ACGTATAGAGATACAC} \\
read_4 & \text{CGTACA} \\
read_5 & \text{ACAGAG} \\
read_6 & \qquad \text{AGATGC}
\end{array}$$

We construct the following matrix M:

$$M = \begin{pmatrix} - & 1 & 0 \\ 0 & 1 & - \\ 0 & - & - \\ 1 & - & - \\ 1 & 0 & - \\ - & 0 & 1 \end{pmatrix},$$

where $m_{i,j} = 0$ tells that the ith read does not support the jth SNP, $m_{i,j} = 1$ tells that the ith read supports the jth SNP, and $m_{i,j} = -$ tells that the ith read does not overlap the jth SNP.

The two haplotypes are 010 and 101, and they correspond to assigning the first three reads (rows) to the first haplotype, and the other three reads (rows) to the second haplotype; that is, $R^0 = \{1, 2, 3\}$ and $R^1 = \{4, 5, 6\}$. In this case, no bit flips in M are needed, since all the reads (rows) are compatible with one of the two haplotypes.

Haplotype phasing was known to be a hard problem even before the development of high-throughput sequencing. Unfortunately, even the above haplotype assembly formulation based on read alignments remains so.

THEOREM 14.3 *Problem 14.4 is NP-hard.*

Proof We reduce from the NP-hard problem MAX-CUT. In this problem, one is given an undirected graph $G = (V, E)$ and is asked to find a subset C of vertices that maximizes the number of edges between C and $V \setminus C$, that is, edges having one endpoint in C and the other one outside C.

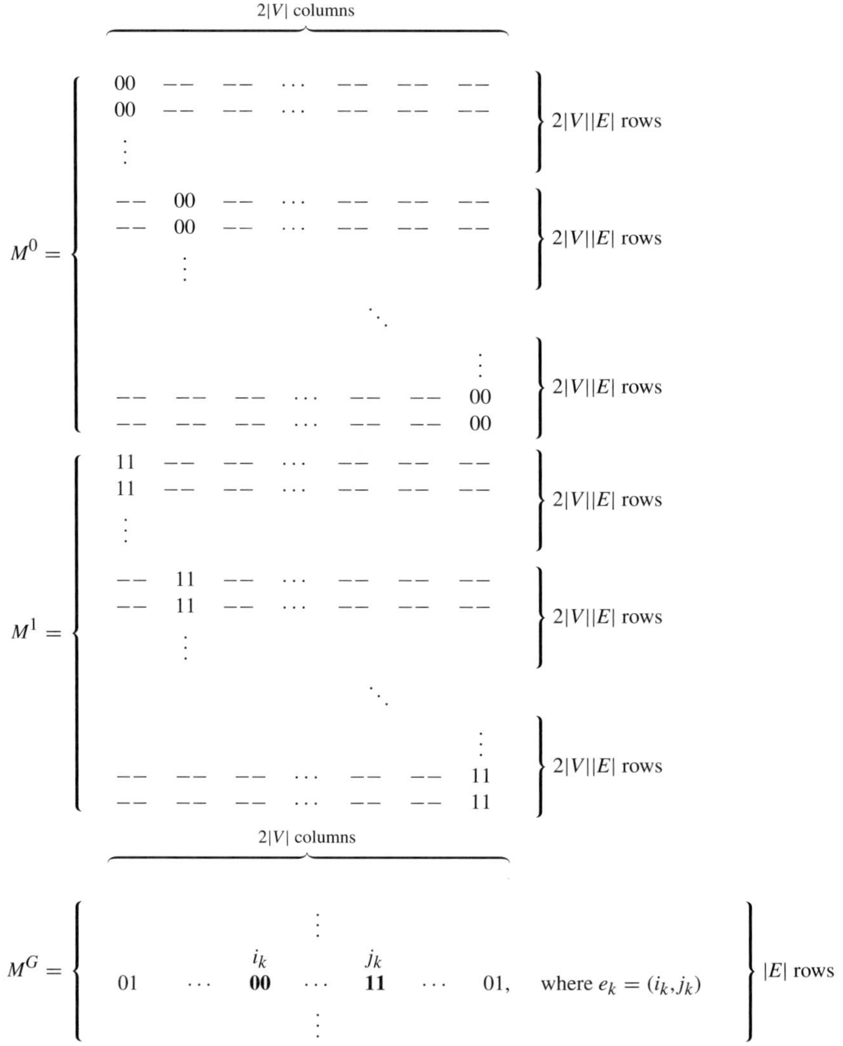

Figure 14.5 The three binary matrices needed in reducing the MAX-CUT problem to Problem 14.4.

Given a graph G as input for the MAX-CUT problem, we construct a matrix M made up of the rows of the three matrices M^0, M^1, and M^G shown in Figure 14.5, (that is, M has $2|V|$ columns and $2|V|^2|E| + 2|V|^2|E| + |E|$ rows).

The matrix M^G has a row for each edge $(i,j) \in E$. To simplify the notation, let $V = \{1, \ldots, |V|\}$ and $E = \{e_1, \ldots, e_{|E|}\}$. For each $e_k \in E$ with $e_k = (i,j)$, $i < j$, we set

- $m^G_{k,2i-1} = m^G_{k,2i} = 0$,
- $m^G_{k,2j-1} = m^G_{k,2j} = 1$,
- $m^G_{k,2t-1} = 0$ and $m^G_{k,2t} = 1$, for all $t \in \{1, \ldots, |V|\} \setminus \{i,j\}$.

We will show that $G = (V, E)$ admits a subset $C \subseteq V$ with t edges between C and $V \setminus C$ if and only if the number of flips in an optimal solution for M is $|V||E| - 2t$. Thus, finding a maximum cut in G is equivalent to finding a solution with the minimum number of flips for M.

For the forward implication, let $C \subseteq V$ be such that there are t edges between C and $V \setminus C$. Define the haplotypes A and B as

$$A[2i - 1] = \begin{cases} A[2i] = 0, B[2i - 1] = B[2i] = 1, & \text{if vertex } i \in C, \\ A[2i] = 1, B[2i - 1] = B[2i] = 0, & \text{otherwise.} \end{cases}$$

Observe first that each row of M^0 and M^1 is compatible with either A or B, without flipping any bits. Moreover, for every edge $e_k = (i, j)$, with $i < j$, of G the following statements apply.

- If e_k is between C and $V \setminus C$, then the kth row of M^G needs exactly $|V| - 2$ bit flips to be compatible with one of A or B, one bit for exactly one of the positions $m_{k,2t-1}$ or $m_{k,2t}$, for all $t \in \{1, \ldots, |V|\} \setminus \{i, j\}$.
- If e_k is not between C and $V \setminus C$, then assume, without loss of generality, that $i, j \in C$ (the case $i, j \in V \setminus C$ is symmetric). Since by construction $m_{k,2i-1} = m_{k,2i} = 0$, and $m_{k,2j-1} = m_{k,2j} = 1$, but $A[2i - 1] = A[2i] = 0 = A[2j - 1] = A[2j]$, we need to flip $m_{k,2j-1}$ and $m_{k,2j}$. Therefore, the kth row of M^G needs $|V|$ bit flips, one bit for exactly one of the positions $m_{k,2t-1}$ or $m_{k,2t}$, for all $t \in \{1, \ldots, |V|\} \setminus \{i, j\}$, and two more bit flips for $m_{k,2j-1}$ and $m_{k,2j}$.

Thus, we have constructed a solution for M requiring in total $|V||E| - 2t$ bit flips.

The reverse implication is left as an exercise for the reader (Exercise 14.7) □

We will now present a practical algorithm for solving Problem 14.4. Onwards, we are concerned with finding an optimal partition (R^0, R^1) of the rows of M, since, as explained on page 316, we can easily derive the two optimal haplotypes from an optimal partition.

We start with the simple strategy of trying out all possible ways of partitioning the rows of M, and choosing the partition requiring the minimum number of bit flips. Given such a partition (R^0, R^1), let $D(R^0, R^1, j)$ denote the minimum number of bit flips needed for making the first j columns of M compatible with two haplotypes. Using the notation $c(R^0, R^1, j)$ introduced on page 316, we can clearly express D column-wise as

$$D(R^0, R^1, j) = D(R^0, R^1, j - 1) + c(R^0, R^1, j). \tag{14.5}$$

Such an algorithm has complexity $O(2^r rs)$, since we try out all of the 2^r partitions (R^0, R^1) of the rows of M, and then evaluate them by computing $D(R^0, R^1, s)$ in time $O(rs)$ (computing $c(R^0, R^1, j)$ takes time $O(r)$).

Let us see now how such an algorithm can be refined for the practical features of the haplotype assembly problem. Observe that the number of reads overlapping a given SNP j (that is, having a 0 or a 1 on column j) is much smaller than the total number r of reads. We will obtain an algorithm of complexity $O(2^K(2^K + r)s)$, where K is the maximum coverage of an SNP. Such an algorithm is called *fixed-parameter tractable*, with the maximum coverage as a parameter, since, if we assume K to be bounded by a constant, then we have a polynomial-time algorithm.

The cost of making a column j of M compatible with two haplotypes does not depend on the rows of M having a $-$. Call such rows *inactive* at column j, and call the rows having a 0 or a 1 *active* at column j. Therefore, for each column j, we can enumerate over all partitions only of its active rows. The cost of making the jth column, whose active rows are partitioned as (S^0, S^1), compatible with two haplotypes is obtained just as before, and we analogously denote it as $c(S^0, S^1, j)$.

We also analogously denote by $D(S^0, S^1, j)$ the minimum number of bit flips needed to make the first j columns of M compatible with two haplotypes, with the constraint that the active rows at the jth column are partitioned as (S^0, S^1). In order to express $D(S^0, S^1, j)$ depending on $D(S^0, S^1, j-1)$, observe that some rows are active both in column j and in column $j-1$. Denote by $U(j)$ the rows of M active both in column $j-1$ and in column j. Thus, the cost $D(S^0, S^1, j)$ depends only on those partitions (Q^0, Q^1) of the $(j-1)$th column which partition the rows in $U(j)$ in the same manner as (S^0, S^1). We can thus write

$$D(S^0, S^1, j) = \min\{D(Q^0, Q^1, j-1) \mid U(j) \cap S^0 \subseteq Q^0 \text{ and } U(j) \cap S^1 \subseteq Q^1\}$$
$$+ c(S^0, S^1, j), \tag{14.6}$$

where we take $D(S^0, S^1, 1) = c(S^0, S^1, 1)$. The number of bit flips in a solution to Problem 14.4 is obtained as (see Example 14.6)

$$\min\{D(S^0, S^1, s) \mid (S^0, S^1) \text{ is a partition of the rows active at the last column } s\}. \tag{14.7}$$

An optimal solution can be traced back by starting with the partition minimizing (14.7), and then iteratively adding to the two sets of this partition the sets in the partition minimizing (14.6). Since computing $c(S^0, S^1, j)$ takes $O(r)$ time, the number of partitions of the rows active at any column is at most 2^K, and, for every such partition in a column j, we consider all $O(2^K)$ partitions of the rows active in column $j-1$, we have the following result.

THEOREM 14.4 *Problem 14.4 can be solved in $O(2^K(2^K + r)s)$ time, where K is the maximum coverage of an SNP.*

Exercise 14.8 asks the reader to show that this algorithm can be implemented to run in time $O(2^K s)$. Observe that a direct implementation of this algorithm requires memory $O(2^K s)$, so that every partition of the active rows is stored for every column; see Exercise 14.9 for an implementation of this algorithm using memory $O(2^K \sqrt{s})$. Exercise 14.10 asks the reader to modify this algorithm for solving Problem 14.4 in time $O(K^T s)$, where T denotes the maximum number of bit flips required in any column in an optimal solution. Observe that in practice T is usually rather small.

Moreover, Problem 14.4 can be generalized so that every cell of M also has a cost of flipping (in practice, the confidence that the prediction is correct). Exercise 14.11 asks the reader which changes are needed to the above algorithm for solving this weighted problem.

Example 14.6 Consider the matrix

$$
M = \begin{pmatrix}
- & 1 & 0 \\
0 & 1 & - \\
0 & - & - \\
1 & - & - \\
1 & 1 & 1 \\
- & - & 1
\end{pmatrix},
$$

which is obtained from the one in Example 14.5 by changing the entries $m_{5,2}$, $m_{5,3}$, and $m_{6,2}$. Note that $U(2) = \{2,5\}$ and $U(3) = \{1,5\}$. For the first column, we have

$$
\begin{aligned}
D(\emptyset, \{2,3,4,5\}, \mathbf{1}) &= 2 = D(\{2,3,4,5\}, \emptyset, \mathbf{1}), \\
D(\{2\}, \{3,4,5\}, \mathbf{1}) &= 1 = D(\{3,4,5\}, \{2\}, \mathbf{1}), \\
D(\{3\}, \{2,4,5\}, \mathbf{1}) &= 1 = D(\{2,4,5\}, \{3\}, \mathbf{1}), \\
D(\{4\}, \{2,3,5\}, \mathbf{1}) &= 1 = D(\{2,3,5\}, \{4\}, \mathbf{1}), \\
D(\{5\}, \{2,3,4\}, \mathbf{1}) &= 1 = D(\{2,3,4\}, \{5\}, \mathbf{1}), \\
D(\{2,3\}, \{4,5\}, \mathbf{1}) &= 0 = D(\{4,5\}, \{2,3\}, \mathbf{1}), \\
D(\{2,4\}, \{3,5\}, \mathbf{1}) &= 2 = D(\{3,5\}, \{2,4\}, \mathbf{1}), \\
D(\{2,5\}, \{3,4\}, \mathbf{1}) &= 2 = D(\{3,4\}, \{2,5\}, \mathbf{1}).
\end{aligned}
$$

For the second column, we have

$$
\begin{aligned}
D(\emptyset, \{1,2,5\}, \mathbf{2}) &= \min(D(\emptyset, \{2,3,4,5\}, \mathbf{1}), \\
&\qquad D(\{3\}, \{2,4,5\}, \mathbf{1}), \\
&\qquad D(\{4\}, \{2,3,5\}, \mathbf{1}), \\
&\qquad D(\{3,4\}, \{2,5\}, \mathbf{1})) + 0 = 1 = D(\{1,2,5\}, \emptyset, \mathbf{2}), \\
D(\{5\}, \{1,2\}, \mathbf{2}) &= \min(D(\{3,4,5\}, \{2\}, \mathbf{1}), \\
&\qquad D(\{5\}, \{2,3,4\}, \mathbf{1}), \\
&\qquad D(\{4,5\}, \{2,3\}, \mathbf{1}), \\
&\qquad D(\{3,5\}, \{2,4\}, \mathbf{1})) + 1 = 1 = D(\{1,2\}, \{5\}, \mathbf{2}), \\
D(\{1\}, \{2,5\}, \mathbf{2}) &= 1 + 1 = D(\{2,5\}, \{1\}, \mathbf{2}), \\
D(\{2\}, \{1,5\}, \mathbf{2}) &= 0 + 1 = D(\{1,5\}, \{2\}, \mathbf{2}).
\end{aligned}
$$

For the third column, we have

$$D(\emptyset, \{1,5,6\}, \mathbf{3}) = \min(D(\emptyset, \{1,2,5\}, \mathbf{2}),$$
$$D(\{2\}, \{1,5\}, \mathbf{2})) + 1 = 2 = D(\{1,5,6\}, \emptyset, \mathbf{2}),$$
$$D(\{1\}, \{5,6\}, \mathbf{3}) = \min(D(\{1,2\}, \{5\}, \mathbf{2}),$$
$$D(\{1\}, \{2,5\}, \mathbf{2})) + 0 = 1 = D(\{5,6\}, \{1\}, \mathbf{2}),$$
$$D(\{5\}, \{1,6\}, \mathbf{3}) = 1 + 1 = D(\{1,6\}, \{5\}, \mathbf{3}),$$
$$D(\{6\}, \{1,5\}, \mathbf{3}) = 1 + 1 = D(\{1,5\}, \{6\}, \mathbf{3}).$$

The optimal partition $(\{1,2,3\}, \{4,5,6\})$ has cost $D(\{1\}, \{5,6\}, \mathbf{3}) = 1$, from flipping entry $m_{5,2}$.

14.4   Literature

We omitted algorithms for motif discovery related to the peak area analysis of Insight 14.1. The foundations of such algorithms are covered in a dedicated book by Parida (2007). A similar approach to ChIP-seq peak detection (using more complex distributions) is proposed in Qin *et al.* (2010): see the references therein for alternative approaches.

The basic approach of read filtering, read alignment, local re-alignment, variant calling from read pileup, and variant annotation, is more or less the standard workflow applied in biomedical laboratories. As is typical for such workflows, the choice of the software for implementing each subroutine is an art of engineering, with best-practice recommendations for each study case in question. From an algorithmic perspective, the key contributions are inside each subtask, and the literature for those can be found from previous chapters. For an overview of the best practices, there are numerous surveys and experimental tool comparisons available: see for example Pabinger *et al.* (2013).

Calling larger variants has been more on the focus of the algorithmically oriented bioinformatics community. Our formulation differs slightly from the literature. As far as we know, the covering assignment problem has not been formulated before, but instead some other routes to the set cover relaxation exist (Hormozdiari *et al.* 2009). See also the monograph by Vazirani (2001) for further references to the set cover problem. The hypotheses we formulated are usually defined as maximal cliques in an interval graph or some other graph defined by read alignment overlaps; see Hormozdiari *et al.* (2009), Marschall *et al.* (2012), and the references therein.

The pan-genome variant calling covered here breaks the standardized workflow, giving also algorithmically more interesting questions to study. As far as we know, variant calling from alignments to multiple genomes has not been seriously considered in the literature, except for some experiments in articles focusing on the alignment problem on multiple genomes. Our approach follows that of ongoing work (Valenzuela *et al.* 2015).

Haplotyping is a rich area of algorithm research. Here, we focused on the haplotype assembly formulation, since it is the closest to high-throughput sequencing. The NP-hardness proof is from Cilibrasi *et al.* (2005), and the dynamic programming algorithm is from Patterson *et al.* (2014); notice that they obtain a better complexity, see Exercise 14.8.

Exercises

14.1 Calculate the probability of an SNP given a read pileup taking into account the measurement errors.

14.2 *Variant annotation* is the process of assigning a function for each detected variant. For example, a 1-base deletion inside an exon creates a frame-shift and may cause an abnormal protein product to be translated. Consider different variants that might appear and think about why some variants can be called *silent mutations*. Browse the literature to find out what *nonsense mutations*, *missense mutations*, and *regulatory variants* are.

14.3 A greedy algorithm to solve Problem 14.1 is to start with empty E', choose $h \in H$ with most incoming edges not in E', add those unmatched edges to E', and iterate the process until all $r \in R$ are incident to an edge in E'. Show that this can produce a solution whose size is $O(|H|)$ times the optimal solution size.

14.4 Consider the following algorithm for the minimum-cost set cover problem. Start with an empty C and iteratively add to C the most cost-effective set H_i, that is, the set H_i maximizing the ratio between $c(H_i)$ and the number of un-covered elements of $\{r_1, \ldots, r_m\}$ that H_i covers. This is repeated until all elements are covered by some set in C. Show that this algorithm always produces a solution whose cost is $O(\log m)$ times the cost of an optimal solution.

14.5 Show that Problem 14.3 can be reduced to the particular shortest-path problem on DAGs from Exercise 4.16.

14.6 Under what assumptions does the shortest-detour algorithm of Section 6.1.2 applied to aligning haploid prediction to labeled DAG of diploid ground-truth give $O(dn)$ running time, where d is the resulting edit distance and n is the maximum of the DAG size and haploid length?

14.7 Complete the proof of Theorem 14.3, by proving that, if A and B are two haplotypes obtained with $|V||E| - 2t$ bit flips from M, then the set $C = \{i \mid A[2i - 1] = A[2i] = 0\} \subseteq V$ has the property that there are t edges between C and $V \setminus C$. Obtain this proof by showing the following intermediary properties:

- $A[2i - 1] = A[2i]$, for all $i \in \{1, \ldots, |V|\}$;
- $B[2i - 1] = B[2i]$, for all $i \in \{1, \ldots, |V|\}$;
- A is the bit-wise complement of B.

14.8   Show that the algorithm given for the haplotype assembly in the framework of the minimum error correction problem can be optimized to run in $O((2^K + r)s)$ time. How could you get the time complexity down to $O(2^K s)$?

14.9   Show how the algorithm given for the haplotype assembly in the minimum error correction problem can be implemented to use $O(2^K \sqrt{s})$ memory, while maintaining the same asymptotic time complexity. *Hint.* Use the same idea as in Exercise 7.2.

14.10   Let T denote the maximum number of bit flips needed in any column of the matrix M in an optimal solution to Problem 14.4. Show that the algorithm given for this problem can be modified to run in time $O((K^T + r)s)$, and even in time $O(K^T s)$ (see also Exercise 14.8 above).

14.11   Consider the problem of haplotype assembly under minimum error correction in which each cell of M has also a cost of flipping, and one is looking for a partition of the rows of M that is compatible with two haplotypes and minimizes the sum of the costs of all bit flips. What changes to the algorithm we presented are needed in order to solve this more general problem with the same time complexity?

15 Transcriptomics

Recall from Chapter 1 that, through transcription and alternative splicing, each gene produces different RNA transcripts. Depending on various factors, such as the tissue the cell is in, owing to disease, or in response to some stimuli, the RNA transcripts of a gene and the number of copies produced (their *expression level*) can be different.

In this chapter we assume that we have a collection of reads from all the different (copies of the) transcripts of a gene. We also assume that these reads have been aligned to the reference genome, using for example techniques from Section 10.6; in addition, Section 15.4 shows how to exploit the output of genome analysis techniques from Chapter 11 to obtain an aligner for long reads of RNA transcripts. Our final goal is to assemble the reads into the different RNA transcripts, and to estimate the expression level of each transcript. The main difficulty of this problem, which we call *multi-assembly*, arises from the fact that the transcripts share identical substrings.

We illustrate different scenarios, and corresponding multi-assembly formulations, stated and solved for each individual gene. In Section 15.1 we illustrate the simplest one, in which the gene's transcripts are known in advance, and we need only find their expression levels from the read alignments. In Section 15.2 we illustrate the problem of assembling the RNA reads into the different unknown transcripts, without estimating their levels of expression. In Section 15.3 we present a problem formulation for simultaneous assembly and expression estimation.

As just mentioned, in this chapter we assume that we have a reference genome, and thus that we are in a so-called *genome-guided* setting. De novo multi-assembly is in general a rather hard task. Thus, we prefer here to stick to genome-guided multi-assembly, which admits clean problem formulations and already illustrates the main algorithmic concepts. Nevertheless, in Insight 15.1 we briefly discuss how the least-squares method from Section 15.1 could be applied in a de novo setting.

15.1 Estimating the expression of annotated transcripts

Recall from Chapter 1 that each gene can be seen as partitioned into exons (substrings that appear in at least one transcript) and introns (substrings that are removed through transcription and do not appear in any transcript). In this section we assume that we have so-called *gene annotation*, meaning that we have a list T_i of exon starting and ending positions, for each of the k transcripts of a gene. Given a collection of reads sequenced from some of these annotated transcripts, we want to discover the *expression level* of each transcript, that is, how many copies of it appear in the sample.

First, we align the reads to the annotated transcripts, either by concatenating the annotated exons and aligning the reads to the resulting genomic sequences (using techniques from Chapter 10), or by split alignment of the reads directly to the genome (using techniques from Section 10.6).

Second, we construct a directed graph G whose vertices are the annotated exons, and whose arcs correspond to pairs of exons consecutive in some annotated transcript. The resulting graph G is called a *splicing graph*, because it models the alternative splicing of the transcripts. Moreover, G is a DAG, since the direction of its arcs is according to the position in the reference genome of its two endpoint exons. The annotated transcripts $T_1, \ldots, T_k$ correspond to paths $P_1, \ldots, P_k$ in G. For each annotated exon, namely for each vertex of G, we can compute an average read coverage. Similarly, we can compute the number of reads overlapping each splice-site alternative, that is, each position where two exons are concatenated, and thus each arc of G.

A reasonable assumption on how transcript sequencing behaves is that the coverage does not depend on the position inside the transcript. Observe, however, that, if the reads are of length m, the transcript coverage in the first and last $m - 1$ positions of each transcript presents an upward and downward slope, respectively. In this chapter we assume that such artifacts have been appropriately addressed, so that each annotated transcript T_i has an unknown average coverage e_i, its expression level. In the following two sections, where we assume that gene annotation is not available, this upward/downward slope is actually helpful in identifying potential start and end exons.

We can formulate this problem as follows. Given paths $P_1, \ldots, P_k$ in G (the annotated transcripts), find the expression level e_i of each path P_i, minimizing the sum of squared differences between the observed coverage of a vertex or an arc and the sum of the expression levels of the paths using that vertex or arc.

We make a simplifying reduction from this problem to one in which the input graph has coverages associated only with its arcs, as was done also in Section 5.4.2. This can be achieved by subdividing each vertex v into two vertices v_{in} and v_{out}, and adding the arc (v_{in}, v_{out}) having the same coverage as v. We also make all former in-neighbors of v in-neighbors of v_{in}, and all former out-neighbors of v out-neighbors of v_{out}. We use this simplified problem input henceforth, and later in Section 15.3.

Problem 15.1   Annotated transcript expression estimation

Given a directed graph $G = (V, E)$, a weight function $w : E \to \mathbb{Q}_+$, and a collection of k directed paths $P_1, \ldots, P_k$ in G, find expression levels $e_1, \ldots, e_k$ for each path, minimizing the following sum of errors:

$$\sum_{(x,y) \in E} \left(w(x, y) - \sum_{j \,:\, (x,y) \in P_j} e_j \right)^2. \tag{15.1}$$

Problem 15.1 is a *least-squares problem*, and can be solved analytically. The idea is to reduce the problem to solving a system of *linear* equations, which reduction is possible thanks to the particular definition of the objective function, as a sum of *squared* errors. See Insight 15.3 on page 342 for a discussion on different error functions.

We introduce some further notation here. We assume that the set E of arcs of G is $\{a_1, \ldots, a_m\}$, and that each a_i has weight w_i. Moreover, for every $i \in \{1, \ldots, m\}$ and $j \in \{1, \ldots, k\}$, let $\alpha_{i,j}$ be indicator variables defined as

$$\alpha_{i,j} = \begin{cases} 1, & \text{if arc } a_i \text{ belongs to path } P_j, \\ 0, & \text{otherwise.} \end{cases}$$

Using the indicator variables $\alpha_{i,j}$, we can rewrite (15.1) as

$$\begin{aligned} f(e_1, \ldots, e_k) = & \left(w_1 - \alpha_{1,1}e_1 - \alpha_{1,2}e_2 - \cdots - \alpha_{1,k}e_k\right)^2 \\ & + \left(w_2 - \alpha_{2,1}e_1 - \alpha_{2,2}e_2 - \cdots - \alpha_{2,k}e_k\right)^2 \\ & \cdots \\ & + \left(w_m - \alpha_{m,1}e_1 - \alpha_{m,2}e_2 - \cdots - \alpha_{m,k}e_k\right)^2. \end{aligned} \tag{15.2}$$

The function f is a function in k variables $e_1, \ldots, e_k$; it is quadratic in each variable. See Figure 15.1 for an example in two variables. It attains its global minimum in any point satisfying the system of equations obtained by equating to 0 each partial derivative of $f(e_1, \ldots, e_k)$ with respect to each variable. We thus obtain the following system of equations:

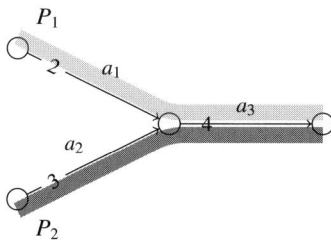

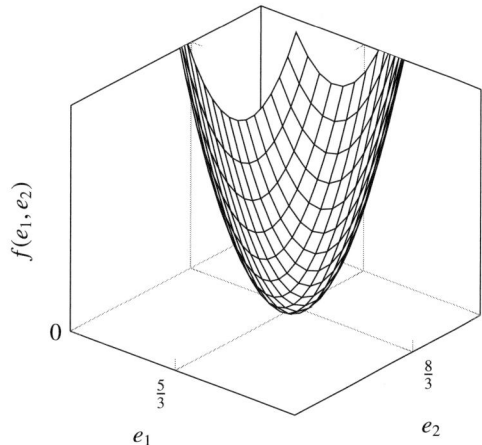

Figure 15.1 On the left, a directed graph G with arcs a_1, a_2, and a_3, having weights $2, 3$, and 4, respectively. P_1 and P_2 are two directed paths in G. The indicator matrix A is shown, as well as the corresponding system of equations (15.4). On the right is shown the function $f(e_1, e_2)$, having a unique minimum at $\left(\frac{5}{3}, \frac{8}{3}\right)$, the unique solution to the system (15.4).

$$e_1 \sum_{j=1}^{m} \alpha_{j,1}\alpha_{j,1} + \cdots + e_k \sum_{j=1}^{m} \alpha_{j,1}\alpha_{j,k} = w_1\alpha_{1,1} + \cdots + w_m\alpha_{m,1},$$

$$e_1 \sum_{j=1}^{m} \alpha_{j,2}\alpha_{j,1} + \cdots + e_k \sum_{j=1}^{m} \alpha_{j,2}\alpha_{j,k} = w_1\alpha_{1,2} + \cdots + w_m\alpha_{m,2},$$

$$\cdots \tag{15.3}$$

$$e_1 \sum_{j=1}^{m} \alpha_{j,k}\alpha_{j,1} + \cdots + e_k \sum_{j=1}^{m} \alpha_{j,k}\alpha_{j,k} = w_1\alpha_{1,k} + \cdots + w_m\alpha_{m,k}.$$

(Exercise 15.1 asks the reader to fill in the derivations needed for obtaining this system.)

We rewrite (15.3) more compactly using matrix notation. Let $A_{1..m,1..k}$ be a matrix such that $A[i,j] = \alpha_{i,j}$, for all $i \in \{1,\ldots,m\}$, $j \in \{1,\ldots,k\}$. Let A^{T} be its transpose, namely A^{T} is the $k \times m$ matrix such that $A[i,j] = \alpha_{j,i}$, for all $i \in \{1,\ldots,k\}, j \in \{1,\ldots,m\}$. A solution $(e_1,\ldots,e_k)$ for Problem 15.1 is any solution to the system of equations

$$A^{\mathrm{T}}A \begin{pmatrix} e_1 \\ \vdots \\ e_k \end{pmatrix} = A^{\mathrm{T}} \begin{pmatrix} w_1 \\ \vdots \\ w_m \end{pmatrix}, \tag{15.4}$$

(recall that A and $w_1,\ldots,w_m$ are known). Observe that system (15.4), and thus Problem 15.1, always admits a solution since the matrix $A^{\mathrm{T}}A$ is a square matrix of size k. Moreover, the solution is unique if and only if $\det(A^{\mathrm{T}}A) \neq 0$.

The matrix $A^{\mathrm{T}}A$ can be computed in time $O(mk^2)$, since each of the k^2 elements of $A^{\mathrm{T}}A$ can be computed in time $O(m)$. The system (15.4) can be solved for example by Gaussian elimination, which requires in total $O(k^3)$ arithmetic operations.

Insight 15.1 De novo annotated transcript expression estimation

Observe that the solution for Problem 15.1 does not require the input directed graph to be acyclic. Therefore, we could try applying it to a de novo setting, where G is the de Bruijn graph of a set of reads. For each arc, representing a k-mer, we can store the number of reads containing that k-mer. The annotated transcripts that we now take as input are strings, which need to be aligned to G, combining methods from Section 9.7 and Chapter 10; namely, one can slide a k-mer window through a transcript sequence, navigate the de Bruijn graph, and branch in the search to allow a limited amount of alignment errors.

Therefore, from the alignment of each transcript T_i we can again obtain a path P_i picking up the vertices belonging to the alignment (P_i can possibly have repeated vertices) in G. Thus, we are back in the same setting as in Problem 15.1, with the exception that the input graph might not be acyclic.

Observe that the complexity $O(mk^2)$ (where m now denotes the number of distinct k-mers) of computing the matrix $A^{\mathrm{T}}A$ is still linear in m, and finding a solution to the corresponding system of equations has a complexity independent of m.

15.2   Transcript assembly

We now study the problem of finding the most likely transcripts of a gene, given only the RNA split-read alignments. Thus, in contrast to the previous section, we assume here that we do not have any gene annotation. We also assume that we just need to find the transcripts, without estimating their expression levels.

Even without existing gene annotation, the exons of a gene can be discovered from split-read alignments, by observing that

- coverage information helps in discovering the start position of the first exon in a transcript and the end position of the last exon of a transcript, which can be marked as virtual splice-sites;
- the exons and introns of a gene are then contiguous stretches of the reference genome between two consecutive splice-sites; the exons are those stretches with coverage above a given confidence threshold.

These two steps can be formalized and implemented in the same manner as in Insight 14.1 for peak detection using HMMs. Exercise 15.2 asks the reader to formalize this exon detection phase using HMMs. The exons give the set of vertices of the splicing graph. The arcs of the splicing graph are similarly derived from reads overlapping two exons, indicating that they are consecutive in some transcript. Henceforth we assume that we have already constructed a splicing directed acyclic graph G.

The main difference in this section with respect to the rest of this chapter is that we do not try to explain the observed coverages of the splicing graph in terms of a least-squares model, but instead associate a cost with each arc between two vertices u and v, as the belief that u and v originate from *different* transcripts. Accordingly, we want to find a collection of paths that will cover all of the exons, at minimum total cost.

We should underline that the algorithms which we will present in this section depend only on the acyclicity of the splicing graph. As a matter of fact, the input could also be an overlap graph (whose vertices stand for reads and whose arcs stand for overlaps between reads – recall Section 13.2.3), since in this case the graph is also acyclic, because the reads have been aligned to the reference.

15.2.1   Short reads

In this section we assume that we have only a collection of "short" reads, in the sense that they overlap at most two consecutive exons. If a read happens to overlap more exons, then we can consider each overlap over two consecutive exons as a separate short read.

It is common practice to impose a *parsimony* principle in this multi-assembly problem, and to require the simplest interpretation that can explain the data. Thus, among all possible collections of transcripts covering all the exons of a gene, one with the minimum number of transcripts is preferred. Among all such minimum solutions, the one with the minimum total cost is preferred.

This problem is that of the minimum-cost minimum path cover, studied in Section 5.4.2 as Problem 5.11. However, we make one amendment to it here. Observe that, from the splicing graph construction phase, the read coverage indicates exons that are likely candidates to be start vertices of the solution paths. Likewise, we also know which exons are candidates for the end vertices of the solution paths. Therefore, we have the following problem.

Problem 15.2   Transcript assembly with short reads

Given a DAG $G = (V, E)$, a set $S \subseteq V$ of possible start vertices, a set $T \subseteq V$ of possible end vertices, and cost function $c : E \rightarrow \mathbb{Q}_+$, find the minimum number k of paths $P_1, \ldots, P_k$ such that

- every $v \in V$ belongs to some P_i (that is, $P_1, \ldots, P_k$ form a path cover of G),
- every P_i starts in some vertex of S, and ends in some vertex of T, and

among all path covers with k paths satisfying the above condition, they minimize

$$\sum_{i=1}^{k} \sum_{(x,y) \in P_i} c(x, y).$$

Henceforth we assume that all sources of G belong to S and all sinks of G belong to T, since otherwise Problem 15.2 admits no solution. Problem 15.2 can be solved as was done in Section 5.4.2: we construct the same flow network $N = (G^*, \ell, u, c, q)$, with the only modification that the global source s^* is connected to every vertex v_{in}, with $v \in S$ (instead of every source v of G), and every vertex v_{out}, with $v \in T$ (instead of every sink v of G), is connected to the global sink t^*. Computing the minimum-cost flow on N and splitting it into paths gives the solution paths. Thus, we analogously obtain the following theorem.

THEOREM 15.1   *The problem of transcript assembly with short reads on a DAG G with n vertices, m arcs, and positive costs is solvable in time $O(nm + n^2 \log n)$.*

15.2.2   Long reads

A splicing graph, through its arcs, models constraints only on pairs of exons that must be consecutive in some transcript. However, if the reads are long enough (or if some exons are short enough) a read can overlap multiple exons. We will consider in Section 15.4 how to find such multiple-exon read overlaps. Such overlaps give rise to further constraints on the admissible solutions to a transcript assembly problem. One way to model this problem of transcript assembly with long reads is to receive in the input also a collection of paths in the splicing graph (the long reads) and to require that every such path be completely contained in some solution path. Thus, we extend Problem 15.2 to Problem 15.3 below. See Figure 15.2 for an example.

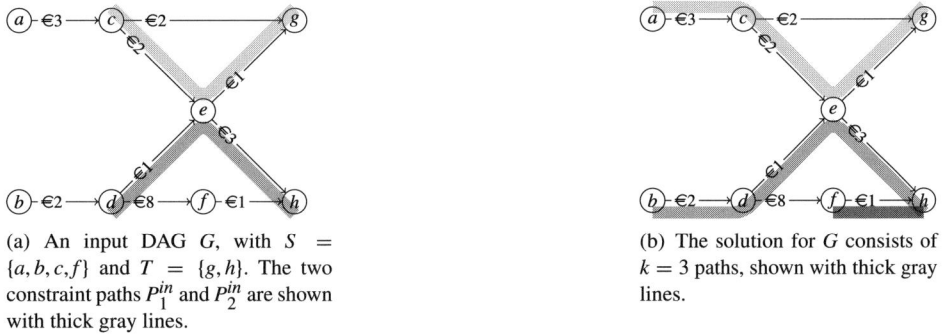

(a) An input DAG G, with $S = \{a,b,c,f\}$ and $T = \{g,h\}$. The two constraint paths P_1^{in} and P_2^{in} are shown with thick gray lines.

(b) The solution for G consists of $k = 3$ paths, shown with thick gray lines.

Figure 15.2 An example for the problem of transcript assembly with long reads.

Problem 15.3   Transcript assembly with long reads

Given a DAG $G = (V, E)$, a set $S \subseteq V$ of possible start vertices, a set $T \subseteq V$ of possible end vertices, a cost function $c : E \to \mathbb{Q}_+$, and a collection $\mathcal{P}^{in} = \{P_1^{in}, \ldots, P_t^{in}\}$ of directed paths in G, find the minimum number k of paths $P_1^{sol}, \ldots, P_k^{sol}$ forming a path cover of G, such that

- every P_i^{sol} starts in some vertex of S, and ends in some vertex of T,
- every path $P^{in} \in \mathcal{P}^{in}$ is entirely contained in some P_i^{sol}, and

among all path covers with k paths satisfying the above conditions, they minimize

$$\sum_{i=1}^{k} \sum_{(x,y) \in P_i^{sol}} c(x, y).$$

We solve Problem 15.3 also by reducing it to a network flow problem. The idea is to represent each path constraint P^{in} starting in a vertex u and ending in a vertex v by an arc (u, v) with demand 1 in the corresponding flow network N. The other arcs of N corresponding to vertices of P^{in} receive demand 0, since, if we cover P^{in}, we have already covered all of its vertices. However, the problematic cases (see Figure 15.3) are when a path constraint is completely contained in another one, or when the suffix of a path constraint equals the prefix of another path constraint. This is because adding such arcs between their endpoints increases the number k of paths in a solution for Problem 15.3.

We solve this by preprocessing the input constraints with the following two steps.

Step 1. While there are paths $P_i^{in}, P_j^{in} \in \mathcal{P}^{in}$ such that P_i^{in} is contained in P_j^{in}, remove P_i^{in} from $\mathcal{P}^{in}$.

If a path P_i^{in} is completely contained in another path P_j^{in}, then if we cover P_j^{in} with a solution path, this solution already covers P_i^{in}. This is key also for the correctness of the next step.

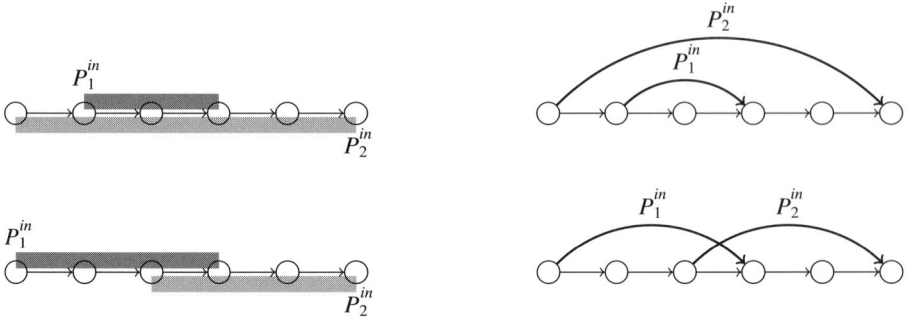

Figure 15.3 Problematic cases in reducing the problem of transcript assembly with long reads to a network flow problem. They can be solved after preprocessing the collection $\mathcal{P}^{in}$ of path constraints by Steps 1 and 2. Top, a path constraint P_1^{in} completely included in P_2^{in}; bottom, path constraints P_1^{in} and P_2^{in} have a suffix–prefix overlap.

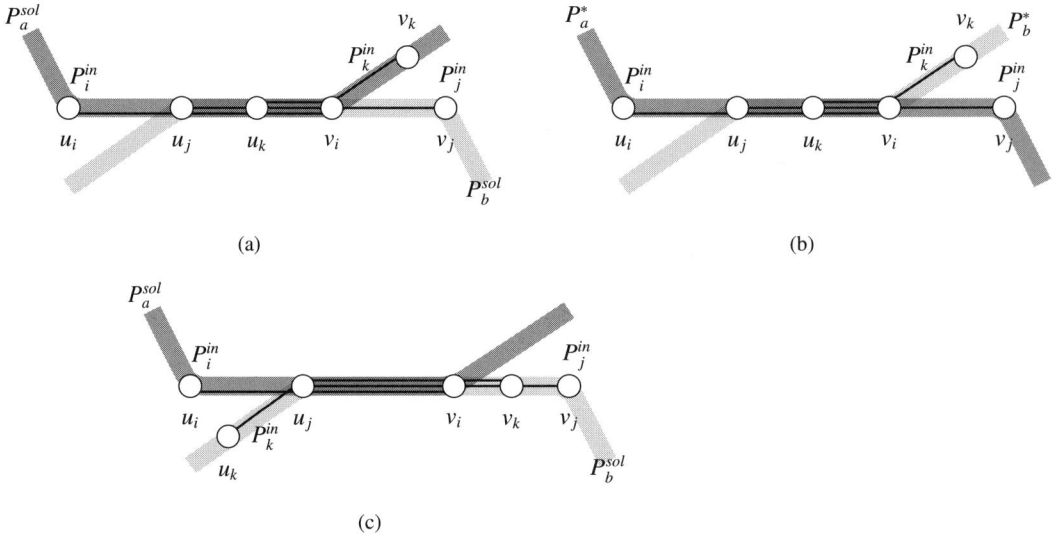

Figure 15.4 A visual proof of Lemma 15.2. Subpath constraints are drawn with thin black lines, solution paths are drawn with thick gray lines.

Step 2. While there are paths $P_i^{in}, P_j^{in} \in \mathcal{P}^{in}$ such that a suffix of P_i^{in} is a prefix of P_j^{in} do the following:

- let $P_i^{in}, P_j^{in} \in \mathcal{P}^{in}$ be as above, with the *longest* suffix–prefix overlap;
- let $P_{new}^{in} :=$ the path $P_i^{in} \cup P_j^{in}$ which starts as P_i^{in} and ends as P_j^{in};
- $\mathcal{P}^{in} := \left(\mathcal{P}^{in} \setminus \{P_i^{in}, P_j^{in}\} \right) \cup \{P_{new}^{in}\}$.

In this step, we merge paths having a suffix–prefix overlap. We do this iteratively, at each step merging that pair of paths for which the suffix–prefix overlap is the longest possible.

The correctness of these two steps is guaranteed by the following lemma.

LEMMA 15.2   *If the problem of transcript assembly with long reads for an instance* $(G, \mathcal{P}^{in})$ *admits a solution with k paths, then also the problem instance transformed by applying Steps 1 and 2 admits a solution with k paths, and this solution has the same cost and also satisfies the original constraints* $\mathcal{P}^{in}$.

Proof The correctness of Step 1 is clear. Assume that $\mathcal{P}^{in}$ has been transformed by Steps 1 and 2, and let $P_i^{in}, P_j^{in} \in \mathcal{P}^{in}$ be such that their suffix–prefix overlap is the *longest* possible. Suppose that the original problem admits a solution $\mathcal{P}^{sol} = \{P_1^{sol}, \dots, P_k^{sol}\}$ such that P_i^{in} and P_j^{in} are covered by different solution paths, say P_a^{sol} and P_b^{sol}, respectively. We show that the transformed problem admits a solution $\mathcal{P}^* = \left(\{P_1^{sol}, \dots, P_k^{sol}\} \setminus \{P_a^{sol}, P_b^{sol}\}\right) \cup \{P_a^*, P_b^*\}$ having the same cardinality as $\mathcal{P}$, in which P_i^{in} and P_j^{in} are covered by the same path P_a^*, and $\mathcal{P}^*$ also satisfies all of the original constraints $\mathcal{P}^{in}$.

Suppose that P_i^{in} starts with vertex u_i and ends with vertex v_i, and that P_j^{in} starts with vertex u_j and ends with vertex v_j.

- Let P_a^* be the path obtained as the concatenation of the path P_a^{sol} taken from its starting vertex until v_i with the path P_b^{sol} taken from v_i until its end vertex (so that P_a^* covers both P_i^{in} and P_j^{in}).
- Let P_b^* be the path obtained as the concatenation of the path P_b^{sol} taken from its starting vertex until v_i with the path P_a^{sol} taken from v_i until its end vertex.

See Figures 15.4(a) and (b). We have to show that the collection of paths $\mathcal{P}^* = \left(\{P_1^{sol}, \dots, P_k^{sol}\} \setminus \{P_a^{sol}, P_b^{sol}\}\right) \cup \{P_a^*, P_b^*\}$ covers all the vertices of G, has the some cost as $\mathcal{P}^{sol}$ and satisfies the original constraints $\mathcal{P}^{in}$. Since P_a^* and P_b^* use exactly the same arcs as P_a^{sol} and P_b^{sol}, $\mathcal{P}^*$ is a path cover of G, having the some cost as $\mathcal{P}^{sol}$.

The only two problematic cases are when there is a subpath constraint P_k^{in} that has v_i as internal vertex and is satisfied only by P_a^{sol}, or it is satisfied only by P_b^{sol}. Denote, analogously, by u_k and v_k the endpoints of P_k^{in}. From the fact that the input was transformed by Step 1, P_i^{in} and P_j^{in} are not completely included in P_k^{in}.

Case 1. P_k^{in} is satisfied only by P_a^{sol} (Figures 15.4(a) and (b)). Since P_i^{in} is not completely included in P_k^{in}, u_k is an internal vertex of P_i^{in}; thus, a suffix of P_i^{in} is a prefix also of P_k^{in}. From the fact that the overlap between P_i^{in} and P_j^{in} is the longest possible, we have that vertices u_j, u_k, and v_i appear in this order in P_i^{in}. Thus, P_k^{in} is also satisfied by P_b^*, since u_k appears after u_j on P_i.

Case 2. P_k^{in} is satisfied only by P_b^{sol}, and it is not satisfied by P_a^* (Figure 15.4(c)). This means that P_k^{in} starts on P_b^{sol} before u_j and, since it contains v_i, it ends on P_b^{sol} after v_i. From the fact that P_j^{in} is not completely included in P_k^{in}, v_k is an internal vertex of P_j^{in}, and thus a suffix of P_k^{in} equals a prefix of P_j^{in}. This common part is now longer than the common suffix/prefix between P_i^{in} and P_j^{in}, which contradicts the maximality of the suffix/prefix between P_i^{in} and P_j^{in}. This proves the lemma.   □

We are now ready to reduce this problem to a minimum-cost flow problem. As was done in Section 5.4.2 for the minimum path cover problem, and in the previous section, we construct a flow network $N = (G^*, \ell, u, c, q)$ (see Figure 15.5).

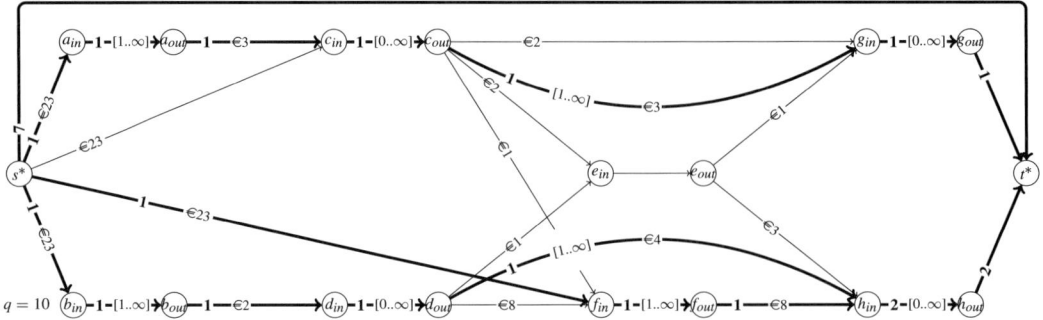

Figure 15.5 The flow network N constructed from the graph in Figure 15.2(a). Arcs without cost labels have null cost; arcs without demand labels have null demand. Arcs with non-null value in a minimum-cost flow over N are highlighted.

- For every vertex x of G, we introduce two vertices x_{in} and x_{out} in G^*, and add the arc (x_{in}, x_{out}) to G^*, with null cost;
- for every arc (y, x) incoming into x, we add the arc (y_{out}, x_{in}) to G^*, with cost $c(y, x)$;
- for every arc (x, y) outgoing from x, we add the arc (x_{out}, y_{in}) to G^* with cost $c(x, y)$.

We add a global source s^* with arcs to every vertex x_{in}, if $x \in S$. Likewise, we add a global sink t^* and arcs (x_{out}, t^*), for every sink $x \in T$.

For every path $P^{in} \in \mathcal{P}^{in}$ ($\mathcal{P}^{in}$ is transformed by Steps 1 and 2), starting in u and ending in v, we add the arc (u_{out}, v_{in}). We set its cost $c(u_{out}, v_{in})$ as the sum of the costs of all arcs of P^{in}, and we set its demand to 1. Moreover, for every arc (v_{in}, v_{out}) of G^* such that v belongs to no path constraint in $\mathcal{P}^{in}$, we set its demand to 1. All other arcs have demand 0.

We set the value of the flow to $q = |V| + |P^{in}|$, since at most $|V| + |P^{in}|$ paths are needed in any solution for Problem 15.3. To account for the supplementary flow, we also add the arc (s^*, t^*) with cost 0 and demand 0.

Finally, we set the cost of each arc (s^*, x_{in}) to 1 plus the sum of all other costs of G^*, since we want a solution with the minimum number of paths. A minimum-cost flow f induces a path cover in G in the following way. Since G^* is acyclic, we can consider a decomposition of f minus the arc (s^*, t^*) into the k weighted paths $P_1, \ldots, P_k$ from s^* to t^* (all having weight 1, from the minimality of f). We remove the vertices s^* and t^* from each P_i, and contract the arcs of the form (x_{in}, x_{out}) back into x. We obtain a collection $Q_1, \ldots, Q_k$ of paths in G^*. Moreover, whenever some Q_i uses an arc (u_{out}, v_{in}) corresponding to a constraint path P^{in}, we replace it in Q_i with P^{in}. This gives a collection $R_1, \ldots, R_k$ of paths in G, which is the solution to Problem 15.3.

Therefore, we have the following result.

THEOREM 15.3 *The problem of transcript assembly with long reads on a graph with n vertices, m arcs, and c subpath constraints, with K being the sum of subpath constraint* *lengths, can be solved by reducing it to a minimum-cost circulation problem on a network with $O(n)$ vertices and $O(m + c)$ arcs, with demands only. This network can be computed in time $O(K + c^2 + m)$, and the complexity of the problem of transcript assembly with long reads becomes $O(K + n^2 \log n + n(m + c))$.*

Proof Network N has $O(n)$ vertices and $O(m + c)$ arcs. The minimum-cost flow problem on N can be reduced to a minimum-cost circulation problem on a network obtained from N by adding the arc (t^*, s^*), with cost 0 and demand 0. This network has demands only, and they all equal 1, thus the minimum-cost circulation can be computed in time $O(n^2 \log n + n(m + c))$.

We now analyze the complexity of the preprocessing phase. If implemented naively, it takes $O(c^2 n^2)$ time. However, this can be lowered to $O(K + \text{output})$, where $\text{output} \leq c^2$ is the number of pairs having an overlap.

Step 1 can be solved by first building a (generalized) suffix tree on the concatenation of path constraints with a distinct symbol $\#_i$ added after each constraint sequence P_i^{in}. This can be done in $O(K)$ time even on our alphabet of size $O(n)$ by Theorem 8.15 on page 143. Then one can do as follows during depth-first traversal of the tree: if a leaf corresponding to the suffix starting at the beginning of subpath constraint P_i^{in} has an incoming arc labeled only by $\#_i$ and its parent still has other children, then the constraint is a substring of another constraint and must be removed (together with the leaf).

For Step 2, we compute all pairs of longest suffix–prefix overlaps between the subpath constraints using the $O(K + \text{output})$ time algorithm in Theorem 8.22. The output can be put into the form of a list L containing elements of the form (i, j, len) in decreasing order of the overlap length, len, between constraints P_i^{in} and P_j^{in}. Notice that each constraint must take part in only one merge as the first element, and in one merge as the second element, since merging cannot produce completely new overlaps; otherwise some constraint would be included in another constraint, which is not possible after Step 1. We can thus maintain two bitvectors $F[1..c]$ and $S[1..c]$ initialized to zeros, such that, once P_i^{in} and P_j^{in} have been merged, we set $F[i] = 1$ and $S[j] = 1$. That is, we scan list L from beginning to end, apply the merge if the correspondong bits are not set in F and S, and set the bits after each merge.

Merging itself requires a linked list structure: all the constraints are represented as doubly-linked lists with vertex numbers as elements. Merging can be done by linking the doubly-linked lists together, removing the extra overlapping part from the latter list, and redirecting its start pointer to point inside the newly formed merged list. When one has finished with merging, the new constraints are exactly those old constraints whose start pointers still point to the beginning of a vertex list. The complexity of merging is thus $O(K)$.   □

15.2.3   Paired-end reads

As paired-end reads are sequenced from ends of the same fragment of RNA, they also impose constraints on sequences of vertices that must appear together in the same assembled transcript. Thus, we can further generalize Problems 15.2 and 15.3 into

Problem 15.4 below, in which we are given a collection of pairs of paths, and we require that both paths in a pair are fully covered by the same solution path.

Problem 15.4   Transcript assembly with paired-end reads

Given a DAG $G = (V, E)$, a set $S \subseteq V$ of possible start vertices, a set $T \subseteq V$ of possible end vertices, a cost function $c : E \rightarrow \mathbb{Q}_+$, and a collection $\mathcal{P}^{in} = \{(P_{1,1}^{in}, P_{1,2}^{in}), \dots, (P_{t,1}^{in}, P_{t,2}^{in})\}$ of pairs of directed paths in G, find the minimum number k of paths $P_1^{sol}, \dots, P_k^{sol}$ forming a path cover of G, such that

- every P_i^{sol} starts in some vertex of S, and ends in some vertex of T,
- for every pair $(P_{j,1}^{in}, P_{j,2}^{in}) \in \mathcal{P}^{in}$, there exists P_i^{sol} such that both $P_{j,1}^{in}$ and $P_{j,2}^{in}$ are entirely contained in P_i^{sol}, and

among all path covers with k paths satisfying the above conditions, they minimize

$$\sum_{i=1}^{k} \sum_{(x,y) \in P_i^{sol}} c(x, y).$$

However, this problem of transcript assembly with paired-end reads becomes NP-hard. We show this by reducing from the NP-hard problem of deciding whether a graph G is 3-colorable, that is, if $V(G)$ can be partitioned into three sets V_A, V_B, V_C such that for no edge of G do both of its endpoints belong to the same set. Such a partition of $V(G)$ is called a *proper coloring* of G.

Let $G = (V, E)$ with $V = \{v_1, \dots, v_n\}$ and let $E = \{e_1, \dots, e_m\}$ be an undirected graph. We create the DAG $P(G)$ drawn in Figure 15.6, which consists of a first stage of n blocks corresponding to the n vertices of G, and a second stage of m blocks corresponding to each edge $e_k = (v_{i_k}, v_{j_k})$ of G, $k \in \{1, \dots, m\}$. Only some of the vertices and arcs of $P(G)$ have been labeled; when an arc is labeled $[L]$, we mean that in the family of paired subpath constraints we have the constraint $(L, [L])$. Also, we set all costs to 0.

THEOREM 15.4   *The problem of transcript assembly with paired-end reads is NP-hard.*

Proof   We show that an undirected graph $G = (V, E)$ is 3-colorable if and only if the DAG $P(G)$ drawn in Figure 15.6 admits a solution to Problem 15.4 with three paths.

For the forward implication, observe that we need at least three paths to solve $P(G)$, since the three arcs v_1, X_1, Y_1 exiting from vertex 0 cannot be covered by the same path, and each of them appears in some constraint. By definition, G is 3-colorable if and only if V can be partitioned into three sets V_A, V_B, V_C such that no arc of G is contained in any of them. We use these three sets to create the three solution paths for the problem of transcript assembly with paired-end reads as follows:

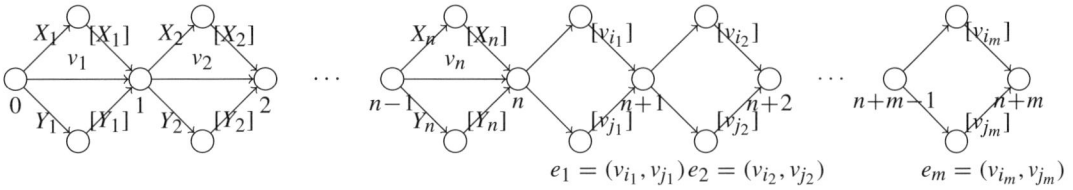

Figure 15.6 A reduction from the graph 3-coloring problem to the problem of transcript assembly with paired-end reads.

- for all $X \in \{A, B, C\}$, in the first stage (until vertex n) path P_X picks up all arcs labeled with a vertex in V_X, and no arc labeled with a vertex in $V \setminus V_X$;
- in the second stage (from vertex n until vertex $n + m$), P_X picks up those arcs labeled with $[v_{i_k}]$ such that v_{i_k} belongs to P_X.

This construction is possible since no edge $e_k = (v_{i_k}, v_{j_k})$ is contained in the same color class, and consequently the two arcs of $P(G)$ labeled v_{i_k} and v_{j_k} do not belong to the same path among $\{P_A, P_B, P_C\}$. Thus, $[v_{i_k}]$ and $[v_{j_k}]$ do not have to be both covered by the same solution path. Therefore, the three paths P_A, P_B, P_C satisfy all path constraints, and they constitute a solution to Problem 15.4.

For the backward implication, suppose that the DAG $P(G)$ drawn in Figure 15.6 admits a solution to Problem 15.4 with three paths P_A, P_B, P_C. Then, we partition V into three color classes A, B, C by setting $v_i \in X$ if and only if the arc of $P(G)$ labeled by v_i (in the first stage from vertex 0 to vertex n) belongs to P_X, for all $X \in \{A, B, C\}$. To see that $\{A, B, C\}$ is indeed a partition of V, observe that in each block k of the first stage of $P(G)$ no two paths in $\{P_A, P_B, P_C\}$ can share an arc, since all three arcs labeled v_k, X_k, Y_k appear in some constraint. Therefore, each arc v_k appears in exactly one of $\{P_A, P_B, P_C\}$. The proof that the partition $\{A, B, C\}$ is also a proper coloring of G follows as in the forward implication. For every arc (v_{i_k}, v_{j_k}) of G, its endpoints belong to different paths among P_A, P_B, P_C, because of the path constraints $(v_{i_k}, [v_{i_k}])$ and $(v_{j_k}, [v_{j_k}])$. □

15.3 Simultaneous assembly and expression estimation

In this section we study the problem in which we need to do simultaneous transcript assembly and expression estimation. We assume that we have already constructed a splicing graph, for example by methods described at the beginning of Section 15.2.

Problem 15.5 Transcript assembly and expression estimation (least squares)

Given a DAG $G = (V, E)$, a set $S \subseteq V$ of possible start vertices, a set $T \subseteq V$ of possible end vertices, and a weight function $w : E \to \mathbb{Q}_+$, find a collection of paths $P_1, \ldots, P_k$ in G, and their corresponding expression levels $e_1, \ldots, e_k$, such that

- every P_i starts in some vertex of S and ends in some vertex of T, and the following sum of errors is minimized:

$$\sum_{(x,y)\in E} \left(w(x,y) - \sum_{j\,:\,(x,y)\in P_j} e_j \right)^2. \tag{15.5}$$

The problem formulation we give in Problem 15.5 above coincides with the problem of annotated transcript expression estimation (Problem 15.1), with the amendment that we also need to find the paths starting in some vertex of S and ending in some vertex of T which minimize the least-squared error. Recall that each vertex of the splicing graph is considered as being subdivided as in Section 15.1, so that we have coverages associated only with the arcs of the splicing graph. Observe also that we do not impose an upper bound on the number of paths in a solution. Insight 15.2 discusses some practical strategies for limiting the number of solution paths. See Figure 15.7 for an example.

Problem 15.5 can be solved by a minimum-cost flow problem with convex cost functions (recall Insight 5.2), as follows. Construct the flow network $N = (G^*, \ell, u, c, q)$, where G^* is obtained from G by adding a global source s^* connected to all vertices in S, and a global sink t^* to which all vertices in T are connected. The demand of every arc is set to 0, and the capacity of every arc is unlimited. The cost function for every arc (x, y) of G is $c(x, y, z) = (w(x,y) - z)^2$, which is a convex function in z, and the cost of all arcs incident to s^* or t^* is set to 0. Exercise 15.6 asks the reader to work out the formal details showing that any decomposition into weighted paths of the minimum-cost flow on N gives an optimal solution for Problem 15.5.

We now modify Problem 15.5 into Problem 15.6 by replacing the square function with the absolute value function. See Insight 15.3 for a brief discussion on the differences between the two functions.

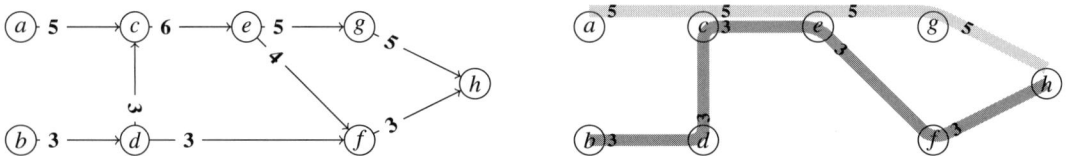

Figure 15.7 On the left, a weighted DAG G for Problem 15.5 or 15.6, with $S = \{a, b\}$ and $T = \{h\}$. On the right, one optimal solution for both problems, consisting of two paths with expression levels 5 and 3, respectively. Their cost is $|6 - 8|^\beta + |3 - 0|^\beta + |4 - 3|^\beta$, for $\beta \in \{1, 2\}$, from arcs (c, e), (d, f), and (e, f).

Problem 15.6 Transcript assembly and expression estimation (least absolute values)

Given a DAG $G = (V, E)$, a set $S \subseteq V$ of possible start vertices, a set $T \subseteq V$ of possible end vertices, and a weight function $w : E \to \mathbb{Q}_+$, find a collection of paths $P_1, \ldots, P_k$ in G, and their corresponding expression levels $e_1, \ldots, e_k$, such that

- every P_i starts in some vertex of S and ends in some vertex of T,

and the following sum of errors is minimized:

$$\sum_{(x,y) \in E} \left| w(x, y) - \sum_{j : (x,y) \in P_j} e_j \right|. \tag{15.6}$$

This admits a reduction to a minimum-cost flow problem with linear cost functions (in fact, null or unit costs), which we describe next. Suppose that $P_1, \ldots, P_k$, with expression levels $e_1, \ldots, e_k$, are some paths in G. Let $g : E \to \mathbb{Q}_+$ be the flow on G induced by these paths, namely $g(x, y) = \sum_{j : (x,y) \in P_j} e_j$. Using g, we can rewrite (15.6), and call it the *error* of g, as

$$\mathsf{error}(g) = \sum_{(x,y) \in E} |w(x, y) - g(x, y)|.$$

If $P_1, \ldots, P_k$, with expression levels $e_1, \ldots, e_k$, are optimal for Problem 15.6, then the induced flow g minimizes $\mathsf{error}(g)$. Vice versa, if g is a flow on G minimizing $\mathsf{error}(g)$, then any decomposition of g into weighted paths (obtained by Theorem 5.2) gives an optimal solution to Problem 15.6. Therefore, for solving Problem 15.6, it suffices to find a flow g on G minimizing $\mathsf{error}(g)$ and split it into (an arbitrary number of) weighted paths.

In order to find such an optimal g, we build a flow network N, which we call an *offset network*, such that the flow on an arc (x, y) of N models the quantity $|w(x, y) - g(x, y)|$. In other words, a minimum-cost flow in N models the offsets (positive or negative) between the weights of arcs of G and their expression levels in an optimal solution to Problem 15.6. We achieve this by considering, for every vertex $y \in V(G)$, the quantity

$$\Delta(y) = \sum_{x \in N^+(y)} w(y, x) - \sum_{x \in N^-(y)} w(x, y).$$

If $\Delta(y) > 0$, then we need to force $\Delta(y)$ units of flow to go from vertex y to a global sink t^*. If $\Delta(y) < 0$, then we need to force $-\Delta(y)$ units of flow to go from a global source s^* to vertex y.

Formally, the offset network $N = (G^*, \ell, u, c, q)$ is constructed as follows (see Figure 15.8 for an example).

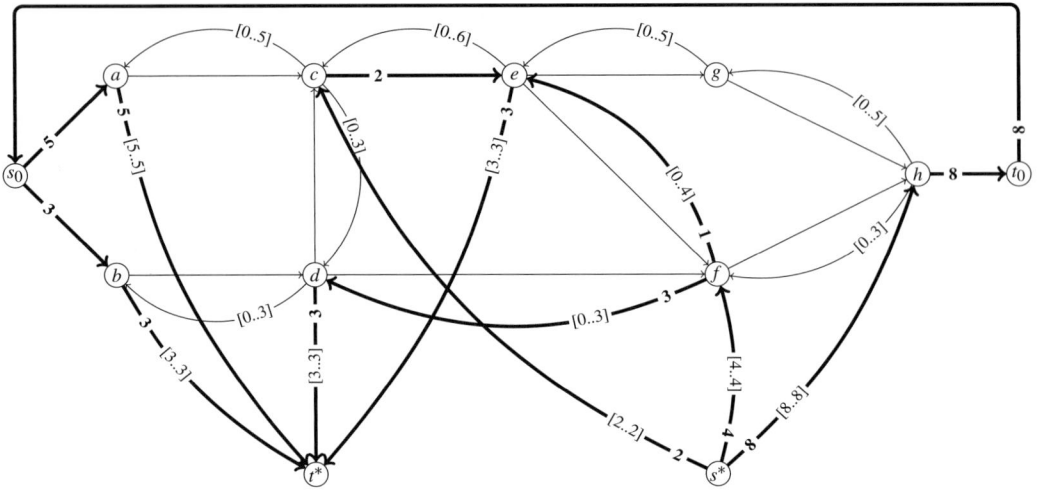

Figure 15.8 The flow network N constructed from the input in Figure 15.7 (left). The arcs between vertices of G have cost 1; all other arcs have cost 0. Arcs with non-null value in a minimum-cost flow f over N are highlighted; f induces the flow drawn in Figure 15.7 (right).

- We add to G^* all vertices and arcs of G, together with
  - a new source s_0 and a new sink t_0, with arcs (s_0, s) for every $s \in S$, arcs (t, t_0), for every $t \in T$, and arc (t_0, s_0), all with infinite capacity, and with null demand and cost;
  - vertices s^* and t^*.
- For every arc $(x, y) \in E(G)$,
  - we set its demand $\ell(x, y) = 0$, its capacity $u(x, y) = \infty$, and its cost $c(x, y) = 1$;
  - we add the reverse arc (y, x) with $\ell(y, x) = 0$, $u(y, x) = w(x, y)$, $c(y, x) = 1$.
- Starting with initial flow value $q := 0$, for every $y \in V(G)$
  - let $\Delta(y) := \sum_{x \in N^+(y)} w(y, x) - \sum_{x \in N^-(y)} w(x, y)$;
  - if $\Delta(y) > 0$ then
    - we add the arc (y, t^*) with $\ell(y, t^*) = u(y, t^*) = \Delta(y)$, $c(y, t^*) = 0$;
    - we update $q := q + \Delta(y)$;
  - if $\Delta(y) < 0$ then
    - we add the arc (s^*, y) with $\ell(s^*, y) = u(s^*, y) = -\Delta(y)$, $c(s^*, y) = 0$.

Given a minimum-cost flow f on N, we construct the desired flow $g : E(G) \to \mathbb{Q}_+$ as follows: given an arc $(x, y) \in E(G)$,

$$g(x, y) := \begin{cases} w(x, y) + f(x, y), & \text{if } f(x, y) > 0, \\ w(x, y) - f(y, x), & \text{if } f(y, x) > 0, \\ w(x, y), & \text{otherwise.} \end{cases} \tag{15.7}$$

Observe that g is a well-defined function, by the fact that $f(y,x) \leq u(y,x) = w(x,y)$, for every arc $(x,y) \in E(G)$, and by Lemma 15.5 below. Exercise 15.7 asks the reader to show that g is indeed a flow on G. Since, for every arc $(x,y) \in E(G)$, we set $c(x,y) = c(y,x) = 1$ in N, it holds that

$$\text{error}(g) = \text{cost}(f).$$

Thus, since f is of minimum cost, g minimizes $\text{error}(g)$.

LEMMA 15.5   *If f is a minimum-cost flow in N, then at most one of $f(x,y)$ and $f(y,x)$ is non-null, for every arc $(x,y) \in E(G)$.*

Proof   We argue by contradiction, and suppose that, for an arc $(x,y) \in E$, we have $\min(f(x,y),f(y,x)) > 0$. Then, we show that we can find a flow f' on N of strictly smaller cost than f, as follows. Flow f' is defined as coinciding with f on all arcs, except for $f'(x,y)$, which equals $f(x,y) - \min(f(x,y),f(y,x))$ and $f'(y,x)$, which equals $f(y,x) - \min(f(x,y),f(y,x))$. Note that f' is a flow on N. Since we assumed $\min(f(x,y),f(y,x)) > 0$, and the cost of both arcs (x,y) and (y,x) is 1, f' has a strictly smaller cost than the minimum-cost flow f.   □

Insight 15.2 Limiting the number of transcripts

In Problems 15.5 and 15.6 we did not impose a constraint on the number of paths in an optimal solution. However, in practice, according to the same parsimony principle as was invoked in Section 15.2.1, one prefers a solution with a small number of solution paths.

One way this can be achieved is to split the resulting flow into the minimum number of paths. Since this is an NP-hard problem (Theorem 5.3), one can apply some of the heuristics from Insight 5.1.

However, we can ask for an exact algorithm: Problems 15.5 and 15.6 can be reformulated by receiving in the input also an upper bound on the number k of solution paths. Exercise 15.11 asks the reader to show that both of these problems become NP-hard. Exercise 15.12 asks you to show that these extensions can be solved with an algorithm based on dynamic programming.

Another common practical modification consists of changing the objective function in Problems 15.5 and 15.6 by adding a regularization term $\lambda \sum_{i=1}^{k} e_i$, for some given $\lambda \geq 0$:

$$\sum_{(x,y) \in E} \left(w(x,y) - \sum_{j\,:\,(x,y) \in P_j} e_j \right)^2 + \lambda \sum_{i=1}^{k} e_i.$$

Exercise 15.10 asks the reader to modify the minimum-cost flow reductions presented in this section for solving this generalization. Just as before, splitting the resulting flow into the minimum number of paths is also NP-hard, but in practice decompositions of it based on heuristics from Insight 5.1 can lead to solutions with fewer paths.

Therefore, the following theorem holds.

THEOREM 15.6   *The problem of transcript assembly and expression estimation (least absolute values) can be solved in polynomial time.*

Proof   Given an input for the problem of transcript assembly and expression estimation (least absolute values), we construct the flow network N and compute in polynomial time a minimum-cost flow f on N. The flow f induces a flow g on G, defined in (15.7). Using Theorem 5.2 we can split the flow g into weighted paths in time linear in the number of vertices and arcs of the input graph. From the minimality of f, and the fact that $\mathrm{error}(g) = \mathrm{error}(f)$, these paths are optimal for the problem of transcript assembly and expression estimation (least absolute values).   □

15.4   Transcript alignment with co-linear chaining

In Section 6.5 we studied a dynamic programming solution for aligning a eukaryote gene sequence to the genome, allowing a limited number of free gaps, modeling the introns. Applying such an algorithm, for example, on long reads obtained by RNA-sequencing is infeasible. For this reason, in Section 10.6 we studied some alternatives using index structures. Now, we are ready for our last proposal for implementing split-read alignment, this time combining maximal exact matches with a rigorous dynamic programming approach. This approach gives a scalable solution to find the path constraints on the splicing graph required in Section 15.2.2.

Insight 15.3 Error functions

Throughout this chapter we have used the sum of squared differences and the sum of absolute differences as objective functions. The absolute value function has the advantage that it is more robust with respect to outliers than the square function. However, in contrast to the square function, it leads to unstable solutions, in the sense that small changes in the coverage values can lead to great differences in the optimal solution. For example, one choice that could mitigate both of these issues is the function $\log(1 + x^2)$. However, this function is not convex, so no network flow solution can apply. Nevertheless, the dynamic programming solution asked for in Exercise 15.13 can apply for this function.

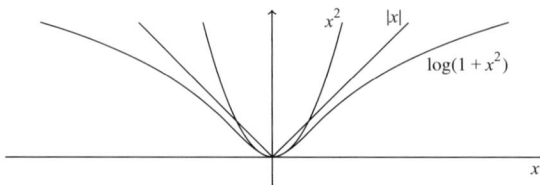

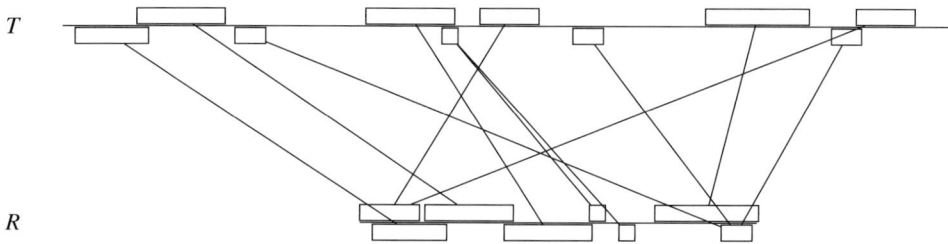

Figure 15.9 Maximal exact matches as input for co-linear chaining for aligning a long read from an RNA transcript to the genome. Boxes cover the area spanned by the matches, and edges show the pairs.

Consider a long read R from an RNA transcript of a gene. Compute the maximal exact matches between R and the genome string T, using the techniques from Section 11.1.3. The alignment result should look like the one illustrated in Figure 15.9.

Co-linear chaining aims to collect a chain of maximal exact matches (or any other local alignment anchors) so that the order of the selected pairs satisfies the order in which they appear in both T and R, and the collected chain covers the maximum area of R, among all such feasible chains. This problem statement is formalized below.

Problem 15.7   Co-linear chaining

Let T and R be two strings, and let V be a set of N tuples (x, y, c, d) with the property that $T_{x..y}$ matches $R_{c..d}$. Find a sequence $S = s_1 s_2 \cdots s_p$ of tuples, where each $s_i \in V$, and

- $s_j.y > s_{j-1}.y$ and $s_j.d > s_{j-1}.d$, for all $1 \leq j \leq p$, and
- the *ordered* coverage of R, defined as

$$\mathrm{coverage}(R, S) = |\{i \in [1..|R|] \mid i \in [s_j.c..s_j.d] \text{ for some } 1 \leq j \leq p\}|,$$

  is maximized.

The solution for Problem 15.7 uses dynamic programming and the invariant technique studied in Chapter 6.

First, sort the tuples in V by the coordinate y into the sequence $v_1 v_2 \cdots v_N$. Then, fill a table $C[1..N]$ such that $C[j]$ gives the maximum ordered coverage of $R[1..v_j.d]$ using the tuple v_j and any subset of tuples from $\{v_1, v_2, \ldots, v_{j-1}\}$. Hence, $\max_j C[j]$ gives the total maximum ordered coverage of R. It remains to derive the recurrence for computing $C[j]$. For any tuple $v_{j'}$, with $j' < j$, we consider two cases:

(a) $v_{j'}$ does not overlap v_j in R (that is, $v_{j'}.d < v_j.c$), or
(b) $v_{j'}$ overlaps v_j in R (that is, $v_j.c \leq v_{j'}.d \leq v_j.d$).

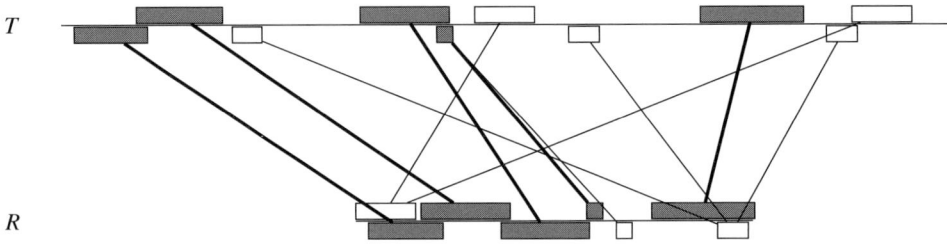

Figure 15.10 Optimal solution of co-linear chaining visualized by the shaded boxes and thicker edges.

The best ordered coverage that uses the tuple v_j and a subset of tuples from $\{v_1, v_2, \ldots, v_{j-1}\}$ with none of them overlapping v_j (case (a) above) can thus be obtained by the recurrence

$$C^a[j] = \max_{j' \,:\, v_{j'}.d < v_j.c} C[j'] + (v_j.d - v_j.c + 1). \tag{15.8}$$

Similarly, the best ordered coverage that uses the tuple v_j and a subset of tuples from $\{v_1, v_2, \ldots, v_{j-1}\}$ with at least one of them overlapping v_j (case (b) above) can be obtained by the recurrence

$$C^b[j] = \max_{j' \,:\, v_j.c \le v_{j'}.d \le v_j.d} C[j'] + (v_j.d - v_{j'}.d), \tag{15.9}$$

which works correctly unless there is a tuple $v_{j'}$ satisfying the condition $v_j.d \ge v_{j'}.d \ge v_{j'}.c > v_j.c$. Such containments can, however, be ignored when computing the final value $C[j] = \max(C^a[j], C^b[j])$, because case (a) always gives a better result: see Exercise 15.16.

We can use the invariant technique to implement these recurrence relations with range maximum queries. These queries can be solved using the search tree in Lemma 3.1 on page 20 (in fact, here we need its dual version, with minimum replaced by maximum, which we denote as RMaxQ):

$$C^a[j] = (v_j.d - v_j.c + 1) + \max_{j' \,:\, v_{j'}.d < v_j.c} C[j']$$

$$= (v_j.d - v_j.c + 1) + \mathcal{T}.\text{RMaxQ}(0, v_j.c - 1),$$

$$C^b[j] = v_j.d + \max_{j' \,:\, v_j.c \le v_{j'}.d \le v_j.d} C[j'] - v_{j'}.d$$

$$= v_j.d + \mathcal{I}.\text{RMaxQ}(v_j.c, v_j.d),$$

$$C[j] = \max(C^a[j], C^b[j]).$$

For these to work correctly, we need to have properly updated the trees $\mathcal{T}$ and $\mathcal{I}$ for all $j' \in [1..j-1]$. That is, we need to call $\mathcal{T}.\text{Update}(v_{j'}.d, C[j'])$ and $\mathcal{I}.\text{Update}(v_{j'}.d, C[j'] - v_{j'}.d)$ after computing each $C[j']$. The running time is $O(N \log N)$. The pseudocode is given in Algorithm 15.1.

Algorithm 15.1: Co-linear chaining

Input: A set of tuples V sorted by y-coordinate into sequence $v_1 v_2 \cdots v_N$.

Output: The index j giving $\max_j C[j]$.

1 Initialize search trees $\mathcal{T}$ and $\mathcal{I}$ with keys $v_j.d$, $1 \leq j \leq N$, and with key 0, all keys associated with values $-\infty$;

2 $\mathcal{T}$.Update$(0, 0)$;

3 $\mathcal{I}$.Update$(0, 0)$;

4 **for** $j \leftarrow 1$ *to* N **do**

5 $\quad C^a[j] \leftarrow (v_j.d - v_j.c + 1) + \mathcal{T}$.RMaxQ$(0, v_j.c - 1)$;

6 $\quad C^b[j] \leftarrow v_j.d + \mathcal{I}$.RMaxQ$(v_j.c, v_j.d)$;

7 $\quad C[j] \leftarrow \max(C^a[j], C^b[j])$;

8 $\quad \mathcal{T}$.Update$(v_j.d, C[j])$;

9 $\quad \mathcal{I}$.Update$(v_j.d, C[j] - v_j.d)$;

10 **return** $\operatorname{argmax}_j C[j]$;

Figure 15.10 illustrates the optimal chain on our schematic example. This chain can be extracted by modifying Algorithm 15.1 to store traceback pointers.

THEOREM 15.7 *Co-linear chaining (Problem 15.7) on N input tuples can be solved in $O(N \log N)$ time.*

15.5 Literature

The least-squares method given in Section 15.1 is a standard topic in numerical analysis and statistics (it is attributed to Gauss in 1795). Faster solutions are possible if approximation is allowed and the number of equations is much larger than the number of variables (Drineas *et al.* 2011).

A very similar minimum path cover formulation to the problem of transcript assembly with short reads from Section 15.2.1 was introduced by Trapnell *et al.* (2010). The differences are that Trapnell *et al.* (2010) uses an overlap graph (which is extended to paired-end reads, for a suitably defined notion of overlap), and the costs are assigned only to the edges of the bipartite graph arising from the alternative solution of the minimum path cover problem via a bipartite matching problem (Fulkerson 1956). An even earlier connection between the minimum path cover problem and a transcript assembly problem is considered in Jenkins *et al.* (2006). The problem of transcript assembly with long reads from Section 15.2.2 was initially proposed in the context of transcript assembly by Bao *et al.* (2013), and fully solved in Rizzi *et al.* (2014). Exercise 15.3 asks the reader to rediscover another reduction, from Ntafos & Hakimi (1979), of the unweighted version of this problem to a minimum-cost flow problem.

The problem of transcript assembly with paired-end reads from Section 15.2.3 was proposed in Song & Florea (2013). The unweighted versions of these two problems first appeared in an application to program testing (Ntafos & Hakimi 1979). The latter was proved NP-hard in Ntafos & Hakimi (1979) by a reduction from 3-SAT and in Beeren-winkel *et al.* (2014) and Rizzi *et al.* (2014) by a reduction from graph 3-colorability.

The problem of transcript assembly and expression estimation (least squares) from Section 15.3 was considered in Tomescu *et al.* (2013a), and variants of it appeared before in Li *et al.* (2011a) and Feng *et al.* (2011). Its least absolute values variant appeared in Lin *et al.* (2012), and the solution we presented is adapted from the one given in Tomescu *et al.* (2013a).

The addition of a regularization term mentioned in Insight 15.2 was considered by Li *et al.* (2011b) and Bernard *et al.* (2014). The addition of an input upper bound on the number of predicted transcripts was proposed by Tomescu *et al.* (2013b).

The co-linear chaining solution is analogous to one in Abouelhoda (2007). Exercise 15.17 covers some variants studied in Mäkinen *et al.* (2012).

Exercises

15.1   Work out the calculations for deriving the system of equations (15.3), using the formula

$$\frac{\partial \left(w_i - \alpha_{i,1}e_1 - \alpha_{i,2}e_2 - \cdots - \alpha_{i,k}e_k \right)^2}{\partial e_j} = 2 \left(w_i - \alpha_{i,1}e_1 - \alpha_{i,2}e_2 - \cdots - \alpha_{i,k}e_k \right)$$

$$\cdot \frac{\partial \left(w_i - \alpha_{i,1}e_1 - \alpha_{i,2}e_2 - \cdots - \alpha_{i,k}e_k \right)}{\partial e_j}$$

$$= -2\alpha_{i,j} \left(w_i - \alpha_{i,1}e_1 - \alpha_{i,2}e_2 - \cdots - \alpha_{i,k}e_k \right).$$

15.2   Formalize the exon detection phase described on page 329 using HMMs (recall Insight 14.1 and Chapter 7).

15.3   Consider the following version of Problem 15.3 in which there are no costs associated with the arcs for the DAG and in which we need to cover only a given collection of paths. Given a DAG $G = (V, E)$, a set $S \subseteq V$ of possible start vertices, a set $T \subseteq V$ of possible end vertices, and a collection $\mathcal{P}^{in} = \{P_1^{in}, \ldots, P_t^{in}\}$ of directed paths in G, find the minimum number k of paths $P_1^{sol}, \ldots, P_k^{sol}$ such that

- every P_i^{sol} starts in some vertex of S and ends in some vertex of T,
- every path $P^{in} \in \mathcal{P}^{in}$ is entirely contained in some P_i^{sol}.

Assume also that no P_i^{in} is entirely included in another P_j^{in} (otherwise P_i^{in} can be removed from $\mathcal{P}^{in}$). Show that this problem can be solved by a reduction to a minimum-cost flow problem (without having to iteratively merge paths with longest suffix–prefix overlaps). What features does the resulting flow network have? Can you apply a specialized minimum-cost flow algorithm with a better complexity for such a network?

15.4   Given an input for the problem of transcript assembly with paired-end reads, show that we can decide in polynomial time whether it admits a solution with two paths, and if so, find the two solution paths.

15.5 Using the reduction in the proof of Theorem 15.4, conclude that there exists for no $\varepsilon > 0$ an algorithm returning a solution with k paths for the problem of transcript assembly with paired-end reads, where k is greater than the optimal number of paths by a multiplicative factor $\frac{4}{3} - \varepsilon$.

15.6 Argue that the reduction of Problem 15.5 to a network flow problem with convex costs presented on page 338 is correct.

15.7 Show that the function g defined in (15.7) on page 340 is a flow on G, namely that

- for all $y \in V(G) \setminus (S \cup T)$, $\sum_{x \in N^-(y)} g(x, y) = \sum_{x \in N^+(y)} g(y, x)$;
- $\sum_{s \in S} \sum_{y \in N^+(s)} g(s, y) = \sum_{t \in T} \sum_{x \in N^-(t)} g(x, t)$.

15.8 Explain what changes need to be made to the flow network N, if in Problem 15.6 we also get in the input a coefficient $\alpha(x, y)$ for every arc (x, y) of G, and we are asked to find the paths and their expression levels which minimize

$$\sum_{(x,y)\in E} \alpha(x, y) \left| w(x, y) - \sum_{j\,:\,(x,y)\in P_j} e_j \right|.$$

15.9 Consider a variant of Problem 15.6 in which we want to minimize the absolute differences between the total coverage (instead of the average coverage) of an exon and its predicted coverage. Explain how this problem can be reduced to the one in Exercise 15.8 above, for an appropriately chosen function α.

15.10 Adapt the reduction to a minimum-cost flow problem from Section 15.3 to solve the following problem. Given a DAG $G = (V, E)$, a set $S \subseteq V$ of possible start vertices, a set $T \subseteq V$ of possible end vertices, and a weight function $w : E \to \mathbb{Q}_+$, find a collection of paths $P_1, \ldots, P_k$ in G, and their corresponding expression levels $e_1, \ldots, e_k$, such that

- every P_i starts in some vertex of S and ends in some vertex of T,

and the following function is minimized:

$$\sum_{(x,y)\in E} \left| w(x, y) - \sum_{j\,:\,(x,y)\in P_j} e_j \right| + \lambda \sum_{i=1}^{k} e_i.$$

What can you say about the problem in which we use the squared difference, instead of the absolute value of the difference?

15.11 Show that the following problem is NP-hard, for any fixed $\beta \geq 1$. Given a DAG $G = (V, E)$, a set $S \subseteq V$ of possible start vertices, a set $T \subseteq V$ of possible end vertices, a weight function $w : E \to \mathbb{Q}_+$, and an integer k, find a collection of paths $P_1, \ldots, P_k$ in G, and their corresponding expression levels $e_1, \ldots, e_k$, such that

- every P_i starts in some vertex of S and ends in some vertex of T, and the following function is minimized:

$$\sum_{(x,y)\in E} \left| w(x,y) - \sum_{j\,:\,(x,y)\in P_j} e_j \right|^{\beta}.$$

Hint. Use Theorem 5.3 or its proof.

15.12   Show that, given $k \geq 1$ and given $e_1, \ldots, e_k$, the following problem is solvable in time $O(n^{2k+2})$. Given a DAG $G = (V, E)$, a set $S \subseteq V$ of possible start vertices, a set $T \subseteq V$ of possible end vertices, a weight function $w : E \to \mathbb{Q}_+$, and an integer k, find a collection of paths $P_1, \ldots, P_k$ in G, such that

- every P_i starts in some vertex of S and ends in some vertex of T,

and the following function is minimized:

$$\sum_{(x,y)\in E} \left| w(x,y) - \sum_{j\,:\,(x,y)\in P_j} e_j \right|. \tag{15.10}$$

Hint. Use dynamic programming and consider the optimal k paths ending in every tuple of k vertices.

15.13   Show that, given an error function $\delta : \mathbb{Q}_+ \times \mathbb{Q}_+ \to \mathbb{Q}_+$, the problem considered in Exercise 15.12 in which the objective function (15.10) is replaced with

$$\sum_{(x,y)\in E} \delta \left(w(x,y), \sum_{j\,:\,(x,y)\in P_j} e_j \right) \tag{15.11}$$

is solvable with the same time complexity $O(n^{2k+2})$ (assuming δ is computable in time $O(1)$).

15.14   Consider the following problem of *transcript assembly and expression estimation with outliers*, in which the weights of the arcs not belonging to any solution path do not contribute in the objective function. Given $k \geq 1$, a DAG $G = (V, E)$, a set $S \subseteq V$ of possible start vertices, a set $T \subseteq V$ of possible end vertices, a weight function $w : E \to \mathbb{Q}_+$, and an integer k, find a collection of paths $P_1, \ldots, P_k$ in G, such that

- every P_i starts in some vertex of S and ends in some vertex of T,

and the following function is minimized:

$$\sum_{(x,y)\,:\,\text{exists } P_i \text{ with } (x,y)\in P_i} \left| w(x,y) - \sum_{j\,:\,(x,y)\in P_j} e_j \right|.$$

Show that this problem is NP-hard. *Hint.* Use Theorem 5.3 or its proof.

15.15   Reduce the problem of transcript assembly and expression estimation with outliers from Exercise 15.14 to the generalized problem from Exercise 15.13, by showing an appropriate error function δ.

15.16 Prove that the co-linear chaining algorithm works correctly even when there are tuples containing other tuples in T or in R, that is, tuples of type (x, y, c, d) and (x', y', c', d') such that either $x < x' \leq y' < y$ or $c < c' \leq d' < d$ (or both).

15.17 Modify the co-linear chaining algorithm to solve the following variations of the ordered coverage problem.

(a) Find the maximum ordered coverage of R such that all the tuples involved in the coverage must overlap in R.

(b) Find the maximum ordered coverage of R such that the distance in R between two consecutive tuples involved in the coverage is at most a given threshold value α.

(c) Find the maximum ordered coverage of R such that the distance in T between two consecutive tuples involved in the coverage is at most a given threshold value β.

15.18 Co-linear chaining gives a rough alignment between the transcript and the genome as a list of subregion correspondences. Consider how this rough alignment can be fine-grained into a sequence-level alignment.

16   Metagenomics

Assume that a drop of seawater contains cells from n distinct species $\mathcal{S} = \{S^1, S^2, \ldots, S^n\}$. Denote the genomes of such species by $\mathcal{G} = \{G^1, G^2, \ldots, G^n\}$, and let $\mathbf{F}[1..n]$ be a vector that contains the number of cells of each species. We use the related vector $\mathbf{P}[i]$ to denote *relative frequencies*: in other words, $\mathbf{P}[i]$ equals $\mathbf{F}[i]$ divided by the total number of cells in the drop of water. In practice the distribution $\mathbf{P}$ is heavy-tailed, with few species at high frequency and many species at low frequency. We call the triplet $(\mathcal{S}, \mathcal{G}, \mathbf{P})$ the *environment*. In most practical cases n is large, we know just a small fraction of $\mathcal{S}$ and $\mathcal{G}$, and we completely ignore $\mathbf{P}$ and $\mathbf{F}$. A *metagenomic sample* is a collection of r strings $\mathcal{R} = \{R^1, R^2, \ldots, R^r\}$, where each R^i, called a *read*, is a substring of a genome in $\mathcal{G}$, possibly perturbed by sequencing errors. Methods and tools for the computational analysis of metagenomic samples are still under active development, and some of the questions raised by this type of heterogeneous, large-scale data remain unresolved. In this chapter we focus on a subset of key problems with a clear algorithmic formulation.

Throughout the chapter we assume that, except for a small variance, all reads in a sample have the same length ℓ, which is significantly smaller than the length of any genome in $\mathcal{G}$. For simplicity, we assume that the process that generated $\mathcal{R}$ from the environment sampled a substring of length ℓ with equal probability from any position of the genome of any cell: in other words, the probability that a read comes from genome G^i is $|G^i| \cdot \mathbf{F}[i] / \sum_{j=1}^{n} |G^j| \cdot \mathbf{F}[j]$. We call the average number of reads that cover a position in G^i, or equivalently the number of reads in $\mathcal{R}$ that were sampled from a cell with genome G^i, multiplied by $\ell/|G^i|$, the *coverage* of a genome $G^i \in \mathcal{G}$.

In practice just a small fraction of all existing genomes has been sequenced: we might know some genomes just as a set of contigs (see Section 13.2), or just as an even smaller set of *genes*. In some applications we are more interested in estimating the biological functions performed by the species in an environment, rather than the relative frequency of the species themselves: for this purpose we might use precomputed *gene families*, collections of sequenced genes with similar functions coming from known species. Such families are typically annotated with a reference alignment or with a hidden Markov model that describes the common features of all sequences in a family (see Section 6.6 and Chapter 7). In this chapter we denote by $\mathcal{D}$ any precomputed database that is used by an algorithm, leaving its details to the context.

We also denote by $\mathcal{T}$ a known *reference taxonomy*, that is, a tree whose leaves are all known species, and whose (possibly unary) internal nodes, called *taxa*, are higher-order groups of species based on evolutionary history or on shared biological features. The depth of a taxon in $\mathcal{T}$ is called its *taxonomic rank*. When it is hard to estimate the composition of a sample at the species level, algorithms focus on immediately smaller ranks, called *genus* and *family*.

16.1   Species estimation

A key problem in metagenome analysis consists in reconstructing the *taxonomic composition* of a sample, or in other words in estimating $\mathcal{S}$ and $\mathbf{P}$ from $\mathcal{R}$. Knowing which taxa are present in a sample, and in which proportion, can help one to study the effects of a drug, of a pathogen, or of a disease on the microbes that live in the human gut, for example. The taxonomic composition can also serve as a succinct fingerprint to compare, classify, and retrieve large collections of metagenomic samples. The methods described in this section can be used almost verbatim with databases of protein families, to estimate the biochemical processes that are at work in an environment, rather than its taxonomic composition.

16.1.1   Single-read methods

Assume that all genomes in $\mathcal{G}$ are stored in a reference database $\mathcal{D}$. Perhaps the most natural way of estimating $\mathcal{S}$ and $\mathbf{P}$ from $\mathcal{R}$ consists in finding the substring W^i of a genome in $\mathcal{D}$ that best matches each read $R^i \in \mathcal{R}$ taken in isolation, for example by using the read-mapping techniques described in Chapter 10. In this section we assume that every substring W of a genome in $\mathcal{D}$ that matches a read R^i has a corresponding *matching score* (or, symmetrically, a *matching cost*, for example the edit distance between R^i and W) that quantifies the quality of the alignment between R^i and W. The search for a best match can be naturally performed in parallel for each read, or for each genome in $\mathcal{D}$. If R^i is long enough, we could first compare its k-mer composition with the k-mer composition of the genomes or of the gene families in $\mathcal{D}$, and then match R^i just to the portions of the database with similar enough k-mer composition (see Section 11.2 for a sampler of similarity measures based on k-mers).

A read R^i can approximately match more than one genome in $\mathcal{D}$, or even the same genome at different positions: let $\mathcal{W}^i$ denote the set of all substrings of $\mathcal{D}$ that match R^i with a high enough score. If $|\mathcal{W}^i| > 1$, it is common practice to assign R^i to the *lowest common ancestor* in $\mathcal{T}$ of all elements in $\mathcal{W}^i$: this effectively builds an approximation of $\mathcal{S}$ and $\mathbf{P}$ in which species and relative frequencies are grouped by $\mathcal{T}$, possibly at different depths. This strategy can be refined by re-mapping the best match W^i to $\mathcal{W}^i$ itself, and by assigning R^i to the lowest common ancestor *of the matches of W^i in $\mathcal{W}^i$* whose score is higher than the matching score between W^i and R^i. Strategies based on the least common ancestor, however, tend to assign reads to taxa with low rank in $\mathcal{T}$, thereby limiting the accuracy of the estimate, and they discard reads with many high-scoring matches when the corresponding lowest common ancestor is close to the root of $\mathcal{T}$.

Marker-based methods

Rather than using entire genomes as a reference database, we could employ a precomputed set of *markers* that are known to uniquely characterize a node in $\mathcal{T}$.

DEFINITION 16.1   *A marker for a node $v \in \mathcal{T}$ is a substring that occurs at most τ times in every genome in v, and that does not occur in any genome not in v.*

We can compute such markers of maximal length by adapting the document counting algorithm described in Section 8.4.2, or alternatively by using the matching statistics and the distinguishing statistics vectors described in Section 11.2.3 (see Exercises 16.11 and 16.12). If we restrict markers to having a specific length k, we can compute all of them in small space, as follows.

LEMMA 16.2   *Let $\mathcal{S} = \{S^1, S^2, \ldots, S^n\}$ and $\mathcal{T} = \{T^1, T^2, \ldots, T^m\}$ be two sets of strings on alphabet $[1..\sigma]$. Given the bidirectional BWT index of S^i for all $i \in [1..n]$, the bidirectional BWT index of T^j for all $j \in [1..m]$, and two integers $k > 1$ and $\tau \geq 1$, we can compute all k-mers that occur between two and τ times in every string of $\mathcal{S}$, and that do not occur in any string of $\mathcal{T}$, in $O((x + y) \log \sigma \cdot (|\mathcal{S}| + |\mathcal{T}|))$ time and $3x + o(x) + y + O(\sigma \log^2 y)$ bits of working space, where $x = \min\{|S^i| : i \in [1..n]\}$ and $y = \max\{|X| : X \in \mathcal{S} \cup \mathcal{T}\}$.*

Proof   Let S^* be a string of minimum length in $\mathcal{S}$. We build a bitvector `intervals` $[1..|S^*| + 1]$ such that `intervals`$[i] = 1$ if and only if i is the starting position of a k-mer interval in $\mathsf{BWT}_{S^*\#}$, and `intervals`$[i] = 0$ otherwise. Recall from Section 9.7.3 that this vector can be built from the bidirectional BWT index of S^* in $O(|S^*| \log \sigma)$ time, by enumerating the internal nodes of the suffix tree of S^*. We also initialize to zero a bitvector `intersection`$[1..|S^*| + 1]$, and to one a bitvector `found`$[1..|S^*| + 1]$. Then, we scan `intervals` sequentially, and we set `intersection`$[p] = 1$ for every starting position p of an interval of size at most τ in `intervals`. We index `intervals` to support `rank` queries in constant time, as described in Section 3.2.

Then, we consider every other string $S^i \in \mathcal{S} \setminus \{S^*\}$. First, we build a bitvector `intervals`$^i[1..|S^i| + 1]$ marking the intervals of all k-mers in $\mathsf{BWT}_{S^i\#}$. Then, we use the bidirectional BWT index of S^* and the bidirectional BWT index of S^i to enumerate the internal nodes of their generalized suffix tree, as described in Algorithm 11.3. Assume that we reach a node v of the generalized suffix tree whose depth is at least k: let $\overset{\leftrightarrow}{v}_i$ and $\overset{\leftrightarrow}{v}_*$ be the intervals of v in $\mathsf{BWT}_{S^i\#}$ and in $\mathsf{BWT}_{S^*\#}$, respectively, and let $\overset{\leftarrow}{W}_i$ and $\overset{\leftarrow}{W}_*$ be the intervals of the length-k prefix W of $\ell(u)$ in $\mathsf{BWT}_{S^i\#}$ and in $\mathsf{BWT}_{S^*\#}$, respectively. If $\overset{\leftrightarrow}{v}_i$ coincides with the interval of a k-mer in $\mathsf{BWT}_{S^i\#}$, we compute $\overset{\leftarrow}{W}_*$ in constant time from $\overset{\leftrightarrow}{v}_*$ and `intervals`, and we set `found`$[p] = 1$, where p is the starting position of $\overset{\leftarrow}{W}_*$ in $\mathsf{BWT}_{S^*\#}$. At the end of this process, we set `intersection`$[i] = 0$ for all $i \in [1..|S^*|+1]$ such that `intersection`$[i] = 1$ and `found` $= 0$. A similar algorithm can be applied to unmark from `intersection` the starting position of all k-mers that occur in a string of $\mathcal{T}$.

Then, we use again the bidirectional BWT index of S^* and the bidirectional BWT index of S^i to enumerate the internal nodes of their generalized suffix tree. Assume that we reach a node v of the generalized suffix tree, at depth at least k, such that $\vec{v}_*$ coincides with the interval of a k-mer in $\mathsf{BWT}_{S^*\#}$, and such that $\texttt{intersection}[p] = 1$, where p is the starting position of $\vec{v}_*$. Then we set $\texttt{intersection}[p] = 0$ if $\vec{v}_i$ coincides with the interval of a k-mer in $\mathsf{BWT}_{T^i\#}$, and if $|\vec{v}_i| > \tau$. At the end of this process, the bits set to one in $\texttt{intersection}$ mark the starting positions of the intervals of the distinct k-mers of S^* that occur between two and τ times in every string of $\mathcal{S}$.

Once all of the strings in $\mathcal{S}$ and $\mathcal{T}$ have been processed in this way, we invert $\mathsf{BWT}_{S^*\#}$, and we return the length-k prefix of the current suffix of $S^*\#$ whenever we reach a marked position in $\texttt{intersection}$. □

It is easy to generalize Lemma 16.2 to k-mers that occur between *one* and τ times.

For concreteness, assume that markers have been precomputed for every genome in $\mathcal{G}$. Recall that the probability that a read in $\mathcal{R}$ is sampled from a genome G^i is $(|G^i| \cdot \mathbf{F}[i])/(\sum_{j=1}^{n} |G^j| \cdot \mathbf{F}[j])$. The coverage of G^i, or equivalently the number of reads that cover a position of G^i, is thus

$$C(i) = |\mathcal{R}|\ell \cdot \frac{\mathbf{F}[i]}{\sum_{j=1}^{n} |G^j| \cdot \mathbf{F}[j]}$$

and the sum of all coverages is

$$\sum_{i=1}^{n} C(i) = |\mathcal{R}|\ell \cdot \frac{\sum_{i=1}^{n} \mathbf{F}[i]}{\sum_{j=1}^{n} |G^j| \cdot \mathbf{F}[j]}.$$

It follows that $\mathbf{P}[i] = C(i)/\sum_{i=1}^{n} C(i)$. Let M^i be the concatenation of all markers of genome G^i, and let $\mathcal{R}^i$ be the set of reads in $\mathcal{R}$ whose best match is located inside a marker of G^i. We can estimate $C(i)$ by $|\mathcal{R}^i|\ell/|M^i|$ for every G^i, and thus derive an estimate of $\mathbf{P}[i]$.

In practice aligning reads to markers can be significantly faster than aligning reads to entire genomes, since the set of markers is significantly smaller than the set of genomes, and the probability that a read matches a marker is lower than the probability that a read matches a substring of a genome. This method allows one also to detect taxa that are not in $\mathcal{T}$, since the difference between the relative frequency of a taxon v and the sum of the relative frequencies of all children of v in $\mathcal{T}$ can be assigned to an unknown child below v.

16.1.2   Multi-read and coverage-sensitive methods

Rather than assigning each read to a taxon in isolation, we could combine the taxonomic assignments of multiple reads to increase both the number of classified reads and the accuracy of the estimate. For example, we could first use the set of all mappings of all reads to reference genomes in $\mathcal{D}$ to determine *which taxa in $\mathcal{T}$ are present in the sample*, and only afterwards try to assign reads to taxa. Specifically, we could compute the smallest set of taxa that can annotate all reads, at each taxonomic rank i of $\mathcal{T}$: this is an instance of the set cover problem described in Section 14.1, in which the taxa at rank i are sets, and a taxon v *covers* a read R^j if R^j has a high-scoring match in some genomes that belong to v. Then, read R^j can be assigned to a largest set that covers it, or it can be declared unclassified if the assigned taxon at rank i is not a child of the assigned taxon at rank $i - 1$.

Another way of combining information from multiple reads consists in assembling reads into contigs (see Section 13.2), and then mapping such long contigs to the reference genomes in $\mathcal{D}$. Longer contigs reduce the number of false matches, and potentially allow one to annotate reads that could not be annotated when mapped to the database in isolation. Contigs could be preliminarily clustered according to their k-mer composition before being matched to the database: all contigs in the same cluster X could then be assigned to the lowest common ancestor of the best matches of all contigs in X. Similar strategies can be applied to clusters of reads built without using an assembler, as described in Section 16.2.

A third way of using global information consists in exploiting the assumption that *the probability of sampling a read from a given position in a given genome is the same for all positions in the genome*. We call a method that exploits this assumption *coverage-sensitive*. In what follows, we denote by $\mathcal{D}' = \{D^1, \dots, D^m\}$ the subset of $\mathcal{D}$ that contains all genomes with a high-scoring match to at least one read in $\mathcal{R}$. We want to select exactly one mapping location for each read (or to declare the read unmapped), so that the resulting coverage is as uniform as possible along the sequence of every genome in $\mathcal{D}'$, and so that the sum of all matching costs is as small as possible: this provides an estimate of the number of reads c_i sampled from each genome $D^i \in \mathcal{D}'$, which can be immediately converted to its relative frequency $\mathbf{P}[i]$. It is natural to express this problem with the formalism of minimum-cost flows, as follows.

Mapping reads using candidate abundances

Assume first that we already have an estimate of the number of reads $c_1, \dots, c_m$ sampled from each genome in $\mathcal{D}'$, which we call the *candidate abundance*. We will deal later with the problem of finding an optimal set of such abundances. If each genome D^i has uniform coverage, then its uniform *candidate coverage* is $c_i/(|D^i| - \ell + 1)$. In order to model the uniformity constraint, we partition every genome D^i into substrings of a fixed length L, which we will refer to as *chunks*. Denote by s_i the number of chunks that each genome D^i is partitioned into. We construct a bipartite graph $G = (A \cup B, E)$, such that the vertices of A correspond to reads, and the vertices of B correspond to the chunks of all genomes $D^1, \dots, D^m$. Specifically, for every chunk j of genome D^i, we introduce a vertex $y_{i,j}$, and we add an edge between a read $x \in A$ and vertex $y_{i,j} \in B$ if there is a match of read x inside chunk j of genome D^i. This edge is assigned the cost of the match, which we denote here by $c(x, y_{i,j})$. The uniformity assumption can now be modeled by requiring that every genome chunk receives close to c_i/s_i read mappings. In order to model the fact that reads can originate from unknown species whose genome is not present in $\mathcal{D}'$, we introduce an "unknown" vertex z in B, with edges from every read vertex $x \in A$ to z, and with an appropriately initialized cost $c(x, z)$.

In the problem of coverage-sensitive abundance evaluation stated below, the task is to select exactly one edge for every vertex $x \in A$ (a mapping of the corresponding read), either to a chunk of some genome D^i, or to the unknown vertex z, which at the same time minimizes the sum, over all chunks of every genome D^i, of the absolute difference between c_i/s_i and the number of mappings it receives, and also minimizes the sum of all selected edge costs. We can combine these two optimization criteria into a single objective function by receiving in the input also a parameter $\alpha \in (0, 1)$, which controls the relative contribution of the two objectives. Given a set M of edges in a graph G, recall from page 53 that we denote by $d_M(x)$ the number of edges of M incident to a vertex x of G. Formally, we have the following problem.

Problem 16.1   Coverage-sensitive mapping and abundance evaluation

Given a bipartite graph $G = (A \cup B, E)$, where A is a set of n reads and $B = \{y_{1,1}, \ldots, y_{1,s_1}, \ldots, y_{m,1}, \ldots, y_{m,s_m}\} \cup \{z\}$ is a set of genome chunks; a cost function $c : E \to \mathbb{Q}$; a constant $\alpha \in (0, 1)$; and a candidate tuple $c_1, \ldots, c_m$ of reads per genome, find a many-to-one matching M that satisfies the condition

- for every $x \in A$, $d_M(x) = 1$ (that is, every read is covered by exactly one edge of M),

and minimizes the following quantity:

$$(1 - \alpha) \sum_{(x,y) \in M} c(x, y) + \alpha \sum_{i=1}^{m} \sum_{j=1}^{s_i} \left| \frac{c_i}{s_i} - d_M(y_{i,j}) \right|. \tag{16.1}$$

This problem can be solved in polynomial time by a reduction to a minimum-cost flow problem, as was done for the problem of many-to-one matching with optimal residual load factors from Section 5.3.3. The only difference here is the introduction of the vertex z, which does not have any required load factor. Exercise 16.1 asks the reader to work out the details of this reduction.

Finding optimal abundances
We still need to find the *optimal* tuple $c_1, \ldots, c_N$ of abundances which minimizes (16.1) in Problem 16.1 over all possible tuples of abundances. We extend the formulation of Problem 16.1 by requiring to find the optimal such abundances as well.

Problem 16.2   Coverage-sensitive mapping and abundance estimation

Given a bipartite graph $G = (A \cup B, E)$, where A is a set of n reads, and $B = \{y_{1,1}, \ldots, y_{1,s_1}, \ldots, y_{m,1}, \ldots, y_{m,s_m}\} \cup \{z\}$ is a set of genome chunks; a cost function $c : E \to \mathbb{Q}$; and constant $\alpha \in (0, 1)$, find a many-to-one matching M that satisfies the condition

- for every $x \in A$, $d_M(x) = 1$ (that is, every read is covered by exactly one edge of M), and find a tuple of abundances $c_1, \ldots, c_N$ that minimizes the following quantity:

$$(1 - \alpha) \sum_{(x,y) \in M} c(x, y) + \alpha \sum_{i=1}^{m} \sum_{j=1}^{s_i} \left| \frac{c_i}{s_i} - d_M(y_{i,j}) \right|. \tag{16.2}$$

Unfortunately, this problem is NP-hard, as shown in Theorem 16.3. However, in practice, since Problem 16.1 is solvable in polynomial time, we can use ad-hoc algorithms to guess different candidate tuples of abundances, and then we can evaluate their performance using Problem 16.1.

THEOREM 16.3   *Problem 16.2 is NP-hard.*

Proof   We reduce from the problem of exact cover with 3-sets (X3C). In this problem, we are given a collection S of 3-element subsets $S_1, \ldots, S_m$ of a set $U = \{1, \ldots, n\}$, and we are required to decide whether there is a subset $C \subseteq S$, such that every element of U belongs to exactly one $S_i \in C$.

Given an instance of problem X3C, we construct the bipartite graph $G = (A \cup B, E)$, where $A = U = \{1, \ldots, n\}$ and B corresponds to S, in the following sense:

- $s_i = 3$, for every $i \in \{1, \ldots, m\}$;
- for every $S_j = \{i_1 < i_2 < i_3\}$, we add to B the three vertices $y_{j,1}, y_{j,2}, y_{j,3}$ and the edges $\{i_1, y_{j,1}\}, \{i_2, y_{j,2}\}, \{i_3, y_{j,3}\}$, each with cost 0.

For completeness, we also add vertex z to B, and edges of some positive cost between it and every vertex of A.

We now show that, for any $\alpha \in (0, 1)$, an instance for problem X3C is a "yes" instance if and only if Problem 16.2 admits on this input a many-to-one matching M of cost 0. Observe that, in any solution M of cost 0, the genome abundances are either 0 or 3, and that vertex z has no incident edges in M.

For the forward implication, let C be an exact cover with 3-sets. We assign the abundances $c_1, \ldots, c_m$ as follows:

$$c_j = \begin{cases} 3, & \text{if } S_j \in C, \\ 0, & \text{if } S_j \notin C; \end{cases}$$

and we construct M as containing, for every $S_j = \{i_1 < i_2 < i_3\} \in C$, the three edges $\{i_1, y_{j,1}\}, \{i_2, y_{j,2}\}, \{i_3, y_{j,3}\}$. Clearly, this M gives a solution of cost 0 to Problem 16.2.

Vice versa, if Problem 16.2 admits a solution M of cost 0, in which the genome coverages are either 0 or 3 and z has no incident edges, then we can construct an exact cover with 3-sets C by taking

$$C = \{S_i \mid d_M(y_{i,1}) = d_M(y_{i,2}) = d_M(y_{i,3}) = 1\}.$$

$\square$

16.2 Read clustering

When reads in $\mathcal{R}$ cannot be successfully mapped to known genomes, we might want to assemble them into the corresponding, unknown source genomes, for example by using the algorithms described in Chapter 13. This corresponds to estimating just $\mathcal{G}$ and $\mathbf{P}$, but not $\mathcal{S}$. However, the size of metagenomic samples, the large number of species they contain, and the sequence similarity of such species, make assembly a slow and inaccurate process in practice. It might thus be useful to preliminarily group reads together on the basis of their k-mer composition: the resulting clusters loosely correspond to long strings that occur in one or in multiple genomes of $\mathcal{G}$, and in the best case to entire genomes. Assembly can then be carried out inside each cluster, possibly in parallel. The number of clusters can also be used to estimate $\mathbf{P}$ and derived measures of the biodiversity of an environment, and the k-mer composition of a cluster can be used to position it approximately inside a known taxonomy, without assembling its reads. Like the frequency of species, the number and k-mer composition of clusters can also be used as a succinct fingerprint to compare and classify large collections of metagenomic samples.

Clustering reads becomes more difficult as the number of species increases, and as the frequency vector $\mathbf{P}$ diverges from a uniform distribution. To improve accuracy, it is becoming more common to combine multiple metagenomic samples that are believed to contain approximately the same species, possibly with different relative frequencies. The input to this problem is then even larger than the already massive metagenomic samples taken in isolation. In this section we describe a space-efficient clustering pipeline that is based on the bidirectional BWT index described in Section 9.4, where R is the concatenation of all reads in $\mathcal{R}$, each terminated by a distinct binary string on the artificial characters $\{\#_0, \#_1\} \notin [1..\sigma]$. The main steps of the pipeline are summarized in Figure 16.1.

16.2.1 Filtering reads from low-frequency species

As mentioned at the beginning of this chapter, the distribution of relative frequencies $\mathbf{P}$ is heavy-tailed, with few species at high frequency and many species at low frequency. In practice reads from low-frequency species tend to degrade the quality of the clusters of high-frequency species: a preliminary step consists thus in isolating reads from low-frequency species and processing them in a separate phase. This has the additional advantages of removing reads with many errors and of reducing the size of the input of each phase.

Specifically, a read is filtered out if and only if *all its k_1-mers occur fewer than τ_1 times in $\mathcal{R}$*, where k_1 and τ_1 are user-defined parameters that depend on the error rate, coverage, and read length.

LEMMA 16.4 *Given the bidirectional BWT index of R, we can detect in $O(|R| \log \sigma)$ time and in $2|R| + o(|R|) + \sigma \log^2 |R|$ bits of working space all of the reads in R that contain at least one k_1-mer that occurs at least $\tau_1 \geq 2$ times in R.*

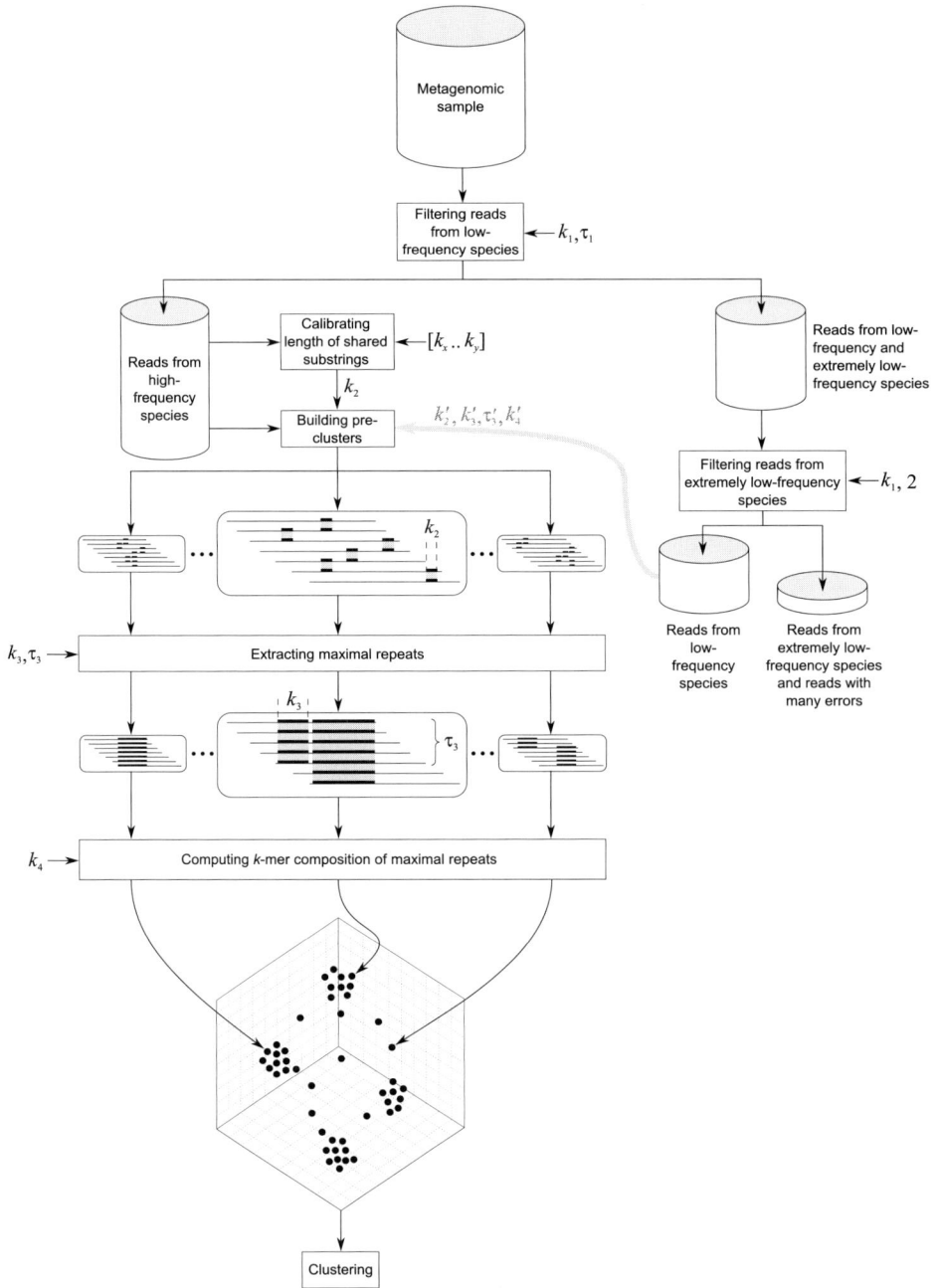

Figure 16.1 The main steps for clustering metagenomic reads. See Section 16.2 for a detailed description of each step and parameter. The light gray arc indicates that reads from low-frequency species can be processed with the same pipeline as reads from high-frequency species, but using different parameters.

Proof We build a bitvector of intervals `intervals[1..|R|]` such that `intervals` $[i] = 1$ if and only if i is the starting or the ending position of the interval of a k_1-mer in BWT_R, and `intervals[i]` $= 0$ otherwise. Recall from Section 9.7.3 that by enumerating the internal nodes of the suffix tree of R in $O(|R|\log\sigma)$ time, using the bidirectional BWT index of R, we can build a bitvector first_{k_1} that is zero everywhere, except at the first position of the interval of every k_1-mer in BWT_R. We can easily derive `intervals` by a linear scan of first_{k_1}. During this scan, we avoid marking the intervals of k_1-mers that are smaller than τ_1. Finally, we preprocess `intervals` in linear time so that it supports `rank` queries in constant time (see Section 3.2). We then build a temporary bitvector `keep[1..r]` initialized to zeros, with a bit for each read in $\mathcal{R}$, and we invert BWT_R: at each step of the inversion, we know a position i in R, the read q this position belongs to, and the corresponding position j in BWT_R. If j belongs to an interval in `intervals`, we set `keep[q]` $= 1$. □

Note that, during the enumeration of the internal nodes of the suffix tree of R, we can waive the trick described in Algorithm 9.3 to keep the depth of the stack bounded by $\log|R|$, and we can instead perform a standard pre-order traversal of the suffix-link tree of R: indeed, the depth of the stack is bounded by the read length ℓ, which is a small constant in practice.

From now on, we assume that the reads from low-frequency species are stored on disk, that R denotes the concatenation of all reads from high-frequency species only, and that the bidirectional BWT index is built on this new version of R. Note that, rather than computing the Burrows–Wheeler transforms of the new R and $\underline{R}$ from scratch, we could derive them from their previous versions: see Exercise 16.6.

16.2.2 Initializing clusters

Let W_1^i and W_2^i be two, sufficiently long substrings of the same genome G^i, and let W^j be a substring of a different genome G^j. It is typically the case that the k-mer composition of W_1^i is more similar to the k-mer composition of W_2^i than to the k-mer composition of W^j for suitably small k. If the reads are long enough, we can thus compute their k-mer compositions and use this directly to build clusters that approximate genomes. If the reads are short, however, we cannot estimate their k-mer composition reliably: a possible workaround consists in grouping similar reads into *pre-clusters*, estimating the k-mer composition of such pre-clusters, and iteratively merging pre-clusters according to their k-mer compositions.

Perhaps the most natural criterion to create pre-clusters consists in merging two reads if they share a sufficiently long substring: this makes pre-clusters loosely correspond to unassembled contigs (see Section 13.2), or to substrings of a genome. Creating this kind of pre-cluster has the additional advantage of assigning approximately the same number of pre-clusters to species with very different frequencies. Indeed, consider two species $\{G^i, G^j\} \subseteq \mathcal{G}$ such that $\mathbf{P}[i]$ is very different from $\mathbf{P}[j]$: since pre-clusters loosely correspond to substrings of a genome, we expect that all the pre-clusters associated with G^i contain approximately the same number of reads c_i, and that all the pre-clusters associated with G^j contain approximately the same number of reads c_j, with c_i very different from c_j. However, the number of pre-clusters associated with each genome should depend only on its length, and the variation in length among different species in the same metagenomic sample is significantly lower than the variation in their frequency.

But how can we choose the length of the shared substrings to be used to create pre-clusters? In practice it is common to start from a range of *candidate lengths*, say $[k_x..k_y]$. Such a range could have been estimated from known genomes, so that a substring of length $k \in [k_x..k_y]$ occurs with high probability at most once in every known genome. Alternatively, the range could have been computed from some estimate of the error rate and coverage, to minimize the probability that two reads from different genomes share a substring of length inside the range. Note that this probability decreases as k increases, but setting k too large might prevent reads from the same genome being clustered together, because of sequencing errors. We thus want to derive, *from the read set $\mathcal{R}$ itself*, the maximum $k_2 \in [k_x..k_y]$ such that *the majority of distinct k_2-mers of R will occur at least twice in R*: in other words, most such k_2-mers are not likely to contain sequencing errors. This can be done in a single traversal of the internal nodes of the suffix tree of R, using the techniques described in Section 11.2.1: see Exercise 11.15.

Once we have calibrated k_2, we can merge two reads if they share at least one k_2-mer:

LEMMA 16.5   *Given the bidirectional BWT index of R and an integer k_2, we can build in $O(|R| \log \sigma)$ time and in $\max\{|R| + \sigma \log^2 |R|, 2|R|, |R| + o(|R|) + K \log r, r \log r\}$ bits of additional space all pre-clusters of R induced by k_2-mers, where K is the number of distinct repeating k_2-mers of R.*

Proof   We compute again the bitvector $\mathtt{first}_{k_2}$ using Lemma 9.22 in Section 9.7.3, and we compute from $\mathtt{first}_{k_2}$ another bitvector, $\mathtt{intervals}$, which marks the first and the last position of all the disjoint k_2-mer intervals of size at least two in BWT_R. Note that we can avoid marking in $\mathtt{intervals}$ the boundaries of the intervals of other k_2-mers as well. Indeed, consider the interval of k_2-mer W in BWT_R, and extend W to the left by one character in all possible ways: we can discard any of the resulting intervals if it happens to be itself the interval of a k_2-mer. Such a test can be performed during a linear scan of BWT_R: see Exercise 16.7. We then index $\mathtt{intervals}$ to support $\mathtt{rank}$ queries in constant time (see Section 3.2). Let K be the number of distinct k_2-mers that survive these filters, and let $\mathtt{cluster}[1..K]$ be a vector of read identifiers that takes $K \log r$ bits of space. We also initialize a disjoint-set data structure with one cluster for each read, taking $O(r \log r)$ bits of space (see Insight 16.1).

Insight 16.1 Disjoint-set forests

Given a *fixed* set of n elements $\mathcal{A}$, assume that we want to maintain a *dynamic* collection $\mathcal{C}$ of disjoint subsets of $\mathcal{A}$ that supports the following operations:

- $\mathtt{find}(x)$: given an element $x \in \mathcal{A}$, return a unique identifier of the set in $\mathcal{C}$ that contains x;

- union(X, Y): given the identifiers of two sets X and Y in $\mathcal{C}$, remove X and Y from $\mathcal{C}$ and add $X \cup Y$ to $\mathcal{C}$.

A *disjoint-set forest* represents a set $X \in \mathcal{C}$ as a tree T_X, in which every node corresponds to an element of X and has a pointer to its parent in T_X. The root of T_X is used as the unique identifier of set X. To answer query find(x) for some $x \in X$ and $X \in \mathcal{C}$, it suffices to follow parent pointers from x to the root of T_X, and to return the root of T_X. To implement union(X, Y), where sets X and Y are such that $|Y| \leq |X|$, it suffices to set the parent pointer of the root of T_Y to the root of T_X: this guarantees that find takes $O(\log n)$ time (see Exercise 16.14).

A second optimization to speed up find queries consists in resetting to y the parent pointer of every node that is traversed while answering a query find(x), where y is the result of such query. This makes the tree flatter, reducing the time taken to answer queries about the ancestors of x and their descendants. It is possible to show that this technique achieves $O(\alpha(n))$ amortized time per operation, where $\alpha(n)$ is the inverse of the fast-growing Ackermann function $A(n, n)$. For most practical settings of n, $A(n, n)$ is less than 5.

Then, we invert BWT_R: at each step we know a position i in R, the read p this position belongs to, and the corresponding position j in BWT_R. We use the bitvector intervals to detect whether j belongs to the interval of a repeating k_2-mer that survived the filtering, and to get the identifier q of such a k_2-mer. If we have reached k_2-mer q for the first time, then cluster[q] is null, and we set it to the read identifier p. Otherwise, we ask the disjoint-set data structure to merge clusters cluster[q] and p, and we set cluster[q] to the root of the tree created by the disjoint-set data structure after this union.

Note that the disjoint-set data structure does not need to be present in memory during this process: we can just stream to disk all queries to the disjoint-set data structure, free the memory, initialize a new disjoint-set data structure, and update it by streaming the queries back into memory. □

This algorithm can be easily adapted to cluster reads that share a k_2-mer *or its reverse complement*: see Exercise 16.9. More advanced approaches could merge two reads if they share at least two distinct k_2-mers (see Exercise 16.8), or if the regions that surround a shared k_2-mer in the two reads are within a specified edit distance. Note that the probability of erroneously merging two reads is directly proportional to the size of pre-clusters in practice, thus pre-clusters are not allowed to grow beyond a user-specified upper bound. We could then process k_2-mers *in order of increasing frequency*, stopping when the pre-clusters become too large: indeed, sharing a rare substring more likely implies that two reads come from the same genome.

Since k_2 is large, the number of occurrences of a repeating k_2-mer is approximately equal to the coverage γ of high-frequency species. Thus, the number of bits taken by the vector cluster is approximately $(|R|/\gamma) \log r$. This space can be further reduced by using *maximal repeats of length at least k_2* (see Section 8.4.1 for a definition of maximal repeat). Indeed, let $\mathcal{M}_R$ denote the set of all maximal repeats of R of length at least k_2. Every repeating k_2-mer is a substring of a maximal repeat in $\mathcal{M}_R$, and distinct repeating k_2-mers might occur in R only as substrings of the same maximal repeat $W \in \mathcal{M}_R$: thus, considering just W produces an equivalent set of queries to the disjoint-set data structure. More generally, every substring V of a maximal repeat $W \in \mathcal{M}_R$ occurs in R wherever W occurs, and possibly at other positions, therefore the union operations induced by W are a subset of the union operations induced by V, and we can safely disregard W for clustering. We are thus interested in the following subset of maximal repeats of R.

DEFINITION 16.6 *Let R be a string and k be an integer. A repeat of R is called k-submaximal if it is a maximal repeat of R of length at least k, and if it does not contain any maximal repeat of length at least k as a substring.*

We denote by $\mathcal{M}_R^* \subseteq \mathcal{M}_R$ the set of all k_2-submaximal repeats of R. Note that $\mathcal{M}_R^*$ is at most as big as the set of distinct repeating k_2-mers of R: see Exercise 16.13.

LEMMA 16.7 *Given the bidirectional BWT index of R and an integer k_2, we can mark the intervals in BWT_R of all k_2-submaximal repeats of R in $O(|R| \log \sigma)$ time and in $|R| + o(|R|) + \sigma \log^2 |R|$ bits of working space.*

Proof Let intervals$[1..|R|]$ be a bitvector in which we mark the starting position and the ending position of all intervals in BWT_R of maximal repeats in $\mathcal{M}_R^*$. First, we mark only the intervals in BWT_R that correspond to maximal repeats that do not contain another maximal repeat *as a suffix*. The intervals of such repeats in BWT_R are not contained in the interval of any other maximal repeat. To do so, we traverse the suffix-link tree of R depth-first using the bidirectional BWT index of R (see Algorithm 9.3), and, as soon as we meet an internal node v of the suffix tree of R with string depth k_2, we mark in intervals the block associated with v in BWT_R and we stop traversing the subtree rooted at v. We denote by $\mathcal{M}_R'$ the set of maximal repeats that results from this phase. Since $\mathcal{M}_R'$ is suffix-free, at most one repeat of $\mathcal{M}_R'$ ends at every position of R, therefore $|\mathcal{M}_R'| \leq |R|$.

In practice we may want to discard also maximal repeats that contain another maximal repeat *as a prefix*. To do so, we traverse the suffix-link tree of $\underline{R}$ in the same way as before, exchanging the role of BWT_R and $\mathsf{BWT}_{\underline{R}}$. Note that this is possible because the intervals in BWT_R and in $\mathsf{BWT}_{\underline{R}}$ are always synchronized by the bidirectional BWT index during the traversal. At the end of this process, we index intervals to support rank operations in constant time (see Section 3.2).

Then, we backward-search R *from left to right* in $\mathsf{BWT}_{\underline{R}}$, keeping the current position p in R, the corresponding position q in $\mathsf{BWT}_{\underline{R}}$, and an initially empty interval $[i..j]$. Note that this corresponds to reconstructing R *from left to right*. We use vector intervals to detect whether q belongs to the interval of a maximal repeat W: if so, we set $i = \overleftarrow{W}$ and $j = \overrightarrow{W}$, and we continue updating q by a backward step with character $R[p+1]$ and updating interval $[i..j]$ by backward-searching string $W \cdot R[p+1]$. Assume that, at some point, q belongs again to the interval of a maximal repeat V. If $\overleftarrow{V} \subseteq [i..j]$, then the occurrence of V that ends at position p in R includes the previous occurrence of W, therefore W is a substring of V and the interval of V can be unmarked from `intervals`. Otherwise, if $[i..j] \subset \overleftarrow{V}$, then the occurrence of V that ends at position p in R cannot include the previous occurrence of W: we thus reset $[i..j]$ to $\overleftarrow{V}$, and continue. $\square$

We can then use the bitvector `intervals` to cluster reads as described in Lemma 16.5. Other approaches build pre-clusters using long suffix–prefix overlaps (see Section 8.4.4), or long maximal exact matches (see Section 11.1.3): note that both of these structures are maximal repeats of R.

16.2.3   Growing clusters

As mentioned, the pre-clusters created in Section 16.2.2 typically correspond to short, possibly noncontiguous substrings of a single genome: we thus need to merge pre-clusters that belong to the same genome. Before proceeding, we can clearly filter out pre-clusters with too few reads, or such that the string which would result from assembling the reads in the pre-cluster would be too short: the reads in such pre-clusters can be processed in a separate phase. If we have paired-end information, we can also merge two pre-clusters if they contain corresponding paired reads. In general, we want to merge two pre-clusters if they have similar k_4-mer composition, where $k_4 < k_2$ is a user-specified constant estimated from known genomes.

Owing to the existence of repeating k_2-mers inside the same pre-cluster, we estimate the k_4-mer composition not directly from the reads in a pre-cluster, but from long, repeating substrings in the pre-cluster. Specifically, we extract all the maximal repeats of length at least $k_3 > k_2$ that occur at least τ_3 times in a pre-cluster. This operation could be carried out for each pre-cluster separately. However, since $k_3 > k_2$, a maximal repeat of length at least k_3 occurs in exactly one pre-cluster in R: we can thus compute all the maximal repeats of length at least k_3 *in the whole of R* by enumerating the internal nodes of the suffix tree of R just once (see Section 11.1.1), we could mark their intervals in BWT_R, and then we could extract the corresponding strings by inverting BWT_R: the details are left to Exercise 16.15. The parameters k_3 and τ_3 are again computed from estimates of the error rate and coverage, to minimize the probability that a k_3-mer in a genome occurs fewer than τ_3 times in a read set due to sequencing errors, and to minimize the probability that spurious k_3-mers are generated in a read set by sequencing errors.

Finally, we compute the k_4-mer composition of all the maximal repeats of a pre-cluster, and we input all such composition vectors to any clustering algorithm, for example k-means (see Insight 16.2). At the end of this step, all reads sampled from high-frequency species have been clustered. Reads from low-frequency species can be processed with a similar pipeline (see Figure 16.1): first, we filter out all reads that contain only *unique k_1-mers*; that is, we lower to two the filtering threshold τ_1 used in Section 16.2.1. Such reads either have a very high number of errors, or they belong to extremely low-frequency species for which there is too little information in the dataset to perform a reliable clustering. We then pre-cluster the remaining reads, using either

k_2'-mers or maximal repeats of length at least k_2', where $k_2' < k_2$. In practice we might want to organize this step in multiple rounds, in which we use decreasing values of k_2'. The k_4'-mer composition of the resulting pre-clusters (with $k_4' < k_4$) is then used to build new clusters.

Insight 16.2 *k-means clustering*

Let $\mathcal{S} \subset \mathbb{R}^n$ be a set of vectors, and let $\mathcal{S}^1, \mathcal{S}^2, \ldots, \mathcal{S}^k$ be a partitioning of $\mathcal{S}$ into k disjoint subsets, called *clusters*. Consider the cost function $\sum_{i=1}^{k} \sum_{\mathbf{X} \in \mathcal{S}^i} ||\mathbf{X} - \mathbf{M}^i||^2$, called the *within-cluster sum of squares*, where $\mathbf{M}^i$ is the mean of all vectors in cluster $\mathcal{S}^i$, and $|| \cdot ||$ is the 2-norm described in Section 11.2. It can be shown that the problem of finding the set of k clusters $\mathcal{S}^1, \mathcal{S}^2, \ldots, \mathcal{S}^k$ that minimize the within-cluster sum of squares is NP-hard. However, given an initial set $\mathbf{M}^1, \mathbf{M}^2, \ldots, \mathbf{M}^k$ of k mean vectors, we could find a local minimum of the cost function by iteratively assigning each vector $\mathbf{X} \in \mathcal{S}$ to a mean vector $\mathbf{M}^i$ that minimizes $|\mathbf{X} - \mathbf{M}^i|^2$, and updating every mean vector to the average of all vectors assigned to it in the previous iteration. This iterative approach is called *k-means clustering*.

Assume that each vector $\mathbf{X} \in \mathcal{S}$ is the composition vector of a corresponding string $X \in [1..\sigma]^+$, with one component for every distinct substring of a given length ℓ. In other words, $\mathbf{X}[W]$ is the frequency of ℓ-mer W in X, divided by the length of X (see Section 11.2.1). Assume also that we have the bidirectional BWT index of every string X in the set. We can perform k-means clustering *without building the composition vectors* $\mathbf{X} \in \mathcal{S}$ *explicitly*, as follows.

We use the bidirectional BWT index of X to mark in a bitvector $\texttt{intervals}_X[1..|X|]$ the first position of the interval in $\mathsf{BWT}_{X\#}$ of every ℓ-mer that occurs in X, as described in Section 9.7.3. Then, we index $\texttt{intervals}_X$ to support rank queries in constant time, and we initialize another bitvector, $\texttt{used}_X[1..L_X]$, where L_X is the number of distinct ℓ-mers in X. Computing the distance between $\mathbf{X}$ and $\mathbf{M}^i$ amounts to inverting $\mathsf{BWT}_{X\#}$: the first time we reach the interval of an ℓ-mer W of X, we compute $|\mathbf{X}[W] - \mathbf{M}^i[W]|^2$ and we flag the interval of W in the bitvector $\texttt{used}$. The contribution of the ℓ-mers with a non-zero component in $\mathbf{M}^i$ and a zero component in $\mathbf{X}$ can be computed using a bitvector that flags the non-zero components of $\mathbf{M}^i$ that have been found in X. Updating $\mathbf{M}^i$ also amounts to inverting $\mathsf{BWT}_{X\#}$ for each X that has been assigned to $\mathbf{M}^i$ in the previous phase: the first time we reach the interval of an ℓ-mer W in $\mathsf{BWT}_{X\#}$, we add to its component in a new vector $\mathbf{M}^i$ (initially set to zero) the size of the interval divided by the length of X, and we flag the interval of W in the bitvector $\texttt{used}$.

16.3 Comparing metagenomic samples

Assume that we have a database $\mathcal{D}$ of known metagenomes. Given a new metagenome $\mathcal{R}$, we want to determine which metagenomes in $\mathcal{D}$ are most similar to $\mathcal{R}$. One way of doing this could be to use the algorithms described in Section 16.1 to estimate the set of taxa $\mathcal{S}$ present in $\mathcal{R}$ and their relative frequencies **P**: assuming that all metagenomes in $\mathcal{D}$ have been annotated in this way, we could use such taxonomic compositions to retrieve similar metagenomes.

16.3.1 Sequence-based methods

When the species in $\mathcal{R}$ and in $\mathcal{D}$ are mostly unknown, we need to compare metagenomes using their sequence composition rather than their taxonomic composition. One way of doing this is by using the k-mer and substring kernels described in Section 11.2.

Note that the relative frequency of a k-mer W in $\mathcal{R}$ is affected both by the frequency of W in the distinct genomes of the sample and by the relative frequencies **P** of such genomes. Therefore, it could happen that samples containing very different sets of *distinct genomes* display a very similar k-mer composition, because of the relative frequencies of the genomes in the samples. We might thus be interested in estimating the k-mer composition vectors *of the distinct genomes in a metagenome*, for example by preliminarily clustering or assembling the reads as described in Section 16.2 and Chapter 13.

16.3.2 Read-based methods

Rather than comparing two metagenomic samples at the substring level, we could compare them *at the read level*, by estimating the proportion of reads they have in common.

DEFINITION 16.8 *Let $\mathcal{R} = \{R^1, R^2, \ldots, R^r\}$ and $\mathcal{Q} = \{Q^1, Q^2, \ldots, Q^q\}$ be two metagenomic samples. We say that a read $R^i \in \mathcal{R}$ is* similar *to a read $Q^j \in \mathcal{Q}$ if there are at least τ, not necessarily distinct, substrings of length k that occur in R^i and in Q^j in the same order and without overlaps, where $k < \ell$ and $\tau \geq 1$ are two user-specified parameters.*

Note that the same read R^i can be similar to multiple reads in $\mathcal{Q}$. Denote by $\mathcal{R} \otimes \mathcal{Q}$ the set of reads in $\mathcal{R}$ that are similar to at least one read in $\mathcal{Q}$, and denote by $\mathcal{Q} \otimes \mathcal{R}$ the set of reads in $\mathcal{Q}$ that are similar to at least one read in $\mathcal{R}$. We could estimate the similarity between $\mathcal{R}$ and $\mathcal{Q}$ by computing $f(\mathcal{R}, \mathcal{Q}) = (|(\mathcal{R} \otimes \mathcal{Q})| + |(\mathcal{Q} \otimes \mathcal{R})|)/(|\mathcal{R}| + |\mathcal{Q}|)$, or equivalently the proportion of reads that are similar to another read in the union of the two read sets. This can be done by adapting the approach of Exercise 16.8.

In order to scale to larger datasets, we might want to waive the requirement of computing $\mathcal{R} \otimes \mathcal{Q}$ and $\mathcal{Q} \otimes \mathcal{R}$ exactly. For example, we could approximate $\mathcal{R} \otimes \mathcal{Q}$ with the set $\mathcal{R} \odot \mathcal{Q}$ that contains all reads $R^i \in \mathcal{R}$ with at least τ nonoverlapping substrings of length k that occur also in $\mathcal{Q}$, but not necessarily in the same read Q^i and not necessarily in the same order. Setting k long enough makes $\mathcal{R} \odot \mathcal{Q}$ sufficiently close to $\mathcal{R} \otimes \mathcal{Q}$ in practice. One way of computing $\mathcal{R} \odot \mathcal{Q}$ consists in building the matching statistics vector MS_{R^i} of each read $R^i \in \mathcal{R}$ with respect to the string that results from concatenating all reads in $\mathcal{Q}$, as described in Section 11.2.3. Then, we can create a directed acyclic graph $G = (V, E)$ for each read R^i, whose vertices correspond to all positions $p \in [1..\ell]$ with $\mathsf{MS}_{R^i}[p] \geq k$, and whose arcs $(v_p, v_q) \in E$ correspond to pairs of positions $p < q$ such that $q - p \geq k$. A longest path in G is a set of nonoverlapping substrings of length k of maximum size, and the length of such a path can be computed in $O(|V| + |E|)$ time by dynamic programming: see Exercise 16.16.

16.3.3   Multi-sample methods

Assume that we have multiple metagenomic samples of type 1 and multiple metagenomic samples of type 2: a substring that is common to all samples of type 1 and that does not occur in any sample of type 2 could be used to discriminate between type 1 and type 2, and possibly to build a model of the two classes. Recall the *document counting problem* described on page 149 (Problem 8.4): given a set of d strings, we want to report, for every maximal repeat in the set, the number of strings that include the repeat as a substring. One can extend this problem and its linear-time solution described in Theorem 8.21 to consider two sets, one containing type 1 metagenomic samples and the other containing type 2 metagenomic samples, and to report, for every maximal repeat in the union of the two sets, the number of times it occurs in type 1 samples and in type 2 samples as a substring. Note that this method is not restricted to a fixed length k, and it can be made scalable since document counting can be solved space-efficiently: see Exercise 11.3 on page 257.

16.4   Literature

The species estimation strategy that combines read mapping with a lowest common ancestor is described in Huson *et al.* (2011) and references therein, and some of its variants using contigs and read clusters are detailed in Wang *et al.* (2014a). The idea of re-mapping the best hit of a read to the set of all its hits is from Haque *et al.* (2009). The variant of lowest common ancestor based on the set cover problem is described in Gori *et al.* (2011). Further heuristics based on detecting open reading frames in reads are detailed in Sharma *et al.* (2012). The idea of using taxonomic markers to speed up frequency estimation is described in Liu *et al.* (2011) and Segata *et al.* (2012), where markers are entire genes rather than arbitrary substrings, and in Edwards *et al.* (2012), where k-mers are used to mark protein families. The idea of computing markers from matching statistics and from the BWT of genomes was developed for this book. Markers are related to the notion of *cores* (substrings that are present in all genomes of a taxon v) and of *crowns* (cores of a taxon v that are not cores of `parent`(v), or equivalently strings that can separate v from its siblings in a taxonomy): see Huang *et al.* (2013).

Problem 16.1 is adapted from Lo *et al.* (2013) and its generalization to Problem 16.2 is from Sobih *et al.* (2015). The NP-hardness proof for Problem 16.2 is from R. Rizzi (personal communication, 2014). This problem formulation has been applied to species estimation in Sobih *et al.* (2015). A similar objective of obtaining a predefined read coverage, expressed again with the formalism of minimum-cost flows, appears in Medvedev *et al.* (2010) in connection with a copy-number variation problem.

The read clustering pipeline in Section 16.2 is from Wang *et al.* (2012a): this paper contains additional details on estimating the probability of merging two reads from different genomes used in Section 16.2.2, and on the expected error for a specific setting of parameters k_3 and τ_3 used in Section 16.2.3. The space-efficient implementation of the pipeline and the notion of k-submaximal repeats were developed for this book, and are detailed in Alanko *et al.* (2014). The idea of performing k-means clustering without storing the composition vectors of the input strings explicitly was developed for this book. Filtering out low-frequency species to improve the clustering accuracy has been proposed in Wang *et al.* (2012b). An alternative clustering criterion based on rare or unique k-mers appears in Haubold *et al.* (2005) and Tanaseichuk *et al.* (2011). Using maximal repeats to cluster reads has been described in Baran & Halperin (2012); conversely, using contigs produced by an assembler has been described in Alneberg *et al.* (2013) and Wang *et al.* (2014a). More details on the approach of combining multiple metagenomic samples that contain the same species before clustering appear in Baran & Halperin (2012). Alneberg *et al.* (2013) use the coverage of a contig or of a cluster in each sample of a collection, in addition to its k-mer composition, to improve clustering accuracy.

A method for comparing metagenomic samples using their estimated taxonomic composition, and its efficient implementation on GPU, is described in Su *et al.* (2013). More information about the k-mer composition of metagenomic samples can be found in Willner *et al.* (2009), and studies on the performance of a number of similarity measures based on taxonomic and k-mer composition can be found in Su *et al.* (2012), Jiang *et al.* (2012), and Wang *et al.* (2014b). The idea of comparing metagenomes at the read level using operator $\odot$ is from Maillet *et al.* (2012). This paper solves Exercise 16.16.

The space-efficient approach for comparing multiple metagenomic samples using document counting is from Fischer *et al.* (2008). As multi-samples become larger, even such a space-efficient solution might not be scalable in practice, unless executed on a machine with several terabytes of main memory. The distributed algorithm described in Välimäki & Puglisi (2012) scales this approach to a cluster of standard computers: this solution was applied to metagenomic samples in Seth *et al.* (2014).

Exercises

16.1 Show that Problem 16.1 can be reduced to a minimum-cost network flow problem, by extending the reduction constructed in Section 5.3.3 for Problem 5.7.

16.2 Show that you can reduce Problem 16.1 to one in which every read vertex in A has at least two incident edges.

16.3 Discuss how to modify the statement of Problem 16.1 if the probability of sequencing a read from a certain location inside each genome is no longer uniform for all locations, but follows a different distribution function, which is known beforehand.

16.4 Explain what changes you need to make to the reduction constructed in Exercise 16.1 if the objective function in Problem 16.1 is

$$(1 - \alpha) \sum_{(x,y) \in M} c(x, y) + \alpha \sum_{i=1}^{m} \sum_{j=1}^{s_i} \left(\frac{c_i}{s_i} - d_M(y_{i,j}) \right)^2. \tag{16.3}$$

Hint. Recall Exercise 5.16.

16.5 What can you say about the complexity of Problem 16.2 for $\alpha = 0$? What about $\alpha = 1$?

16.6 Describe an algorithm to compute the BWT of the reads that are kept and of the reads that are filtered out in Section 16.2.1, by reusing the BWT of the original file. Recall that reads in the input file R are separated by distinct binary strings on the artificial characters $\#_0$ and $\#_1$ which do not belong to $[1..\sigma]$.

16.7 Given the concatenation R of all reads in a metagenomic sample as defined in Section 16.2, consider the interval of a k_2-mer W in BWT_R. Section 16.2.2 claims that, during the creation of pre-clusters, we can discard any k_2-mer V whose interval in BWT_R coincides with the interval of aW for some $a \in [1..\sigma]$. Prove this claim, and describe an algorithm that implements it in linear time using just one scan of the bitvector `intervals`.

16.8 With reference to Section 16.2.2, describe an algorithm to create pre-clusters in which two reads are merged if they share *at least two* distinct k_2-mers. Assume that you have $|R| \log r$ bits of main memory in addition to the space required by Lemma 16.5, and assume that you have an algorithm that sorts tuples in external memory as a black box. Describe how to implement additional conditions on top of your algorithm, for example that the two shared k_2-mers are in the same order in both reads, that they do not overlap, or that they are offset by exactly one position.

16.9 With reference to Section 16.2.2, describe an algorithm to create pre-clusters in which two reads are merged if they share a k_2-mer *or its reverse complement*. Note that in this case we cannot discard k_2-mers that occur exactly once in R, since they could occur twice in $R \cdot \underline{R}$. Describe a solution that takes $K + K' \log r$ bits of space in addition to the bidirectional BWT index of R, where K is the number of distinct k_2-mers in R, and K' is the number of distinct k_2-mers that occur at least twice in $R \cdot \underline{R}$. This algorithm should perform $k_2 + 1$ operations for every position of the input file. Adapt your solution to filter out reads such that all their k_1-mers occur less than τ_1 times in $R \cdot \underline{R}$, as described in Section 16.2.1. Finally, describe another solution for creating pre-clusters that uses string $R \cdot \underline{R}$ as input and that takes approximately twice the time and the space of Lemma 16.5.

16.10 Describe a way to parallelize the algorithm in Lemma 16.5 by using samples of SA_R (see Section 9.2.3).

16.11 Adapt the algorithm for document counting described in Section 8.4.2 such that, given a reference taxonomy $\mathcal{T}$, it computes the markers of every taxon $v \in \mathcal{T}$.

16.12 Let $MS_{S,T,\tau}$ be the following generalization of the matching statistics vector described in Section 11.2.3: $MS_{S,T,\tau}[i]$ is the longest prefix of suffix $S[i..|S|]\#$ that occurs at least τ times in T. Similarly, let $SUS_{T,\tau}[i]$ be the shortest prefix of $T[i..|T|]\#$ that occurs at most τ times in T. Describe an algorithm that computes the markers of every taxon v in a reference taxonomy $\mathcal{T}$ using the shortest unique substring vector $SUS_{S,\tau}$, the matching statistics vector $MS_{S,T,\tau+1}$, and a bitvector $\mathtt{flags}_{S,T,\tau}[1..|S|]$ such that $\mathtt{flags}_{S,T,\tau}[i] = 1$ if and only if $S[i..i + MS_{S,T,1}[i] - 1]$ occurs at most τ times in T, for every genome $T \neq S$ in the database.

16.13 Prove that the number of k-submaximal repeats of a string S is upper-bounded by the number of distinct k-mers that occur at least twice in S.

16.14 With reference to Insight 16.1, prove that attaching the root of the smallest subtree to the root of the largest subtree guarantees $O(\log n)$ time for $\mathtt{find}$.

16.15 With reference to Section 16.2.3, give the pseudocode of an algorithm that extracts all the maximal repeats of length at least k_4 from all pre-clusters, using a single enumeration of all the internal nodes of the suffix tree of R.

16.16 With reference to Section 16.3.2, describe a way to approximate $\mathcal{R} \otimes \mathcal{Q}$ that is more accurate than just computing $\mathcal{R} \odot \mathcal{Q}$. *Hint.* Use the operator $\odot$ iteratively on suitable subsets of $\mathcal{R}$ and $\mathcal{Q}$. Then, give the pseudocode of an algorithm that, given vector MS_{R^i} for a read R^i, finds a set of nonoverlapping substrings of length k of maximum size.

References

Abouelhoda, M. I. (2007), A chaining algorithm for mapping cDNA sequences to multiple genomic sequences, in *14th International Symposium on String Processing and Information Retrieval*, Vol. 4726 of Lecture Notes in Computer Science. Berlin: Springer, pp. 1–13.

Adelson-Velskii, G. & Landis, E. M. (1962), 'An algorithm for the organization of information', *Proceedings of the USSR Academy of Sciences* **146**, 263—266 [in Russian]. English translation by Myron J. Ricci in *Soviet Mathematics Doklady*, **3**, 1259–1263 (1962).

Aho, A. V. & Corasick, M. J. (1975), 'Efficient string matching: An aid to bibliographic search', *Communications of the ACM* **18**(6), 333–340.

Ahuja, R., Goldberg, A., Orlin, J. & Tarjan, R. (1992), 'Finding minimum-cost flows by double scaling', *Mathematical Programming* **53**, 243–266.

Ahuja, R. K., Magnanti, T. L. & Orlin, J. B. (1993), *Network Flows: Theory, Algorithms, and Applications*. Upper Saddle River, NJ: Prentice-Hall, Inc.

Alanko, J., Belazzougui, D., Cunial, F. & Mäkinen, V. (2015), Scalable clustering of metagenomic reads. To be published.

Alneberg, J., Bjarnason, B. S., de Bruijn, I., Schirmer, M., Quick, J., Ijaz, U. Z., Loman, N. J., Andersson, A. F. & Quince, C. (2013), 'CONCOCT: Clustering cONtigs on COverage and ComposiTion', arXiv:1312.4038.

Alstrup, S., Husfeldt, T. & Rauhe, T. (1998), Marked ancestor problems, in *Proceedings of the 39th Annual Symposium on Foundations of Computer Science*, IEEE, pp. 534–543.

Apostolico, A. (2010), Maximal words in sequence comparisons based on subword composition, in *Algorithms and Applications*. Berlin: Springer, pp. 34–44.

Apostolico, A. & Bejerano, G. (2000), 'Optimal amnesic probabilistic automata or how to learn and classify proteins in linear time and space', *Journal of Computational Biology* **7**(3–4), 381–393.

Apostolico, A. & Denas, O. (2008), 'Fast algorithms for computing sequence distances by exhaustive substring composition', *Algorithms for Molecular Biology* **3**, 13.

Apostolico, A. & Lonardi, S. (2002), 'A speed-up for the commute between subword trees and DAWGs', *Information Processing Letters* **83**(3), 159–161.

Arge, L., Fischer, J., Sanders, P. & Sitchinava, N. (2013), On (dynamic) range minimum queries in external memory, in *Algorithms and Data Structures*, Vol. 8037 of Lecture Notes in Computer Science. Berlin: Springer, pp. 37–48.

Arlazarov, V., Dinic, E., Kronrod, M. & Faradzev, I. (1970), 'On economic construction of the transitive closure of a directed graph', *Doklady Akademii Nauk SSSR* **194**(11), 487–488 [in Russian]. English translation in *Soviet Mathematics Doklady* **11**, 1209–1210 (1975).

Babenko, M., Gawrychowski, P., Kociumaka, T. & Starikovskaya, T. (2015), Wavelet trees meet suffix trees, in *Symposium on Discrete Algorithms, SODA 2015*, pp. 572–591.

Baeza-Yates, R. A. & Ribeiro-Neto, B. A. (2011), *Modern Information Retrieval – The Concepts and Technology behind Search*, 2nd edn. New York, NY: Addison-Wesley.

Baker, B. S. (1993), On finding duplication in strings and software, Technical report, AT&T Bell Laboratories, New Jersey.

Bang-Jensen, J. & Gutin, G. (2008), *Digraphs: Theory, Algorithms and Applications*, 2nd edn. Springer Monographs in Mathematics. Berlin: Springer.

Bao, E., Jiang, T. & Girke, T. (2013), 'BRANCH: boosting RNA-Seq assemblies with partial or related genomic sequences', *Bioinformatics* **29**(10), 1250–1259.

Baran, Y. & Halperin, E. (2012), 'Joint analysis of multiple metagenomic samples', *PLoS Computational Biology* **8**(2), e1002373.

Baum, L. E. (1972), An inequality and associated maximization technique in statistical estimation for probabilistic functions of Markov processes, in *Inequalities III: Proceedings of the Third Symposium on Inequalities*. New York: Academic Press, pp. 1–8.

Bayer, R. (1972), 'Symmetric binary B-trees: Data structure and maintenance algorithms', *Acta Informatica* **1**, 290–306.

Beerenwinkel, N., Beretta, S., Bonizzoni, P., Dondi, R. & Pirola, Y. (2014), Covering pairs in directed acyclic graphs, in *Language and Automata Theory and Applications*, Vol. 8370 of Lecture Notes in Computer Science. Berlin: Springer, pp. 126–137.

Behnam, E., Waterman, M. S. & Smith, A. D. (2013), 'A geometric interpretation for local alignment-free sequence comparison', *Journal of Computational Biology* **20**(7), 471–485.

Belazzougui, D. (2014), Linear time construction of compressed text indices in compact space, in *Symposium on Theory of Computing, STOC 2014*, pp. 148–193.

Belazzougui, D., Cunial, F., Kärkkäinen, J. & Mäkinen, V. (2013), Versatile succinct representations of the bidirectional Burrows-Wheeler transform, in *21st Annual European Symposium on Algorithms (ESA 2013)*, Vol. 8125 of Lecture Notes in Computer Science. Berlin: Springer, pp. 133–144.

Belazzougui, D. & Puglisi, S. (2015), Range predecessor and Lempel–Ziv parsing. To be published.

Beller, T., Berger, K. & Ohlebusch, E. (2012), Space-efficient computation of maximal and supermaximal repeats in genome sequences, in *19th International Symposium on String Processing and Information Retrieval (SPIRE 2012)*, Vol. 7608 of Lecture Notes in Computer Science. Berlin: Springer, pp. 99–110.

Beller, T., Gog, S., Ohlebusch, E. & Schnattinger, T. (2013), 'Computing the longest common prefix array based on the Burrows–Wheeler transform', *Journal of Discrete Algorithms* **18**, 22–31.

Bellman, R. (1958), 'On a routing problem', *Quarterly of Applied Mathematics* **16**, 87–90.

Bender, M. A. & Farach-Colton, M. (2000), The LCA problem revisited, in *4th Latin American Symposium on Theoretical Informatics (LATIN 2000)*, Vol. 1776 of Lecture Notes in Computer Science, Berlin: Springer, pp. 88–94.

Bernard, E., Jacob, L., Mairal, J. & Vert, J.-P. (2014), 'Efficient RNA isoform identification and quantification from RNA-seq data with network flows', *Bioinformatics* **30**(17), 2447–2455.

Blazewicz, J. & Kasprzak, M. (2003), 'Complexity of DNA sequencing by hybridization', *Theoretical Computer Science* **290**(3), 1459–1473.

Blumer, A., Blumer, J., Haussler, D., Ehrenfeucht, A., Chen, M.-T. & Seiferas, J. (1985), 'The smallest automation recognizing the subwords of a text', *Theoretical Computer Science* **40**, 31–55.

Bowe, A., Onodera, T., Sadakane, K. & Shibuya, T. (2012), Succinct de Bruijn graphs, in *Algorithms in Bioinformatics*, Vol. 7534 of Lecture Notes in Computer Science. Berlin: Springer, pp. 225–235.

Brodal, G. S., Davoodi, P. & Rao, S. S. (2011), Path minima queries in dynamic weighted trees, in *Algorithms and Data Structures*, Vol. 6844 of Lecture Notes in Computer Science. Berlin: Springer, pp. 290–301.

Burrows, M. & Wheeler, D. (1994), A block sorting lossless data compression algorithm, Technical report 124, Digital Equipment Corporation.

Cancedda, N., Gaussier, E., Goutte, C. & Renders, J. M. (2003), 'Word sequence kernels', *The Journal of Machine Learning Research* **3**, 1059–1082.

Chairungsee, S. & Crochemore, M. (2012), 'Using minimal absent words to build phylogeny', *Theoretical Computer Science* **450**, 109–116.

Chaisson, M., Pevzner, P. A. & Tang, H. (2004), 'Fragment assembly with short reads', *Bioinformatics* **20**(13), 2067–2074.

Chan, H.-L., Lam, T.-W., Sung, W.-K., Tam, S.-L. & Wong, S.-S. (2006), A linear size index for approximate pattern matching, in *Proceedings of the Annual Symposium on Combinatorial Pattern Matching*, Vol. 4009 of Lecture Notes in Computer Science. Berlin: Springer, pp. 49–59.

Chan, R. H., Chan, T. H., Yeung, H. M. & Wang, R. W. (2012), 'Composition vector method based on maximum entropy principle for sequence comparison', *IEEE/ACM Transactions on Computational Biology and Bioinformatics* **9**(1), 79–87.

Charikar, M., Lehman, E., Liu, D., Panigrahy, R., Prabhakaran, M., Rasala, A., Sahai, A. & Shelat, A. (2002), Approximating the smallest grammar: Kolmogorov complexity in natural models, in *Proceedings of the Thirty-Fourth Annual ACM Symposium on Theory of Computing*, ACM, pp. 792–801.

Chazelle, B. (1988), 'A functional approach to data structures and its use in multidimensional searching', *SIAM Journal on Computing* **17**(3), 427–462.

Chikhi, R., Limasset, A., Jackman, S., Simpson, J. & Medvedev, P. (2014), 'On the representation of de Bruijn graphs', in *Proceeedings of the Annual International Conference on Research in Computational Molecular Biology*, Vol. 8394 of Lecture Notes in Computer Science. Berlin: Springer, pp. 35–55.

Chikhi, R. & Rizk, G. (2012), Space-efficient and exact de Bruijn graph representation based on a Bloom filter, in B. J. Raphael & J. Tang, eds., *Algorithms in Bioinformatics*, Vol. 7534 of Lecture Notes in Computer Science. Berlin: Springer, pp. 236–248.

Cilibrasi, R., Iersel, L., Kelk, S. & Tromp, J. (2005), On the complexity of several haplotyping problems, in R. Casadio & G. Myers, eds., *Algorithms in Bioinformatics*, Vol. 3692 of Lecture Notes in Computer Science. Berlin: Springer, pp. 128–139.

Cilibrasi, R. & Vitányi, P. M. (2005), 'Clustering by compression', *IEEE Transactions on Information Theory* **51**(4), 1523–1545.

Clark, D. (1996), Compact Pat Trees, PhD thesis, University of Waterloo, Canada.

Cole, R., Gottlieb, L. A. & Lewenstein, M. (2004), Dictionary matching and indexing with errors, in *Proceedings of the Symposium on Theory of Computing*, pp. 91–100.

Coleman, J. R., Papamichail, D., Skiena, S., Futcher, B., Wimmer, E. & Mueller, S. (2008), 'Virus attenuation by genome-scale changes in codon pair bias', *Science* **320**(5884), 1784–1787.

Cormen, T. H., Leiserson, C. E., Rivest, R. L. & Stein, C. (2009), *Introduction to Algorithms*, 3rd edn. Cambridge, MA: MIT Press.

Cristianini, N. & Shawe-Taylor, J. (2000), *An Introduction to Support Vector Machines*. New York, NY: Cambridge University Press.

Crochemore, M., Landau, G. & Ziv-Ukelson, M. (2002), A sub-quadratic sequence alignment algorithm for unrestricted cost matrices, in *Proceedings of the 13th ACM–SIAM Symposium on Discrete Algorithms (SODA 2002)*, pp. 679–688.

Crochemore, M., Mignosi, F. & Restivo, A. (1998), 'Automata and forbidden words', *Information Processing Letters* **67**(3), 111–117.

Crochemore, M., Mignosi, F., Restivo, A. & Salemi, S. (2000), 'Data compression using antidictionaries', *Proceedings of the IEEE* **88**(11), 1756–1768.

Crochemore, M. & Rytter, W. (2002), *Jewels of Stringology*. Singapore: World Scientific.

Crochemore, M. & Vérin, R. (1997a), Direct construction of compact directed acyclic word graphs, in *Proceeding of the 8th Annual Symposium on Combinatorial Pattern Matching (CPM)*, Vol. 1264 of Lecture Notes in Computer Science. Berlin: Springer, pp. 116–129.

Crochemore, M. & Vérin, R. (1997b), On compact directed acyclic word graphs, in *Structures in Logic and Computer Science*. Berlin: Springer, pp. 192–211.

Davoodi, P. (2011), Data Structures: Range Queries and Space Efficiency, PhD thesis, Department of Computer Science, Aarhus University.

Dayarian, A., Michael, T. P. & Sengupta, A. M. (2010), 'SOPRA: Scaffolding algorithm for paired reads via statistical optimization', *BMC Bioinformatics* **11**, 345.

de Berg, M., van Kreveld, M., Overmars, M. & Schwarzkopf, O. (2000), *Computational Geometry – Algorithms and Applications*, Vol. 382 of Lecture Notes in Computer Science. Berlin: Springer.

Delcher, A. L., Kasif, S., Fleischmann, R. D., Peterson, J., White, O. & Salzberg, S. L. (1999), 'Alignment of whole genomes', *Nucleic Acids Research* **27**(11), 2369–2376.

Deorowicz, S. & Grabowski, S. (2011), 'Robust relative compression of genomes with random access', *Bioinformatics* **27**(21), 2979–2986.

Dietz, P. F. (1989), Optimal algorithms for list indexing and subset rank, in *Algorithms and Data Structures (WADS' 89)*, Vol. 382 of Lecture Notes in Computer Science. Berlin: Springer, pp. 39–46.

Dilworth, R. P. (1950), 'A decomposition theorem for partially ordered sets', *The Annals of Mathematics* **51**(1), 161–166.

Do, H. H., Jansson, J., Sadakane, K. & Sung, W.-K. (2012), Fast relative Lempel–Ziv self-index for similar sequences, in *Joint International Conference on Frontiers in Algorithmics and Algorithmic Aspects in Information and Management (FAW-AAIM)*, Vol. 7285 of Lecture Notes in Computer Science. Berlin: Springer, pp. 291–302.

Donmez, N. & Brudno, M. (2013), 'SCARPA: Scaffolding reads with practical algorithms', *Bioinformatics* **29**(4), 428–434.

Drineas, P., Mahoney, M. W., Muthukrishnan, S. & Sarlós, T. (2011), 'Faster least squares approximation', *Numerische Mathematik* **117**(2), 219–249.

Durbin, R., Eddy, S. R., Krogh, A. & Mitchison, G. (1998), *Biological Sequence Analysis: Probabilistic Models of Proteins and Nucleic Acids*. Cambridge: Cambridge University Press.

Eddy, S. R. (2011), 'Accelerated profile HMM searches', *PLoS Computational Biology* **7**(10), e1002195.

Edmonds, J. & Karp, R. (1972), 'Theoretical improvements in algorithmic efficiency for network flow problems', *Journal of the Association for Computing Machinery* **19**, 248–264.

Edwards, R. A., Olson, R., Disz, T., Pusch, G. D., Vonstein, V., Stevens, R. & Overbeek, R. (2012), 'Real time metagenomics: Using k-mers to annotate metagenomes', *Bioinformatics* **28**(24), 3316–3317.

Elias, P. (1975), 'Universal codeword sets and representations of the integers', *IEEE Transactions on Information Theory* **21**(2), 194–203.

Eppstein, D., Galil, Z., Giancarlo, R. & Italiano, G. F. (1992a), 'Sparse dynamic programming I: Linear cost functions', *Journal of the ACM* **39**(3), 519–545.

Eppstein, D., Galil, Z., Giancarlo, R. & Italiano, G. F. (1992b), 'Sparse dynamic programming II: Convex and concave cost functions', *Journal of the ACM* **39**(3), 546–567.

Farach, M. (1997), Optimal suffix tree construction with large alphabets, in *Proceedings of the 38th IEEE Symposium on Foundations of Computer Science (FOCS)*, pp. 137–143.

Farach, M., Noordewier, M., Savari, S., Shepp, L., Wyner, A. & Ziv, J. (1995), On the entropy of DNA: Algorithms and measurements based on memory and rapid convergence, in *Proceedings of the Sixth Annual ACM–SIAM Symposium on Discrete Algorithms*, Society for Industrial and Applied Mathematics, pp. 48–57.

Feng, J., Li, W. & Jiang, T. (2011), 'Inference of isoforms from short sequence reads', *Journal of Computational Biology* **18**(3), 305–321.

Ferrada, H., Gagie, T., Hirvola, T. & Puglisi, S. J. (2014), 'Hybrid indexes for repetitive datasets', *Philosophical Transactions of the Royal Society A* **372**(2016), 20130137.

Ferragina, P., Luccio, F., Manzini, G. & Muthukrishnan, S. (2009), 'Compressing and indexing labeled trees, with applications', *Journal of the ACM* **57**(1), Article 4.

Ferragina, P. & Manzini, G. (2000), Opportunistic data structures with applications, in *Proceedings of the 41st IEEE Symposium on Foundations of Computer Science (FOCS)*, IEEE, pp. 390–398.

Ferragina, P. & Manzini, G. (2005), 'Indexing compressed texts', *Journal of the ACM* **52**(4), 552–581.

Ferragina, P., Nitto, I. & Venturini, R. (2009), On the bit-complexity of Lempel-Ziv compression, in *Proceedings of the Twentieth Annual ACM–SIAM Symposium on Discrete Algorithms*, Society for Industrial and Applied Mathematics, pp. 768–777.

Ferragina, P., Nitto, I. & Venturini, R. (2013), 'On the bit-complexity of Lempel–Ziv compression', *SIAM Journal on Computing* **42**, 1521–1541.

Fischer, J. & Heun, V. (2011), 'Space-efficient preprocessing schemes for range minimum queries on static arrays', *SIAM Journal on Computing* **40**(2), 465–492.

Fischer, J., Mäkinen, V. & Välimäki, N. (2008), Space-efficient string mining under frequency constraints, in *Proceedings of the Eighth IEEE International Conference on Data Mining (ICDM 2008)*, IEEE Computer Society, pp. 193–202.

Ford, L. R. (1956), Network flow theory, Technical Report Paper P-923, The RAND Corporation, Santa Monica, CA.

Ford, L. R. & Fulkerson, D. R. (1956), 'Maximal flow through a network', *Canadian Journal of Mathematics* **8**, 399–404.

Fredman, M. L. (1975), 'On computing the length of longest increasing subsequences', *Discrete Mathematics* **11**(1), 29–35.

Fredman, M. L. & Willard, D. E. (1994), 'Trans-dichotomous algorithms for minimum spanning trees and shortest paths', *Journal of Computer and System Sciences* **48**(3), 533–551.

Fulkerson, D. R. (1956), 'Note on Dilworth's decomposition theorem for partially ordered sets', *Proceedings of the American Mathematical Society* **7**(4), 701–702.

Gabow, H. N. (1990), Data structures for weighted matching and nearest common ancestors with linking, in *Proceedings of the First Annual ACM–SIAM Symposium on Discrete Algorithms, SODA '90*, Society for Industrial and Applied Mathematics, Philadelphia, PA, pp. 434–443.

Gabow, H. N., Bentley, J. L. & Tarjan, R. E. (1984), Scaling and related techniques for geometry problems, in *Proceedings of the 16th Annual ACM Symposium on Theory of Computing (STOC 1984)*, ACM, pp. 135–143.

Gabow, H. N. & Tarjan, R. E. (1989), 'Faster scaling algorithms for network problems', *SIAM Journal on Computing* **18**(5), 1013–1036.

Gagie, T., Gawrychowski, P., Kärkkäinen, J., Nekrich, Y. & Puglisi, S. J. (2012), A faster grammar-based self-index, in *6th International Conference on Language and Automata Theory and Applications (LATA 2012)*, Vol. 7183 of Lecture Notes in Computer Science. Berlin: Springer, pp. 240–251.

Gallé, M. (2011), Searching for Compact Hierarchical Structures in DNA by Means of the Smallest Grammar Problem, PhD thesis, Université Rennes 1.

Gao, S., Sung, W.-K. & Nagarajan, N. (2011), 'Opera: Reconstructing optimal genomic scaffolds with high-throughput paired-end sequences', *Journal of Computational Biology* **18**(11), 1681–1691.

Garcia, S. P. & Pinho, A. J. (2011), 'Minimal absent words in four human genome assemblies', *PLoS One* **6**(12), e29344.

Garcia, S. P., Pinho, A. J., Rodrigues, J. M., Bastos, C. A. & Ferreira, P. J. (2011), 'Minimal absent words in prokaryotic and eukaryotic genomes', *PLoS One* **6**(1), e16065.

Garey, M. R. & Johnson, D. S. (1979), *Computers and Intractability: A Guide to the Theory of NP-Completeness*. New York, NY: W. H. Freeman & Co.

Giancarlo, R., Rombo, S. E. & Utro, F. (2014), 'Compressive biological sequence analysis and archival in the era of high-throughput sequencing technologies', *Briefings in Bioinformatics* **15**(3), 390–406.

Göke, J., Schulz, M. H., Lasserre, J. & Vingron, M. (2012), 'Estimation of pairwise sequence similarity of mammalian enhancers with word neighbourhood counts', *Bioinformatics* **28**(5), 656–663.

Goldberg, A. & Tarjan, R. (1990), 'Finding minimum-cost circulations by successive approximation', *Mathematics of Operations Research* **15**, 430–466.

Goldberg, A. V. & Tarjan, R. E. (1987), Solving minimum-cost flow problems by successive approximation, in A. V. Aho, ed., *STOC '87 Proceedings of the Nineteenth Annual ACM Symposium on Theory of Computing*, ACM, pp. 7–18.

Goldberg, A. V. & Tarjan, R. E. (1988), Finding minimum-cost circulations by canceling negative cycles, in J. Simon, ed., *Proceedings of the 20th Annual ACM Symposium on Theory of Computing*, ACM, pp. 388–397.

Goldberg, A. V. & Tarjan, R. E. (1989), 'Finding minimum-cost circulations by canceling negative cycles', *Journal of the ACM* **36**(4), 873–886.

Gonnet, G., Baeza-Yates, R. & Snider, T. (1992), *Information Retrieval: Data Structures and Algorithms*. Upper Saddle River, NJ: Prentice-Hall, Chapter 3: New indices for text: Pat trees and Pat arrays, pp. 66–82.

Gori, F., Folino, G., Jetten, M. S. & Marchiori, E. (2011), 'MTR: Taxonomic annotation of short metagenomic reads using clustering at multiple taxonomic ranks', *Bioinformatics* **27**(2), 196–203.

Gotoh, O. (1982), 'An improved algorithm for matching biological sequences', *Journal of Molecular Biology* **162**(3), 705–708.

Grossi, R., Gupta, A. & Vitter, J. (2003), High-order entropy-compressed text indexes, in *Proceedings of the 14th Annual ACM–SIAM Symposium on Discrete Algorithms (SODA)*, pp. 841–850.

Grossi, R. & Vitter, J. (2000), Compressed suffix arrays and suffix trees with applications to text indexing and string matching, in *Proceedings of the 32nd ACM Symposium on Theory of Computing (STOC)*, pp. 397–406.

Grossi, R. & Vitter, J. (2006), 'Compressed suffix arrays and suffix trees with applications to text indexing and string matching', *SIAM Journal on Computing* **35**(2), 378–407.

Guisewite, G. & Pardalos, P. (1990), 'Minimum concave-cost network flow problems: Applications, complexity, and algorithms', *Annals of Operations Research* **25**(1), 75–99.

Gusfield, D. (1997), *Algorithms on Strings, Trees and Sequences: Computer Science and Computational Biology*. Cambridge: Cambridge University Press.

Gusfield, D., Landau, G. M. & Schieber, B. (1992), 'An efficient algorithm for the all pairs suffix-prefix problem', *Information Processing Letters* **41**(4), 181–185.

Hampikian, G. & Andersen, T. (2007), Absent sequences: nullomers and primes, in *Pacific Symposium on Biocomputing*, Vol. 12, pp. 355–366.

Haque, M. M., Ghosh, T. S., Komanduri, D. & Mande, S. S. (2009), 'SOrt-ITEMS: Sequence orthology based approach for improved taxonomic estimation of metagenomic sequences', *Bioinformatics* **25**(14), 1722–1730.

Harel, D. & Tarjan, R. E. (1984), 'Fast algorithms for finding nearest common ancestors', *SIAM Journal on Computing* **13**(2), 338–355.

Hartman, T., Hassidim, A., Kaplan, H., Raz, D. & Segalov, M. (2012), How to split a flow?, in A. G. Greenberg & K. Sohraby, eds., *Proceedings of INFOCOM 2012*, IEEE, pp. 828–836.

Haubold, B., Pierstorff, N., Möller, F. & Wiehe, T. (2005), 'Genome comparison without alignment using shortest unique substrings', *BMC Bioinformatics* **6**(1), 123.

Haussler, D. (1999), Convolution kernels on discrete structures, Technical report UCSC-CRL-99-10, UC Santa Cruz.

Herold, J., Kurtz, S. & Giegerich, R. (2008), 'Efficient computation of absent words in genomic sequences', *BMC Bioinformatics* **9**(1), 167.

Hirschberg, D. S. (1975), 'A linear space algorithm for computing maximal common subsequences', *Communications of the ACM* **18**(6), 341–343.

Hoare, C. A. R. (1962), 'Quicksort', *Computer Journal* **5**(1), 10–15.

Hon, W.-K. & Sadakane, K. (2002), Space-economical algorithms for finding maximal unique matches, in *Proceedings of the 13th Annual Symposium on Combinatorial Pattern Matching (CPM '02)*, Vol. 2373 of Lecture Notes in Computer Science. Berlin: Springer, pp. 144–152.

Hon, W.-K., Sadakane, K. & Sung, W.-K. (2003), Breaking a time-and-space barrier in constructing full-text indices, in *Proceedings of the 44th IEEE Symposium on Foundations of Computer Science (FOCS)*, IEEE, pp. 251–260.

Hoobin, C., Puglisi, S. J. & Zobel, J. (2011), 'Relative Lempel–Ziv factorization for efficient storage and retrieval of web collections', *Proceedings of the VLDB Endowment* **5**(3), 265–273.

Hopcroft, J. E. & Karp, R. M. (1973), 'An $n^{5/2}$ algorithm for maximum matchings in bipartite graphs.', *SIAM Journal on Computing* **2**(4), 225–231.

Hormozdiari, F., Alkan, C., Eichler, E. E. & Sahinalp, S. C. (2009), 'Combinatorial algorithms for structural variation detection in high-throughput sequenced genomes', *Genome Research* **19**(7), 1270–1278.

Hu, T. C. (1966), 'Minimum-cost flows in convex-cost networks', *Naval Research Logistics Quarterly* **13**(1), 1–9.

Huang, K., Brady, A., Mahurkar, A., White, O., Gevers, D., Huttenhower, C. & Segata, N. (2013), 'MetaRef: A pan-genomic database for comparative and community microbial genomics', *Nucleic Acids Research* **42**(D1), D617–D624.

Huang, L., Popic, V. & Batzoglou, S. (2013), 'Short read alignment with populations of genomes', *Bioinformatics* **29**(13), 361–370.

Hui, L. C. K. (1992), Color set size problem with application to string matching, in *Proceedings of the Annual Symposium on Combinatorial Pattern Matching (CPM)*, Vol. 644 of Lecture Notes in Computer Science. Berlin: Springer, pp. 230–243.

Hunt, J. W. & Szymanski, T. G. (1977), 'A fast algorithm for computing longest common subsequences', *Communications of the ACM* **20**(5), 350–353.

Huson, D. H., Mitra, S., Ruscheweyh, H.-J., Weber, N. & Schuster, S. C. (2011), 'Integrative analysis of environmental sequences using MEGAN4', *Genome Research* **21**(9), 1552–1560.

Huson, D. H., Reinert, K. & Myers, E. W. (2001), The greedy path-merging algorithm for sequence assembly, in *Proceedings of RECOMB 2001*, pp. 157–163.

Hyyrö, H. (2001), Explaining and extending the bit-parallel approximate string matching algorithm of Myers, Technical report A2001-10, Department of Computer and Information Sciences, University of Tampere, Finland.

Jacobson, G. (1989), Space-efficient static trees and graphs, in *Proceedings of the 30th IEEE Symposium on Foundations of Computer Science (FOCS)*, pp. 549–554.

Jenkins, P., Lyngsø, R. B. & Hein, J. (2006), How many transcripts does it take to reconstruct the splice graph?, in *Proceedings of Algorithms in Bioinformatics, 6th International Workshop, WABI 2006*, Vol. 4175 of Lecture Notes in Computer Science. pp. 103–114.

Jiang, B., Song, K., Ren, J., Deng, M., Sun, F. & Zhang, X. (2012), 'Comparison of metagenomic samples using sequence signatures', *BMC Genomics* **13**, 730.

Kärkkäinen, J. & Na, J. C. (2007), Faster filters for approximate string matching, in *Proceedings of the 9th Workshop on Algorithm Engineering and Experiments (ALENEX07)*, SIAM, pp. 84–90.

Kärkkäinen, J., Sanders, P. & Burkhardt, S. (2006), 'Linear work suffix array construction', *Journal of the ACM* **53**(6), 918–936.

Kasai, T., Lee, G., Arimura, H., Arikawa, S. & Park, K. (2001), Linear-time longest-common-prefix computation in suffix arrays and its applications, in *Proceedings of the Annual Symposium on Combinatorial Pattern Matching (CPM)*, Vol. 2089 of Lecture Notes in Computer Science. Berlin: Springer, pp. 181–192.

Kececioglu, J. (1991), Exact and Approximation Algorithms for DNA Sequence Reconstruction, PhD thesis, The University of Arizona.

Kececioglu, J. D. & Myers, E. W. (1993), 'Combinatorial algorithms for DNA sequence assembly', *Algorithmica* **13**(1–2), 7–51.

Kim, D., Sim, J., Park, H. & Park, K. (2005), 'Constructing suffix arrays in linear time', *Journal of Discrete Algorithms* **3**(2–4), 126–142.

Knuth, D. E., Morris, J. & Pratt, V. R. (1977), 'Fast pattern matching in strings', *SIAM Journal of Computing* **6**(2), 323–350.

Ko, P. & Aluru, S. (2005), 'Space efficient linear time construction of suffix arrays', *Journal of Discrete Algorithms* **3**(2–4), 143–156.

Krogh, A., Brown, M., Mian, I. S., Sjölander, K. & Haussler, D. (1994), 'Hidden Markov models in computational biology: Applications to protein modeling', *Journal of Molecular Biology* **235**(5), 1501–1531.

Kuksa, P. P., Huang, P.-h. & Pavlovic, V. (2008), V.: Scalable algorithms for string kernels with inexact matching, in *Proceedings of the Neural Information Processing Systems 21 (NIPS 2008)*, pp. 881–888.

Kuksa, P. P. & Pavlovic, V. (2012), Efficient evaluation of large sequence kernels, in *Proceedings of the 18th ACM SIGKDD International Conference on Knowledge Discovery and Data Mining*, ACM, pp. 759–767.

Kulekci, M. O., Vitter, J. S. & Xu, B. (2012), 'Efficient maximal repeat finding using the Burrows–Wheeler transform and wavelet tree', *IEEE/ACM Transactions on Computational Biology and Bioinformatics* **9**(2), 421–429.

Kuruppu, S., Puglisi, S. J. & Zobel, J. (2011a), Optimized relative Lempel–Ziv compression of genomes, in *Proceedings of the Thirty-Fourth Australasian Computer Science Conference*, Australian Computer Society, Inc., pp. 91–98.

Kuruppu, S., Puglisi, S. J. & Zobel, J. (2011b), Reference sequence construction for relative compression of genomes, in *String Processing and Information Retrieval*, Vol. 7024 of Lecture Notes in Computer Science. Berlin: Springer, pp. 420–425.

Lam, T. W., Sung, W. K., Tam, S. L., Wong, C. K. & Yiu, S. M. (2008), 'Compressed indexing and local alignment of DNA', *Bioinformatics* **24**(6), 791–797.

Landau, G. & Vishkin, U. (1988), 'Fast string matching with k differences', *Journal of Computer and System Sciences* **37**, 63–78.

Langmead, B., Trapnell, C., Pop, M. & Salzberg, S. L. (2009), 'Ultrafast and memory-efficient alignment of short DNA sequences to the human genome', *Genome Biology* **10**(3), R25.

Lee, C., Grasso, C. & Sharlow, M. F. (2002), 'Multiple sequence alignment using partial order graphs', *Bioinformatics* **18**(3), 452–464.

Lehman, E. & Shelat, A. (2002), Approximation algorithms for grammar-based compression, in *Proceedings of the Thirteenth Annual ACM–SIAM Symposium on Discrete Algorithms*, Society for Industrial and Applied Mathematics, pp. 205–212.

Leslie, C. & Kuang, R. (2003), Fast kernels for inexact string matching, in *Learning Theory and Kernel Machines*, Vol. 2777 of Lecture Notes in Computer Science. Berlin: Springer, pp. 114–128.

Levenshtein, V. (1966), 'Binary codes capable of correcting deletions, insertions and reversals', *Soviet Physics Doklady* **10**, 707.

Li, H. & Durbin, R. (2009), 'Fast and accurate short read alignment with Burrows–Wheeler transform', *Bioinformatics* **25**(14), 1754–1760.

Li, J. J., Jiang, C. R., Brown, J. B., Huang H. & Bickel, P. J. (2011a), 'Sparse linear modeling of next-generation mRNA sequencing (RNA-Seq) data for isoform discovery and abundance estimation', *Proceedings of the National Academy of Sciences* **108**(50), 19867–19872.

Li, M., Chen, X., Li, X., Ma, B. & Vitányi, P. M. (2004), 'The similarity metric', *IEEE Transactions on Information Theory* **50**(12), 3250–3264.

Li, M. & Vitányi, P. M. (2008), *An Introduction to Kolmogorov Complexity and its Applications*. Berlin: Springer.

Li, R., Yu, C., Li, Y., Lam, T.-W., Yiu, S.-M., Kristiansen, K. & Wang, J. (2009), 'SOAP2: An improved ultrafast tool for short read alignment', *Bioinformatics* **25**(15), 1966–1967.

Li, W., Feng, J. & Jiang, T. (2011b), 'IsoLasso: A LASSO regression approach to RNA-Seq based transcriptome assembly', *Journal of Computational Biology* **18**(11), 1693–1707.

Lifshits, Y., Mozes, S., Weimann, O. & Ziv-Ukelson, M. (2009), 'Speeding up HMM decoding and training by exploiting sequence repetitions', *Algorithmica* **54**(3), 379–399.

Lin, Y.-Y., Dao, P., Hach, F., Bakhshi, M., Mo, F., Lapuk, A., Collins, C. & Sahinalp, S. C. (2012), CLIIQ: Accurate comparative detection and quantification of expressed isoforms in a population, in B. J. Raphael & J. Tang, eds., *Algorithms in Bioinformatics*, Vol. 7534 of Lecture Notes in Computer Science. Berlin: Springer, pp. 178–189.

Lindsay, J., Salooti, H., Zelikovsky, A. & Măndoiu, I. (2012), Scalable genome scaffolding using integer linear programming, in *Proceedings of the ACM Conference on Bioinformatics, Computational Biology and Biomedicine, BCB '12*, ACM, New York, NY, pp. 377–383.

Liu, B., Gibbons, T., Ghodsi, M., Treangen, T. & Pop, M. (2011), 'Accurate and fast estimation of taxonomic profiles from metagenomic shotgun sequences', *BMC Genomics* **12**(Suppl. 2), S4.

Lo, C., Kim, S., Zakov, S. & Bafna, V. (2013), 'Evaluating genome architecture of a complex region via generalized bipartite matching', *BMC Bioinformatics* **14**(Suppl. 5), S13.

Lodhi, H., Saunders, C., Shawe-Taylor, J., Cristianini, N. & Watkins, C. (2002), 'Text classification using string kernels', *The Journal of Machine Learning Research* **2**, 419–444.

Löytynoja, A., Vilella, A. J. & Goldman, N. (2012), 'Accurate extension of multiple sequence alignments using a phylogeny-aware graph algorithm', *Bioinformatics* **28**(13), 1684–1691.

Maier, D. (1978), 'The complexity of some problems on subsequences and supersequences', *Journal of the ACM* **25**(2), 322–336.

Maillet, N., Lemaitre, C., Chikhi, R., Lavenier, D. & Peterlongo, P. (2012), 'Compareads: Comparing huge metagenomic experiments', *BMC Bioinformatics* **13**(Suppl. 19), S10.

Mäkinen, V. & Navarro, G. (2007), 'Rank and select revisited and extended', *Theoretical Computer Science* **387**(3), 332–347.

Mäkinen, V., Navarro, G., Síren, J. & Välimäki, N. (2009), Storage and retrieval of individual genomes, in *Proceedings of the 13th Annual International Conference on Research in Computational Molecular Biology (RECOMB 2009)*, Vol. 5541 of Lecture Notes in Bioinformatics. Berlin: Springer, pp. 121–137.

Mäkinen, V., Navarro, G., Sirén, J. & Välimäki, N. (2010), 'Storage and retrieval of highly repetitive sequence collections', *Journal of Computational Biology* **17**(3), 281–308.

Mäkinen, V., Navarro, G. & Ukkonen, E. (2003), Algorithms for transposition invariant string matching, in *Proceedings of the 20th International Symposium on Theoretical Aspects of Computer Science (STACS '03)*, Vol. 2607 of Lecture Notes in Computer Science. Berlin: Springer, pp. 191–202.

Mäkinen, V., Salmela, L. & Ylinen, J. (2012), 'Normalized N50 assembly metric using gap-restricted co-linear chaining', *BMC Bioinformatics* **13**, 255.

Manber, U. & Myers, G. (1993), 'Suffix arrays: A new method for on-line string searches', *SIAM Journal on Computing* **22**(5), 935–948.

Marschall, T., Costa, I. G., Canzar, S., Bauer, M., Klau, G. W., Schliep, A. & Schönhuth, A. (2012), 'CLEVER: Clique-enumerating variant finder', *Bioinformatics* **28**(22), 2875–2882.

Masek, W. & Paterson, M. (1980), 'A faster algorithm for computing string edit distances', *Journal of Computer and System Sciences* **20**(1), 18–31.

Mathé, C., Sagot, M.-F., Schiex, T. & Rouzé, P. (2002), 'Current methods of gene prediction, their strengths and weaknesses', *Nucleic Acids Research* **30**(19), 4103–4117.

McCreight, E. (1976), 'A space-economical suffix tree construction algorithm', *Journal of the ACM* **23**(2), 262–272.

Medvedev, P., Fiume, M., Dzamba, M., Smith, T. & Brudno, M. (2010), 'Detecting copy number variation with mated short reads', *Genome Research* **20**(11), 1613–1622.

Medvedev, P., Georgiou, K., Myers, G. & Brudno, M. (2007), Computability of models for sequence assembly, in *Workshop on Algorithms in Bioinformatics (WABI 2007)*, Vol. 4645 of Lecture Notes in Computer Science. Berlin: Springer, pp. 289–301.

Minoux, M. (1986), Solving integer minimum cost flows with separable convex cost objective polynomially, in G. Gallo & C. Sandi, eds, *Netflow at Pisa*, Vol. 26 of Mathematical Programming Studies. Berlin: Springer, pp. 237–239.

Moore, E. (1959), The shortest path through a maze, in H. Aiken, ed., *Proceedings of an International Symposium on the Theory of Switching, 2–5 April 1957, Part II*. Cambridge, MA: Harvard University Press, pp. 285–292.

Morris, J. & Pratt, V. R. (1970), A linear pattern-matching algorithm, Technical Report 40, University of California, Berkeley.

Munro, I. (1996), Tables, in *Proceedings of the 16th Conference on Foundations of Software Technology and Theoretical Computer Science (FSTTCS)*, Vol. 1180 of Lecture Notes in Computer Science. Berlin: Springer, pp. 37–42.

Myers, E. W. (2005), 'The fragment assembly string graph', *Bioinformatics* **21**(Suppl. 2), ii79–ii85.

Myers, G. (1999), 'A fast bit-vector algorithm for approximate string matching based on dynamic programming', *Journal of the ACM* **46**(3), 395–415.

Nagarajan, N. & Pop, M. (2009), 'Parametric complexity of sequence assembly: Theory and applications to next generation sequencing', *Journal of Computational Biology* **16**(7), 897–908.

Navarro, G. (2001), 'A guided tour to approximate string matching', *ACM Computing Surveys* **33**(1), 31–88.

Navarro, G. (2012), Wavelet trees for all, in J. Kärkkäinen & J. Stoye, eds., *Combinatorial Pattern Matching*, Vol. 7354 of Lecture Notes in Computer Science. Berlin: Springer, pp. 2–26.

Navarro, G. & Mäkinen, V. (2007), 'Compressed full-text indexes', *ACM Computing Surveys* **39**(1), Article 2.

Navarro, G. & Raffinot, M. (2002), *Flexible Pattern Matching in Strings – Practical On-Line Search Algorithms for Texts and Biological Sequences*. Cambridge: Cambridge University Press.

Needleman, S. B. & Wunsch, C. D. (1970), 'A general method applicable to the search for similarities in the amino acid sequence of two proteins', *Journal of Molecular Biology* **48**(3), 443–453.

Nevill-Manning, C. G. (1996), Inferring Sequential Structure, PhD thesis.

Ntafos, S. & Hakimi, S. (1979), 'On path cover problems in digraphs and applications to program testing', *IEEE Transactions on Software Engineering* **5**(5), 520–529.

Nykänen, M. & Ukkonen, E. (2002), 'The exact path length problem', *Journal of Algorithms* **42**(1), 41–52.

Ohlebusch, E., Gog, S. & Kügel, A. (2010), Computing matching statistics and maximal exact matches on compressed full-text indexes, in *String Processing and Information Retrieval*, Vol. 6393 of Lecture Notes in Computer Science. Berlin: Springer, pp. 347–358.

Onodera, T., Sadakane, K. & Shibuya, T. (2013), Detecting superbubbles in assembly graphs, in *Algorithms in Bioinformatics*, Vol. 8126 of Lecture Notes in Computer Science. Berlin: Springer, pp. 338–348.

Orlin, J. (1993), 'A faster strongly polynomial minimum cost flow algorithm', *Operations Research* **41**, 338–350.

Orlin, J. B. (1988), A faster strongly polynominal minimum cost flow algorithm, in J. Simon, ed., *Proceedings of the 20th Annual ACM Symposium on Theory of Computing*, ACM, pp. 377–387.

Pabinger, S., Dander, A., Fischer, M., Snajder, R., Sperk, M., Efremova, M., Krabichler, B., Speicher, M. R., Zschocke, J. & Trajanoski, Z. (2013), 'A survey of tools for variant analysis of next-generation genome sequencing data', *Briefings in Bioinformatics* **15**(2), 256–278.

Parida, L. (2007), *Pattern Discovery in Bioinformatics: Theory and Algorithms*. New York, NY: Chapman & Hall/CRC.

Patterson, M., Marschall, T., Pisanti, N., van Iersel, L., Stougie, L., Klau, G. W. & Schönhuth, A. (2014), WhatsHap: Haplotype assembly for future-generation sequencing reads, in *18th Annual International Conference on Research in Computational Molecular Biology (RECOMB 2014)*, Vol. 8394 of Lecture Notes in Computer Science. Berlin: Springer, pp. 237–249.

Pell, J., Hintze, A., Canino-Koning, R., Howe, A., Tiedje, J. M. & Brown, C. T. (2012), 'Scaling metagenome sequence assembly with probabilistic de Bruijn graphs', *Proceedings of the National Academy of Sciences* **109**(33), 13272–13277.

Pellicer, J., Fay, M. F. & Leitch, I. J. (2010), 'The largest eukaryotic genome of them all?', *Botanical Journal of the Linnean Society* **164**(1), 10–15.

Pevzner, P. (1989), 'ℓ-tuple DNA sequencing: Computer analysis', *Journal of Biomolecular Structure and Dynamics* **7**(1), 63–73.

Pevzner, P. A., Tang, H. & Waterman, M. S. (2001), A new approach to fragment assembly in DNA sequencing, in *Proceedings of RECOMB 2001*, pp. 256–267.

Pijls, W. & Potharst, R. (2013), 'Another note on Dilworth's decomposition theorem', *Journal of Discrete Mathematics* **2013**, 692645.

Policriti, A. & Prezza, N. (2014a), 'Fast randomized approximate string matching with succinct hash data structures', *BMC Bioinformatics*, in press.

Policriti, A. & Prezza, N. (2014b), Hashing and indexing: Succinct data structures and smoothed analysis, in *ISAAC 2014 – 25th International Symposium on Algorithms and Computation*, Vol. 8889 of Lecture Notes in Computer Science. Berlin, Springer, pp. 157–168.

Policriti, A., Tomescu, A. I. & Vezzi, F. (2012), 'A randomized numerical aligner (rNA)', *Journal of Computer and System Sciences* **78**(6), 1868–1882.

Puglisi, S. J., Smyth, W. F. & Turpin, A. (2007), 'A taxonomy of suffix array construction algorithms', *ACM Computing Surveys* **39**(2).

Qi, J., Wang, B. & Hao, B.-I. (2004), 'Whole proteome prokaryote phylogeny without sequence alignment: A k-string composition approach', *Journal of Molecular Evolution* **58**(1), 1–11.

Qin, Z. S., Yu, J., Shen, J., Maher, C. A., Hu, M., Kalyana-Sundaram, S., Yu, J. & Chinnaiyan, A. M. (2010), 'HPeak: An HMM-based algorithm for defining read-enriched regions in ChIP-Seq data', *BMC Bioinformatics* **11**, 369.

Rabin, M. O. & Scott, D. (1959), 'Finite automata and their decision problems', *IBM Journal of Research and Development* **3**(2), 114–125.

Raffinot, M. (2001), 'On maximal repeats in strings', *Information Processing Letters* **80**(3), 165–169.

Räihä, K.-J. & Ukkonen, E. (1981), 'The shortest common supersequence problem over binary alphabet is NP-complete', *Theoretical Computer Science* **16**, 187–198.

Reinert, G., Chew, D., Sun, F. & Waterman, M. S. (2009), 'Alignment-free sequence comparison (I): Statistics and power', *Journal of Computational Biology* **16**(12), 1615–1634.

Rizzi, R., Tomescu, A. I. & Mäkinen, V. (2014), 'On the complexity of minimum path cover with subpath constraints for multi-assembly', *BMC Bioinformatics* **15**(Suppl. 9), S5.

Rødland, E. A. (2013), 'Compact representation of k-mer de Bruijn graphs for genome read assembly', *BMC Bioinformatics* **14**(1), 313.

Rousu, J., Shawe-Taylor, J. & Jaakkola, T. (2005), 'Efficient computation of gapped substring kernels on large alphabets', *Journal of Machine Learning Research* **6**(9), 1323–1344.

Rytter, W. (2003), 'Application of Lempel–Ziv factorization to the approximation of grammar-based compression', *Theoretical Computer Science* **302**(1), 211–222.

Sacomoto, G. (2014), Efficient Algorithms for de novo Assembly of Alternative Splicing Events from RNA-seq Data, PhD thesis, Université Claude Bernard Lyon 1, Lyon.

Sadakane, K. (2000), Compressed text databases with efficient query algorithms based on the compressed suffix array, in *Proceedings of the 11th International Symposium on Algorithms and Computation (ISAAC)*, Vol. 1969 of Lecture Notes in Computer Science. Berlin: Springer, pp. 410–421.

Sahlin, K., Street, N., Lundeberg, J. & Arvestad, L. (2012), 'Improved gap size estimation for scaffolding algorithms', *Bioinformatics* **28**(17), 2215–2222.

Sakharkar, M. K., Chow, V. T. & Kangueane, P. (2004), 'Distributions of exons and introns in the human genome', *In Silico Biology* **4**(4), 387–393.

Salikhov, K., Sacomoto, G. & Kucherov, G. (2013), Using cascading Bloom filters to improve the memory usage for de Brujin graphs, in *Algorithms in Bioinformatics*, Vol. 8126 of Lecture Notes in Computer Science. Berlin: Springer, pp. 364–376.

Salmela, L., Mäkinen, V., Välimäki, N., Ylinen, J. & Ukkonen, E. (2011), 'Fast scaffolding with small independent mixed integer programs', *Bioinformatics* **27**(23), 3259–3265.

Salmela, L., Sahlin, K., Mäkinen, V. & Tomescu, A. I. (2015), Gap filling as exact path length problem, in *Proceedings of RECOMB 2015, 19th International Conference on Research in Computational Molecular Biology*. To be published.

Salmela, L. & Schröder, J. (2011), 'Correcting errors in short reads by multiple alignments', *Bioinformatics* **27**(11), 1455–1461.

Sankoff, D. (1975), 'Minimal mutation trees of sequences', *SIAM Journal on Applied Mathematics* **28**, 35–42.

Schnattinger, T., Ohlebusch, E. & Gog, S. (2010), Bidirectional search in a string with wavelet trees, in *21st Annual Symposium on Combinatorial Pattern Matching (CPM 2010)*, Vol. 6129 of Lecture Notes in Computer Science. Berlin: Springer, pp. 40–50.

Schneeberger, K., Hagmann, J., Ossowski, S., Warthmann, N., Gesing, S., Kohlbacher, O. & Weigel, D. (2009), 'Simultaneous alignment of short reads against multiple genomes', *Genome Biology* **10**, R98.

Schrijver, A. (2003), *Combinatorial Optimization. Polyhedra and Efficiency. Vol. A*, Vol. 24 of Algorithms and Combinatorics. Berlin: Springer.

Segata, N., Waldron, L., Ballarini, A., Narasimhan, V., Jousson, O. & Huttenhower, C. (2012), 'Metagenomic microbial community profiling using unique clade-specific marker genes', *Nature Methods* **9**(8), 811–814.

Seth, S., Välimäki, N., Kaski, S. & Honkela, A. (2014), 'Exploration and retrieval of whole-metagenome sequencing samples', *Bioinformatics* **30**(17), 2471–2479.

Sharma, V. K., Kumar, N., Prakash, T. & Taylor, T. D. (2012), 'Fast and accurate taxonomic assignments of metagenomic sequences using MetaBin', *PLoS One* **7**(4), e34030.

Shawe-Taylor, J. & Cristianini, N. (2004), *Kernel Methods for Pattern Analysis*. Cambridge: Cambridge University Press.

Shimbel, A. (1955), Structure in communication nets, in *Proceedings of the Symposium on Information Networks (New York, 1954)*. Brooklyn, NY: Polytechnic Press of the Polytechnic Institute of Brooklyn, pp. 199–203.

Simpson, J. T. & Durbin, R. (2010), 'Efficient construction of an assembly string graph using the FM-index', *Bioinformatics [ISMB]* **26**(12), 367–373.

Sims, G. E., Jun, S.-R., Wu, G. A. & Kim, S.-H. (2009), 'Alignment-free genome comparison with feature frequency profiles (FFP) and optimal resolutions', *Proceedings of the National Academy of Sciences* **106**(8), 2677–2682.

Sirén, J., Välimäki, N. & Mäkinen, V. (2011), Indexing finite language representation of population genotypes, in *11th International Workshop on Algorithms in Bioinformatics (WABI 2011)*, Vol. 6833 of Lecture Notes in Computer Science. Berlin: Springer, pp. 270–281.

Sirén, J., Välimäki, N. & Mäkinen, V. (2014), 'Indexing graphs for path queries with applications in genome research', *IEEE/ACM Transactions on Computational Biology and Bioinformatics* **11**(2), 375–388.

Smith, T. & Waterman, M. (1981), 'Identification of common molecular subsequences', *Journal of Molecular Biology* **147**(1), 195–197.

Smola, A. J. & Vishwanathan, S. (2003), Fast kernels for string and tree matching, in S. Becker, S. Thrun & K. Obermayer, eds., *Advances in Neural Information Processing Systems 15*. Cambridge, MA: MIT Press, pp. 585–592.

Sobih, A., Belazzougui, D. & Cunial, F. (2015), Space-efficient substring kernels. To be published.

Sobih, A., Tomescu, A. I. & Mäkinen, V. (2015), High-throughput whole-genome metagenomic microbial profiling. To be published.

Song, K., Ren, J., Zhai, Z., Liu, X., Deng, M. & Sun, F. (2013), 'Alignment-free sequence comparison based on next-generation sequencing reads', *Journal of Computational Biology* **20**(2), 64–79.

Song, K., Ren, J., Reinert, G., Deng, M., Waterman, M. S. & Sun, F. (2014), 'New developments of alignment-free sequence comparison: Measures, statistics and next-generation sequencing', *Briefings in Bioinformatics* **15**(3), 343–353.

Song, L. & Florea, L. (2013), 'CLASS: Constrained transcript assembly of RNA-seq reads', *BMC Bioinformatics* **14**(Suppl. 5), S14.

Spang, R., Rehmsmeier, M. & Stoye, J. (2002), 'A novel approach to remote homology detection: Jumping alignments', *Journal of Computational Biology* **9**(5), 747–760.

Su, C.-H., Wang, T.-Y., Hsu, M.-T., Weng, F. C.-H., Kao, C.-Y., Wang, D. & Tsai, H.-K. (2012), 'The impact of normalization and phylogenetic information on estimating the distance for metagenomes', *IEEE/ACM Transactions on Computational Biology and Bioinformatics* **9**(2), 619–628.

Su, X., Wang, X., Xu, J. & Ning, K. (2013), GPU-meta-storms: Computing the similarities among massive microbial communities using GPU, in *Proceedings of the 7th International Conference on Systems Biology (ISB)*, IEEE, pp. 69–74.

Tanaseichuk, O., Borneman, J. & Jiang, T. (2011), Separating metagenomic short reads into genomes via clustering, in *Algorithms in Bioinformatics*, Vol. 6833 of Lecture Notes in Computer Science. Berlin: Springer, pp. 298–313.

Tardos, E. (1985), 'A strongly polynomial algorithm to solve combinatorial linear programs', *Combinatorica* **5**, 247–255.

Teo, C. H. & Vishwanathan, S. (2006), Fast and space efficient string kernels using suffix arrays, in *Proceedings of the 23rd International Conference on Machine Learning*, ACM, pp. 929–936.

Tomescu, A. I., Kuosmanen, A., Rizzi, R. & Mäkinen, V. (2013a), 'A novel min-cost flow method for estimating transcript expression with RNA-Seq', *BMC Bioinformatics* **14**(Suppl. 5), S15.

Tomescu, A. I., Kuosmanen, A., Rizzi, R. & Mäkinen, V. (2013b), A novel combinatorial method for estimating transcript expression with RNA-Seq: Bounding the number of paths, in *Algorithms in Bioinformatics*, Vol. 8126 of Lecture Notes in Computer Science. Berlin: Springer, pp. 85–98.

Trapnell, C., Pachter, L. & Salzberg, S. L. (2009), 'TopHat: Discovering splice junctions with RNA-Seq', *Bioinformatics* **25**(9), 1105–1111.

Trapnell, C., Williams, B. A., Pertea, G., Mortazavi, A., Kwan, G., van Baren, M. J., Salzberg, S. L., Wold, B. J. & Pachter, L. (2010), 'Transcript assembly and quantification by RNA-Seq reveals unannotated transcripts and isoform switching during cell differentiation', *Nature Biotechnology* **28**(5), 511–515.

Tutte, W. (1954), 'A short proof of the factor theorem for finite graphs', *Canadian Journal of Mathematics* **6**(1954), 347–352.

Ukkonen, E. (1985), 'Algorithms for approximate string matching', *Information and Control* **64**(1–3), 100–118.

Ukkonen, E. (1995), 'On-line construction of suffix trees', *Algorithmica* **14**(3), 249–260.

Ulitsky, I., Burstein, D., Tuller, T. & Chor, B. (2006), 'The average common substring approach to phylogenomic reconstruction', *Journal of Computational Biology* **13**(2), 336–350.

Valenzuela, D., Välimäki, N. & Mäkinen, V. (2015), Variation calling over pan-genomes. To be published.

Välimäki, N. (2012), Least random suffix/prefix matches in output-sensitive time, in *23rd Annual Symposium on Combinatorial Pattern Matching (CPM 2012)*, Vol. 7354 of Lecture Notes in Computer Science. Berlin: Springer, pp. 269–279.

Välimäki, N., Ladra, S. & Mäkinen, V. (2012), 'Approximate all-pairs suffix/prefix overlaps', *Information and Computation* **213**, 49–58.

Välimäki, N. & Puglisi, S. J. (2012), Distributed string mining for high-throughput sequencing data, in *12th International Workshop on Algorithms in Bioinformatics (WABI 2012)*, Vol. 7534 of Lecture Notes in Computer Science. Berlin: Springer, pp. 441–452.

Välimäki, N. & Rivals, E. (2013), Scalable and versatile k-mer indexing for high-throughput sequencing data, in *Bioinformatics Research and Applications*, Vol. 7875 of Lecture Notes in Computer Science. Berlin: Springer, pp. 237–248.

van Emde Boas, P. (1977), 'Preserving order in a forest in less than logarithmic time and linear space', *Information Processing Letters* **6**(3), 80–82.

Vatinlen, B., Chauvet, F., Chrétienne, P. & Mahey, P. (2008), 'Simple bounds and greedy algorithms for decomposing a flow into a minimal set of paths', *European Journal of Operational Research* **185**(3), 1390–1401.

Vazirani, V. V. (2001), *Approximation Algorithms*. New York, NY: Springer.

Vinga, S. & Almeida, J. (2003), 'Alignment-free sequence comparison – a review', *Bioinformatics* **19**(4), 513–523.

Viterbi, A. (1967), 'Error bounds for convolutional codes and an asymptotically optimum decoding algorithm', *IEEE Transactions on Information Theory* **13**(2), 260–269.

Wagner, R. A. & Fischer, M. J. (1974), 'The string-to-string correction problem', *Journal of the Association for Computing Machinery* **21**, 168–173.

Wan, L., Reinert, G., Sun, F. & Waterman, M. S. (2010), 'Alignment-free sequence comparison (II): Theoretical power of comparison statistics', *Journal of Computational Biology* **17**(11), 1467–1490.

Wandelt, S., Starlinger, J., Bux, M. & Leser, U. (2013), 'RCSI: Scalable similarity search in thousand(s) of genomes', *PVLDB* **6**(13), 1534–1545.

Wang, Y., Leung, H. C., Yiu, S.-M. & Chin, F. Y. (2012a), 'MetaCluster 4.0: A novel binning algorithm for NGS reads and huge number of species', *Journal of Computational Biology* **19**(2), 241–249.

Wang, Y., Leung, H. C., Yiu, S.-M. & Chin, F. Y. (2012b), 'MetaCluster 5.0: A two-round binning approach for metagenomic data for low-abundance species in a noisy sample', *Bioinformatics* **28**(18), i356–i362.

Wang, Y., Leung, H. C., Yiu, S. M. & Chin, F. Y. (2014a), 'MetaCluster-TA: Taxonomic annotation for metagenomic data based on assembly-assisted binning', *BMC Genomics* **15**(Suppl. 1), S12.

Wang, Y., Liu, L., Chen, L., Chen, T. & Sun, F. (2014b), 'Comparison of metatranscriptomic samples based on k-tuple frequencies', *PLoS One* **9**(1), e84348.

Weiner, P. (1973), Linear pattern matching algorithm, in *Proceedings of the 14th Annual IEEE Symposium on Switching and Automata Theory*, pp. 1–11.

Weintraub, A. (1974), 'A primal algorithm to solve network flow problems with convex costs', *Management Science* **21**(1), 87–97.

West, D. B. (2000), *Introduction to Graph Theory*, 2nd edn. Upper Saddle River, NJ: Prentice-Hall, Inc.

Willner, D., Thurber, R. V. & Rohwer, F. (2009), 'Metagenomic signatures of 86 microbial and viral metagenomes', *Environmental Microbiology* **11**(7), 1752–1766.

Yang, X., Chockalingam, S. P. & Aluru, S. (2012), 'A survey of error-correction methods for next-generation sequencing', *Briefings in Bioinformatics* **14**(1), 56–66.

Zerbino, D. & Birney, E. (2008), 'Velvet: Algorithms for de novo short read assembly using de Bruijn graphs', *Genome Research* **18**, 821–829.

Ziv, J. & Lempel, A. (1977), 'A universal algorithm for sequential data compression', *IEEE Transactions on Information Theory* **23**(3), 337–343.

Ziv, J. (2008), 'On finite memory universal data compression and classification of individual sequences', *IEEE Transactions on Information Theory* **54**(4), 1626–1636.

Index